# FISH SCIENCES

# FISH SCIENCES

*Editor*

**Feray Marangoz**

scitus
academics

**Fish Sciences**

Edited by **Feray Marangoz**



ISBN: 9781681179520

Library of Congress Control Number: 2017949087


SCITUS Academics LLC,
616, Corporate Way, Suite 2, 4766,
Valley Cottage, NY 10989

www.scitusacademics.com



Printed in United States of America.

# Preface

The study of fishes is known as Ichthyology. The study of fish dates from the Upper Paleolithic Revolution (with the advent of "high culture"). The science of ichthyology was developed in several interconnecting epochs, each with various significant advancements. The study of fish receives its origins from human's desire to feed, clothe, and equip themselves with useful implements. Early cultures manifested these insights in abstract and identifiable artistic expressions. A thorough knowledge of fish diversity, distributions, habitat requirements and life histories is essential to the management of fisheries, and conservation of species and the aquatic environment. Fish is a significant source of protein for millions of people in the world. Human population growth has placed a significant demand on fish as a food source. It is estimated that approximately 25 percent of all protein consumed by people comes from fish. For some developing nations fish provides a large percentage of protein for its own people as well as a source of income as an export product. Fish research has become an increasingly important study as fish populations decline throughout the world.

This book 'Fish Sciences' brings research studies to give in a concise form an account of the principal facts relating to the structure, classification, and life-history of Fishes. Research in ichthyology helps policy makers determine what course of action to take in order to maintain fish populations throughout the world. National and international policies are determined based on the data provided by fish research. It is intended to meet the requirements of those who are desirous of studying the elements of Ichthyology; to serve as a book of reference to zoologists generally; and, finally, to supply those who, like travelers, have frequent opportunities of observing fishes, with a ready means of obtaining information.

# Table of Contents

# CHAPTER 1

# Fish Cytokines and Immune Response

**Sebastián Reyes-Cerpa[1], Kevin Maisey[2], Felipe Reyes-López[3], Daniela Toro-Ascuy[1], Ana María Sandino[4] and Mónica Imarai[2]**

*[1]Dept. Biology, Laboratory of Virology, University of Santiago of Chile, Santiago, Chile*

*[2]Dept. Biology, Laboratory of Immunology, University of Santiago of Chile, Santiago, Chile*

*[3]Dept. Cell Biology, Physiology & Immunology, Unit of Animal Physiology, Autonomous University of Barcelona, Barcelona, Spain*

*[4]Dept. Biology, Laboratory of Virology, University of Santiago of Chile, Santiago, Chile; Activaq S.A., Santiago, Chile*

## 1. INTRODUCTION

The immune system can be defined as a complex system that protects the organism against organisms or substances that might cause infection or disease. One of the most fascinating characteristics of the immune system is its capability to recognize and respond to pathogens with significant specificity. Innate and adaptive immune responses are able to recognize foreign structures and trigger different molecular and cellular mechanisms for antigen elimination. The immune response is critical to all individuals; therefore numerous changes have taken place during evolution to generate variability and specialization, although the immune system has conserved some important features over millions of years of evolution that are common for all species. The emergence of new taxonomic categories coincided with the diversification of the immune response. Most notably, the emergence of vertebrates coincided with the development of a novel type of immune response. Apparently, vertebrates inherited innate immunity from their invertebrate ancestors [1].

In higher vertebrates, the immune system consists of primary and secondary lymphoid organs with distinct compartments and morphology located in anatomically distinct sites. The thymus and bone marrow constitute the primary lymphoid organs, while the spleen, lymph nodes, and mucosal associated lymphoid tissue (MALT) comprise the secondary lymphoid organs [2].

Fish are a heterogeneous group divided into three classes: Agnatha (jawless fish such as the hagfish and lampreys), Chrondrichthyes (cartilaginous fish such as sharks, rays and skates) and Osteichthyes (bony fish) [3]. As in all vertebrates, fish have cellular and humoral immune responses and organs, the main function of which is immune defence. Most generative and secondary lymphoid organs in mammals are also found in fish, except for lymphatic nodules and bone marrow [3].

The head kidney or pronephros has hematopoietic functions [3, 4], and unlike in higher vertebrates, it is the immune organ involved in phagocytosis [5], antigen processing, production of IgM [6, 7] and immune memory through melanomacrophagic centres [8, 9]. The thymus, another lymphoid organ situated near the opercular cavity in teleosts, produces T lymphocytes involved in allograft rejection, stimulation of phagocytosis and antibody production by B cells [10, 11]. The spleen is a large, blood-filtering organ that undergoes increasing structural complexity in order to augment its efficiency in trapping and processing antigens [12-15]. Melanomacrophage centres are present for clearance of ingested material and can be surrounded by immunoglobulin-positive cells, especially after immunization [8]. Proliferation of granular cells has also been observed in association with ellipsoids and melanomacrophage centres after immunization [16].

## 1.1. Innate and Adaptive Immune Response

The development of an immune system is essential for the survival of living organisms. In vertebrates, immunity can be divided into two components, the innate immune response and the adaptive immune response. The innate immune response is the initial line of defence against infection, which includes physical barriers and cellular response. The adaptive immune response is capable of specific antigen recognition and is responsible for the secondary immune response.

The innate immune system recognizes conserved molecular structures common to pathogenic microorganisms such as polysaccharides, lipopolysaccharides (LPS), peptidoglycans, bacterial DNA, and double-strand viral RNA, among others, through their interaction with specific receptors like toll receptors (TLRs). These mechanisms of recognition may lead directly to successful removal of pathogens, for instance by phagocytosis, or may trigger additional protective responses through induction of adaptive immune responses [17]. Cells of the innate immune system have a diverse array of functions. Some cells are phagocytic, allowing them to engulf and degrade pathogenic particles. Other cells produce and secrete cytokines and chemokines that can stimulate and help guide the migration of cells and further direct the immune response [18].

The adaptive system recognizes foreign structures by means of two cellular receptors, the B cell receptor (BCR) and the T cell receptor (TCR). Adaptive immunity is highly regulated by several mechanisms. It increases with antigen exposure and produces immunological memory, which is the basis of vaccine development and the preventive function of vaccines [19, 20]. The adaptive response generally starts days after infection and is capable of recognizing specific protein motifs of peptides, which leads to a response that increases in both speed and magnitude with each successive exposure [21]. The main effector cells of the adaptive immune response are the lymphocytes, specifically B cells and T cells. When B cells are activated, they are capable of differentiating into plasma cells that can secrete antibodies. Upon activation T cells differentiate into either helper T cells or cytotoxic T cells. Helper T cells are capable of activating other cells of the adaptive immune response such as B cells and macrophages, while cytotoxic T cells upon activiation are able to kill cells that have been infected [22].

## 1.2. Fish Immune Response

Immune responses in fish have not been as well characterized as they have in higher vertebrates. Consequently, there is not enough information about the components of the fish immune system and its function and regulation. Key immune mammalian homologous genes have been identified in several fish species, suggesting that the fish immune system shares many features with the mammalian system. For example, the identification of $\alpha$ and $\beta$ T cell receptor genes (TCR) [23], key T cell markers such as CD3, CD4, CD8, CD28, CD40L, and a great number of cytokines and chemokines [24-26] suggest that T helper (Th)1, Th2 and Th17 and the regulatory subset Treg are present in fish. Some cell subsets have been better studied mainly because their activity can be easily differentiated and measured, as in the case of cytotoxic cells [27] and macrophages [28, 29]. Finally, B cells have been much more studied due to the availability of monoclonal antibodies that have been isolated and identified by a number of techniques [30, 31]. Phenotypic characterization of leukocytes has been hampered mainly by the lack of membrane cell markers [32, 33]. Researchers anticipate developing antibodies for cell lineage markers of fish immunocompetent cells that can be used to isolate and characterize immune cells to obtain insights into their regulation and role in immune response [34-36].

Antibodies in teleosts play a key role in the immune response. In general, IgM is the main immunoglobulin in teleosts that can elicit effective specific humoral responses against various antigens. For IgM, one gene alone can generate as many as six structural isoforms. Therefore, diversity is the result of structural organization rather than genetic variability [37]. Recently, several reports have provided evidence for the existence of IgD/IgZ/IgT in fish [38-41]. Interestingly, B cells from rainbow trout and salmon have high phagocytic capacity, suggesting a transition in B lymphocyte during evolution in which a key cell type of the innate immunity and phagocytosis evolved into a highly specialized component of the adaptive immune response in higher vertebrates [42, 43].

## 1.3. Fish Cytokines

Cytokines are secreted proteins with growth, differentiation, and activation functions that regulate the nature of immune responses. Cytokines are involved in several steps of the immune response, from induction of the innate response to the generation of cytotoxic T cells and the production of antibodies. In higher vertebrates, the combination of cytokines that are secreted in response to an immune stimulation induces the expression of immune-related genes through multiple signalling pathways, which contributes to the initiation of the immune response. Cytokines can modulate immune responses through an autocrine or paracrine manner upon binding to their corresponding receptors [44].

Cytokines have overlapping and sometimes contradictory pleiotropic functions that make their classification difficult. Cytokines are produced by macrophages, lymphocytes, granulocytes, DCs, mast cells, and epithelial cells, and can be divided into interferons (IFNs), interleukins (ILs), tumor necrosis factors (TNFs), colony stimulating factors, and chemokines [45]. They are secreted by activated immune-related cells upon induction by various pathogens, such as parasitic, bacterial, or viral components [46]. Macrophages can secret IL-1, IL-6, IL-12, TNFα, and chemokines such as IL-8 and MCP-1, all of which are indispensable for macrophage, neutrophil, and lymphocyte recruitment to the infected tissues and their activation as pathogen eliminators [47]. Meanwhile, cytokines released by phagocytes in tissues can also induce acute phase proteins, including mannose-binding lectin (MBL) and C-reactive protein (CRP), and promote migration of DCs [48].

Fish appear to possess a repertoire of cytokines similar to those of mammals. To date several cytokine homologues and suppressors have been cloned in fish species [24, 25, 49]. Some cytokines described in fish are TNFα, IL-1β, IL-6 or IFN.

Current knowledge of fish cytokines is based on mammal models of the cytokines network and their complex interactions. In this review we included the pro-inflammatory cytokines associated with innate and adaptive immunity, regulatory cytokines and anti-inflammatory cytokines.

## 1.4. Pro-Inflammatory Fish Cytokines

### 1.4.1. Tumour Necrosis Factor A (TNFA)

TNFα (tumour necrosis factor alpha) is a pro-inflammatory cytokine that plays an important role in diverse host responses, including cell proliferation, differentiation, necrosis, apoptosis, and the induction of other cytokines. TNFα can induce either NF-kB mediated survival or apoptosis, depending on the cellular context [50]. TNFα mediates powerful anti-microbial responses, including inducing apoptosis, killing infected cells, inhibiting intracellular pathogen replication, and up-regulating diverse host response genes. Many viruses have evolved strategies to neutralize TNFα by direct binding and inhibition of the ligand or its receptor or modulation of various downstream signalling events [51].

TNFα has been identified, cloned, and characterized in several bony fish, including Japanese flounder [52], rainbow trout [53, 54], gilthead seabream [55], carp [56] catfish [57], tilapia [58], turbot [59] and goldfish [60]. These studies have revealed the existence of some obvious differences from their mammalian counterpart, such as the presence of multiple isoforms of TNFα in some teleost species [54, 56] the high constitutive expression of this gene in different tissues of healthy fish and its relatively poor up-regulation by immune challenge *in vitro* and *in vivo* [53, 55, 57]. However, the most unexpected and interesting difference between fish and mammal TNFα concerns the weak *in vitro* effects of TNFα on phagocyte activation in goldfish [60], rainbow trout [57], turbot [59] and gilthead seabram [61]. This weak *in vitro* activity of fish TNFα sharply contrasts with the powerful actions exerted by the i.p. injection of recombinant TNFα in gilthead seabream, which includes the recruitment of phagocytes to the injection site, with a concomitant strong increase in their respiratory burst [61]. Apparently endothelial cells are the main target cells of fish TNFα, suggesting that TNFα is mainly involved in the recruitment of leukocytes to the inflammatory foci rather than in their activation [62]. Despite the above, differential expression has been observed in studies with rainbow trout leucocytes, which have shown increased response to different pro-inflammatory stimuli, as human recombinant TNFα [63], LPS [53, 64], zimosan and muramyl dipeptide as a peptidoglycan constituent of both gram-positive and gram-negative bacteria [64]. Moreover, it is known that Infectious Pancreatic Necrosis Virus (IPNV)-mediated up-regulation of TNFα regulates both the Bad/Bid-mediated apoptotic pathway and the RIP1 (receptor-interacting protein-1)/ROS-mediated secondary necrosis pathway [65].

### 1.4.2. Interleukin 1 Family

In mammals, the 11 members of the Interleukin-1 family include IL-1α (IL-1F1), IL-1β (IL-1F2), IL-1 receptor antagonist (IL-1ra/IL-1F3), IL-18 (IL-1F4), IL-1F5-10 and IL-33 (IL-1F11). These molecules tend to be either pro-inflammatory or act as antagonists that inhibit the activities of particular family members [66]. Despite these semantic issues, to date only two clear homologues of these molecules have been discovered in fish, IL-1β and IL-18 [24].

#### 1.4.2.1. Interleukin 1β

IL-1β is one of the earliest expressed pro-inflammatory cytokines and enables organisms to respond promptly to infection by inducing a cascade of reactions leading to inflammation. Many of the effector roles of IL-1β are mediated through the up- or down-regulation of expression of other cytokines and chemokines [67]. Mammalian IL-1β is produced by a wide variety of cells, but mainly by blood monocytes and tissue macrophages. IL-1β was the first interleukin to be characterized in fish and has since been identified in a number of fish species, such as rainbow trout [68], carp [69], sea bass [70], gilt head seabram [71], haddock [72], tilapia [73]. A second IL-1β gene (IL-1beta2] has been identified in trout [74]

In mammals pro-IL-1β remains cytosolic and requires cellular proteases to release the mature peptide. It is known that the peptide is cleaved by the IL-1β converting enzyme (ICE)[75]. However, the aspartic acid residue for which this enzyme has specificity is not present in all fish genes sequenced to date. Nevertheless, using a combination of multiple alignments and analysis of the N-terminal sequences of known mature peptides, it is possible to predict fish gene cutting sites. In trout, this gives a mature peptide of 166 and 165 aminoacids for IL-1β1 and IL-1β2 [76].

Like its mammalian counterpart, teleost IL-1β has been found to be regulated in response to various stimuli, such as LPS or poly I:C [68, 70-74, 77-81]. The biological activity of recombinant IL-1β (rIL-1β) has been studied in several fish species, indicating that fish IL-1β is involved in the regulation of immune relevant genes, lymphocyte activation, migration of leucocytes, phagocytosis and bactericidal activities [77, 81-84].

#### 1.4.2.2. Interleukin 18

In mammals, IL-18 is mainly produced by activated macrophages. It is an important cytokine with multiple functions in innate and acquired immunity [85-87]. One of its primary biological properties is to induce interferon gamma (IFNγ) synthesis in Th1 and NK cells in synergy with IL-12 [88, 89]. It promotes T and NK cell maturation, activates neutrophils and enhances Fas ligand-mediated cytotoxicity [90-92]. Like IL-1β, it is synthesized as an inactive precursor of approximately 24 kDa and is stored intracellularly. Activation and secretion of IL-18 is mainly effected through specific cleavage of the precursor after D35 by caspase 1, also termed the IL-1β-converting enzyme (ICE), which is believed to be one of the key processes regulating IL-18 bioactivity [93, 94]. Some other enzymes, including caspase 3 and neutrophil proteinase 3, also cleave the IL-18 precursor to generate active or inactive mature molecules [95, 96].

IL-18 was discovered in fish by analysis of sequenced fish genomes (fugu) and EST databases (medaka) [97, 98]. An alternative splicing form of the IL-18 mRNA was discovered in trout that may have an important role in regulating IL-18 expression and processing in this species. This form shows a lower constitutive expression relative to the full length transcript, but unlike the full length transcript, it increases in response to LPS and polyI:C stimulation in the RTG-2 fibroblast cell line [98]. The expression level of the full length transcript can increase in response to LPS plus IL- 1b in head kidney leucocyte cultures, and by IFNγ in RTS-11 cells [99].

### 1.4.3. Other Pro-Inflammatory Cytokines

#### *1.4.3.1. INTERLEUKIN 6*

A number of other interleukins are considered pro-inflammatory, some of which are released during the cytokine cascade that follows bacterial infection. Of these IL-6 is one of the best known, and is itself a member of the IL-6 family of cytokines that includes IL-11 and IL-31, as well as cytokines such as mammalian CNTF, LIF, OSM, CT-1 and CT-2 [24]. Whilst the homology of known fish molecules with many of these IL-6 family members is not conclusive [100], true homologues appear to be present in at least in the cases of IL-6 and IL-11 [24]. IL-6 is produced by a diverse group of cells including T lymphocytes, macrophages, fibroblasts, neurons, endothelial and glial cells. The pleiotropic effects of IL-6 are mediated by a 2-subunit receptor [101] and include the regulation of diverse immune and neuro-endocrine processes. IL-6 has been implicated in the control of immunoglobulin production, lymphocyte and monocyte differentiation, chemokine secretion and migration of leukocytes to inflammation sites [102-104].

IL-6 was first discovered in fugu by analysis of the genome sequence [105] and subsequently in other species as part of EST analysis of immune gene-enriched cDNA libraries [106-108]. However, little is known about the function and signalling pathways of IL-6 in fish. Interestingly, trout IL-6 expression in macrophages is reported to be induced by LPS, poly I:C and IL-1β in the macrophage cell line RTS-11, as well as in head kidney macrophages [109]. Moreover, IL-6 induces the expression of itself, so it can act in an autocrine and paracrine fashion to increase its expression, with the potential to both amplify and exacerbate the inflammatory response. However, IL-6 can significantly down-regulate the expression of trout TNFα1, TNFα2, and IL-1β, suggesting a potential role of trout IL-6 in limiting host damage during inflammation [109].

#### *1.4.3.2. INTERLEUKIN 11*

In mammals, IL-11 is produced by many cell types throughout the body. Basal and inducible IL-11 mRNA expression can be detected in fibroblasts, epithelial cells, chondrocytes, synoviocytes, keratinocytes, endothelial cells, osteoblasts and certain tumour cells and cell lines [110]. Viral [111] and bacterial infection [112] and cytokine stimulation (IL-1, TNFα and TGF-β1] induce IL-11 expression. IL-11 acts on multiple cell types, including hemotopoietic cells, hepatocytes, adipocytes, intestinal epithelial cells, tumour cells, macrophages, and both osteoblasts and osteoclasts. In the hematopoietic compartment IL-11 supports multilineage and committed progenitors, contributing to myeloid, erythroid, megakaryocyte and lymphoid lineages [113]. IL-11 is also an anti-inflammatory cytokine that inhibits the production of pro-inflammatory cytokines from lipopolysaccharde (LPS)-stimulated macrophages [114]. In combination with its trophic effects on the gastrointestinal epithelium, IL-11 plays an important role in the protection and restoration of gastrointestinal mucosa [115, 116].

The teleostean IL-11 orthologue has been found to consist of duplicate IL-11 genes, named IL-11a and IL-11b [117], with expression patterns indicating that both divergent forms of teleostean IL-11 play roles in antibacterial and antiviral defence mechanisms of fish [117-119]. In trout, IL-11 molecule is grouped with IL-11a and is constitutively expressed in intestine and gills and is highly up-regulated at other immune sites (spleen, head kidney, liver) following bacterial infection. *In vitro*, the macrophage-like RTS-11 cell line has shown enhanced IL-11 expression in response to LPS, bacteria, poly I:C and rIL-1β [118]. In carp, IL-11a is modulated by LPS, ConA and peptidoglycan in head kidney macrophages [117, 120] and cortisol has been found to inhibit IL-11 expression on its own and in combination with LPS [117]. In contrast to carp IL-11a, which shows low levels of constitutive expression in blood leucocytes, IL-11b in Japanese flounder shows higher expression at this site, and strong up-regulation was found in response to rhabdovirus infection in kidney cells [119]. This suggests that these paralogues have some complementarity of function related to their differential expression, although study of both forms in a single experiment is still required [24].

## 1.5. Chemokines

Chemokines are a superfamily of approximately 40 different small secreted cytokines that direct the migration of immune cells to infection sites. Their activity is coordinated by binding to G-protein-linked receptors with seven transmembrane domains. Four distinct subgroups make up the chemokine superfamily. These are designated as CXC (or a), CC (or b), C (or g) and $CX_3C$ (or d), which are defined by the arrangement of the first two cysteine residues within their peptide structure. The CC subfamily can be further subdivided according to the total number of cysteine residues, as some members of this group contain four cysteines whilst the remainder possesses six (and are known as the C6-b group). Similarly, the CXC subfamily contains two subgroups based on whether or not the first two cysteines are preceded by a Glu-Leu-Arg (ELR) motif associated with specificity to neutrophils [76, 121].

### 1.5.1. Interleukin 8

An important chemokine related to the pro-inflammatory process is CXCL-8, also called interleukin 8, this chemokine is a member of the CXC chemokine subfamily and attracts neutrophils, T lymphocytes and basophils *in vitro,* but not macrophages or monocytes [122]. Many cell-types, including macrophages, produce IL-8 in response to a variety of stimuli (LPS, cytokines and viruses). The neutrophil-attracting ability of IL-8 can be attributed to the presence of the ELR motif adjacent to the CXC motifs at its N-terminus, presumably by affecting its binding to specific receptors [123, 124]. In contrast, CXC chemokines lack an ELR motif and specifically attract lymphocytes but not neutrophils. The biological effects of IL-8 on neutrophils include increased cytosolic calcium levels, respiratory burst, a change in neutrophil shape and chemotaxis[125].

The fish IL-8 has been found in flounder [126], trout [125, 127], catfish [128], and lamprey [129]. *In vitro* stimulation of a trout macrophage cell line (RTS-11) [125] or *in vivo* intraperitoneal challenge [78] with either LPS or poly I:C did result in clear up-regulation of IL-8 expression. Moreover, induction of IL-8 expression in primary cultures of rainbow trout leukocytes stimulated for 24 hours with LPS and TNFα confirms that this fish chemokine is associated with inflammatory response, as has been suggested in mammals [127]. Interestingly, the ELR motif associated with the neutrophil-attracting ability is absent from the lamprey molecule and it is similar in flounder, where CXCL8 also lacks the ELR motif and appears to be regulated by a bacterial mechanism, since its transcript has only been detected in the major immune organs (spleen and head kidney) of an LPS stimulated flounder. The case of the trout is different, although there is also no ELR preceding the CXC motif, it has a very similar motif (DLR) in this position [130]. The human CXCL8 molecule, where the ELR motif has been mutated to DLR, retains neutrophil-attracting ability, albeit at lower potency [123]. Consequently, it is possible that the trout molecule has similar chemotactic activity to that of mammalian CXCL8 [130].

## 1.6. The Interleukin 2 Family

The IL-2 subfamily of cytokines signals via the common gamma chain (gC or CD132), a member of the type I cytokine receptor family expressed in most leucocytes.These cytokines in mammals include IL-2, IL-4, IL-7, IL-9, IL-15 and IL-21. IL-2, IL-4, IL-9 and IL-21 are all cytokines released from Th cells, which affect their responses [24], whilst IL-7 and IL-15 are particularly important for the maintenance of T cell memory [131]. To date molecules with homology to all of these have been found in fish, except IL-9 [24].

### 1.6.1. Interleukin 2

Interleukin-2 (IL-2 is an important immunomodulatory cytokine that primarily promotes proliferation, activation and differentiation of T cells [132]. IL-2, initially known as T-cell growth factor (TCGF), is synthesized and secreted mainly by Th1 cells that have been activated by stimulation by certain mitogens or by interaction of the T-cell receptor with the antigen/MHC complex on the surface of antigen-presenting cells [133-135]. Although CD4 T cells are the major source of IL-2 production in response to TCR stimulation, transient induction of IL-2 mRNA and production of the protein has been detected in murine dendritic cells activated by gram-negative bacteria [136]. IL-2 can also be produced by B cells in certain situations [137, 138]. The produced IL-2 promotes the expansion and survival of activated T cells and is also required for the activation of natural killer (NK) cells [139] and for immunoglobulin (Ig) synthesis by B cells [140].

The IL-2 gene has been detected only recently in fish by analysis of the fugu genome sequence, which also identified IL-21 as a neighbouring gene, as in mammals, providing the first direct evidence for the existence of a true IL-2 homologue in bony fish [141]. The gene has a 4 exon/3 intron organisation, as in mammals, and showed no constitutive expression in a range of tissues examined. However, injection of Fugu with poly I:C induced expression of IL-2 in the gut and gills [141]. Moreover, IL-2 could be induced in head kidney cell cultures stimulated with PHA, and in T-cell enriched cultures isolated from PBL when stimulated with B7-H3 or B7- H4 Ig fusions proteins in the presence of PHA [24, 142]. IL-2 has since been cloned in rainbow trout [143, 144]. The trout IL-2 was significantly up-regulated in head kidney leucocytes by the T cell mitogen PHA and in classical mixed leucocyte reactions and *in vivo* following infection with bacteria (*Y. ruckeri*) or the parasite *Tetracapsuloides bryosalmonae*. More importantly, the recombinant trout IL-2 produced in *Escherichia coli* was shown to induce expression of two transcription factors (STAT5 and Blimp-1) known to be involved in IL-2 signalling in mammals [143], as well as interferon-g (IFNγ) and IL-2 itself, and a CXC chemokine known to be induced by IFNγ, termed a IFNγ-inducible protein (γIP) [145].

### 1.6.2. Interleukin 4

Interleukin-4 IL-4 is a pleiotropic cytokine produced by T cells, mast cells, and basophils and is known to regulate an array of functions in B cells, T cells, macrophages, hematopoietic and non-hematopoietic cells [146, 147]. IL-4 serves as a key cytokine in driving Th2 differentiation and mediating humoral immunity, allergic responses and certain autoimmune diseases [148]. The IL-4 gene is conserved evolutionally in the animal kingdom and has been isolated from various animals including humans [149], mice [150, 151] and bovines [152], in which the IL-4 locus has been mapped in a region adjacent to those of IL-5 and IL-13 on the same chromosome [153, 154].

Teleost fish have two genes of the IL-4/13 family, IL-4/13A and IL-4/13B, which are situated on separate chromosomes in regions that duplicated during the fish-specific whole genome duplication (FS-WGD) around 350 million years ago [155, 156]. A few IL-4-like genes have been found in fish to date. The first was discovered by searching the *Tetraodon nigroviridis* genome [157]. In this work, IL-4 was constitutively expressed in head kidney, spleen, liver, brain, gill, muscle and heart. The ubiquitous expression of IL-4 is consistent with a postulated role in immune cytokines regulation. Stimulating the fish with a mixed stimulant containing ConA, PHA and PMA significantly up-regulated the expression of IL-4, which suggests that IL-4 is involved in the immune inflammatory responses triggered by mitogens [157], as in mammals, where it has been observed that this mitogen increases IL-4 expression [158]. However, the homology (amino acid identity) of this molecule was very low [12–15%), making it difficult to be sure it is an IL-4 homologue, although clearly related to Th2-type cytokines [24]. In fugu, T cell enriched PBL was found to express more IL-4/13A and IL-4/13B after stimulation with recombinant B7 molecules [142]. In zebrafish a recombinant IL4/13B was shown to increase the number of IgT-positive and CD209-positive cells in blood [159, 160], and in zebrafish spleen the expression of IL-4/13B and transcription factor related to Th2 immune response as GATA-3, and STAT6 was simultaneously enhanced after PHA stimulation [161]. The IL-4/13A gene was identified in trout and salmon [162], where the tissue distribution of salmonid IL-4/13A and GATA-3 expression were compared to the expression of IL-4, IL-13, and GATA-3 in mice. High levels of these transcripts were found in both salmonid and murine thymus, while constitutive IL-4/13A richness of skin and respiratory tissue was found in salmonids but not in mice. Experiments with isolated cells from gill and pronephros (head kidney) indicated that trout IL-4/13A is mainly expressed by surface IgM-negative cells, readily inducible by PHA but not by poly I:C, and regulated differently from the Th1 cytokine IFNγ gene. In mammals, IL-5 is also considered a Th-2 type cytokine and along with IL-3 and GM-CSF it signals through receptors with a common γ-chain (γC). None of these cytokines have been discovered in fish to date [24].

### 1.6.3. Interleukin 7

The cytokine IL-7 plays several important roles during lymphocyte development, survival, and homeostatic proliferation [163]. It is produced by many different stromal cell types, including epithelial cells of the thymus and the intestine [164-166]. There is only one report on IL-7 in fish, for the fugu molecule that was discovered using a gene synteny approach by searching with the mammalian IL-7 gene neighbours C8orf70 and PKIA. Fugu IL-7 shows constitutive expression in head kidney, spleen, liver, intestine, gill and muscle, with expression shown to increase in head kidney cultures stimulated with LPS, poly I:C or PHA [24, 167].

### 1.6.4. Interleukin 15

The central action of IL-15 cytokine is on T-cells, dendritic cells and NK cells. IL- 15 is an important regulator of the innate immune response to infection and autoimmune disease conditions. This gene shares activities with IL-2 and utilizes IL-2R β and γ units [45].

Two genes with homology to IL-15 have been discovered in fish. One shows similar gene organisation and synteny to mammalian and chicken IL-15, and has been termed IL-15. The second gene, which has a 4-exon structure and is in a different genome location, has been termed IL-15-like [168-170]. They show differential expression patterns in terms of the tissues where constitutive expression is apparent, and in terms of inducibility in PBL, with IL-15L being refractory to induction [168]. Two alternative splice variants of IL-15L (IL-15La and IL-15Lb) have also been described [170]. Trout IL-15, which has subsequently been cloned and sequenced, was strongly induced by rIFNγ in two trout cell lines (RTS-11 and RTG-2). rIL-15 could up-regulate IFNγ expression in splenic leucocytes, suggesting a positive feedback loop exists in fish between these two cytokines. Interestingly, unstimulated head kidney leucocytes were not responsive to rIL-15, at least in terms of the IFNγ expression level [171].

### 1.6.5. Interleukin 21

Interleukin 21 (IL-21) is a newly recognized member of IL-2 cytokine family that utilizes the common γ-chain receptor subunit for signal transduction [172-174]. In humans and other mammals, IL-21 is produced by both Th1 and Th2 cells [172, 175, 176]. IL-21 has pleiotropic effects on both innate and adaptive immune responses and can act on CD4+ and CD8+ T cells, B cells, NK cells, dendritic cells (DC), myeloid cells, and other tissue cells. IL-21 enhances the proliferation of anti–CD3-stimulated T cells and acts in concert with other γc cytokines to enhance the growth of CD4+ T cells [177]. IL-21–producing CD4+ T cells exhibit a stable phenotype of IL-21 production in the presence of IL-6 but retain the potential to produce IL-4 under Th2-polarizing conditions and IL-17A under Th17-polarizing conditions [178]. IL-21 stimulates CD8+ T cell proliferation and synergizes with IL-15 in promoting CD8+ T cell expansion in vitro and their antitumor effects in vivo [177, 179]. B cells that encounter IL-21 in the context of Ag-specific (BCR) stimulation and T cell co/stimulation undergo class-switch recombination and differentiate into Ab-producing plasma cells. In contrast, B cells encountering IL-21 during nonspecific TLR stimulation or without proper T cell help undergo apoptosis [180].

Since its discovery in fugu as a gene neighbour of IL-2 [141], IL-21 has been reported in tetraodon [181, 182] and rainbow trout [182]. Fugu IL-21 shows low constitutive expression. However, stimulation of isolated kidney leucocytes with PHA induced IL-21 expression. IL-21 was also up-regulated at mucosal sites as gill and gut when fish were injected with LPS or poly I:C [141]. Similarly, in tetraodon IL/21 expression is low but detectable in the gut, gonad and gills of healthy fish, and is induced in the kidney, spleen and skin following LPS injection [181]. In trout IL-21 expression is highest in gills and intestine, and is induced *in vivo* by bacterial (*Y. ruckeri*) and viral (VHSV) infection [182]. Relative to IL-2, induction of IL-21 expression in head kidney cells appears more rapidly but has shorter duration after stimulation. The trout rIL-21 has also been produced and shown to increase the expression of IL-10, IL-22 and IFNγ, and to a lesser extent IL-21, and to maintain the expression levels of key lymphocyte markers in primary cultures [182]. Thus, IL-21 may act as a survival factor for fish T and B cells [24].

## 1.7. The Interleukin 10 Family

Interleukin-IL-10 is an anti-inflammatory cytokine and a member of the class II cytokine family that also includes IL-19, IL-20, IL-22, IL-24, IL-26 and the interferons [183]. Although the predicted helical structure of these homodimeric molecules is conserved, certain receptor-binding residues are variable and define the interaction with specific heterodimers of different type-2 cytokine receptors. This leads to diverse biological effects through the activation of signal transducer and activator of transcription (STAT) factors [184].

### 1.7.1. Interleukin 10

Interleukin-10 (IL-10) was discovered initially as an inhibitory factor for the production of Th1 cytokines. Subsequently, pleiotropic inhibitory and stimulatory effects on various types of blood cells were described for IL-10, including its role as a survival and differentiation factor for B cells. IL-10, which is produced by activated monocytes, T cells and other cell types like keratinocytes, appears to be a crucial factor for at least some forms of peripheral tolerance and a major suppressor of the immune response and inflammation. The inhibitory function of IL-10 is mediated by the induction of regulatory T cells [185].

IL-10 was discovered in fish by searching the fugu genome. The translation showed 42–45% similarity to mammalian molecules with very low constitutive expression in tissues [186]. IL-10 has since been cloned in several other fish species including carp [187] zebrafish [188], rainbow trout [189], sea bass [190, 191] and cod [79]. Such studies have shown that IL-10 expression can be increased by LPS stimulation, by bacterial infection, by bath administration of immunostimulants [192] and by IPNV infection which may be associated with mechanisms of immune evasion [78].

### 1.7.2. Interleukin 20 (Il-20like)

In mammals, IL-20 was discovered as a new member of the IL-10 family of cytokines. IL-20 shares the highest amino-acid sequence identity with IL-10, IL-24 and IL-19. It is secreted by immune cells and activated epithelial cells like keratinocytes. A high expression of the corresponding IL-20 receptor chains has been detected on epithelial cells. In terms of function, IL-20 might therefore mediate crosstalk between epithelial cells and tissue-infiltrating immune cells under inflammatory conditions [193].

In fish, the gene of IL-20 has been described in putterfish [183], zebrafish [194] and trout [195]. In the latter work, the IL-20 gene, called IL-20-like (IL-20L) has been described as having a high level of expression in immune related tissues and in the brain, suggesting an important role of the fish IL-20L molecule in both the immune and nervous systems. Although the exact cell types expressing IL-20L have yet to be defined, macrophages express IL-20L. Moreover, IL-20L expression in the macrophage cell line RTS-11 is modulated by pro-inflammatory cytokines, signalling pathway activators, microbial mimics and the immuno-suppressor dexamethasone. These data suggest that trout IL-20L plays an important role in the cytokine network. The increased expression of IL-20L was only detected at late stages (4–24

h) of LPS stimulation in RTS-11 cells and in spleen 24–72 h after infection with *Yersinia ruckeri*, which suggests that the increased expression of IL-20L by LPS and infection is via the rapid increase of pro-inflammatory cytokines (e.g., IL-1β) and other factors known to occur [195].

### 1.7.3. Interleukin 22/26

In mammals, interleukin-22 is secreted by Th17 cells [196], as well as by a subset of NK cells, designated as NK22 [197]; and even by some Th1 cells [198]. Studies have suggested there is a distinct Th22 cell lineage [199, 200]. Many of the same cytokines that induce differentiation and proliferation of IL-17-producing cells also lead to the secretion of IL-22 by Th17 cells, NK22 cells, and putative Th22 cells, including IL-6, IL-23, IL- 1β, TGF-β, and TNFα [201]. IL-17 and IL-22 are therefore frequently produced together in response to infections [202]. Interleukin-22 interacts with a heterodimeric receptor, IL-10R2/IL-22R1 [203], which is expressed on a variety of non-lymphoid cells, especially epithelial cells. Ligation of this receptor leads to both protective and detrimental effects. In synergy with IL-17, IL-22 induces pro-inflammatory cytokines in human bronchial epithelial cells against *Klebsiella pneumoniae* infection [204] and in colonic myofibroblasts [205]. Independently or in synergy with IL-17, IL-22 acts in defence against intestinal infection of mice with *Citrobacter rodentium* [206]. Moreover, IL-22 has been implicated in intestinal homeostasis keeping commensal bacteria contained in anatomical niches, which is key to our symbiotic relationship and normal intestinal physiology. However, the mechanisms that restrict colonization to specific niches are unclear. David Artis and colleagues have described a crucial role for IL-22-producing innate lymphoid cells (ILCs) in preventing lymphoid-resident commensal bacteria from escaping their niche and causing inflammation [207].

IL-26 can be produced by primary T cells, NK cells and T cell clones following stimulation with specific antigen or mitogenic lectins. IL-26 was initially shown by several groups to be co-expressed with IL-22 [208]. IL-26 is co-expressed with IFNγ and IL-22 by human Th1 clones, but not by Th2 clones. It was subsequently found that IL-26 is co-expressed with IL-17 and IL-22 by Th17 cells, an important subset of CD4+ T-helper cells that are distinct from Th1 and Th2 cells [209-211]. More recently, a novel subset of CD56+ NKp44+ NK cells was identified that co-expresses IL-22 and IL-26, especially following treatment with IL-23 [212]. Furthermore, a different subset of immature NK cells was described that do not express CD56 or NKp44 but do express CD117 and CD161 and constitutively express IL-22 and IL-26 [213].

The mechanisms that regulate transcription of the human IL-26 gene are so far largely undefined. It is possible and perhaps likely that expression of the

IL-26 gene is induced in an IL-23-dependent manner because IL-23 is known to induce differentiation of Th17 cells, and IL-23 amplifies expression of IL-17 and IL-22 by Th17 cells [214].

In fish, the IFNγ locus was discovered using a gene synteny approach, and was first reported for fugu [215]. It contained a homologue of IL-22/26, that later studies of the zebrafish genome revealed to be two genes, one with clear homology to IL-22 and one with somewhat less clear homology to IL-26 [216]. The IL-22 gene was expressed constitutively in intestine and gills in all the treated and non-treated tissues. The gene was also expressed in kidney and spleen in LPS and PolyI:C-treated tissues, respectively, while IL-26 was expressed only in intestine treated with PolyI:C without expression [216]. IL-22 expression has been correlated with disease resistance in haddock vaccinated against *V. anguillarum,* with a strong constitutive expression in gills in vaccinated fish but not in control fish 24 hours post bath challenge, resulting in complete protection in fish vaccinated [217]. Moreover, IL-22, a cytokine released by Th-17 cells in mammals, is also interesting, and such responses are thought to be crucial for protection against extracellular microbes and at mucosal sites [218]. This coupled with the recent discovery of novel gill-associated immune tissue in fish [219] may provide a clue to a potential mechanism of resistance elicited by the *V. anguillarum* vaccination [24].

## 1.8. The Interleukin 17 Family

Interleukin-17 and a related family of genes are known to have pro-inflammatory actions and are associated with diseases [220]. After the discovery of the human IL- 17 gene [221], five cellular paralogs of IL-17 were identified, namely IL-17B, C, D, E and F [222-227]. These paralogs, identified by ESTs, genomics and proteomic databases, share identities of 20–50% with IL-17A gene. Human IL-17 A and F are present in tandem in opposite transcriptional orientation on the same chromosome 6p12, while IL-17B (Chr 5q24), IL-17C (Chr 16q24), IL-17D (Chr 13q11) and IL-17E (Chr 14q11) are dispersed. The structural similarities lead to the classification of IL-17 A, B, C, D, E, and F genes to a larger IL-17 sub-family [45]. Several IL-17 family members have been discovered in teleost fish, but homology to mammalian genes has not always been easy to assign. Two IL-17A or F homologue genes (IL-17A/F) have been found on the same chromosome. However, it has been difficult to determine which gene codes IL-17A and F. This gene in zebrafish was named IL-17A/F1 and 2. Furthermore, another IL-17A or F homologue gene (IL-17A/F3) has been found in zebrafish localized on a chromosome different from that of IL-17A/F1 and 2 [228]. In addition to those in zebrafish, IL-17A or F homologue genes have been found in rainbow trout [229], Atlantic salmon

[230], pufferfish (IL-17A/F1, 2 and 3) [231], and medaka (IL-17A/F1, 2 and 3) [232].

The tissue distribution of the fugu IL-17 gene family also differs. In particular, IL-17 family genes are highly expressed in the head kidney and gills. Moreover, expression of IL-17 family genes is significantly up-regulated in the lipopolysaccharide-stimulated head kidney, suggesting that Fugu IL-17 family members are involved in inflammatory responses [231]. In Atlantic salmon IL-17D expression is widely distributed in tissues, with the highest levels of expression in testis, ovary and skin. Infection with *A. salmonicida* by injection increases IL-17D expression levels in the head kidney (but not the spleen) in a time-dependent manner. Skin and kidney showed an increased IL-17D expression level in fish given a cohabitation challenge with *A. salmonicida* [230]. The two trout IL-17C genes show some degree of differential expression within tissues, with IL-17C1 being more dominant in the gills and skin, whilst IL- 17C2 is more dominant in the spleen, head kidney and brain. Expression of both genes increases significantly with bacterial infection, although the increased expression of IL-17C2 is greater in terms of fold change. Similarly, both genes could be up-regulated in the trout RTS-11 cell line by LPS, poly I:C, calcium ionophore and rIL-1β, with IL-17C2 showing higher fold increases in all cases [229].

## 1.9. Interleukin 12

IL-12 is a heterodimeric cytokine composed of p35 and p40 subunits. It can mediate a number of different activities, including stimulation of IFNγ secretion from resting lymphocytes, NK cell stimulation and cytolytic T cell maturation. Perhaps most crucially, IL-12 also affects the progression of uncommitted T cells to either the Th1 lineage, which in general is characterized by secretion of lymphokines associated with cell-mediated rather than humoral immunity [233].

The p35 and p40 subunits were discovered in fish by analysis of the fugu genome [234]. The p35 locus is quite well conserved, with Schip1 being the immediate neighbour in all cases. This association has allowed p35 to be cloned by gene walking from Schip1 from fish species for which no genome sequence is available [24, 235]. The p40 subunit in fugu is constitutively expressed in all the tissues examined, except muscle, and no increases in expression were seen 3 h after injection with poly I:C or LPS. This constitutive and broad expression distribution of the p40 subunit suggests that it may be expressed in most cell types. The expression of the p35 subunit is more limited in its tissue expression and is induced after injection with poly I:C in the head kidney and the spleen, but not after injection with LPS. These results show that there are differences from the mammalian data in fugu IL-12 subunit expression. Further investigation will be required to show whether this is unique to fugu, if IL-12 is involved more in antiviral defence in fish and if the two subunits are regulated differently from their regulation in the mammalian system [234].

## 1.10. Transforming Growth Factor Β (Tgf-Β)

TGF-β is a pleiotropic cytokine that regulates cell development, proliferation, differentiation, migration, and survival in various leukocyte lineages including lymphocytes, dendritic cells, NK cells, macrophages and granulocytes [236, 237]. In the mammalian immune system, TGF-β1 is a well-known suppressive cytokine and its dominant role is to maintain immune tolerance and suppress autoimmunity [238, 239]. The potent immunosuppressive effects of TGF-β1 are mediated predominantly through its multiple effects on T cells: TGF-β1 suppress Th1 and Th2 cell proliferation, while it promotes T regulatory cell generation by inducing Foxp3 expression. On the other hand, TGF-β also promotes immune responses by inducing the generation of Th17 cells [236, 240, 241]. Therefore, the regulatory roles of TGF-β as a positive or negative control device in immunity are widely acknowledged in mammals [238, 240, 241].

In teleost, despite the lack of extensive investigation on the functional role of TGF-β, some recent studies have revealed that TGF-β1 also exterts powerful immune depressing effects on activated leukocytes, as it does in mammals. For instance, TGF-β1 significantly blocks TNFα-induced activation of macrophage in goldfish and common carp, but induces the proliferation of the goldfish fibroblast cell line CCL71 [242, 243]. In grass carp, TGF-β1 down-regulates LPS/PHA-stimulated the proliferation of peripheral blood lymphocyte by contrast with the stimulatory effect of TGF-b1 alone in the same cells [244]. In red sea bream, similar phenomenon was observed during leukocyte migration under TGF-β1 treatment, with or without LPS challenges [245]. These findings not only define TGF-β1 as an immune regulator in teleost, but also indicate that TGF-β1 may have retained similar functions in immunity during the evolution of vertebrates [246].

## 1.11. Interferons

Interferons genes are involved in mediating cellular resistance against viral pathogens and modulating innate and adaptive immune systems. Broadly, IFNs are classified into two main groups called type I and type II [45]. Type I IFN includes the classical IFNα/β, which is induced by viruses in most cells, whereas type II IFN is only composed of a single gene called IFNγ and is produced by NK cells (NK cells) and T lymphocytes in response to interleukin-12 (IL-12), IL-18, mitogens or antigens [247]. Structurally both IFN types belong to the class II a-helical cytokine family, but have different 3-dimensional structures and bind to different receptors [248].

Two IFNs (IFNα1 and IFNα2) have been cloned from Atlantic salmon and characterized with respect to sequence, gene structure, promoter, antiviral activity and induction of ISGs [249-252]. Salmon IFNα1 induces both Mx and ISG15 proteins in TO cells and thus has properties similar to mammalian IFNα/β and IFNλ [251, 252]. Furthermore, salmon IFNα1 induces potent antiviral activity against the IPNV *in vitro* [251], but this protection has not been observed *in vivo,* despite a high level of expression of IFNα detected in spleen and head kidney of Atlantic salmon challenged intraperitoneally with IPNV [78].

At least three type I IFNs have been discovered in rainbow trout. The IFN1 (rtIFN1) and rtIFN2 show high sequence similar to Atlantic salmon IFNα1 and IFNα2, which contains two cysteines. On the other hand, rtIFN3 contains four cysteines, which further confirms the relationship between mammalian IFNα and fish IFNs. Recombinant rtIFN1 and rtIFN2 have both been shown to up-regulate expression of Mx and inhibit VHSV replication in RTG-2 cells. In contrast, recombinant rtIFN3 has been found to be a poor inducer of Mx and antiviral activity. Interestingly, the three rtIFNs show differential expression in cells and tissues [253]. This suggests that the three trout IFNs have different functions in the immune system of fish, which is an interesting subject for further research [254].

IFNγ has been identified in several fish species, including rainbow trout and Atlantic salmon [215, 216, 248, 255-257]. In contrast to the type I IFNs, fish and mammalian IFNγ are similar in exon/intron structure and display gene synteny. However, some fish species also possess a second IFNγ subtype named IFN gamma rel, which is quite different from the classical IFNγ [258]. Rainbow trout and carp IFNγ have several functional properties in common with mammalian IFNγ, including the ability to enhance respiratory burst activity, nitric oxide production, and phagocytosis of bacteria in macrophages [257-259]. Far less is known about the antiviral properties of fish IFNγ. However, it has been reported that it induces antiviral activity against both IPNV and the Salmon Alpha Virus (SAV) in salmon cell lines [260]

## 1.12. Tools For Fish Cytokine Analysis

The major strategy of functional genomics is to identify the types of responses to specific pathogens based on cytokines expression as a predictor of profile immune response, which began by using suppressive subtractive hybridization as major tools at the beginning of the immunogenomics and upgrade to platforms of wide screening that allow identify thousands of EST's that are differentially regulated in their expression and that allow identifying potential candidates as biomarkers in the progression of the immune response at differential environmental conditions, not only against pathogens, but also in captivity stress conditions that affect the fisheries production.

## 1.13. Suppressive Subtractive Hybridization (SSH)

One of the most important biological processes in higher eukaryotes against external stimuli is the response mediated by differential gene expression. To understand the molecular regulation of these processes, the relevant subsets of differentially expressed genes of interest must be identified, cloned, and studied in detail using specific molecular techniques. In this matter, subtractive cDNA hybridization has been a powerful approach to identify and isolate cDNAs of differentially expressed genes [261-263]. Numerous cDNA subtraction methods have been reported. In general, they involve hybridization of cDNA from one population (tester) to excess of mRNA (cDNA) from another population (driver) and then separation of the unhybridized fraction (target) from the common hybridized sequences. One of these tools is a PCR-based technique called representational difference analysis, which does not require physical separation of single-strand (ss) and double-strand (ds) cDNAs. Representational difference analysis has been applied to enrich genomic fragments that differ in size or representation [264] and to clone differentially expressed cDNAs [265]. However, representational difference analysis has the problem of the wide differences in abundance of individual mRNA species so that multiple rounds of subtraction are needed [265]. Other strategies, such as mRNA differential display [266] and RNA fingerprinting by arbitrary primed PCR [267], are potentially faster methods for identifying differentially expressed genes, but both of these methods have high levels of false positives [268] that bias high-copy-number mRNA [269], which can inappropriate in experiments where only a few genes are expected to vary [268]. One of the techniques most often used to establish differential expression pattern between two conditions is suppression subtractive hybridization (SSH), which selectively amplifies target cDNA fragments (differentially expressed) and simultaneously suppresses non-target DNA amplification. The method is based on the fact that long inverted terminal repeats attached to DNA fragments can selectively suppress amplification of undesirable sequences in PCR procedures [270]. This method overcomes the problem of differences in mRNA abundance by incorporating a hybridization step that normalizes (equalizes) sequence abundance during the course of subtraction by standard hybridization kinetics [271]. Two types of SSH are possible: forward SSH, when the reaction involves the hybridization of cDNA from one population indicated as the evaluated phenotype (tester) to excess of mRNA (cDNA) from a control phenotype (driver); and reverse SSH, when the conditions described above are inverted. Together, the two processes are called reciprocal SSH.

Different works have been done with SSH to evaluate fish immune response at the gene expression level against challenges with bacteria-derived pathogen-associated molecular patterns (PAMP) like LPS [272, 273] and whole bacteria like *Aeromonas salmonicida* [274, 275], *Listonella anguillarum* [276], *Edwardsiella tarda* [277], and *Vibrio parahaemolyticus* [278] (Table 1).

A critical step in any immune response is the recognition of invading organisms. This is mediated by many proteins, including pattern recognition receptors (PRR), which recognize and bind to molecules present on the surface of microorganisms. LPS is an essential cell wall component of gram-negative bacteria and is recognized by PRR, triggering a series of responses that lead to the activation of the host defence system. These PRRs include a number of toll-like receptors, as well as other cell-surface and cytosolic receptors that, upon stimulation, modulate immunity [279, 280]. In LPS-stimulated yellow grouper spleen a subtracted cDNA library was constructed using SSH. The contigs and singlets obtained were analyzed and a low number of immune-related genes were found [272]. In Asian seabass the up-regulation of differentially expressed genes like pro-inflammatory cytokines and related receptors, such as TNF receptor super family member 14 (TNFRSF14), IL-31 receptor A (IL31RA), chemokine receptor-like 1 (CMKLR1), chemokine (C-X-C motif) receptor 3 (CXCR3), chemokine (C-C motif) receptor 7 (CCR7) and chemokine (C-C motif) ligand 25 (CCL25), was identified at 24h post-challenge by bacterial LPS in spleen Complement components were also identified [273]. These genes are a solid basis for a better understanding of immunity in Asian seabass and for developing effective strategies for immune protection against infections in that species.

Infection of Atlantic salmon by *A. salmonicida* was observed to stimulate an acute-phase response (APR) as part of the innate immune defence system to infection, whose gene expression pattern was remarkably observed in liver at 7 days post-infection [275] indicating that the liver appears to be the main source of APPs in fish, as in mammals. Not surprisingly, the liver gene expression pattern observed in other fish species against *L. anguillarum* [276], *E. tarda* [277], and *V. parahaemolyticus* [278]. The APR is characterized by alterations in the levels of plasma proteins referred to as acute-phase proteins (APPs), as well as the secretion of some other innate defence molecules important for innate immunity, such as complementary systems [281-283]. In Atlantic cod stimulated with atypical *A. salmonicida* (formalin-killed) interleukin-1β (IL-1- β), interleukin-8 (IL-8),

CC chemokine type 3, interferon regulatory factor 1 (IRF1), ferritin heavy subunit, cathelicidin, and hepcidin were identified in the forward spleen SSH library. Atlantic cod IRF1 was constitutively expressed at low levels, and expression was significantly elevated in spleen and head kidney at 24 h following A. salmonicida stimulation, with the highest levels of induction observed in the spleen [274]. The target IRFI genes, as well as their importance in innate immune responses in fish, have not yet been determined, although the expression of IRF1 in teleost macrophages can be induced by both IFNγ and IL-1β, with IFNγ being a much more potent inducer of IRF1 than IL-1β [99].

**TABLE 1.** Transcriptomics studies on fish after treatments with bacteria or virus *in vivo* analyzed with SSH. LPS: Lipopolysaccharide; ISKNV: Infectious spleen and kidney necrosis virus; poly I:C: polyriboinosinic polyribocytidylic acid; SBNNV: Sea bass nervous necrosis virus; ACNNV: Atlantic cod nervous necrosis virus; SGIV: Singapure grouper iridovirus; PBL: Peripheral blood leukocytes.

| **Microorganism** | **Fish** | **Pathogen** | **Tissue/ Cell type** | **Infection route** | **Reference** |
|---|---|---|---|---|---|
| Bacteria | Atlantic salmon | *Aeromonas salmonicida* A449 | Liver, head kidney and spleen | Intraperitoneal | Tsoi et al., 2004 |
| | Yellow grouper | *LPS (E. coli*) | Spleen | Intraperitoneal | Wang et al., 2007 |
| | Atlantic cod | *Aeromonas salmonicida (formalin-killed)* | Head kidney, Spleen | Intraperitoneal | Feng et al., 2009 |
| | Asian seabass | *LPS (E. coli*) | Spleen | Intraperitoneal | Xia and Yue 2010 |
| | Ayu | *Listonella anguillarum* | Liver | Intraperitoneal | Li et al., 2011 |
| | Japanese flounder | *Edwardsiella tarda* (NE8003) | PBL | Intraperitoneal | Matsuyama et al., 2011 |

| | | | | | |
|---|---|---|---|---|---|
| | Marine medaka | *Vibrio parahaemolyticus* | Liver | Intraperitoneal | Bo et al., 2012 |
| Virus | Mandarian fish | ISKNV | Spleen | Intramuscular | He et al., 2006 |
| | Sea bream | Nodavirus (strain 475-9/99) | Brain | Intramuscular | Dios et al., 2007 |
| | Atlantic cod | poly I:C | Spleen | Intraperitoneal | Rise et al., 2008 |
| | Sea bass | SBNNV | Head kidney | Intramuscular | Poisa-Beiro et al., 2009 |
| | Atlantic cod | ACNNV | Brain | Intraperitoneal | Rise et al., 2010 |
| | Orange-spotted grouper | SGIV | Spleen | Intraperitoneal | Xu et al., 2010 |

SSH has been in several investigations to evaluate fish gene expression patterns against challenges with PAMPs, such as polyriboinosinic polyribocytidylic acid (poly I:C) [284], Infectious Spleen and Kidney Necrosis Virus (ISKNV) [285], Nodavirus [286-288], and Singapore grouper iridovirus (SGIV) [289].

Spleen gene expression in mandarin fish at 4 days post-infection with ISKNV of Mx protein, interferon-inducible protein Gig-2, and viperin (interferon-inducible and antiviral protein) was up-regulated, suggesting IFN pathway stimulation after ISKNV infection [285]. Also, two inflammatory cytokine genes, CC chemokine and IL-8, were found in the forward SSH library, whereas the CD59/Neurotoxin/Ly-6-like protein gene was down-regulated. In mammals, CD59 is a complement regulatory protein, which can inhibit complement activation and membrane attack complex (MAC) formation on autologous cells [290], suggesting that down-regulation in the ISKNV-infected host cells may make these cells more sensitive to complement attack, mounting an anti-virus mechanism of the host [285].

In orange-spotted grouper after 5 days of infection with Singapore grouper iridovirus (SGIV) novel genes were annotated as immune-related, such as C-type lectin, epinecidin, and complement components C3 and C9.

Interestingly, the most abundant clone was C-type lectin, and the microarray results at 1, 5 and 9 days post-infection indicated that its expression was up-regulated in liver, spleen and kidney [289]. Lectins are multivalent carbohydrate-binding proteins that function as important pattern-recognition receptors (PRR) and have been isolated and characterized in fish [291-294]. C-type lectin represents a very large family, most members of which are able to bind PAMP and microorganisms themselves through sugar moieties and play important roles in non-self recognition and clearance of invading microorganisms. The up-regulation of C-type lectin in different organs with immunological functions confirmed as SSH as microarrays suggest an important role in the development of control strategies against SGIV infection.

The SSH method was used to generate a subtracted cDNA library enriched in gene transcripts differentially expressed after 1 day post-infection in the brains of sea bream infected with nodavirus. Most of the expressed sequence tags (ESTs) differentially expressed in infected tissues fell into gene categories related to cell structure, transcription, cell signalling or different metabolic routes. Other interesting putative homologies corresponded to genes expressed in stress responses, such as heat shock proteins (Hsp-70) and to immune-related genes such as the Fms-interacting protein, TNFα-induced protein, interferon-induced with helicase C domain protein (mda-5), which in mammals play an important role in the synthesis and secretion of IFN type I [295]. Another nodavirus, sea bass nervous necrosis virus (SBNNV) was studied to identify genes potentially involved in antiviral immune defence in sea bass head kidney using the SSH technique [287]. The results of up-regulated EST from sea bass head kidney SSH showed significant similarities with immune genes, such as β-2 microglobulin, heat shock protein 90 (Hsp-90), IgM, MHC class I and class II, and β-galactoside-binding lectin, identified as a member of the galectin family and closely related to the galectin-1 group (Sbgalectin-1). When the recombinant protein (rSbgalectin-1) was produced and functional assays were conducted, a decrease in IL-1β, TNFα, and Mx expression was observed in the brain of sea bass simultaneously injected with nodavirus and rSbgalectin-1 compared to those infected with the nodavirus alone, suggesting a potential anti-inflammatory protective role of Sbgalectin-1 during viral infection. A similar nodavirus, the Atlantic cod nervous necrosis virus (ACNNV), was studied to evaluate the transcript expression responses in the Atlantic cod (*Gadus morhua*) brain to asymptomatic high nodavirus carrier state [288]. In the forward brain SSH library was identified with significant similarity to genes with immune-relevant functional annotations the interferon stimulated gene 15 (ISG15), IL-8 variant 5, DEXH (Asp-Glu-X-His) box polypeptide 58 (DHX58; LGP2), radical Sadenosyl methionine domain-containing 2 (RSAD2; viperin), β-2-microglobulin (B2M), chemokine CXC-like protein, signal transducer and activator of transcription 1 (STAT1), and CC chemokine type 2. Interestingly, ISG15, DHX58, RSAD2, and sacsin (SACS) transcripts are all strongly upregulated by both high nodavirus carriage and intraperitoneal poly I:C stimulation, suggesting a similar host response is significantly induced in the brain by both nodavirus and poly I:C. This expression pattern is corroborated when the response of Atlantic cod spleen is evaluated against poly I:C stimulation, showing the up-regulation of ISG15, RSAD2, LGP2 and other transcripts such as MHC class I, and IRF1, 7, and 10, indicating that Atlantic cod recognize dsRNA and mount a interferon pathway response [284].

## 1.14. Microarrays

Microarray analysis measures the expression of large numbers of genes in parallel. This methodology, which combines hypotheses-driven and hypotheses-free research strategies, is used to infer molecular mechanisms, classify samples, and diagnose and search for novel biomarkers. With the use of standard platforms, laboratory protocols and procedures for processing of primary data, the results of microarrays analyses are well suited for database management and meta-analysis across multiple experiments, whilst data mining is based on powerful statistical procedures with support from functional and structural annotations of genes [296].

The Atlantic salmon is of particular importance to the global aquaculture industry. Salmonid cDNA microarrays were constructed shortly after large-scale sequencing of salmon and trout cDNA libraries by several research institutes. One of the projects related to salmon sequencing is GRASP (Genomics Research on Atlantic Salmon Project), an initiative funded by Genome Canada that is intended to improve understanding of physiological and evolutionary processes influencing the survival and phenotype of salmonids and other fish in natural and aquaculture environments. The first salmonid GRASP microarray platform (GRASP-1), containing 7356 salmonid elements representing 3557 different cDNAs (3.7K), was obtained from 80,388 ESTs, principally from cDNA libraries [298] of different salmon species such as Atlantic salmon, rainbow trout, Chinook salmon, sockeye salmon, and lake whitefish cDNA libraries. The second version of the GRASP microarray platform (GRASP-2) was developed and contained cDNAs representing 16,006 genes (16K). The genes identified in the array have been stringently selected from Atlantic salmon and rainbow trout EST databases representing a wide variety of different classes of genes [297]. Finally, a new expanded salmonid cDNA microarray (GRASP-3) of 32,000 features (32K) was created where 69% of the total EST collection used was from Atlantic salmon [298]. The Aleksei Krasnov's group designed the rainbow trout microarray (SFA1.0) by identifying a relatively small number of genes (1300 genes; 1.3K) using clones from normalized and subtracted cDNA libraries, as well as genes selected by the functional categories of Gene Ontology for inclusion in a microarray aimed at characterizing transcriptome responses to environmental stressors [299] to maximize the presence of transcripts related directly to immune response in rainbow trout, because of which this platform is also called Immunochip (SFA1.0 immunochip). The updated SFA platform (1.8K; SFA2.0 immunochip) was specially designed for studies of responses to pathogens and stressors and has substantially improved coverage of immune genes [300]. Another cDNA platform in commercial fish species has been designed in Japanese flounder [301] and European flounder [302], turbot [303], and sole [304].However despite impressive achievements, cDNA platforms suffer from limitations and disadvantages. At present most research groups working with salmonids and other aquaculture species do not have full access to clones required for fabrication of cDNA microarrays. Maintenance and PCR amplification of large clone sets is expensive and time consuming, while the risk of errors is high [296]. Probably the most important drawback of cDNA microarrays is their limited ability to discriminate paralogs since long probes cross-hybridize with highly similar transcripts from members of multi-gene families [305]. In salmonids this problem is aggravated by the large number of expressed gene duplicates. These complications can be resolved with oligonucleotide microarrays (ONM) that also provide greater accuracy and reproducibility of analyses. Until recently, the use of ONM platforms was hampered by the cost, but they are now rapidly replacing cDNA platforms. Construction of ONM platforms begins with establishment of mRNA sequence sets for comprehensive coverage of transcriptomes with low redundancy. The next stage is identifying genes by searching protein databases and annotating them according to functions, pathways and structural features. For successful development and use of ONM, it is necessary to define the gene composition and optimum number of spot replicates and to choose criteria for quality assessment [296].

Because of the commercial importance of salmonid species, there is special interest gene expression pattern against different pathogens. Initially salmonid (rainbow trout) ONM contained 1672 elements, representing more than 1400 genes [306]. Currently, one of the most often used ONM platforms to evaluate the response against different conditions and pathogens is the custom salmon ONM (SIQ-3), based on the Agilent Technology system (21K in 4x44K format). Because limited availability of peripheral blood leukocyte (PBL) markers is a well-recognised problem of fish immunology, this platform compares the transcriptomes of PBL and other tissues to search for genes with preferential expression in leukocytes [296], making it a very significant tool to evaluate the response to pathogens in Atlantic salmon. Another ONM platform based on 500K ESTs Atlantic salmon and 250K ESTs rainbow trout [298] is the cGRASP 44K salmonid oligo array (Agilent eArray), although no studies employing this platform have been published yet. Another ONM has been designed in fish model organisms like zebrafish and in commercial fish species such as channel catfish and turbot [307].

Functional genomic studies based on evaluating immune responses, also called immunogenomics, have been conducted *in vivo* to evaluate the response to different pathogens at the systemic level in different organs, especially the liver and head kidney. The functional genomic approach has been used with *Salmo salar* and *Oncorhynchus mykiss*, where PAMPs, whole bacteria and viruses are the most studied pathogens. Here we present different works in fish challenged by bacteria or viruses where differential gene expression profiles were evaluated using microarray platforms with special emphasis on *in vivo* fish immune response (Table 2).

## 1.15. Studies with Bacterial Pathogens, Pamps and Cytokine Network Interactions

One of the most commonly studied bacterial pathogens is *Aeromonas salmonicida*, a gram-negative bacteria and the causative agent of furunculosis. In fact prior to the development of species-specific cDNA microarrays a preliminary study used a human microarray (GENEFILTERS GF211) to explore the liver response in Atlantic salmon infected using a cohabitation model [275]. Only 4 mRNAs were consistently up-regulated ($p < 0.01$) from the 241 positively identified spots with a clearly detectable hybridization signal, none of them related to cytokine expression. This was probably due to the lack of sequence homology, a problem commonly associated with cross-species cDNA hybridization. Thus the creation of species-specific platforms was a key step in fish immunology. Using a custom Atlantic salmon cDNA microarray (NRC-IMB) consisting of over 4,000 different cDNA amplicons, the first results for challenge with Aeromonas salmonicida were reported in 2005 [275]. The study described a cohabitation challenge and identified 16 up-regulated mRNAs in all three tissues studied (spleen, liver and head kidney), whereas 2 and 19 mRNAs were identified as down-regulated in the head kidney and liver, respectively. The authors found that genes related to the acute phase response were up-regulated in spleen and head kidney of infected salmon, indicating that the infected fish underwent a typical acute phase response to infection.

The effects of an *Aeromonas salmonicida* infection were recently reported in turbot, *Scolphtalmus maximus*, [307]. Using a custom designed oligonucleotide-microarray (8x15K), the authors identified a set of 48 differentially regulated mRNAs in the spleen of challenged fish at 3 dpi, mostly related to the acute-phase and the stress/defence immune response. A study using channel and blue catfish explored the effects of a gram-negative bacterial infection on the acute phase response (APR) [308]. The authors showed up-regulation of mRNA transcripts involved in iron homeostasis, transport proteins, complement components and inflammatory and humoral immune response, indicating that conserved APR occurs as part of the innate immune response in both catfish species. Interestingly, a more acute response was observed composed of several immune pathways in the blue but not the channel catfish. More studies are required to elucidate expression patterns resulting from gram-negative bacterial infection of phylogenetically similar and different fish are required to describe common and divergent responses. This could lead to the development of marker systems, consensus on the APR in fish and treatments tailored to certain species, all of which have significant applied interest.

The activity of LPS from gram-negative bacteria, a common membrane-associated PAMP used in immunological studies, has been explored in several fish species. These studies include effects on the spleen in channel catfish [285], rainbow trout head kidney [309], and liver in the Senegalese sole [310]. Using a 19K oligonucleotide microarray (ONM) it was observed that some pro-inflammatory mRNAs in the catfish spleen were up-regulated very quickly, principally between 2 and 4 hours post-injection with LPS, whereas immunoglobulin- (2h post-injection) and antigenic presentation-related mRNA transcripts were repressed 24h post-injection [311]. A similar inhibition was reported in head kidney of rainbow trout, where the suppression of major cellular processes, including immune function and an initial stress reaction, was followed by a proliferative hematopoietic-type/biogenesis response 3 dpi [309]. However, in the Senegalese sole a clear up-regulation of transcripts related to the immune response was reported 24 hpi in the liver [310]. These results collectively highlight the diversity of responses observed at the tissue level and reflect the nature of the immune system that is diffusely located throughout many organ compartments.

**TABLE 2.** Transcriptomics studies on fish after treatments with bacteria or virus in vivo analyzed with microarrays. FDKC: Formaldehyde –killed cells; LPS: Lipopolysaccharide; FKC: Formalin-killed cells; ISAV: Infectious salmon anemia virus; PD: Pancreas disease; CMS: Cardiomyopathy syndrome; HSMI: Heart and skeletal muscle inflammation; IHNV: Infectious hematopoietic necrosis virus; VHSV: Viral hemorrhagic septicemia virus; HIRRV: Hirame rhabdovirus; poly I:C: polyriboinosinic polyribocytidylic acid; PBL: Peripheral blood leukocytes; ONM: oligonucleotide microarray.

| **Microorganism** | **Fish** | **Pathogen** | **Tissue/Cell type** | **Resource** | **Platform** | **Reference** |
|---|---|---|---|---|---|---|
| Bacteria | Atlantic | *Aeromonas salmonicida* ( | Liver | cDNA | GENE-FILTERS | Tsoi et al., 2003 |

| | | | | | |
|---|---|---|---|---|---|
| salmon | A449) | | | GF211 | |
| | *Aeromonas salmonicida* ( A449) | Head kidney/liver/spleen | cDNA | NRC-IMB | Ewart et al., 2005 |
| | *Aeromonas salmonicida* | Liver/spleen | cDNA | SFA-2 | Skugor et al., 2009 |
| | *Aeromonas salmonicida* ( Brivax II) | Liver | cDNA | TRAITS/SPG | Martin et al., 2010 |
| | *Piscirickettsia salmonis* | Head kidney | cDNA | GRASP-1 | Rise et al., 2004 |
| Chinook salmon | *Vibrio anguillarum* | Head kidney | cDNA | GRASP-1 | Ching et al., 2010 |
| Rainbow Trout | *Vibrio anguillarum* ( FDKC) | Liver | ONM | OSUrbt | Gerwick et al., 2007 |
| | LPS (*E.coli* 026:B6 ) | Head kidney | cDNA | SFA-1 | MacKenzie et al., 2008 |
| | LPS (*E. coli* 026:B6) | HK macrophages | cDNA | SFA-1 | Mackenzie et al., 2008 |
| | LPS, PGN (*E. coli* 0111:B4) | HK macrophages | cDNA | SFA-2 | Boltaña et al., 2011 |
| | PGN (*E. coli* 0111:B4, K12) | HK macrophages | cDNA | SFA-2 | Boltaña et al., 2011 |
| | LPS, PGN (*E. coli* 0111:B4) /poly I:C | Erythrocytes | cDNA | SFA-2 | Morera et al., 2011 |
| Brook trout | LPS (*E. coli* 026:B6) | HK macrophages | cDNA | SFA-1 | Mackenzie et al., 2006 |
| Channel catfish | LPS (*E.coli* 0127:B8 ) | Spleen | ONM | UMSMED-1 | Li et al., 2006 |
| | *Edwardsiella ictaluri* (MS-S97-773) | Gills/head kidney/liver/skin /spleen | ONM | UMSMED-2 | Peatman et al., 2007 |
| Blue catfish | *Edwardsiella ictaluri* (MS- | Liver | ONM | UMSMED-2 | Peatman et al., |

| | | | | | | |
|---|---|---|---|---|---|---|
| | | S97-773) | | | | 2008 |
| | Japanese flounder | *Edwardsiella tarda* (NE9505) | Head kidney | cDNA | Japanese flounder custom-3 | Yasuike et al., 2010 |
| | | *Streptococcus iniae* (02; FKC) | Head kidney cells | cDNA | Japanese flounder custom-3 | Dumrongphol et al., 2008 |
| | | *Mycobacterium bovis* (TUMSAT-Msp001) FKC | Kidney cells | cDNA | Japanese flounder custom-3 | Kato el al., 2010 |
| | Zebrafish | *Mycobacterium marinum* (M, E11) | Whole fish | ONM | MWG - Sigma Genosys - Affimetrix | Meijer et al., 2005 |
| | | *Mycobacterium marinum* (E11; Mma20) | Whole fish | ONM | ZF Agilent | Van der Sar et al., 2009 |
| | | *Streptococcus suis* (HA9801) | Whole fish | ONM | Affimetrix Zebrafish GeneChip | Wu et al., 2010 |
| | Turbot | *Aeromonas salmonicida* | Spleen | ONM | Turbot custom Agilent | Millán et al., 2010 |
| | Solea | LPS (*E. coli* 011:B4) | Liver | cDNA | GENIPOL-1 | Osuna-Jimenez et al., 2009 |
| Virus | Atlantic salmon | ISAV (Glesvaer 2/90) | Gills/heart/liver/spleen | cDNA | SFA-2 | Jorgensen et al., 2008 |
| | | ISAV (Glesvaer2/90) | Heart/PBL | ONM | SIQ-3 | Krasnov et al., 2011 |
| | Chinook salmo | ISAV NA-HRP4 (970) | Head kidney | cDNA | GRASP-3 | Leblanc et al., |

| | | | | | |
|---|---|---|---|---|---|
| n | | | | | 2010 |
| Rainbow Trout | IHNV (32/87)/attenuated IHNV | Head kidney | cDNA | SFA-1 | MacKenzie et al., 2008 |
| | IHNV (strain 220-90) | Head kidney | cDNA | GRASP-2 | Purcell et al., 2011 |
| Japanese flounder | VHSV (KRRV9822) | Head kidney cells | cDNA | Japanese flounder custom-1 | Byon et al., 2005 |
| | VHSV (KRRV9822) | Head kidney cells | cDNA | Japanese flounder custom-2 | Byon et al., 2006 |
| | HIRRV (8601H) | Kidney cells | cDNA | Japanese flounder custom-1 | Yasuike et al., 2007 |
| | HIRRV (8601H) | Kidney cells | cDNA | Japanese flounder custom-3 | Yasuike et al., 2010 |
| Zebrafish | VHSV (07.71) | Fin/head kidney/liver/spleen | ONM | ZF Agilent | Encinas et a., 2010 |
| Turbot | Nodavirus (AH95)/poly (I:C) | Kidney | cDNA | Turbot custom | Park et al., 2009 |

For gram-positive infections in fish at the level of transcriptome analyses, infection of zebrafish with *Streptococcus suis* is the only model reported [272]. *Streptococcus suis* is a pathogen associated with zoonosis reported in several countries [312, 313]. The Affymetrix Zebrafish GeneChip was used to identify 125 up-regulated transcripts where the most significant pathways were antigen processing and presentation, leukocyte trans-endothelial migration and the proteosome. The authors suggested that the target list obtained could serve as infection markers for gram-positive infection in fish.

Undoubtedly, the identification of prognostic biomarkers for disease resistance is a major aim for aquaculture. Functional genomics has the potential to identify such potential tools. Disease resistance is normally measured by challenge with the pathogen of interest and assessing the cumulative mortalities. Surviving fish or non-challenged siblings from the same family are then considered 'resistant'. Because this process is costly there is a need for non-lethal methodologies of measuring resistance, ideally based on molecular determinants of resistance. An initial example of this approach used the GRASP 3.7K cDNA array to identify *in vitro* macrophage and *in vivo* head kidney biomarkers in response to *Piscirickettsia salmonis* infection, yielding a number of 11 regulated genes common to both challenges. The researchers proposed 19 highly regulated transcripts as potential biomarkers to evaluate the efficacy of vaccines against *Piscirickettsia salmonis* [314]. C-type lectin 2-1, a gene whose product is involved in endocytosis and the C/EBP-driven inflammatory response [315] was identified and has been identified in almost all reports in which bacterial preparations have been used to challenge live fish [275, 309, 314, 316, 317]. Another study aimed at identifying biomarkers at the transcriptional level described differences between triploid and diploid Chinook salmon under live Vibrio anguillarum challenge using the GRASP 3.7K cDNA microarray [318]. Twelve annotated mRNAs were identified as showing significant differences between diploid and triploid fish. The authors however were unable to provide a description of the underlying mechanisms to explain the observed reduced immune function of triploid salmon.

Individual variation is a major hurdle for the development of prognostic markers as both genetic and epigenetic factors must be taken into account. The utility of, for example, C-type lectin in salmon, and other potential biomarkers in other species for bacterial disease resistance, requires further development. The future publication of several fish genomes coupled to array platforms with a much increased transcript representation could provide an exciting route to further develop this strategy by combining both functional and structural genomics for species of commercial interest with a sequenced genome. Several studies have attempted to correlate gene expression profiles with the activity of bacterins (killed bacteria preparations) used to vaccinate fish in culture. Most studies have concentrated on the rainbow trout and Japanese flounder [262, 277, 319, 320]. In trout, intraperitoneal administration of killed *V. anguillarum* resulted in identifying 36 differentially expressed transcripts [320]. Most of the identified targets are involved in inflammatory response and respond to a broad range of stimuli. This suggests that these targets have little use as markers for vaccination, contrary to previous descriptions in other studies. Both the second and fourth versions of the Japanese flounder cDNA microarray have been used to address vaccination [262, 277, 319]. The results of experimental infection with Gram-negative *E. tarda* indicated that a formalin-killed preparation reduced mortality in vaccinated fish from 90% to 20% [277]. However, a correlation between the transcriptome and the efficacy of vaccination could not be identified.

The effects of a commercial vaccine for Atlantic salmon (a six-component oil-adjuvant vaccine from PHARMAQ) were evaluated to correlate vaccine protection to high and low resistance to furunculosis. The authors did not find any association between either group and suggested that "outcomes of vaccination depend largely on the ability of host to prevent the negative impacts of immune response and to repair damages" [305]. Although this study did not identify correlations between vaccination and gene expression profiles, the potential of a functional genomics approach to evaluate the efficacy and underlying mechanisms of vaccination is highlighted. In terms of the immune response and the resulting complexity in expression patterns resulting from multiple cell types and different tissue responses, the investigator has the potential to obtain a clearer 'image' of the biological response from global expression data. A key objective is therefore to increase the available genomic resources much facilitated by next generation sequencing technologies to form a more robust representation of the immune system among different fish species. Furthermore, the increasing use of ONM platforms will also improve comparison across species as data sets become more easily comparable.

It remains difficult to compare microarray experiments across distinct platforms. In this respect, Meijer et al. 2005, evaluated host transcriptome profiling to *Mycobacterium marinum* infection of adult zebrafish employing three oligonucleotides platforms (MWG, Sigma Genosys, and Affimetrix). At a significance level of $P < 1.00E\text{-}5$, there were differences among the platforms in the total number of more than 2-fold up-regulated genes, whereas the 2-fold down-regulated genes were in a similar range. Evaluation of the distribution of infection-induced genes over different categories reveled some divergence in the set from MWG, probably due to the abundance of genes of the same UniGene cluster. As well, from the total overlap of 4,138 UniGene clusters among the three microarrays, only 66 and 93 genes were up- and down-regulated, respectively [321]. With this antecedent, the same group generated a new platform (Agilent 44K) that includes their 22K probes, a 16K set probes similar to the Sigma-Compugen oligonucleotide library, and 6K set of probes for selected genes of interest indentified by previous data mining of zebrafish transcript and genome databases [322], and they evaluated the transcriptome response to acute and chronic infection by *Mycobacterium marinum*. This important effort in combing different platforms makes it clear that not all relevant genes, including immune-related ones, are represented in all platforms. Consequntly, new efforts are necessary to broaden our understanding of the immune response in fish challenged with a pathogen of interest.

## 1.16. Wide Screening In Fish Challenged With Viral Pathogens

Two studies have reported host responses to IHNV with the SFA and GRASP platforms [309, 323]. The potential mechanisms responsible for host-specific virulence were assessed in rainbow trout infected with high (M) and low virulence (U) strains of IHNV. A marked down-regulation in biological processes, including the immune response, lymphocyte activation, response to stress, transcription and translation, together with a greater viral load (M), suggest that the higher virulence is due to the ability to suppress the immune response via the transcriptional and translational machinery of cell [323]. Furthermore, in rainbow trout was compared the expression profiles of IHNV and attenuated IHNV were compared in rainbow trout over a short time frame of one and three days post-challenge. At 3 dpi, a significant change in the transcriptional program of head kidney revealed an immunological shift orientated toward the activation of adaptive immunity. This shift was IHNV-dependent as determined by differences between the attenuated and virulent IHNV specific expression profiles. The rapid systemic spreading of IHNV inhibited TNFα, MHC class I, and several macrophage and cell cycle/differentiation markers and favored a MHC class II, immunoglobulin and MMP/TBX4-enhanced immune response [309].

Parallel studies were conducted with a cDNA microarray enriched with 213 immune-related genes to study the immune response and the efficacy of DNA vaccines containing the viral G proteins of VHSV and HIRRV administered intramuscularly in the Japanese flounder [301, 324]. As expected, all DNA vaccines containing the viral G glycoprotein conferred specific protection to the challenged fish one month after vaccination. It is suggested that the protection occurs via the IFN type I system due to the number of IFN-related genes up-regulated in both studies, ISG-15, interferon-stimulated gene 56kDa (ISG56) and the Mx protein. In both studies, VHSV and HIRRV in the Japanese flounder, the majority of differentially up-regulated genes were identified between 3 and 7 days post-vaccination (dpv), including the less effective DNA vaccine containing N protein of HIRRV. Interestingly, Mx, an antiviral protein commonly used as a marker for antiviral activity in animal species, was consistently up-regulated across vaccinations [301]. In a similar observation, IRF-3, Mx, Vig-1 and Vig-8 were up-regulated in trout at the site of DNA vaccination against IHNV at 7 dpv [323]. In turbot challenged with nodavirus, both Mx and IFN-inducible proteins were identified 24 hpi [303]. These and the previously described results suggest that both the host-expressed viral glycoprotein and the virulent rhadovirus induce a systemic anti-viral state indicative of non-specific IFN type1 innate immune response and that this canonical response is conserved among all fish. However, the mechanisms to develop a specific cytotoxic T or B lymphocyte-mediated humoral response in fish vaccinated with plasmid DNA-IHNV G that confers protective immunity have not been identified [323].

In direct relation to the above, a significant increase in transcript markers for adaptive immunity was reported in Atlantic salmon during ISA virus (ISAV) infection [325]. Importantly, a progressive increase was observed in IgZ mRNA parallel to a decrease in IgM expression that peaked > 30 days post-infection. This coordinated increase in a group of genes related to B lymphocyte differentiation and maturation and activation of T lymphocyte-mediated immunity, including CD4, TGF-β, CD8α and IFNγ, provides strong evidence for the coordinated regulation of the two arms of the immune system in response to viral infection. An important technological contribution derived from the above study was the development of an ONM for Atlantic salmon (SIQ-3). The first assessment of the performance of these arrays was carried out in Atlantic salmon for the study of virus-responsive genes from samples infected with ISA, salmonid alpha virus/PD-virus, cardiomyopathy syndrome (CMS) agent, heart and skeletal muscle inflammation (HSMI) and PBL from fish infected with ISAV. Some 95 up-regulated transcripts were identified. Most of the regulated transcripts are related directly to the immune response or associated with antiviral response [296]. As previously mentioned, the creation of species-specific platforms has been a key challenge for the study of the immune response against pathogens. Despite impressive achievements, cDNA microarrays suffer from limitations and disadvantages, the most important drawback being the limited ability to discriminate between paralogs as long cDNA probes cross-hybridize with highly similar transcripts from members of multi-gene families [305]. Furthermore this information needs to be supplemented to establish if the increased level of detected transcripts is consistent with specific protein synthesis. Recently, a study employed a combined proteomic and transcriptomic approach to evaluate the immune response against VHS [326]. In the fins of infected fish a series of mRNA transcripts principally related to complement components, immunoglobulin-related proteins, and macrophages were up-regulated (> 2-fold), whereas in parallel using two dimensional differential gel electrophoresis (2D-DIGE), enzymes of the glycolytic pathway and some proteins related to cytoskeletal remodelling and apoptosis (such as annexin A1a) increased with infection. However, very few proteins related to anti-viral response were identified.

## 2. CONCLUDING REMARKS

A complex network exists to regulate the innate and adaptive immune responses of fish from the various cytokines that have been reported. The study of the functional activity of these cytokines is in progress and it will be interesting to know whether mammalian Th1, Th2, Th17 and Treg responses are present in fish, regulating specific cell-mediated immunity. The recombinant production of these cytokines and antibodies against them will be the next challenge in understanding the balance of such immune responses and aid in the effective design of therapeutic strategies to manipulate the fish immune system. towards humoral or cellular immunity in response to specific antigen stimulation, vaccine strategies, functional diets to increase the quality of fishery production and predict the health of cultured fish.

The study of functional genomics in fish has provided substantial data on species of commercial interest. The major aim has been to functionally identify the intensity of responses to specific pathogens and their associated molecular components and to identify transcripts in a whole organism or specific tissue that contribute to such responses. However, the complex biology of the immune response, in which different spatial-temporal expression occurs in multiple cell types at distinct body locations, makes complete mapping of a response difficult and expensive. Moreover, considering that arrays are only as good as the transcripts represented upon them. Thus, the representation of transcripts relevant to the immune response is intimately linked to gene discovery efforts through large-scale sequencing projects, where strategies like SSH contribute not only to understanding transcriptomic response against specific pathogens but also to gene discovery. In this area, access to high-throughput NGS technology has increased in recent years and promises to make an important contribution to understanding immune response in fish. The major task now is the meta-analysis of transcriptomic data to delineate responses common among fish species to specific pathogen groups and highly specific responses. This approach will reveal host specific expression profiles and facilitate the identification of prognostic markers for diseases.

## ACKNOWLEDGMENTS

The authors are grateful for the support of VRID-USACH, INNOVA-CORFO 07CN13PBT-90, INNOVA-CORFO 09MCSS6691, INNOVA-CORFO 09MCSS6698 and CONICYT Fellowships to K. Maisey and D. Toro-Ascuy.

# REFERENCES


1. Flajnik MF, Du Pasquier L. Evolution of the Immune System. 6 th ed. Paul WE, editor: Lippincott Williams & Wilkins; 2008.
2. Akirav EM, Liao S, Ruddle NH. Lymphoid Tissues and Organs. 6 th ed. Paul WE, editor: Lippincott Williams & Wilkins; 2008.
3. Zapata AG, Chibá A, Varas A. Cells and tissues of the immune system of fish. In: Iwama G, Nakanishi T, editors. The fish immune system Organism, pathogen and environment: Academic Press; 1996. p. 1-62.
4. Meseguer J, Lopez-Ruiz A, Garcia-Ayala A. Reticulo-endothelial stroma of the head-kidney from the seawater teleost gilthead seabream (Sparus aurata L.): an ultrastructural and cytochemical study. Anat Rec. 1995 Mar;241(3):303-9.
5. Danneving BH, Lauve A, Press MC, Landsverk T. Receptor-mediated endocytosis and phagocytosis by rainbow trout head kidney sinusoidal cells. Fish Shellfish Immunol. 1994;4:3-18.
6. Brattgjerd S, Evensen O. A sequential light microscopic and ultrastructural study on the uptake and handling of Vibrio salmonicida in phagocytes of the head kidney in experimentally infected Atlantic salmon (Salmo salar L.). Vet Pathol. 1996 Jan;33(1):55-65.
7. Kaattari SL, Irwin MJ. Salmonid spleen and anterior kidney harbor populations of lymphocytes with different B cell repertoires. Dev Comp Immunol. 1985 Summer;9(3):433-44.
8. Herraez MP, Zapata AG. Structure and function of the melano-macrophage centres of the goldfish Carassius auratus. Vet Immunol Immunopathol. 1986 Jun;12(1-4):117-26.
9. Tsujii T, Seno S. Melano-macrophage centers in the aglomerular kidney of the sea horse (teleosts): morphologic studies on its formation and possible function. Anat Rec. 1990 Apr;226(4):460-70.
10. Bowden TJ, Cook P, Rombout JH. Development and function of the thymus in teleosts. Fish Shellfish Immunol. 2005 Nov;19(5):413-27.
11. Zapata AG, Amemiya CT. Phylogeny of lower vertebrates and their immunological structures. Curr Top Microbiol Immunol. 1999;248:67-107.
12. Espenes A, Press CM, Dannevig BH, Landsverk T. Immune-complex trapping in the splenic ellipsoids of rainbow trout (Oncorhynchus mykiss). Cell Tissue Res. 1995 Oct;282(1):41-8.
13. Vallejo AN, Miller NW, Harvey NE, Cuchens MA, Warr GW, Clem LW. Cellular pathway(s) of antigen processing and presentation in fish APC: endosomal involvement and cell-free antigen presentation. Dev Immunol. 1992;3(1):51-65.
14. Vallejo AN, Miller NW, Clem LW. Cellular pathway(s) of antigen processing in fish APC: effect of varying in vitro temperatures on antigen catabolism. Dev Comp Immunol. 1992 Sep-Oct;16(5):367-81.
15. Ellis AE. Antigen trapping in the spleen and kidney of the plaice, Pleuronectes platessa. J Fish Dis. 1980;3:413-26.
16. Agius C, Roberts RJ. Melano-macrophage centres and their role in fish pathology. J Fish Dis. 2003 Sep;26(9):499-509.
17. Medzhitov R. The Innate Immune System. 6 th ed. Paul WE, editor: Lippincott Williams & Wilkins; 2008.
18. Janeway CA, Jr. How the immune system protects the host from infection. Microbes Infect. 2001 Nov;3(13):1167-71.
19. McHeyzer-Williams LJ, McHeyzer-Williams MG. Antigen-specific memory B cell development. Annu Rev Immunol. 2005;23:487-513.
20. Sallusto F, Lanzavecchia A. Heterogeneity of CD4+ memory T cells: functional modules for tailored immunity. Eur J Immunol. 2009 Aug;39(8):2076-82.

21. Dixon B, Stet RJ. The relationship between major histocompatibility receptors and innate immunity in teleost fish. Dev Comp Immunol. 2001 Oct-Dec;25(8-9):683-99.
22. Abbas AK, Lichtman AH, Pillai S. Cells and Tissues of the Adaptive Immune System. 6 th ed: Saunders; 2007.
23. Castro R, Bernard D, Lefranc MP, Six A, Benmansour A, Boudinot P. T cell diversity and TcR repertoires in teleost fish. Fish Shellfish Immunol. 2011 Nov;31(5):644-54.
24. Secombes CJ, Wang T, Bird S. The interleukins of fish. Dev Comp Immunol. 2011 Dec;35(12):1336-45.
25. Wang T, Gorgoglione B, Maehr T, Holland JW, Vecino JL, Wadsworth S, et al. Fish Suppressors of Cytokine Signaling (SOCS): Gene Discovery, Modulation of Expression and Function. J Signal Transduct. 2011;2011:905813.
26. Alejo A, Tafalla C. Chemokines in teleost fish species. Dev Comp Immunol. 2011 Dec;35(12):1215-22.
27. Nakanishi T, Toda H, Shibasaki Y, Somamoto T. Cytotoxic T cells in teleost fish. Dev Comp Immunol. 2011 Dec;35(12):1317-23.
28. Boltana S, Donate C, Goetz FW, MacKenzie S, Balasch JC. Characterization and expression of NADPH oxidase in LPS-, poly(I:C)- and zymosan-stimulated trout (Oncorhynchus mykiss W.) macrophages. Fish Shellfish Immunol. 2009 Apr;26(4):651-61.
29. Hanington PC, Hitchen SJ, Beamish LA, Belosevic M. Macrophage colony stimulating factor (CSF-1) is a central growth factor of goldfish macrophages. Fish Shellfish Immunol. 2009 Jan;26(1):1-9.
30. Salinas I, Zhang YA, Sunyer JO. Mucosal immunoglobulins and B cells of teleost fish. Dev Comp Immunol. 2011 Dec;35(12):1346-65.
31. Zhang YA, Salinas I, Oriol Sunyer J. Recent findings on the structure and function of teleost IgT. Fish Shellfish Immunol. 2011 Nov;31(5):627-34.
32. Randelli E, Buonocore F, Scapigliati G. Cell markers and determinants in fish immunology. Fish Shellfish Immunol. 2008 Oct;25(4):326-40.
33. Rombout JH, Huttenhuis HB, Picchietti S, Scapigliati G. Phylogeny and ontogeny of fish leucocytes. Fish Shellfish Immunol. 2005 Nov;19(5):441-55.
34. Araki K, Akatsu K, Suetake H, Kikuchi K, Suzuki Y. Characterization of CD8+ leukocytes in fugu (Takifugu rubripes) with antiserum against fugu CD8alpha. Dev Comp Immunol. 2008;32(7):850-8.
35. Toda H, Saito Y, Koike T, Takizawa F, Araki K, Yabu T, et al. Conservation of characteristics and functions of CD4 positive lymphocytes in a teleost fish. Dev Comp Immunol. 2011 Jun;35(6):650-60.
36. Takizawa F, Dijkstra JM, Kotterba P, Korytar T, Kock H, Kollner B, et al. The expression of CD8alpha discriminates distinct T cell subsets in teleost fish. Dev Comp Immunol. 2011 Jul;35(7):752-63.
37. Kaattari S, Evans D, Klemer J. Varied redox forms of teleost IgM: an alternative to isotypic diversity? Immunol Rev. 1998 Dec;166:133-42.
38. Wilson M, Bengten E, Miller NW, Clem LW, Du Pasquier L, Warr GW. A novel chimeric Ig heavy chain from a teleost fish shares similarities to IgD. Proc Natl Acad Sci U S A. 1997 Apr 29;94(9):4593-7.
39. Stenvik J, Jorgensen TO. Immunoglobulin D (IgD) of Atlantic cod has a unique structure. Immunogenetics. 2000 May;51(6):452-61.
40. Danilova N, Bussmann J, Jekosch K, Steiner LA. The immunoglobulin heavy-chain locus in zebrafish: identification and expression of a previously unknown isotype, immunoglobulin Z. Nat Immunol. 2005 Mar;6(3):295-302.
41. Hansen JD, Landis ED, Phillips RB. Discovery of a unique Ig heavy-chain isotype (IgT) in rainbow trout: Implications for a distinctive B cell developmental pathway in teleost fish. Proc Natl Acad Sci U S A. 2005 May 10;102(19):6919-24.
42. Li J, Barreda DR, Zhang YA, Boshra H, Gelman AE, Lapatra S, et al. B lymphocytes from early vertebrates have potent phagocytic and microbicidal abilities. Nat Immunol. 2006 Oct;7(10):1116-24.

43. Overland HS, Pettersen EF, Ronneseth A, Wergeland HI. Phagocytosis by B-cells and neutrophils in Atlantic salmon (Salmo salar L.) and Atlantic cod (Gadus morhua L.). Fish Shellfish Immunol. 2010 Jan;28(1):193-204.
44. Wang T, Huang W, Costa MM, Secombes CJ. The gamma-chain cytokine/receptor system in fish: more ligands and receptors. Fish Shellfish Immunol. 2011 Nov;31(5):673-87.
45. Savan R, Sakai M. Genomics of fish cytokines. Comp Biochem Physiol Part D Genomics Proteomics. 2006 Mar;1(1):89-101.
46. Salazar-Mather TP, Hokeness KL. Cytokine and chemokine networks: pathways to antiviral defense. Curr Top Microbiol Immunol. 2006;303:29-46.
47. Svanborg C, Godaly G, Hedlund M. Cytokine responses during mucosal infections: role in disease pathogenesis and host defence. Curr Opin Microbiol. 1999 Feb;2(1):99-105.
48. DeVries ME, Ran L, Kelvin DJ. On the edge: the physiological and pathophysiological role of chemokines during inflammatory and immunological responses. Semin Immunol. 1999 Apr;11(2):95-104.
49. Whyte SK. The innate immune response of finfish--a review of current knowledge. Fish Shellfish Immunol. 2007 Dec;23(6):1127-51.
50. Rahman MM, McFadden G. Modulation of tumor necrosis factor by microbial pathogens. PLoS Pathog. 2006 Feb;2(2):e4.
51. Benedict CA, Banks TA, Ware CF. Death and survival: viral regulation of TNF signaling pathways. Curr Opin Immunol. 2003 Feb;15(1):59-65.
52. Hirono I, Nam BH, Kurobe T, Aoki T. Molecular cloning, characterization, and expression of TNF cDNA and gene from Japanese flounder Paralychthys olivaceus. J Immunol. 2000 Oct 15;165(8):4423-7.
53. Laing KJ, Wang T, Zou J, Holland J, Hong S, Bols N, et al. Cloning and expression analysis of rainbow trout Oncorhynchus mykiss tumour necrosis factor-alpha. Eur J Biochem. 2001 Mar;268(5):1315-22.
54. Zou J, Wang T, Hirono I, Aoki T, Inagawa H, Honda T, et al. Differential expression of two tumor necrosis factor genes in rainbow trout, Oncorhynchus mykiss. Dev Comp Immunol. 2002 Mar;26(2):161-72.
55. Garcia-Castillo J, Pelegrin P, Mulero V, Meseguer J. Molecular cloning and expression analysis of tumor necrosis factor alpha from a marine fish reveal its constitutive expression and ubiquitous nature. Immunogenetics. 2002 Jun;54(3):200-7.
56. Saeij JP, Stet RJ, de Vries BJ, van Muiswinkel WB, Wiegertjes GF. Molecular and functional characterization of carp TNF: a link between TNF polymorphism and trypanotolerance? Dev Comp Immunol. 2003 Jan;27(1):29-41.
57. Zou J, Secombes CJ, Long S, Miller N, Clem LW, Chinchar VG. Molecular identification and expression analysis of tumor necrosis factor in channel catfish (Ictalurus punctatus). Dev Comp Immunol. 2003 Dec;27(10):845-58.
58. Praveen K, Evans DL, Jaso-Friedmann L. Constitutive expression of tumor necrosis factor-alpha in cytotoxic cells of teleosts and its role in regulation of cell-mediated cytotoxicity. Mol Immunol. 2006 Feb;43(3):279-91.
59. Ordas MC, Costa MM, Roca FJ, Lopez-Castejon G, Mulero V, Meseguer J, et al. Turbot TNFalpha gene: molecular characterization and biological activity of the recombinant protein. Mol Immunol. 2007 Jan;44(4):389-400.
60. Grayfer L, Walsh JG, Belosevic M. Characterization and functional analysis of goldfish (Carassius auratus L.) tumor necrosis factor-alpha. Dev Comp Immunol. 2008;32(5):532-43.
61. Garcia-Castillo J, Chaves-Pozo E, Olivares P, Pelegrin P, Meseguer J, Mulero V. The tumor necrosis factor alpha of the bony fish seabream exhibits the in vivo proinflammatory and proliferative activities of its mammalian counterparts, yet it functions in a species-specific manner. Cell Mol Life Sci. 2004 Jun;61(11):1331-40.
62. Roca FJ, Mulero I, Lopez-Munoz A, Sepulcre MP, Renshaw SA, Meseguer J, et al. Evolution of the inflammatory response in vertebrates: fish TNF-alpha is a powerful

activator of endothelial cells but hardly activates phagocytes. J Immunol. 2008 Oct 1;181(7):5071-81.
63. Hardie LJ, Chappell LH, Secombes CJ. Human tumor necrosis factor alpha influences rainbow trout Oncorhynchus mykiss leucocyte responses. Vet Immunol Immunopathol. 1994 Jan;40(1):73-84.
64. Iliev DB, Liarte CQ, MacKenzie S, Goetz FW. Activation of rainbow trout (Oncorhynchus mykiss) mononuclear phagocytes by different pathogen associated molecular pattern (PAMP) bearing agents. Mol Immunol. 2005 Jun;42(10):1215-23.
65. Wang WL, Hong JR, Lin GH, Liu W, Gong HY, Lu MW, et al. Stage-specific expression of TNFalpha regulates bad/bid-mediated apoptosis and RIP1/ROS-mediated secondary necrosis in Birnavirus-infected fish cells. PLoS One. 2011;6(2):e16740.
66. Dinarello C, Arend W, Sims J, Smith D, Blumberg H, O'Neill L, et al. IL-1 family nomenclature. Nat Immunol. 2010 Nov;11(11):973.
67. Dinarello CA. Interleukin-1. Cytokine Growth Factor Rev. 1997 Dec;8(4):253-65.
68. Zou J, Grabowski PS, Cunningham C, Secombes CJ. Molecular cloning of interleukin 1beta from rainbow trout Oncorhynchus mykiss reveals no evidence of an ice cut site. Cytokine. 1999 Aug;11(8):552-60.
69. Fujiki K, Shin DH, Nakao M, Yano T. Molecular cloning and expression analysis of carp (Cyprinus carpio) interleukin-1 beta, high affinity immunoglobulin E Fc receptor gamma subunit and serum amyloid A. Fish Shellfish Immunol. 2000 Apr;10(3):229-42.
70. Scapigliati G, Buonocore F, Bird S, Zou J, Pelegrin P, Falasca C, et al. Phylogeny of cytokines: molecular cloning and expression analysis of sea bass Dicentrarchus labrax interleukin-1beta. Fish Shellfish Immunol. 2001 Nov;11(8):711-26.
71. Pelegrin P, Garcia-Castillo J, Mulero V, Meseguer J. Interleukin-1beta isolated from a marine fish reveals up-regulated expression in macrophages following activation with lipopolysaccharide and lymphokines. Cytokine. 2001 Oct 21;16(2):67-72.
72. Corripio-Miyar Y, Bird S, Tsamopoulos K, Secombes CJ. Cloning and expression analysis of two pro-inflammatory cytokines, IL-1 beta and IL-8, in haddock (Melanogrammus aeglefinus). Mol Immunol. 2007 Feb;44(6):1361-73.
73. Lee DS, Hong SH, Lee HJ, Jun LJ, Chung JK, Kim KH, et al. Molecular cDNA cloning and analysis of the organization and expression of the IL-1beta gene in the Nile tilapia, Oreochromis niloticus. Comp Biochem Physiol A Mol Integr Physiol. 2006 Mar;143(3):307-14.
74. Pleguezuelos O, Zou J, Cunningham C, Secombes CJ. Cloning, sequencing, and analysis of expression of a second IL-1beta gene in rainbow trout (Oncorhynchus mykiss). Immunogenetics. 2000 Oct;51(12):1002-11.
75. Cerretti DP, Kozlosky CJ, Mosley B, Nelson N, Van Ness K, Greenstreet TA, et al. Molecular cloning of the interleukin-1 beta converting enzyme. Science. 1992 Apr 3;256(5053):97-100.
76. Secombes CJ, Wang T, Hong S, Peddie S, Crampe M, Laing KJ, et al. Cytokines and innate immunity of fish. Dev Comp Immunol. 2001 Oct-Dec;25(8-9):713-23.
77. Lu DQ, Bei JX, Feng LN, Zhang Y, Liu XC, Wang L, et al. Interleukin-1beta gene in orange-spotted grouper, Epinephelus coioides: molecular cloning, expression, biological activities and signal transduction. Mol Immunol. 2008 Feb;45(4):857-67.
78. Reyes-Cerpa S, Reyes-Lopez FE, Toro-Ascuy D, Ibanez J, Maisey K, Sandino AM, et al. IPNV modulation of pro and anti-inflammatory cytokine expression in Atlantic salmon might help the establishment of infection and persistence. Fish Shellfish Immunol. 2012 Feb;32(2):291-300.
79. Seppola M, Larsen AN, Steiro K, Robertsen B, Jensen I. Characterisation and expression analysis of the interleukin genes, IL-1beta, IL-8 and IL-10, in Atlantic cod (Gadus morhua L.). Mol Immunol. 2008 Feb;45(4):887-97.
80. Zou J, Holland J, Pleguezuelos O, Cunningham C, Secombes CJ. Factors influencing the expression of interleukin-1 beta in cultured rainbow trout (Oncorhynchus mykiss) leucocytes. Dev Comp Immunol. 2000 Sep-Oct;24(6-7):575-82.

81. Buonocore F, Forlenza M, Randelli E, Benedetti S, Bossu P, Meloni S, et al. Biological activity of sea bass (Dicentrarchus labrax L.) recombinant interleukin-1beta. Mar Biotechnol (NY). 2005 Nov-Dec;7(6):609-17.
82. Hong S, Zou J, Crampe M, Peddie S, Scapigliati G, Bols N, et al. The production and bioactivity of rainbow trout (Oncorhynchus mykiss) recombinant IL-1 beta. Vet Immunol Immunopathol. 2001 Aug 30;81(1-2):1-14.
83. Peddie S, Zou J, Cunningham C, Secombes CJ. Rainbow trout (Oncorhynchus mykiss) recombinant IL-1beta and derived peptides induce migration of head-kidney leucocytes in vitro. Fish Shellfish Immunol. 2001 Nov;11(8):697-709.
84. Peddie S, Zou J, Secombes CJ. Immunostimulation in the rainbow trout (Oncorhynchus mykiss) following intraperitoneal administration of Ergosan. Vet Immunol Immunopathol. 2002 May;86(1-2):101-13.
85. Dinarello CA, Fantuzzi G. Interleukin-18 and host defense against infection. J Infect Dis. 2003 Jun 15;187 Suppl 2:S370-84.
86. Gracie JA, Robertson SE, McInnes IB. Interleukin-18. J Leukoc Biol. 2003 Feb;73(2):213-24.
87. Sugawara I. Interleukin-18 (IL-18) and infectious diseases, with special emphasis on diseases induced by intracellular pathogens. Microbes Infect. 2000 Aug;2(10):1257-63.
88. Micallef MJ, Ohtsuki T, Kohno K, Tanabe F, Ushio S, Namba M, et al. Interferon-gamma-inducing factor enhances T helper 1 cytokine production by stimulated human T cells: synergism with interleukin-12 for interferon-gamma production. Eur J Immunol. 1996 Jul;26(7):1647-51.
89. Okamura H, Nagata K, Komatsu T, Tanimoto T, Nukata Y, Tanabe F, et al. A novel costimulatory factor for gamma interferon induction found in the livers of mice causes endotoxic shock. Infect Immun. 1995 Oct;63(10):3966-72.
90. Dao T, Mehal WZ, Crispe IN. IL-18 augments perforin-dependent cytotoxicity of liver NK-T cells. J Immunol. 1998 Sep 1;161(5):2217-22.
91. Leung BP, Culshaw S, Gracie JA, Hunter D, Canetti CA, Campbell C, et al. A role for IL-18 in neutrophil activation. J Immunol. 2001 Sep 1;167(5):2879-86.
92. Xu D, Trajkovic V, Hunter D, Leung BP, Schulz K, Gracie JA, et al. IL-18 induces the differentiation of Th1 or Th2 cells depending upon cytokine milieu and genetic background. Eur J Immunol. 2000 Nov;30(11):3147-56.
93. Ghayur T, Banerjee S, Hugunin M, Butler D, Herzog L, Carter A, et al. Caspase-1 processes IFN-gamma-inducing factor and regulates LPS-induced IFN-gamma production. Nature. 1997 Apr 10;386(6625):619-23.
94. Gu Y, Kuida K, Tsutsui H, Ku G, Hsiao K, Fleming MA, et al. Activation of interferon-gamma inducing factor mediated by interleukin-1beta converting enzyme. Science. 1997 Jan 10;275(5297):206-9.
95. Akita K, Ohtsuki T, Nukada Y, Tanimoto T, Namba M, Okura T, et al. Involvement of caspase-1 and caspase-3 in the production and processing of mature human interleukin 18 in monocytic THP.1 cells. J Biol Chem. 1997 Oct 17;272(42):26595-603.
96. Sugawara S, Uehara A, Nochi T, Yamaguchi T, Ueda H, Sugiyama A, et al. Neutrophil proteinase 3-mediated induction of bioactive IL-18 secretion by human oral epithelial cells. J Immunol. 2001 Dec 1;167(11):6568-75.
97. Huising MO, Stet RJ, Savelkoul HF, Verburg-van Kemenade BM. The molecular evolution of the interleukin-1 family of cytokines; IL-18 in teleost fish. Dev Comp Immunol. 2004 May 3;28(5):395-413.
98. Zou J, Bird S, Truckle J, Bols N, Horne M, Secombes C. Identification and expression analysis of an IL-18 homologue and its alternatively spliced form in rainbow trout (Oncorhynchus mykiss). Eur J Biochem. 2004 May;271(10):1913-23.
99. Martin SA, Zou J, Houlihan DF, Secombes CJ. Directional responses following recombinant cytokine stimulation of rainbow trout (Oncorhynchus mykiss) RTS-11 macrophage cells as revealed by transcriptome profiling. BMC Genomics. 2007;8:150.

100. Wang T, Secombes CJ. Identification and expression analysis of two fish-specific IL-6 cytokine family members, the ciliary neurotrophic factor (CNTF)-like and M17 genes, in rainbow trout Oncorhynchus mykiss. Mol Immunol. 2009 Jul;46(11-12):2290-8.
101. Taga T, Kishimoto T. Gp130 and the interleukin-6 family of cytokines. Annu Rev Immunol. 1997;15:797-819.
102. Hirano T. Interleukin 6 and its receptor: ten years later. Int Rev Immunol. 1998;16(3-4):249-84.
103. Inoue K, Takano H, Shimada A, Morita T, Yanagisawa R, Sakurai M, et al. Cytoprotection by interleukin-6 against liver injury induced by lipopolysaccharide. Int J Mol Med. 2005 Feb;15(2):221-4.
104. Kaplanski G, Marin V, Montero-Julian F, Mantovani A, Farnarier C. IL-6: a regulator of the transition from neutrophil to monocyte recruitment during inflammation. Trends Immunol. 2003 Jan;24(1):25-9.
105. Bird S, Zou J, Savan R, Kono T, Sakai M, Woo J, et al. Characterisation and expression analysis of an interleukin 6 homologue in the Japanese pufferfish, Fugu rubripes. Dev Comp Immunol. 2005;29(9):775-89.
106. Iliev DB, Castellana B, Mackenzie S, Planas JV, Goetz FW. Cloning and expression analysis of an IL-6 homolog in rainbow trout (Oncorhynchus mykiss). Mol Immunol. 2007 Mar;44(7):1803-7.
107. Nam BH, Byon JY, Kim YO, Park EM, Cho YC, Cheong J. Molecular cloning and characterisation of the flounder (Paralichthys olivaceus) interleukin-6 gene. Fish Shellfish Immunol. 2007 Jul;23(1):231-6.
108. Castellana B, Iliev DB, Sepulcre MP, MacKenzie S, Goetz FW, Mulero V, et al. Molecular characterization of interleukin-6 in the gilthead seabream (Sparus aurata). Mol Immunol. 2008 Jul;45(12):3363-70.
109. Costa MM, Maehr T, Diaz-Rosales P, Secombes CJ, Wang T. Bioactivity studies of rainbow trout (Oncorhynchus mykiss) interleukin-6: effects on macrophage growth and antimicrobial peptide gene expression. Mol Immunol. 2011 Sep;48(15-16):1903-16.
110. Czupryn MJ, McCoy JM, Scoble HA. Structure-function relationships in human interleukin-11. Identification of regions involved in activity by chemical modification and site-directed mutagenesis. J Biol Chem. 1995 Jan 13;270(2):978-85.
111. Bartz H, Buning-Pfaue F, Turkel O, Schauer U. Respiratory syncytial virus induces prostaglandin E2, IL-10 and IL-11 generation in antigen presenting cells. Clin Exp Immunol. 2002 Sep;129(3):438-45.
112. Kernacki KA, Goebel DJ, Poosch MS, Hazlett LD. Early cytokine and chemokine gene expression during Pseudomonas aeruginosa corneal infection in mice. Infect Immun. 1998 Jan;66(1):376-9.
113. Nandurkar HH, Robb L, Begley CG. The role of IL-II in hematopoiesis as revealed by a targeted mutation of its receptor. Stem Cells. 1998;16 Suppl 2:53-65.
114. Trepicchio WL, Bozza M, Pedneault G, Dorner AJ. Recombinant human IL-11 attenuates the inflammatory response through down-regulation of proinflammatory cytokine release and nitric oxide production. J Immunol. 1996 Oct 15;157(8):3627-34.
115. Orazi A, Du X, Yang Z, Kashai M, Williams DA. Interleukin-11 prevents apoptosis and accelerates recovery of small intestinal mucosa in mice treated with combined chemotherapy and radiation. Lab Invest. 1996 Jul;75(1):33-42.
116. Liu Q, Du XX, Schindel DT, Yang ZX, Rescorla FJ, Williams DA, et al. Trophic effects of interleukin-11 in rats with experimental short bowel syndrome. J Pediatr Surg. 1996 Aug;31(8):1047-50; discussion 50-1.
117. Huising MO, Kruiswijk CP, van Schijndel JE, Savelkoul HF, Flik G, Verburg-van Kemenade BM. Multiple and highly divergent IL-11 genes in teleost fish. Immunogenetics. 2005 Jul;57(6):432-43.

118. Wang T, Holland JW, Bols N, Secombes CJ. Cloning and expression of the first nonmammalian interleukin-11 gene in rainbow trout Oncorhynchus mykiss. FEBS J. 2005 Mar;272(5):1136-47.
119. Santos MD, Yasuike M, Kondo H, Hirono I, Aoki T. Teleostean IL11b exhibits complementing function to IL11a and expansive involvement in antibacterial and antiviral responses. Mol Immunol. 2008 Jul;45(12):3494-501.
120. Ribeiro CM, Hermsen T, Taverne-Thiele AJ, Savelkoul HF, Wiegertjes GF. Evolution of recognition of ligands from Gram-positive bacteria: similarities and differences in the TLR2-mediated response between mammalian vertebrates and teleost fish. J Immunol. 2010 Mar 1;184(5):2355-68.
121. Murphy PM, Baggiolini M, Charo IF, Hebert CA, Horuk R, Matsushima K, et al. International union of pharmacology. XXII. Nomenclature for chemokine receptors. Pharmacol Rev. 2000 Mar;52(1):145-76.
122. Mukaida N, Harada A, Matsushima K. Interleukin-8 (IL-8) and monocyte chemotactic and activating factor (MCAF/MCP-1), chemokines essentially involved in inflammatory and immune reactions. Cytokine Growth Factor Rev. 1998 Mar;9(1):9-23.
123. Hebert CA, Vitangcol RV, Baker JB. Scanning mutagenesis of interleukin-8 identifies a cluster of residues required for receptor binding. J Biol Chem. 1991 Oct 5;266(28):18989-94.
124. Clark-Lewis I, Schumacher C, Baggiolini M, Moser B. Structure-activity relationships of interleukin-8 determined using chemically synthesized analogs. Critical role of NH2-terminal residues and evidence for uncoupling of neutrophil chemotaxis, exocytosis, and receptor binding activities. J Biol Chem. 1991 Dec 5;266(34):23128-34.
125. Laing KJ, Zou JJ, Wang T, Bols N, Hirono I, Aoki T, et al. Identification and analysis of an interleukin 8-like molecule in rainbow trout Oncorhynchus mykiss. Dev Comp Immunol. 2002 Jun;26(5):433-44.
126. Lee EY, Park HH, Kim YT, Choi TJ. Cloning and sequence analysis of the interleukin-8 gene from flounder (Paralichthys olivaceous). Gene. 2001 Aug 22;274(1-2):237-43.
127. Sangrador-Vegas A, Lennington JB, Smith TJ. Molecular cloning of an IL-8-like CXC chemokine and tissue factor in rainbow trout (Oncorhynchus mykiss) by use of suppression subtractive hybridization. Cytokine. 2002 Jan 21;17(2):66-70.
128. Chen L, He C, Baoprasertkul P, Xu P, Li P, Serapion J, et al. Analysis of a catfish gene resembling interleukin-8: cDNA cloning, gene structure, and expression after infection with Edwardsiella ictaluri. Dev Comp Immunol. 2005;29(2):135-42.
129. Najakshin AM, Mechetina LV, Alabyev BY, Taranin AV. Identification of an IL-8 homolog in lamprey (Lampetra fluviatilis): early evolutionary divergence of chemokines. Eur J Immunol. 1999 Feb;29(2):375-82.
130. Laing KJ, Secombes CJ. Chemokines. Dev Comp Immunol. 2004 May 3;28(5):443-60.
131. Osborne LC, Abraham N. Regulation of memory T cells by gammac cytokines. Cytokine. 2010 May;50(2):105-13.
132. Sogo T, Kawahara M, Ueda H, Otsu M, Onodera M, Nakauchi H, et al. T cell growth control using hapten-specific antibody/interleukin-2 receptor chimera. Cytokine. 2009 Apr;46(1):127-36.
133. Morgan DA, Ruscetti FW, Gallo R. Selective in vitro growth of T lymphocytes from normal human bone marrows. Science. 1976 Sep 10;193(4257):1007-8.
134. Smith KA. T-cell growth factor. Immunol Rev. 1980;51:337-57.
135. Smith KA. Interleukin-2: inception, impact, and implications. Science. 1988 May 27;240(4856):1169-76.
136. Granucci F, Vizzardelli C, Pavelka N, Feau S, Persico M, Virzi E, et al. Inducible IL-2 production by dendritic cells revealed by global gene expression analysis. Nat Immunol. 2001 Sep;2(9):882-8.
137. Walker E, Leemhuis T, Roeder W. Murine B lymphoma cell lines release functionally active interleukin 2 after stimulation with Staphylococcus aureus. J Immunol. 1988 Feb 1;140(3):859-65.

138. Gaffen SL, Wang S, Koshland ME. Expression of the immunoglobulin J chain in a murine B lymphoma is driven by autocrine production of interleukin 2. Cytokine. 1996 Jul;8(7):513-24.
139. Yu TK, Caudell EG, Smid C, Grimm EA. IL-2 activation of NK cells: involvement of MKK1/2/ERK but not p38 kinase pathway. J Immunol. 2000 Jun 15;164(12):6244-51.
140. Gold MR, DeFranco AL. Biochemistry of B lymphocyte activation. Adv Immunol. 1994;55:221-95.
141. Bird S, Zou J, Kono T, Sakai M, Dijkstra JM, Secombes C. Characterisation and expression analysis of interleukin 2 (IL-2) and IL-21 homologues in the Japanese pufferfish, Fugu rubripes, following their discovery by synteny. Immunogenetics. 2005 Mar;56(12):909-23.
142. Sugamata R, Suetake H, Kikuchi K, Suzuki Y. Teleost B7 expressed on monocytes regulates T cell responses. J Immunol. 2009 Jun 1;182(11):6799-806.
143. Diaz-Rosales P, Bird S, Wang TH, Fujiki K, Davidson WS, Zou J, et al. Rainbow trout interleukin-2: cloning, expression and bioactivity analysis. Fish Shellfish Immunol. 2009 Sep;27(3):414-22.
144. Zhang YA, Hikima J, Li J, LaPatra SE, Luo YP, Sunyer JO. Conservation of structural and functional features in a primordial CD80/86 molecule from rainbow trout (Oncorhynchus mykiss), a primitive teleost fish. J Immunol. 2009 Jul 1;183(1):83-96.
145. Laing KJ, Bols N, Secombes CJ. A CXC chemokine sequence isolated from the rainbow trout Oncorhynchus mykiss resembles the closely related interferon-gamma-inducible chemokines CXCL9, CXCL10 and CXCL11. Eur Cytokine Netw. 2002 Oct-Dec;13(4):462-73.
146. Banchereau J, Bidaud C, Fluckiger AC, Galibert L, Garrone P, Malisan F, et al. Effects of interleukin 4 on human B-cell growth and differentiation. Res Immunol. 1993 Oct;144(8):601-5.
147. Mosmann TR, Cherwinski H, Bond MW, Giedlin MA, Coffman RL. Two types of murine helper T cell clone. I. Definition according to profiles of lymphokine activities and secreted proteins. J Immunol. 1986 Apr 1;136(7):2348-57.
148. Finkelman FD, Urban JF, Jr. The other side of the coin: the protective role of the TH2 cytokines. J Allergy Clin Immunol. 2001 May;107(5):772-80.
149. Yokota T, Otsuka T, Mosmann T, Banchereau J, DeFrance T, Blanchard D, et al. Isolation and characterization of a human interleukin cDNA clone, homologous to mouse B-cell stimulatory factor 1, that expresses B-cell- and T-cell-stimulating activities. Proc Natl Acad Sci U S A. 1986 Aug;83(16):5894-8.
150. Noma Y, Sideras P, Naito T, Bergstedt-Lindquist S, Azuma C, Severinson E, et al. Cloning of cDNA encoding the murine IgG1 induction factor by a novel strategy using SP6 promoter. Nature. 1986 Feb 20-26;319(6055):640-6.
151. Lee F, Yokota T, Otsuka T, Meyerson P, Villaret D, Coffman R, et al. Isolation and characterization of a mouse interleukin cDNA clone that expresses B-cell stimulatory factor 1 activities and T-cell- and mast-cell-stimulating activities. Proc Natl Acad Sci U S A. 1986 Apr;83(7):2061-5.
152. Heussler VT, Eichhorn M, Dobbelaere DA. Cloning of a full-length cDNA encoding bovine interleukin 4 by the polymerase chain reaction. Gene. 1992 May 15;114(2):273-8.
153. Li-Weber M, Krammer PH. Regulation of IL4 gene expression by T cells and therapeutic perspectives. Nat Rev Immunol. 2003 Jul;3(7):534-43.
154. Ansel KM, Djuretic I, Tanasa B, Rao A. Regulation of Th2 differentiation and Il4 locus accessibility. Annu Rev Immunol. 2006;24:607-56.
155. Ohtani M, Hayashi N, Hashimoto K, Nakanishi T, Dijkstra JM. Comprehensive clarification of two paralogous interleukin 4/13 loci in teleost fish. Immunogenetics. 2008 Jul;60(7):383-97.

156. Jaillon O, Aury JM, Brunet F, Petit JL, Stange-Thomann N, Mauceli E, et al. Genome duplication in the teleost fish Tetraodon nigroviridis reveals the early vertebrate proto-karyotype. Nature. 2004 Oct 21;431(7011):946-57.
157. Li JH, Shao JZ, Xiang LX, Wen Y. Cloning, characterization and expression analysis of pufferfish interleukin-4 cDNA: the first evidence of Th2-type cytokine in fish. Mol Immunol. 2007 Mar;44(8):2078-86.
158. Secrist H, Egan M, Peters MG. Tissue-specific regulation of IL-4 mRNA expression in human tonsil. J Immunol. 1994 Feb 1;152(3):1120-6.
159. Lin AF, Xiang LX, Wang QL, Dong WR, Gong YF, Shao JZ. The DC-SIGN of zebrafish: insights into the existence of a CD209 homologue in a lower vertebrate and its involvement in adaptive immunity. J Immunol. 2009 Dec 1;183(11):7398-410.
160. Hu YL, Xiang LX, Shao JZ. Identification and characterization of a novel immunoglobulin Z isotype in zebrafish: implications for a distinct B cell receptor in lower vertebrates. Mol Immunol. 2010 Jan;47(4):738-46.
161. Mitra S, Alnabulsi A, Secombes CJ, Bird S. Identification and characterization of the transcription factors involved in T-cell development, t-bet, stat6 and foxp3, within the zebrafish, Danio rerio. FEBS J. 2010 Jan;277(1):128-47.
162. Takizawa F, Koppang EO, Ohtani M, Nakanishi T, Hashimoto K, Fischer U, et al. Constitutive high expression of interleukin-4/13A and GATA-3 in gill and skin of salmonid fishes suggests that these tissues form Th2-skewed immune environments. Mol Immunol. 2011 Jul;48(12-13):1360-8.
163. Fry TJ, Mackall CL. The many faces of IL-7: from lymphopoiesis to peripheral T cell maintenance. J Immunol. 2005 Jun 1;174(11):6571-6.
164. Alves NL, Richard-Le Goff O, Huntington ND, Sousa AP, Ribeiro VS, Bordack A, et al. Characterization of the thymic IL-7 niche in vivo. Proc Natl Acad Sci U S A. 2009 Feb 3;106(5):1512-7.
165. Mazzucchelli RI, Warming S, Lawrence SM, Ishii M, Abshari M, Washington AV, et al. Visualization and identification of IL-7 producing cells in reporter mice. PLoS One. 2009;4(11):e7637.
166. Repass JF, Laurent MN, Carter C, Reizis B, Bedford MT, Cardenas K, et al. IL7-hCD25 and IL7-Cre BAC transgenic mouse lines: new tools for analysis of IL-7 expressing cells. Genesis. 2009 Apr;47(4):281-7.
167. Kono T, Bird S, Sonoda K, Savan R, Secombes CJ, Sakai M. Characterization and expression analysis of an interleukin-7 homologue in the Japanese pufferfish, Takifugu rubripes. FEBS J. 2008 Mar;275(6):1213-26.
168. Bei JX, Suetake H, Araki K, Kikuchi K, Yoshiura Y, Lin HR, et al. Two interleukin (IL)-15 homologues in fish from two distinct origins. Mol Immunol. 2006 Mar;43(7):860-9.
169. Fang W, Xiang LX, Shao JZ, Wen Y, Chen SY. Identification and characterization of an interleukin-15 homologue from Tetraodon nigroviridis. Comp Biochem Physiol B Biochem Mol Biol. 2006 Mar;143(3):335-43.
170. Gunimaladevi I, Savan R, Sato K, Yamaguchi R, Sakai M. Characterization of an interleukin-15 like (IL-15L) gene from zebrafish (Danio rerio). Fish Shellfish Immunol. 2007 Apr;22(4):351-62.
171. Wang T, Holland JW, Carrington A, Zou J, Secombes CJ. Molecular and functional characterization of IL-15 in rainbow trout Oncorhynchus mykiss: a potent inducer of IFN-gamma expression in spleen leukocytes. J Immunol. 2007 Aug 1;179(3):1475-88.
172. Parrish-Novak J, Dillon SR, Nelson A, Hammond A, Sprecher C, Gross JA, et al. Interleukin 21 and its receptor are involved in NK cell expansion and regulation of lymphocyte function. Nature. 2000 Nov 2;408(6808):57-63.
173. Asao H, Okuyama C, Kumaki S, Ishii N, Tsuchiya S, Foster D, et al. Cutting edge: the common gamma-chain is an indispensable subunit of the IL-21 receptor complex. J Immunol. 2001 Jul 1;167(1):1-5.
174. Vosshenrich CA, Di Santo JP. Cytokines: IL-21 joins the gamma(c)-dependent network? Curr Biol. 2001 Mar 6;11(5):R175-7.

175. Chtanova T, Tangye SG, Newton R, Frank N, Hodge MR, Rolph MS, et al. T follicular helper cells express a distinctive transcriptional profile, reflecting their role as non-Th1/Th2 effector cells that provide help for B cells. J Immunol. 2004 Jul 1;173(1):68-78.
176. Wurster AL, Rodgers VL, Satoskar AR, Whitters MJ, Young DA, Collins M, et al. Interleukin 21 is a T helper (Th) cell 2 cytokine that specifically inhibits the differentiation of naive Th cells into interferon gamma-producing Th1 cells. J Exp Med. 2002 Oct 7;196(7):969-77.
177. Leonard WJ, Zeng R, Spolski R. Interleukin 21: a cytokine/cytokine receptor system that has come of age. J Leukoc Biol. 2008 Aug;84(2):348-56.
178. Suto A, Kashiwakuma D, Kagami S, Hirose K, Watanabe N, Yokote K, et al. Development and characterization of IL-21-producing CD4+ T cells. J Exp Med. 2008 Jun 9;205(6):1369-79.
179. Frederiksen KS, Lundsgaard D, Freeman JA, Hughes SD, Holm TL, Skrumsager BK, et al. IL-21 induces in vivo immune activation of NK cells and CD8(+) T cells in patients with metastatic melanoma and renal cell carcinoma. Cancer Immunol Immunother. 2008 Oct;57(10):1439-49.
180. Konforte D, Simard N, Paige CJ. IL-21: an executor of B cell fate. J Immunol. 2009 Feb 15;182(4):1781-7.
181. Wang HJ, Xiang LX, Shao JZ, Jia S. Molecular cloning, characterization and expression analysis of an IL-21 homologue from Tetraodon nigroviridis. Cytokine. 2006 Aug;35(3-4):126-34.
182. Wang T, Diaz-Rosales P, Costa MM, Campbell S, Snow M, Collet B, et al. Functional characterization of a nonmammalian IL-21: rainbow trout Oncorhynchus mykiss IL-21 upregulates the expression of the Th cell signature cytokines IFN-gamma, IL-10, and IL-22. J Immunol. 2011 Jan 15;186(2):708-21.
183. Lutfalla G, Roest Crollius H, Stange-Thomann N, Jaillon O, Mogensen K, Monneron D. Comparative genomic analysis reveals independent expansion of a lineage-specific gene family in vertebrates: the class II cytokine receptors and their ligands in mammals and fish. BMC Genomics. 2003 Jul 17;4(1):29.
184. Fickenscher H, Hor S, Kupers H, Knappe A, Wittmann S, Sticht H. The interleukin-10 family of cytokines. Trends Immunol. 2002 Feb;23(2):89-96.
185. Moore KW, de Waal Malefyt R, Coffman RL, O'Garra A. Interleukin-10 and the interleukin-10 receptor. Annu Rev Immunol. 2001;19:683-765.
186. Zou J, Clark MS, Secombes CJ. Characterisation, expression and promoter analysis of an interleukin 10 homologue in the puffer fish, Fugu rubripes. Immunogenetics. 2003 Aug;55(5):325-35.
187. Savan R, Igawa D, Sakai M. Cloning, characterization and expression analysis of interleukin-10 from the common carp, Cyprinus carpio L. Eur J Biochem. 2003 Dec;270(23):4647-54.
188. Zhang DC, Shao YQ, Huang YQ, Jiang SG. Cloning, characterization and expression analysis of interleukin-10 from the zebrafish (Danio rerion). J Biochem Mol Biol. 2005 Sep 30;38(5):571-6.
189. Inoue Y, Kamota S, Ito K, Yoshiura Y, Ototake M, Moritomo T, et al. Molecular cloning and expression analysis of rainbow trout (Oncorhynchus mykiss) interleukin-10 cDNAs. Fish Shellfish Immunol. 2005 Apr;18(4):335-44.
190. Buonocore F, Randelli E, Bird S, Secombes C, Facchiano A, Constantini S, et al. Interleukin-10 expression by real-time PCR and homology modelling analysis in the European sea bass (Dicentrarchus Labrax L.). Aquaculture. 2007;270(1-4):512 - 22.
191. Pinto RD, Nascimento DS, Reis MI, do Vale A, Dos Santos NM. Molecular characterization, 3D modelling and expression analysis of sea bass (Dicentrarchus labrax L.) interleukin-10. Mol Immunol. 2007 Mar;44(8):2056-65.
192. Zhang Z, Swain T, Bogwald J, Dalmo RA, Kumari J. Bath immunostimulation of rainbow trout (Oncorhynchus mykiss) fry induces enhancement of inflammatory cytokine transcripts, while repeated bath induce no changes. Fish Shellfish Immunol. 2009 May;26(5):677-84.

193. Wegenka UM. IL-20: biological functions mediated through two types of receptor complexes. Cytokine Growth Factor Rev. 2010 Oct;21(5):353-63.
194. Stein C, Caccamo M, Laird G, Leptin M. Conservation and divergence of gene families encoding components of innate immune response systems in zebrafish. Genome Biol. 2007;8(11):R251.
195. Wang T, Diaz-Rosales P, Martin SA, Secombes CJ. Cloning of a novel interleukin (IL)-20-like gene in rainbow trout Oncorhynchus mykiss gives an insight into the evolution of the IL-10 family. Dev Comp Immunol. 2010 Feb;34(2):158-67.
196. Liang SC, Tan XY, Luxenberg DP, Karim R, Dunussi-Joannopoulos K, Collins M, et al. Interleukin (IL)-22 and IL-17 are coexpressed by Th17 cells and cooperatively enhance expression of antimicrobial peptides. J Exp Med. 2006 Oct 2;203(10):2271-9.
197. Cupedo T, Crellin NK, Papazian N, Rombouts EJ, Weijer K, Grogan JL, et al. Human fetal lymphoid tissue-inducer cells are interleukin 17-producing precursors to RORC+ CD127+ natural killer-like cells. Nat Immunol. 2009 Jan;10(1):66-74.
198. Duhen T, Geiger R, Jarrossay D, Lanzavecchia A, Sallusto F. Production of interleukin 22 but not interleukin 17 by a subset of human skin-homing memory T cells. Nat Immunol. 2009 Aug;10(8):857-63.
199. Eyerich S, Eyerich K, Pennino D, Carbone T, Nasorri F, Pallotta S, et al. Th22 cells represent a distinct human T cell subset involved in epidermal immunity and remodeling. J Clin Invest. 2009 Dec;119(12):3573-85.
200. Fujita H, Nograles KE, Kikuchi T, Gonzalez J, Carucci JA, Krueger JG. Human Langerhans cells induce distinct IL-22-producing CD4+ T cells lacking IL-17 production. Proc Natl Acad Sci U S A. 2009 Dec 22;106(51):21795-800.
201. Wolk K, Haugen HS, Xu W, Witte E, Waggie K, Anderson M, et al. IL-22 and IL-20 are key mediators of the epidermal alterations in psoriasis while IL-17 and IFN-gamma are not. J Mol Med (Berl). 2009 May;87(5):523-36.
202. Feinen B, Russell MW. Contrasting Roles of IL-22 and IL-17 in Murine Genital Tract Infection by Neisseria gonorrhoeae. Front Immunol. 2012;3:11.
203. Wolk K, Kunz S, Witte E, Friedrich M, Asadullah K, Sabat R. IL-22 increases the innate immunity of tissues. Immunity. 2004 Aug;21(2):241-54.
204. Aujla SJ, Chan YR, Zheng M, Fei M, Askew DJ, Pociask DA, et al. IL-22 mediates mucosal host defense against Gram-negative bacterial pneumonia. Nat Med. 2008 Mar;14(3):275-81.
205. Andoh A, Zhang Z, Inatomi O, Fujino S, Deguchi Y, Araki Y, et al. Interleukin-22, a member of the IL-10 subfamily, induces inflammatory responses in colonic subepithelial myofibroblasts. Gastroenterology. 2005 Sep;129(3):969-84.
206. Zheng Y, Valdez PA, Danilenko DM, Hu Y, Sa SM, Gong Q, et al. Interleukin-22 mediates early host defense against attaching and effacing bacterial pathogens. Nat Med. 2008 Mar;14(3):282-9.
207. Bird L. Mucosal immunology: IL-22 keeps commensals in their place. Nat Rev Immunol. 2012 Jul 6.
208. Wolk K, Kunz S, Asadullah K, Sabat R. Cutting edge: immune cells as sources and targets of the IL-10 family members? J Immunol. 2002 Jun 1;168(11):5397-402.
209. Wilson NJ, Boniface K, Chan JR, McKenzie BS, Blumenschein WM, Mattson JD, et al. Development, cytokine profile and function of human interleukin 17-producing helper T cells. Nat Immunol. 2007 Sep;8(9):950-7.
210. Manel N, Unutmaz D, Littman DR. The differentiation of human T(H)-17 cells requires transforming growth factor-beta and induction of the nuclear receptor RORgammat. Nat Immunol. 2008 Jun;9(6):641-9.
211. Pene J, Chevalier S, Preisser L, Venereau E, Guilleux MH, Ghannam S, et al. Chronically inflamed human tissues are infiltrated by highly differentiated Th17 lymphocytes. J Immunol. 2008 Jun 1;180(11):7423-30.
212. Cella M, Fuchs A, Vermi W, Facchetti F, Otero K, Lennerz JK, et al. A human natural killer cell subset provides an innate source of IL-22 for mucosal immunity. Nature. 2009 Feb 5;457(7230):722-5.

213. Hughes T, Becknell B, McClory S, Briercheck E, Freud AG, Zhang X, et al. Stage 3 immature human natural killer cells found in secondary lymphoid tissue constitutively and selectively express the TH 17 cytokine interleukin-22. Blood. 2009 Apr 23;113(17):4008-10.
214. Donnelly RP, Sheikh F, Dickensheets H, Savan R, Young HA, Walter MR. Interleukin-26: an IL-10-related cytokine produced by Th17 cells. Cytokine Growth Factor Rev. 2010 Oct;21(5):393-401.
215. Zou J, Yoshiura Y, Dijkstra JM, Sakai M, Ototake M, Secombes C. Identification of an interferon gamma homologue in Fugu, Takifugu rubripes. Fish Shellfish Immunol. 2004 Oct;17(4):403-9.
216. Igawa D, Sakai M, Savan R. An unexpected discovery of two interferon gamma-like genes along with interleukin (IL)-22 and -26 from teleost: IL-22 and -26 genes have been described for the first time outside mammals. Mol Immunol. 2006 Mar;43(7):999-1009.
217. Corripio-Miyar Y, Zou J, Richmond H, Secombes CJ. Identification of interleukin-22 in gadoids and examination of its expression level in vaccinated fish. Mol Immunol. 2009 Jun;46(10):2098-106.
218. Kolls JK. Th17 cells in mucosal immunity and tissue inflammation. Semin Immunopathol. 2010 Mar;32(1):1-2.
219. Haugarvoll E, Bjerkas I, Nowak BF, Hordvik I, Koppang EO. Identification and characterization of a novel intraepithelial lymphoid tissue in the gills of Atlantic salmon. J Anat. 2008 Aug;213(2):202-9.
220. Moseley TA, Haudenschild DR, Rose L, Reddi AH. Interleukin-17 family and IL-17 receptors. Cytokine Growth Factor Rev. 2003 Apr;14(2):155-74.
221. Yao Z, Timour M, Painter S, Fanslow W, Spriggs M. Complete nucleotide sequence of the mouse CTLA8 gene. Gene. 1996 Feb 12;168(2):223-5.
222. Li H, Chen J, Huang A, Stinson J, Heldens S, Foster J, et al. Cloning and characterization of IL-17B and IL-17C, two new members of the IL-17 cytokine family. Proc Natl Acad Sci U S A. 2000 Jan 18;97(2):773-8.
223. Shi Y, Ullrich SJ, Zhang J, Connolly K, Grzegorzewski KJ, Barber MC, et al. A novel cytokine receptor-ligand pair. Identification, molecular characterization, and in vivo immunomodulatory activity. J Biol Chem. 2000 Jun 23;275(25):19167-76.
224. Hymowitz SG, Filvaroff EH, Yin JP, Lee J, Cai L, Risser P, et al. IL-17s adopt a cystine knot fold: structure and activity of a novel cytokine, IL-17F, and implications for receptor binding. EMBO J. 2001 Oct 1;20(19):5332-41.
225. Lee J, Ho WH, Maruoka M, Corpuz RT, Baldwin DT, Foster JS, et al. IL-17E, a novel proinflammatory ligand for the IL-17 receptor homolog IL-17Rh1. J Biol Chem. 2001 Jan 12;276(2):1660-4.
226. Starnes T, Robertson MJ, Sledge G, Kelich S, Nakshatri H, Broxmeyer HE, et al. Cutting edge: IL-17F, a novel cytokine selectively expressed in activated T cells and monocytes, regulates angiogenesis and endothelial cell cytokine production. J Immunol. 2001 Oct 15;167(8):4137-40.
227. Starnes T, Broxmeyer HE, Robertson MJ, Hromas R. Cutting edge: IL-17D, a novel member of the IL-17 family, stimulates cytokine production and inhibits hemopoiesis. J Immunol. 2002 Jul 15;169(2):642-6.
228. Gunimaladevi I, Savan R, Sakai M. Identification, cloning and characterization of interleukin-17 and its family from zebrafish. Fish Shellfish Immunol. 2006 Oct;21(4):393-403.
229. Wang T, Martin SA, Secombes CJ. Two interleukin-17C-like genes exist in rainbow trout Oncorhynchus mykiss that are differentially expressed and modulated. Dev Comp Immunol. 2010 May;34(5):491-500.
230. Kumari J, Larsen AN, Bogwald J, Dalmo RA. Interleukin-17D in Atlantic salmon (Salmo salar): molecular characterization, 3D modelling and promoter analysis. Fish Shellfish Immunol. 2009 Nov;27(5):647-59.

231. Korenaga H, Kono T, Sakai M. Isolation of seven IL-17 family genes from the Japanese pufferfish Takifugu rubripes. Fish Shellfish Immunol. 2010 May-Jun;28(5-6):809-18.
232. Kono T, Korenaga H, Sakai M. Genomics of fish IL-17 ligand and receptors: a review. Fish Shellfish Immunol. 2011 Nov;31(5):635-43.
233. Huang D, Cancilla MR, Morahan G. Complete primary structure, chromosomal localisation, and definition of polymorphisms of the gene encoding the human interleukin-12 p40 subunit. Genes Immun. 2000 Dec;1(8):515-20.
234. Yoshiura Y, Kiryu I, Fujiwara A, Suetake H, Suzuki Y, Nakanishi T, et al. Identification and characterization of Fugu orthologues of mammalian interleukin-12 subunits. Immunogenetics. 2003 Aug;55(5):296-306.
235. Nascimento DS, do Vale A, Tomas AM, Zou J, Secombes CJ, dos Santos NM. Cloning, promoter analysis and expression in response to bacterial exposure of sea bass (Dicentrarchus labrax L.) interleukin-12 p40 and p35 subunits. Mol Immunol. 2007 Mar;44(9):2277-91.
236. Li MO, Flavell RA. TGF-beta: a master of all T cell trades. Cell. 2008 Aug 8;134(3):392-404.
237. Li MO, Wan YY, Sanjabi S, Robertson AK, Flavell RA. Transforming growth factor-beta regulation of immune responses. Annu Rev Immunol. 2006;24:99-146.
238. Saxena V, Lienesch DW, Zhou M, Bommireddy R, Azhar M, Doetschman T, et al. Dual roles of immunoregulatory cytokine TGF-beta in the pathogenesis of autoimmunity-mediated organ damage. J Immunol. 2008 Feb 1;180(3):1903-12.
239. Zhang L, Yi H, Xia XP, Zhao Y. Transforming growth factor-beta: an important role in CD4+CD25+ regulatory T cells and immune tolerance. Autoimmunity. 2006 Jun;39(4):269-76.
240. Li B, Samanta A, Song X, Furuuchi K, Iacono KT, Kennedy S, et al. FOXP3 ensembles in T-cell regulation. Immunol Rev. 2006 Aug;212:99-113.
241. Wan YY, Flavell RA. 'Yin-Yang' functions of transforming growth factor-beta and T regulatory cells in immune regulation. Immunol Rev. 2007 Dec;220:199-213.
242. Haddad G, Hanington PC, Wilson EC, Grayfer L, Belosevic M. Molecular and functional characterization of goldfish (Carassius auratus L.) transforming growth factor beta. Dev Comp Immunol. 2008;32(6):654-63.
243. Kadowaki T, Yasui Y, Takahashi Y, Kohchi C, Soma G, Inagawa H. Comparative immunological analysis of innate immunity activation after oral administration of wheat fermented extract to teleost fish. Anticancer Res. 2009 Nov;29(11):4871-7.
244. Yang M, Wang Y, Wang X, Chen C, Zhou H. Characterization of grass carp (Ctenopharyngodon idellus) Foxp1a/1b/2: evidence for their involvement in the activation of peripheral blood lymphocyte subpopulations. Fish Shellfish Immunol. 2010 Feb;28(2):289-95.
245. Cai Z, Gao C, Li L, Xing K. Bipolar properties of red seabream (Pagrus major) transforming growth factor-beta in induction of the leucocytes migration. Fish Shellfish Immunol. 2010 Apr;28(4):695-700.
246. Yang M, Wang X, Chen D, Wang Y, Zhang A, Zhou H. TGF-beta1 exerts opposing effects on grass carp leukocytes: implication in teleost immunity, receptor signaling and potential self-regulatory mechanisms. PLoS One. 2012;7(4):e35011.
247. Samuel CE. Antiviral actions of interferons. Clin Microbiol Rev. 2001 Oct;14(4):778-809, table of contents.
248. Robertsen B. The interferon system of teleost fish. Fish Shellfish Immunol. 2006 Feb;20(2):172-91.
249. Bergan V, Steinsvik S, Xu H, Kileng O, Robertsen B. Promoters of type I interferon genes from Atlantic salmon contain two main regulatory regions. FEBS J. 2006 Sep;273(17):3893-906.
250. Kileng O, Brundtland MI, Robertsen B. Infectious salmon anemia virus is a powerful inducer of key genes of the type I interferon system of Atlantic salmon, but is not inhibited by interferon. Fish Shellfish Immunol. 2007 Aug;23(2):378-89.

251. Robertsen B, Bergan V, Rokenes T, Larsen R, Albuquerque A. Atlantic salmon interferon genes: cloning, sequence analysis, expression, and biological activity. J Interferon Cytokine Res. 2003 Oct;23(10):601-12.
252. Rokenes TP, Larsen R, Robertsen B. Atlantic salmon ISG15: Expression and conjugation to cellular proteins in response to interferon, double-stranded RNA and virus infections. Mol Immunol. 2007 Feb;44(5):950-9.
253. Zou J, Tafalla C, Truckle J, Secombes CJ. Identification of a second group of type I IFNs in fish sheds light on IFN evolution in vertebrates. J Immunol. 2007 Sep 15;179(6):3859-71.
254. Robertsen B. Expression of interferon and interferon-induced genes in salmonids in response to virus infection, interferon-inducing compounds and vaccination. Fish Shellfish Immunol. 2008 Oct;25(4):351-7.
255. Milev-Milovanovic I, Long S, Wilson M, Bengten E, Miller NW, Chinchar VG. Identification and expression analysis of interferon gamma genes in channel catfish. Immunogenetics. 2006 Feb;58(1):70-80.
256. Stolte EH, Savelkoul HF, Wiegertjes G, Flik G, Lidy Verburg-van Kemenade BM. Differential expression of two interferon-gamma genes in common carp (Cyprinus carpio L.). Dev Comp Immunol. 2008;32(12):1467-81.
257. Zou J, Carrington A, Collet B, Dijkstra JM, Yoshiura Y, Bols N, et al. Identification and bioactivities of IFN-gamma in rainbow trout Oncorhynchus mykiss: the first Th1-type cytokine characterized functionally in fish. J Immunol. 2005 Aug 15;175(4):2484-94.
258. Grayfer L, Garcia EG, Belosevic M. Comparison of macrophage antimicrobial responses induced by type II interferons of the goldfish (Carassius auratus L.). J Biol Chem. 2010 Jul 30;285(31):23537-47.
259. Arts JA, Tijhaar EJ, Chadzinska M, Savelkoul HF, Verburg-van Kemenade BM. Functional analysis of carp interferon-gamma: evolutionary conservation of classical phagocyte activation. Fish Shellfish Immunol. 2010 Nov;29(5):793-802.
260. Sun B, Skjaeveland I, Svingerud T, Zou J, Jorgensen J, Robertsen B. Antiviral activity of salmonid gamma interferon against infectious pancreatic necrosis virus and salmonid alphavirus and its dependency on type I interferon. J Virol. 2011 Sep;85(17):9188-98.
261. Duguid JR, Dinauer MC. Library subtraction of in vitro cDNA libraries to identify differentially expressed genes in scrapie infection. Nucleic Acids Res. 1990 May 11;18(9):2789-92.
262. Hara E, Kato T, Nakada S, Sekiya S, Oda K. Subtractive cDNA cloning using oligo(dT)30-latex and PCR: isolation of cDNA clones specific to undifferentiated human embryonal carcinoma cells. Nucleic Acids Res. 1991 Dec;19(25):7097-104.
263. Hedrick SM, Cohen DI, Nielsen EA, Davis MM. Isolation of cDNA clones encoding T cell-specific membrane-associated proteins. Nature. 1984 Mar 8-14;308(5955):149-53.
264. Lisitsyn N, Wigler M. Cloning the differences between two complex genomes. Science. 1993 Feb 12;259(5097):946-51.
265. Hubank M, Schatz DG. Identifying differences in mRNA expression by representational difference analysis of cDNA. Nucleic Acids Res. 1994 Dec 25;22(25):5640-8.
266. Liang P, Pardee AB. Differential display of eukaryotic messenger RNA by means of the polymerase chain reaction. Science. 1992 Aug 14;257(5072):967-71.
267. Welsh J, Chada K, Dalal SS, Cheng R, Ralph D, McClelland M. Arbitrarily primed PCR fingerprinting of RNA. Nucleic Acids Res. 1992 Oct 11;20(19):4965-70.
268. Sompayrac L, Jane S, Burn TC, Tenen DG, Danna KJ. Overcoming limitations of the mRNA differential display technique. Nucleic Acids Res. 1995 Nov 25;23(22):4738-9.
269. Bertioli DJ, Schlichter UH, Adams MJ, Burrows PR, Steinbiss HH, Antoniw JF. An analysis of differential display shows a strong bias towards high copy number mRNAs. Nucleic Acids Res. 1995 Nov 11;23(21):4520-3.

270. Siebert PD, Chenchik A, Kellogg DE, Lukyanov KA, Lukyanov SA. An improved PCR method for walking in uncloned genomic DNA. Nucleic Acids Res. 1995 Mar 25;23(6):1087-8.
271. Diatchenko L, Lau YF, Campbell AP, Chenchik A, Moqadam F, Huang B, et al. Suppression subtractive hybridization: a method for generating differentially regulated or tissue-specific cDNA probes and libraries. Proc Natl Acad Sci U S A. 1996 Jun 11;93(12):6025-30.
272. Wang L, Wu X. Identification of differentially expressed genes in lipopolysaccharide-stimulated yellow grouper Epinephelus awoara spleen. Fish Shellfish Immunol. 2007 Aug;23(2):354-63.
273. Xia JH, Yue GH. Identification and analysis of immune-related transcriptome in Asian seabass Lates calcarifer. BMC Genomics. 2010;11:356.
274. Feng CY, Johnson SC, Hori TS, Rise M, Hall JR, Gamperl AK, et al. Identification and analysis of differentially expressed genes in immune tissues of Atlantic cod stimulated with formalin-killed, atypical Aeromonas salmonicida. Physiol Genomics. 2009 May 13;37(3):149-63.
275. Tsoi SC, Ewart KV, Penny S, Melville K, Liebscher RS, Brown LL, et al. Identification of immune-relevant genes from atlantic salmon using suppression subtractive hybridization. Mar Biotechnol (NY). 2004 May-Jun;6(3):199-214.
276. Li CH, Chen J, Shi YH, Lu XJ. Use of suppressive subtractive hybridization to identify differentially expressed genes in ayu (Plecoglossus altivelis) associated with Listonella anguillarum infection. Fish Shellfish Immunol. 2011 Sep;31(3):500-6.
277. Matsuyama T, Fujiwara A, Takano T, Nakayasu C. Suppression subtractive hybridization coupled with microarray analysis to examine differential expression of genes in Japanese flounder Paralichthys olivaceus leucocytes during Edwardsiella tarda and viral hemorrhagic septicemia virus infection. Fish Shellfish Immunol. 2011 Oct;31(4):524-32.
278. Bo J, Giesy JP, Ye R, Wang KJ, Lee JS, Au DW. Identification of differentially expressed genes and quantitative expression of complement genes in the liver of marine medaka Oryzias melastigma challenged with Vibrio parahaemolyticus. Comp Biochem Physiol Part D Genomics Proteomics. 2012 Jun;7(2):191-200.
279. Purcell MK, Smith KD, Hood L, Winton JR, Roach JC. Conservation of Toll-Like Receptor Signaling Pathways in Teleost Fish. Comp Biochem Physiol Part D Genomics Proteomics. 2006 Mar;1(1):77-88.
280. Trinchieri G, Sher A. Cooperation of Toll-like receptor signals in innate immune defence. Nat Rev Immunol. 2007 Mar;7(3):179-90.
281. Bayne CJ, Gerwick L. The acute phase response and innate immunity of fish. Dev Comp Immunol. 2001 Oct-Dec;25(8-9):725-43.
282. Bayne CJ, Gerwick L, Fujiki K, Nakao M, Yano T. Immune-relevant (including acute phase) genes identified in the livers of rainbow trout, Oncorhynchus mykiss, by means of suppression subtractive hybridization. Dev Comp Immunol. 2001 Apr;25(3):205-17.
283. Holland MC, Lambris JD. The complement system in teleosts. Fish Shellfish Immunol. 2002 May;12(5):399-420.
284. Rise ML, Hall J, Rise M, Hori T, Gamperl AK, Kimball J, et al. Functional genomic analysis of the response of Atlantic cod (Gadus morhua) spleen to the viral mimic polyriboinosinic polyribocytidylic acid (pIC). Dev Comp Immunol. 2008;32(8):916-31.
285. He W, Yinlt ZX, Li Y, Huo WL, Guan HJ, Weng SP, et al. Differential gene expression profile in spleen of mandarin fish Siniperca chuatsi infected with ISKNV, derived from suppression subtractive hybridization. Dis Aquat Organ. 2006 Dec 14;73(2):113-22.
286. Dios S, Poisa-Beiro L, Figueras A, Novoa B. Suppression subtraction hybridization (SSH) and macroarray techniques reveal differential gene expression profiles in brain of sea bream infected with nodavirus. Mol Immunol. 2007 Mar;44(9):2195-204.

287. Poisa-Beiro L, Dios S, Ahmed H, Vasta GR, Martinez-Lopez A, Estepa A, et al. Nodavirus infection of sea bass (Dicentrarchus labrax) induces up-regulation of galectin-1 expression with potential anti-inflammatory activity. J Immunol. 2009 Nov 15;183(10):6600-11.
288. Rise ML, Hall JR, Rise M, Hori TS, Browne MJ, Gamperl AK, et al. Impact of asymptomatic nodavirus carrier state and intraperitoneal viral mimic injection on brain transcript expression in Atlantic cod (Gadus morhua). Physiol Genomics. 2010 Jul 7;42(2):266-80.
289. Xu D, Wei J, Cui H, Gong J, Yan Y, Lai R, et al. Differential profiles of gene expression in grouper Epinephelus coioides, infected with Singapore grouper iridovirus, revealed by suppression subtractive hybridization and DNA microarray. J Fish Biol. 2010 Aug;77(2):341-60.
290. Fisicaro N, Aminian A, Hinchliffe SJ, Morgan BP, Pearse MJ, D'Apice AJ, et al. The pig analogue of CD59 protects transgenic mouse hearts from injury by human complement. Transplantation. 2000 Sep 27;70(6):963-8.
291. Janeway CA, Jr., Medzhitov R. Innate immune recognition. Annu Rev Immunol. 2002;20:197-216.
292. Tasumi S, Ohira T, Kawazoe I, Suetake H, Suzuki Y, Aida K. Primary structure and characteristics of a lectin from skin mucus of the Japanese eel Anguilla japonica. J Biol Chem. 2002 Jul 26;277(30):27305-11.
293. Tsutsui S, Iwamoto K, Nakamura O, Watanabe T. Yeast-binding C-type lectin with opsonic activity from conger eel (Conger myriaster) skin mucus. Mol Immunol. 2007 Feb;44(5):691-702.
294. Zhang H, Robison B, Thorgaard GH, Ristow SS. Cloning, mapping and genomic organization of a fish C-type lectin gene from homozygous clones of rainbow trout (Oncorhynchus mykiss). Biochim Biophys Acta. 2000 Nov 15;1494(1-2):14-22.
295. Honda K, Yanai H, Takaoka A, Taniguchi T. Regulation of the type I IFN induction: a current view. Int Immunol. 2005 Nov;17(11):1367-78.
296. Krasnov A, Timmerhaus G, Afanasyev S, Jorgensen SM. Development and assessment of oligonucleotide microarrays for Atlantic salmon (Salmo salar L.). Comp Biochem Physiol Part D Genomics Proteomics. 2011 Mar;6(1):31-8.
297. von Schalburg KR, Rise ML, Cooper GA, Brown GD, Gibbs AR, Nelson CC, et al. Fish and chips: various methodologies demonstrate utility of a 16,006-gene salmonid microarray. BMC Genomics. 2005;6:126.
298. Koop BF, von Schalburg KR, Leong J, Walker N, Lieph R, Cooper GA, et al. A salmonid EST genomic study: genes, duplications, phylogeny and microarrays. BMC Genomics. 2008;9:545.
299. Krasnov A, Koskinen H, Pehkonen P, Rexroad CE, 3rd, Afanasyev S, Molsa H. Gene expression in the brain and kidney of rainbow trout in response to handling stress. BMC Genomics. 2005;6:3.
300. Schiotz BL, Jorgensen SM, Rexroad C, Gjoen T, Krasnov A. Transcriptomic analysis of responses to infectious salmon anemia virus infection in macrophage-like cells. Virus Res. 2008 Sep;136(1-2):65-74.
301. Kurobe T, Yasuike M, Kimura T, Hirono I, Aoki T. Expression profiling of immune-related genes from Japanese flounder Paralichthys olivaceus kidney cells using cDNA microarrays. Dev Comp Immunol. 2005;29(6):515-23.
302. Williams TD, Diab AM, George SG, Godfrey RE, Sabine V, Conesa A, et al. Development of the GENIPOL European flounder (Platichthys flesus) microarray and determination of temporal transcriptional responses to cadmium at low dose. Environ Sci Technol. 2006 Oct 15;40(20):6479-88.
303. Park KC, Osborne JA, Montes A, Dios S, Nerland AH, Novoa B, et al. Immunological responses of turbot (Psetta maxima) to nodavirus infection or polyriboinosinic polyribocytidylic acid (pIC) stimulation, using expressed sequence tags (ESTs) analysis and cDNA microarrays. Fish Shellfish Immunol. 2009 Jan;26(1):91-108.
304. Cerda J, Mercade J, Lozano JJ, Manchado M, Tingaud-Sequeira A, Astola A, et al. Genomic resources for a commercial flatfish, the Senegalese sole (Solea

senegalensis): EST sequencing, oligo microarray design, and development of the Soleamold bioinformatic platform. BMC Genomics. 2008;9:508.
305. Skugor S, Jorgensen SM, Gjerde B, Krasnov A. Hepatic gene expression profiling reveals protective responses in Atlantic salmon vaccinated against furunculosis. BMC Genomics. 2009;10:503.
306. Tilton SC, Gerwick LG, Hendricks JD, Rosato CS, Corley-Smith G, Givan SA, et al. Use of a rainbow trout oligonucleotide microarray to determine transcriptional patterns in aflatoxin B1-induced hepatocellular carcinoma compared to adjacent liver. Toxicol Sci. 2005 Dec;88(2):319-30.
307. Millan A, Gomez-Tato A, Fernandez C, Pardo BG, Alvarez-Dios JA, Calaza M, et al. Design and performance of a turbot (Scophthalmus maximus) oligo-microarray based on ESTs from immune tissues. Mar Biotechnol (NY). 2010 Aug;12(4):452-65.
308. Peatman E, Baoprasertkul P, Terhune J, Xu P, Nandi S, Kucuktas H, et al. Expression analysis of the acute phase response in channel catfish (Ictalurus punctatus) after infection with a Gram-negative bacterium. Dev Comp Immunol. 2007;31(11):1183-96.
309. MacKenzie S, Balasch JC, Novoa B, Ribas L, Roher N, Krasnov A, et al. Comparative analysis of the acute response of the trout, O. mykiss, head kidney to in vivo challenge with virulent and attenuated infectious hematopoietic necrosis virus and LPS-induced inflammation. BMC Genomics. 2008;9:141.
310. Prieto-Alamo MJ, Abril N, Osuna-Jimenez I, Pueyo C. Solea senegalensis genes responding to lipopolysaccharide and copper sulphate challenges: large-scale identification by suppression subtractive hybridization and absolute quantification of transcriptional profiles by real-time RT-PCR. Aquat Toxicol. 2009 Mar 9;91(4):312-9.
311. Li RW, Waldbieser GC. Production and utilization of a high-density oligonucleotide microarray in channel catfish, Ictalurus punctatus. BMC Genomics. 2006;7:134.
312. Fittipaldi N, Gottschalk M, Vanier G, Daigle F, Harel J. Use of selective capture of transcribed sequences to identify genes preferentially expressed by Streptococcus suis upon interaction with porcine brain microvascular endothelial cells. Appl Environ Microbiol. 2007 Jul;73(13):4359-64.
313. Lun ZR, Wang QP, Chen XG, Li AX, Zhu XQ. Streptococcus suis: an emerging zoonotic pathogen. Lancet Infect Dis. 2007 Mar;7(3):201-9.
314. Rise ML, von Schalburg KR, Brown GD, Mawer MA, Devlin RH, Kuipers N, et al. Development and application of a salmonid EST database and cDNA microarray: data mining and interspecific hybridization characteristics. Genome Res. 2004 Mar;14(3):478-90.
315. Matsumoto M, Tanaka T, Kaisho T, Sanjo H, Copeland NG, Gilbert DJ, et al. A novel LPS-inducible C-type lectin is a transcriptional target of NF-IL6 in macrophages. J Immunol. 1999 Nov 1;163(9):5039-48.
316. MacKenzie S, Iliev D, Liarte C, Koskinen H, Planas JV, Goetz FW, et al. Transcriptional analysis of LPS-stimulated activation of trout (Oncorhynchus mykiss) monocyte/macrophage cells in primary culture treated with cortisol. Mol Immunol. 2006 Mar;43(9):1340-8.
317. Martin SA, Blaney SC, Houlihan DF, Secombes CJ. Transcriptome response following administration of a live bacterial vaccine in Atlantic salmon (Salmo salar). Mol Immunol. 2006 Apr;43(11):1900-11.
318. Ching B, Jamieson S, Heath JW, Heath DD, Hubberstey A. Transcriptional differences between triploid and diploid Chinook salmon (Oncorhynchus tshawytscha) during live Vibrio anguillarum challenge. Heredity (Edinb). 2010 Feb;104(2):224-34.
319. Dumrongphol Y, Hirota T, Kondo H, Aoki T, Hirono I. Identification of novel genes in Japanese flounder (Paralichthys olivaceus) head kidney up-regulated after vaccination with Streptococcus iniae formalin-killed cells. Fish Shellfish Immunol. 2009 Jan;26(1):197-200.

320. Gerwick L, Corley-Smith G, Bayne CJ. Gene transcript changes in individual rainbow trout livers following an inflammatory stimulus. Fish Shellfish Immunol. 2007 Mar;22(3):157-71.

321. Meijer AH, Verbeek FJ, Salas-Vidal E, Corredor-Adamez M, Bussman J, van der Sar AM, et al. Transcriptome profiling of adult zebrafish at the late stage of chronic tuberculosis due to Mycobacterium marinum infection. Mol Immunol. 2005 Jun;42(10):1185-203.

322. van der Sar AM, Spaink HP, Zakrzewska A, Bitter W, Meijer AH. Specificity of the zebrafish host transcriptome response to acute and chronic mycobacterial infection and the role of innate and adaptive immune components. Mol Immunol. 2009 Jul;46(11-12):2317-32.

323. Purcell MK, Nichols KM, Winton JR, Kurath G, Thorgaard GH, Wheeler P, et al. Comprehensive gene expression profiling following DNA vaccination of rainbow trout against infectious hematopoietic necrosis virus. Mol Immunol. 2006 May;43(13):2089-106.

324. Byon JY, Ohira T, Hirono I, Aoki T. Use of a cDNA microarray to study immunity against viral hemorrhagic septicemia (VHS) in Japanese flounder (Paralichthys olivaceus) following DNA vaccination. Fish Shellfish Immunol. 2005 Feb;18(2):135-47.

325. Jorgensen SM, Afanasyev S, Krasnov A. Gene expression analyses in Atlantic salmon challenged with infectious salmon anemia virus reveal differences between individuals with early, intermediate and late mortality. BMC Genomics. 2008;9:179.

326. Encinas P, Rodriguez-Milla MA, Novoa B, Estepa A, Figueras A, Coll J. Zebrafish fin immune responses during high mortality infections with viral haemorrhagic septicemia rhabdovirus. A proteomic and transcriptomic approach. BMC Genomics. 2010;11:518.

# CHAPTER 2

# Narrowed Temperature Adaptability in Non-Natal Osmotic Environments of Two Euryhaline Wanderers, Dace and Black Porgy: Implications for Seasonal Habitat Changes

**Masahiro Nakamura[1]·Katsumi Tsukamoto[3]·Reiji Masuda[2] ·Tsuguo Otake[1]**

## ABSTRACT

Many euryhaline wanderer species exhibit seasonal habitat changes, and temperature is suggested to play a key role in this phenomenon. To examine this possibility, we tested the hypothesis that adaptable temperature ranges of euryhaline wanderers are narrower in a non-natal osmotic environment than in their natal osmotic environment using dace *Tribolodon hakonensis*, a freshwater born euryhaline fish, and black porgy *Acanthopagrus schlegelii*, a marine born euryhaline fish. In both species, all individuals survived for 48 h in all the tested temperatures in their natal osmotic environments. In non-natal osmotic environments, however, both low (dace: 6 °C, black porgy: 10 °C) and high temperatures (dace: 30 °C, black porgy: 33 °C) significantly reduced their survival. These results indicate that euryhaline wanderers have a narrowed adaptable temperature range in non-natal osmotic environments. The similar trend observed in these two species of freshwater and marine fishes suggests that this may be a general characteristic in euryhaline wanderers and could be one of the causes of their seasonal movements.

**Keywords:** ***Acanthopagrusschlegelli*** Euryhaline wanderer Japanese black porgy Japanese dace Natal and non-natal osmotic environment Seasonal habitat change Adaptable temperature ***Tribolodon hakonensis***

# INTRODUCTION

Knowledge of the life history and ecology of fishes in estuaries has been accumulating in recent years [1], and it is becoming increasingly clear that many euryhaline wanderer species have common characteristics including seasonal habitat changes. The abundance and species richness of fishes residing in temperate estuaries is substantially reduced in winter [2, 3, 4]. For example, many marine born euryhaline species in the Hudson River and other species in the mid-Atlantic Bight estuaries along the US east coast migrate offshore or southward for overwintering [1, 3]. The Japanese black porgy *Acanthopagrus schlegelii* (family Sparidae), which moves into estuaries from the marine environment, is found in the brackish area of the Nakaumi Lagoon in Shimane Prefecture of Japan only in the summer season [5] and is suggested to move back to marine habitats during the other seasons. Conversely, freshwater born euryhaline wanderers such as the gizzard shad *Dorosoma cepedianum* and the Japanese dace *Tribolodon hakonensis* (family Cyprinidae) migrate back to freshwater habitats during the winter season for overwintering after they entered estuaries or coastal areas in summer [1, 6, 7, 8, 9]. These observations suggest that euryhaline wanderers in the temperate zone migrate back to their natal osmotic environment (hypo-isotonic environment for freshwater species and hyper-isotonic environment for marine species) for wintering. Hereafter in this paper, both the hypertonic environment for freshwater species and the hypotonic environment for marine species are defined as a "non-natal osmotic environment".

Some ecological studies suggest that water temperature of the non-natal osmotic environment may play a key role in shaping the pattern of seasonal movement in euryhaline wanderers. Able and Fahay [1] indicated that the migration out of estuaries might be influenced by temperature, based on the fact that later egress periods or year-round residency can be observed within the same species, such as *Anchoa mitchilli* [10], in more southern estuaries where autumn cooling occurs later. Temperature-influenced habitat changes are also suggested from the geographical differences of migration patterns of Japanese dace. Overwintering migration of dace from coastal or estuarine areas into freshwater is known for the populations of rivers in northern Japan, including the Mukawa River in Hokkaido [7] and three rivers in Niigata Prefecture [9]. However, the population inhabiting the Kamogawa River in Mie Prefecture, in the middle part of Japan, moves back into freshwater only for reproduction in spring, without performing the overwintering migration [11]. Those differences may be caused by the coastal ocean temperatures in the winter season, because the monthly mean sea surface temperature of the coastal area in Hokkaido drops to 1–2 °C and to 10 °C in Niigata Prefecture in winter, while around the coast of Mie Prefecture it rarely drops to less than 15–16 °C (Japan MeteorologicalAgency: http://www.data.jma.go.jp/kaiyou/data/db/kaikyo/monthly/sst_HQ.html, accessed 6 May 2015). It is possible that in temperature conditions lower than 10 °C, Japanese dace might not be able to adapt to seawater, which is a non-natal osmotic environment for them, and thus the northern populations migrate back to freshwater, a natal osmotic environment, in winter. Furthermore, there is an experimental study that has demonstrated that the low temperature tolerance of a euryhaline fish belonging to the family Sciaenidae decreases in the non-natal osmotic environment [12]. The temperature range that individuals of euryhaline wanderer species can adapt to in non-natal osmotic environments may be narrower than that within their natal osmotic environment, and this might be one of the causes of seasonal habitat changes in euryhaline wanderers. However, experimental research has been conducted only for one species, the Atlantic croaker *Micropogonias undulatus*, and only for low temperature conditions [12]. Thus, how general this characteristic is among euryhaline wanderers, and whether this is consistent for high temperature conditions remain unclear.

The objective of this study, therefore, was to empirically determine if the adaptable temperature range of euryhaline wanderer fishes is narrower in a non-natal osmotic environment than that in their natal osmotic environment, and examine the possible reason of their seasonal habitat changes. Two euryhaline wanderer species, namely Japanese dace and Japanese black porgy belonging to families which have not yet been tested, were used for the experiment. Two groups of each species, one acclimated to the natal osmotic environment and the other to the non-natal osmotic environment, were tested for their survival rate when exposed to either high or low temperature. We also examined the relationships between body size and thermal stress tolerance in each osmotic environment.

# MATERIALS AND METHODS

## Fish used for experiments

The Japanese dace used in this study were collected in October of 2013 with a throw net in the pure freshwater area of the lower Enokigawa River near the estuary adjacent to Maizuru Bay. After being captured, individuals were kept in a 20-l plastic bucket with a lid, which was filled with river water taken from the sampling site, and were transported to the Maizuru Fisheries Research Station (MFRS) of Kyoto University located in Maizuru, Kyoto Prefecture (35°29N, 135°22E). It took approximately 15 min to transport them. The standard length (SL) of the dace used in the experiments ranged between 64.6 and 146.5 mm. Japanese black porgy used in this study were hatchery-born and reared in filtered seawater at MFRS under a natural photoperiod and water temperature. The SL of individuals ranged between 47.9 and 57.5 mm.

About 200 dace and 80 black porgy were kept in 500-l clear polycarbonate tanks (two tanks for dace and one for black porgy) filled with freshwater and seawater, respectively. The seawater tanks for black porgy were supplied with filtered seawater at a rate of 4 l/min and with enough aeration (600 ml/min). Tap water dechlorinated with sodium thiosulfate (0.6 mg/l) was used for rearing the dace in freshwater. Those three tanks were placed outdoors, covered with a roof, and exposed to the natural photoperiod. Fish were fed to satiation with commercial pellets (dace: Kawasakana-no-esa, GEX Co., Ltd., Japan; black porgy: Otohime S2, Marubeni Nisshin Feed Co., Ltd., Japan) once a day for about 10 days until being separated into different acclimation tanks.

## Acclimation of fish and the procedure of challenge tests

Dace and black porgy were acclimated to both natal and non-natal osmotic water for more than 2 weeks prior to the experiment (dace: 80 individuals each; black porgy: 30 individuals each). Salinities of freshwater and seawater used for acclimation were 0.1–0.2 ppt and 32–34 ppt, respectively. Two 500-l clear polycarbonate tanks for each species, one for each osmotic condition, were used for the acclimation, and temperature was kept around 20 °C using a heater equipped with a thermostat. Since direct transfer to a non-natal osmotic environment can be lethal, fish were acclimated through several steps as follows. For acclimating the dace to seawater, they were first transferred to 15 ppt seawater from freshwater. Three days after this first transference, they were transferred to full strength seawater (32–34 ppt). Since black porgy seemed to be less tolerant to a non-natal osmotic environment than dace in a preliminary experiment, they were first transferred to 10 ppt seawater from full strength seawater. After 1 day of acclimation, the salinity of the tank water was reduced to 5 ppt, followed by a 2 ppt degradation each day until the salinity reached 1 ppt. Fish were acclimated at 1 ppt for 72 h and then transferred to freshwater (0.1–0.2 ppt). The two seawater tanks (one for each species) were supplied with filtered seawater at a rate of 4 l/min and well aerated (600 ml/min). Fishes in the two freshwater tanks (one for each species) were kept in water filtrated with three air-driven corner filters. About 1/3 of the water in the freshwater tanks were exchanged every 4 days.

Challenge tests were conducted in the laboratory from 7 November to 3 December 2013 using two fish groups acclimated to natal and non-natal osmotic waters. Only fish with no injury and no sign of abnormal behavior were used in the experiments. Two water baths (width: 1800 mm, length: 900 mm, depth: 400 mm) equipped with four chillers and two heaters were used to regulate the water temperature of the experimental tanks. Water in each water bath was circulated with two water pumps to make the temperature homogeneous throughout the bath. Six acrylic fiber experimental tanks (width: 450 mm, length: 300 mm, depth: 300 mm) were set in each water bath. Three of them were filled with freshwater and the other three filled with full strength seawater. Thus, triplicates for the natal osmotic environment and non-natal osmotic environment treatments were prepared for each experimental water temperature. Temperatures to which the fish would be transferred were 6, 10, 15, 20 (no change), and 30 °C for dace and 10, 20 (no change), and 33 °C for black porgy; thus, experiments with the two osmotic environments with three replications of each treatment for one species were conducted at five (dace) or three (black porgy) temperature treatments. Black porgy was examined at fewer temperature treatments compared to dace because of the limited number of fish available for use in the experiment. Each tank was aerated and held under light conditions of 12L–12D. Two experiments were conducted simultaneously using the two water baths.

Experimental tanks were set in the water baths for at least 24 h before the start of the experiment to adjust the water temperature of tanks to each test temperature. At the time of the experiment, 3–5 individuals in the acclimation tank were transferred to an acrylic fiber container (width: 280 mm, depth: 170 mm, height: 160 mm) filled with acclimation water. The container was then put into the experimental tank with fish and acclimation water. Thus, the ambient water temperature of fishes did not change abruptly, and so we prevented death of fish by sudden thermal change. The number of fish in each tank was adjusted to between 3 and 5 individuals to make the biomass of each experimental tank approximately the same within each species. Therefore, the total numbers of individuals used in each experimental temperature change treatment were 13–15 for dace and 11–12 for black porgy.

It took 0 min (20 °C treatment) to 1–3 h for the temperature of the water in the containers to reach the targeted temperatures. Fish were released into the water of the experimental tanks after the water temperature of the container reached each experimental temperature. Water temperature of the container was checked constantly with a thermometer during this period and all transfers were started around 7:00 a.m. Containers were removed from the experimental tank as soon as fish were released and a lid was put on each experimental tank to prevent evaporation of water and the fish jumping out of the tank. Experiments were conducted for 48 h and tanks were checked for dead fish every 12 h. When dead individuals were found, they were immediately removed from the tank and the standard length (SL) and body weight (BW) were measured. Fish were defined as dead when no movement could be seen in any part of their body. The survival rate of each tank was calculated for 24 and 48 h after fish were exposed to the test temperature. The SL of dace in each experimental tank at the end of the experiment were not the same (Welch's *t*-test, $P < 0.01$), presumably because of the low survival rate of small individuals during seawater acclimation. SL composition of black porgy was the same among the experimental tanks (Welch's *t*-test, $P > 0.05$).

## Data analysis

Prior to the statistical comparison of survival rate among treatments, variation in the triplicated data was tested. Because there was no significant difference in survival rates within each triplicated data (Fisher's exact test, $P > 0.05$), data from each triplicated trial were pooled. Effects of SL and BW on the survival rate in the non-natal osmotic environments were examined by comparing the SL and BW between surviving and dead individuals. Three individuals or more from both surviving and dead groups were needed for this analysis, so only data of time points that fulfilled this requirement were used for the analysis. In the case of dace, data of the 24-h time point of 6 °C (5 individuals survived and 10 individuals were dead) and 30 °C (3 individuals survived and 9 individuals were dead) were used for the analysis. However, in the case of 10 °C, not enough dead individuals were obtained before the time of 24 h, so the data of 48 h (11 individuals survived and 3 individuals were dead) was used for the analysis. For black porgy, data of 24 h at 10 °C (6 individuals survived and 6 individuals were dead) and 33 °C (4 individuals survived and 8 individuals were dead) were used for analysis.

Survival rates were compared between natal and non-natal osmotic environment acclimated groups at the same test temperature at each time point using Fisher's exact test. SL and BW were compared between surviving and dead individuals in the non-natal osmotic environment using Welch's *t*-test. The significance level of all the statistical tests was set at $\alpha = 0.01$. Welch's *t*-tests were performed using Ekuseru-Toukei 2010 (Social Survey Research Information Co., Ltd., Japan) and Fisher's exact tests were performed using the statistical program R (R Development Core Team, 2014; http://www.R-project.org).

# RESULTS

No significant differences in SL and BW of black porgy were observed between the fish used in the natal and non-natal osmotic environment treatments (Table 1, Welch's *t*-test, $P > 0.05$). However, dace in the seawater treatment were significantly larger in SL and BW, with the exception of the 6 °C temperature treatment (Table 1, Welch's *t*-test, $P < 0.01$). This was due to the mortality of small individuals in full-strength seawater during the acclimation period before the experimental trials.

**Table 1** Standard length (SL), body weight (BW) and the total number of individual fish used in each experimental temperature treatment for Japanese dace *Tribolodon hakonensis* and Japanese black porgy *Acanthopagrusschlegelli*

| Species | Temperature (°C) | Sea water | | | Fresh water | | |
|---|---|---|---|---|---|---|---|
| | | SL | BW | *n* | SL | BW | *n* |
| *Tribolodon hakonensis* | 6 | 116.3 ± 12.4 | 18.9 ± 6.0 | 15 | 104.4 ± 19.5 | 16.8 ± 8.3 | 14 |
| | 10 | 119.1 ± 8.4* | 20.5 ± 4.4* | 15 | 101.6 ± 15.2 | 14.8 ± 5.4 | 13 |
| | 15 | 123.3 ± 9.2* | 23.1 ± 4.7* | 14 | 100.1 ± 19.3 | 15.2 ± 9.5 | 15 |
| | 20 | 119.9 ± 9.0* | 20.4 ± 6.2* | 14 | 88.32 ± 26.1 | 12.23 ± 7.3 | 15 |
| | 30 | 102.0 ± 17.2* | 17.2 ± 6.5* | 14 | 80.3 ± 30.5 | 11.2 ± 7.4 | 14 |
| *Acanthopagrus schlegelli* | 10 | 65.1 ± 13.9 | 7.2 ± 4.4 | 12 | 65.7 ± 10.2 | 10.1 ± 4.2 | 12 |
| | 20 | 66.5 ± 15.0 | 8.6 ± 5.6 | 12 | 67.2 ± 13.4 | 9.6 ± 5.8 | 11 |
| | 33 | 68.0 ± 16.1 | 9.3 ± 5.8 | 11 | 63.2 ± 8.0 | 8.4 ± 2.7 | 12 |

* Seawater acclimated individuals were significantly larger compared to those acclimated to freshwater (Welch's *t*-test, $P < 0.01$). SL and BW is expressed with mean ± standard error

In dace, all individuals tested in freshwater survived at all experimental temperatures during the 48 h of the test period after being transferred to the new temperature (Fig. 1b). Also, all individuals tested in seawater survived at temperatures of 20 °C and 15 °C for 48 h. However, about 20 %

of the fish died at 10 °C, although there was no significant difference in survival rate between the natal and non-natal osmotic environments (Fisher's exact test, $P > 0.05$). Most fish died at 6 °C and 30 °C after 24 h in seawater, showing significantly lower survival rates compared to the freshwater acclimated groups at the same test temperatures (Fisher's exact test, $P < 0.01$). The survival rate in seawater markedly decreased after 48 h at 6 °C and 30 °C (Fisher's exact test, $P < 0.01$), with no fish surviving at 6 °C (Fig. 1a, b).

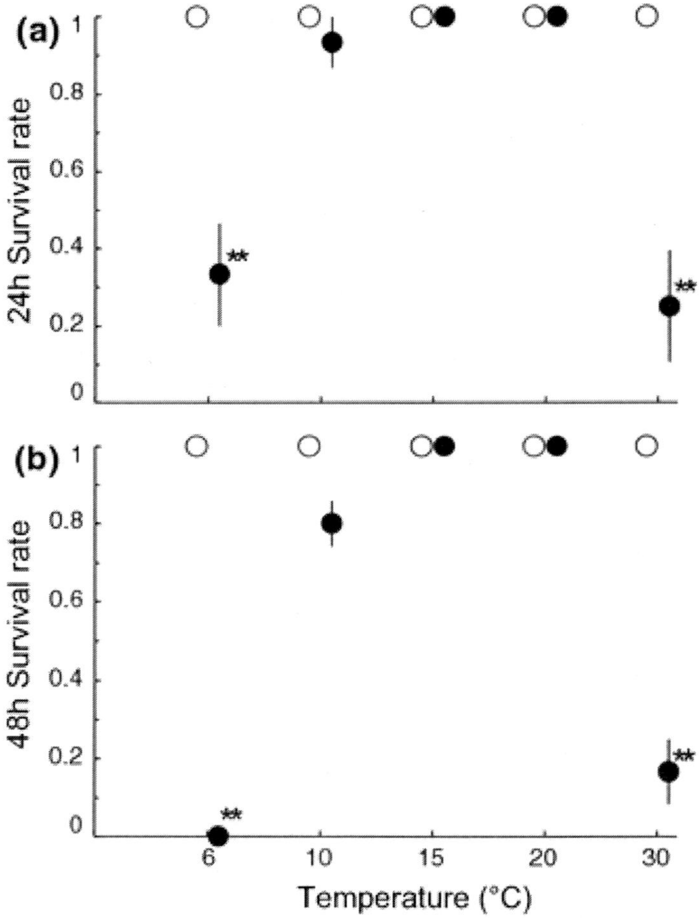


**Figure. 1** Survival rate of Japanese dace *Tribolodon hakonensis* acclimated to natal (freshwater, *open circle*) and non-natal (seawater, *closed circle*) osmotic environments at **A** 24 h and **B** 48 h after the transference to test water temperatures. *Vertical lines* indicate standard error (SE). *Asterisks* indicate $P < 0.01$ level of significant differences between the natal and non-natal osmotic environment acclimated treatments at each temperature

In black porgy, all individuals tested in seawater survived all experimental temperature changes. However, at 10 and 33 °C, more than half of the fish acclimated to freshwater died after 24 h and showed a significantly lower survival rate compared to seawater acclimated groups at the same temperatures (Fisher's exact test, $P < 0.01$). The decrease in survival rate of freshwater acclimated groups at 10 and 33 °C became more notable at 48 h (Fisher's exact test, $P < 0.01$) (Fig. 2a, b).

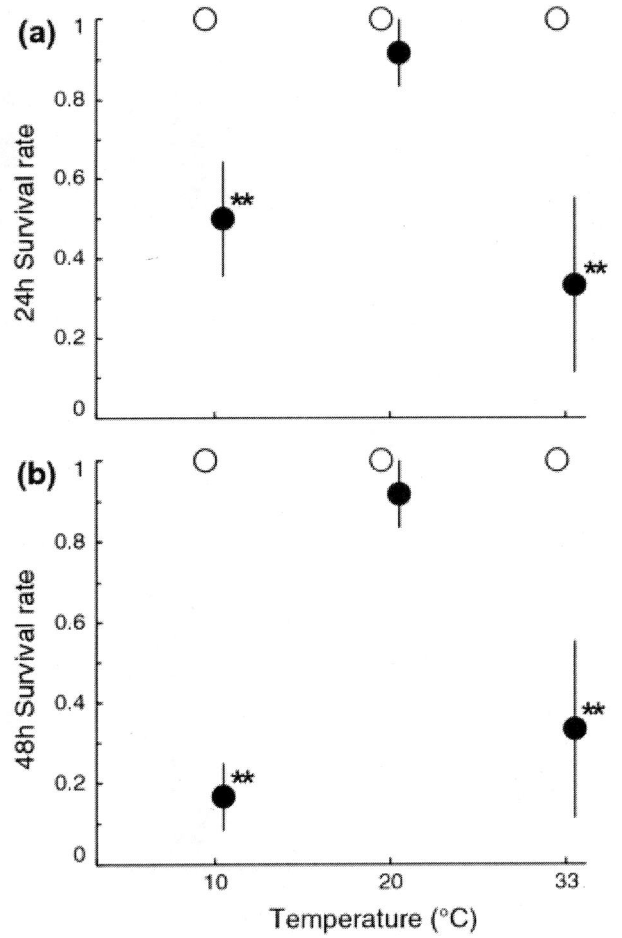


**Figure. 2** Survival rate of Japanese black porgy *Acanthopagrusschlegelli* acclimated to natal (seawater, *open circle*) and non-natal (freshwater, *closed circle*) osmotic environments at **A** 24 h and **B** 48 h after the transference to test water temperatures. *Vertical lines* indicate standard error (SE). *Asterisks* indicate $P < 0.01$ level of significant differences between the natal and non-natal osmotic environment acclimated treatments at each temperature

In dace, there was no significant difference in SL and BW between surviving and dead fish in the non-natal osmotic environment in any of the treatments (Fig. 3a, b). In black porgy, surviving individuals were significantly larger and heavier than dead ones at 10 °C (Welch's *t*-test, $P < 0.01$), and the same tendency was observed at 33 °C, although the difference was not statistically significant (Welch's *t*-test, $P > 0.05$) (Fig. 4a, b).

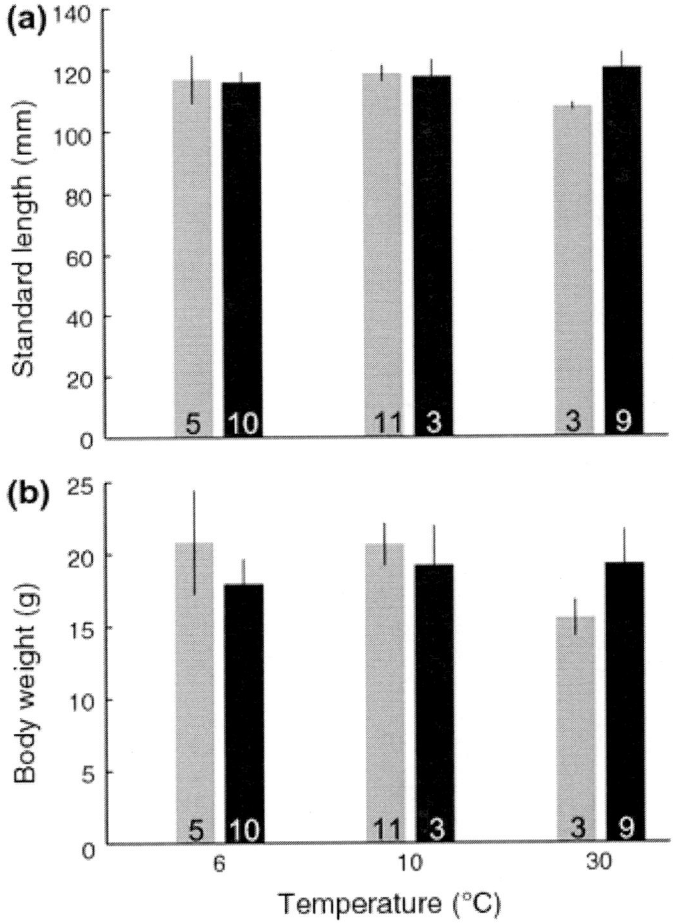


**Figure. 3** Standard length (**A**) and body weight (**B**) of surviving (*grey*) and dead (*black*) individuals in non-natal osmotic environments of Japanese dace *Tribolodon hakonensis*. *Lines on the bars* show SE and *numbers in columns* indicate the total number of individuals used for analysis

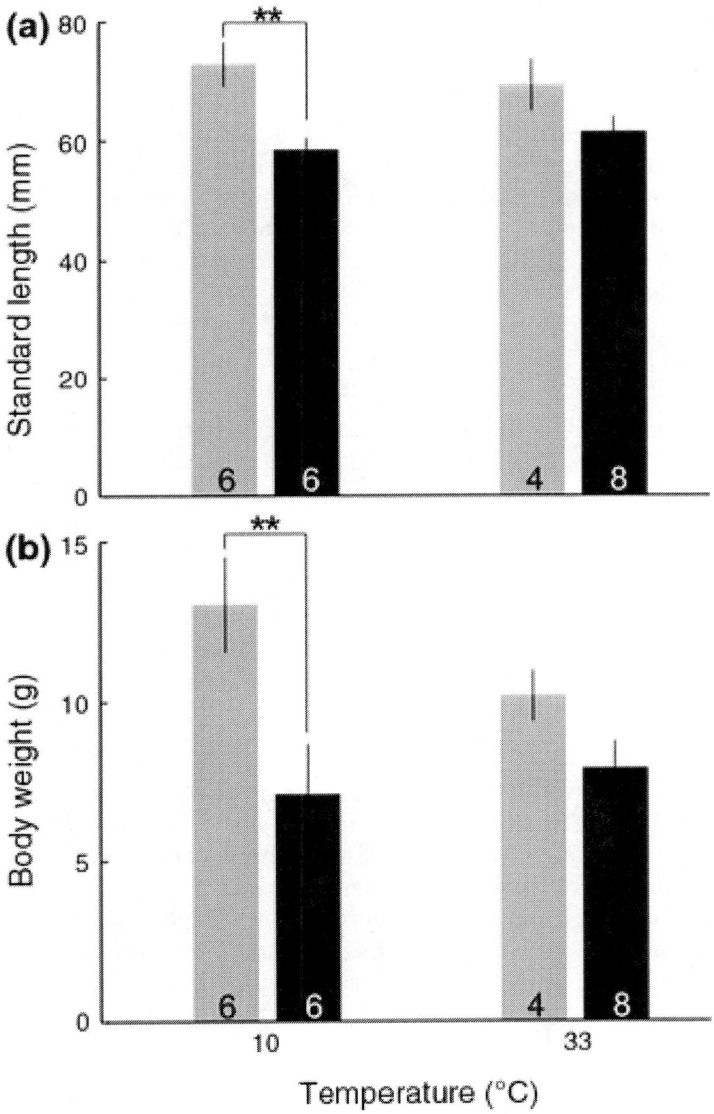


**Figure. 4** Standard length (**A**) and body weight (**B**) of surviving (*grey*) and dead (*black*) individuals in non-natal osmotic environments of Japanese black porgy *Acanthopagrus schlegelii*. *Lines on the bars* show SE and *numbers incolumns* indicate the total number of individuals used for analysis. *Asterisks* indicate $P < 0.01$ level of significant differences between dead and surviving individuals at the same temperature

# DISCUSSION

This study revealed that the survival rates of fish acclimated to non-natal osmotic environments were markedly reduced at both high and low water temperatures, suggesting that the temperature ranges at which they can survive in the non-natal osmotic environments were narrower than those in the natal osmotic environments. The non-natal osmotic environment acclimated group of dace used for the challenge test was significantly larger than those used for the natal osmotic environment. The size difference was due to the mortality of small individuals during the 2-week pre-acclimation period in seawater, which suggests that small fish are less tolerant of living in seawater.

For the dace used in the challenge test, there was no difference in body size between those that survived and died in the non-natal osmotic environment. Also, there was no mortality in the freshwater treatment group. These results indicate that there is no clear relationship between temperature tolerance and body size in dace used in the experiment. Thus, although there was a significant size difference between those used in natal and non-natal osmotic environment treatments in dace, the high mortality observed in the seawater acclimated group can apparently be attributed to a narrowed adaptable temperature range in the non-natal osmotic environment, which was also the case for black porgy that had no size difference between treatments.

Low temperature tolerance in fishes during osmotic changes is generally governed by the ability to maintain ionic balances at both the cellular and organismal level [13, 14, 15]. Thus, individuals that died at low temperature may have failed in osmoregulation. Nevertheless, the direct cause of the high mortality observed both at high and low temperature conditions in non-natal osmotic environments remains unknown. Measuring the plasma osmolarity of individuals used for the experiment would help reveal this. Indeed, a recent study has reported that ion transport greatly decreases in cold temperatures for mummichog *Fundulus heteroclitus* in seawater [16], so this may also be a factor for the species in our study.

The size-selective mortality of smaller individuals of black porgy under a non-natal osmotic environment during the temperature change trials may also be related to osmoregulatory factors. Allanson et al. [17] and Martin [18] reported a similar tendency in which smaller fish were more vulnerable to temperature-induced osmotic stress than larger ones. This may be because smaller individuals have larger gill surface area/body mass ratio compared to larger individuals. In contrast, we did not observe any size-selective mortality in dace used for the challenge tests. Hurst and

Conover [19] also found that there was no effect of body size on the temperature and salinity tolerance of young-of-the-year striped bass. Allometry of gill surface area to body mass might influence the tolerance to osmotic stress only in certain species or body sizes. Further study would be required to reveal the physiological mechanisms underlying the relationships between tolerance to temperature-induced osmotic stress and body size.

To our knowledge, our study is the first to demonstrate that the adaptability of euryhaline wanderers to non-natal osmotic environments decreases not only in low temperature conditions but also in high temperature conditions. Although there is no evidence yet that dace or black porgy avoid the non-natal osmotic environment in extremely high temperature conditions, such cases are known for other species. Spotted hake *Urophycis regia*, a euryhaline wanderer marine species, is known to egress from the estuary in summer as water temperature increases [1, 20, 21], possibly indicating that this species avoids hypotonic conditions under high temperature conditions. Perhaps future studies may find evidence that dace and black porgy might avoid the non-natal osmotic environment in extremely high temperature conditions, especially considering the warming trend in coastal waters [22, 23].

Lankford and Targett [12] reported that Atlantic croaker, a marine species, has a reduced low temperature tolerance in low salinity water. The fact that dace, black porgy and Atlantic croaker, which belong to various taxa, show the same trend may suggest that narrowed temperature adaptability in non-natal osmotic environments is a general characteristic in euryhaline wanderers and could be one of the causes of their seasonal movements.

This is insightful from a phylogenetic perspective because recent studies suggest that the diadromous species of salmonids, anguillids and osmerids are derived from species living in a single osmotic environment via euryhaline wanderer species [24, 25, 26]. For example, ancestral euryhaline wanderer species may have succeeded in expanding their adaptable temperature range to non-natal osmotic environments and became able to stay there for the whole year at latitudes with seasonal temperature changes, which is a common migratory pattern among diadromous species with multiple-year life spans. In fact, some evidence of residual characteristics similar to those of Japanese dace, Japanese black porgy and Atlantic croaker may be that primitive salmonid species such as brook trout *Salvelinus fontinalis* are suggested to have inferior low temperature tolerance in seawater compared to the more recently diverged [27] Atlantic salmon *Salmo salar* and rainbow trout *Oncorhynchus mykiss* [28]. In addition, primitive salmonid species including brook trout are known to be in the sea only during the spring-summer season [29].

Therefore, comparative studies about the adaptable temperature range to non-natal osmotic environments between primitive and recently diverged salmonids as well as additional euryhaline wanderer species might shed light on the physiological processes influencing the evolution of diadromous fishes. Other types of future research should examine the tolerances of euryhaline wanderer fishes to temperature changes over longer time periods, since our study only examined the effect of rapid changes. Studies on changes in individual plasma osmolality should also be conducted to determine the relationships between survival and osmoregulatory ability while under temperature stress.

## ACKNOWLEDGMENTS

We sincerely thank Dr. Kohji Takahashi (Nagasaki University) for technical support in collection and rearing of fishes. We are also grateful to Drs. Mari Kuroki (The University of Tokyo), Michael J. Miller (Nihon University), two anonymous reviewers, and Dr. Yutaka Kurita for constructive comments on the manuscript.

## REFERENCES

1. Able KW, Fahay MP (2010) Ecology of estuarine fishes. The Johns Hopkins University Press, Maryland
2. Day JW, Hall CAS, Kemp WM, Yanez-Arancibia A (1989) Estuarine ecology. John Wiley & Sons, New York
3. Able KW, Fahay MP (1998) The first year in the life of estuarine fishes in the Middle Atlantic Bight. Rutgers University Press, New Brunswick NJ
4. Ishitobi Y, Hiratsuka J, Kuwabara H, Yamamuro M (2000) Comparison of fish fauna in three areas of adjacent eutrophic estuarine lagoons with different salinities. J Mar Syst 26:171–181
5. Yokoo T, Mito T, Iwasaki T, Sasaki T, Michine T, Aranishi F (2010) Spatial and temporal variation of fish assemblages surveyed by set-nets in Nakaumi Lagoon. Jap J Limnol 71:11–18 **(IN JAPANESE WITH ENGLISH ABSTRACT)**
6. Hildebrand SF, Schroeder WC (1928) Fishes of Chesapeake Bay. Bull US Bur Fish 43:1–366
7. Sakai H (1995) Life-histories and genetic divergence in three species of *Triborodon* (Cypriniodae). Mem Fac Fish Hokkaido Univ 42:1–98
8. Murdy EO, Birdsong RS, Musick JA (1997) Fishes of Chesapeake Bay. Smithsonian Institute Press, Washington DC
9. Katano O, Nakamura T, Abe SI, Baba Y (2010) Population density, growth and migration via the sea to different streams of Japanese dace *Tribolodon hakonensis* in lower reaches of small streams. Ichthyol Res 57:1–9
10. Vouglitois JJ, Able KW, Kurtz RJ, Tighe KA (1987) Life history and population dynamics of the bay anchovy in New Jersey. T Am Fish Soc 116:141–153
11. Ishizaki D, Otake T, Sato T, Yodo T, Yoshioka M, Kashiwagi M (2009) Use of otolith microchemistry to estimate the migratory history of Japanese dace *Tribolodon*

*hakonensis* in the Kamo River, Mie Prefecture. Nippon Suisan Gakkaishi 75:419–424 **(IN JAPANESE WITH ENGLISH ABSTRACT)**

12. Lankford TE, Targett TE (2001) Low-temperature tolerance of age-0 Atlantic croakers: recruitment implications for US mid-Atlantic estuaries. T Am Fish Soc 130:236–249
13. Maetz J, Evans DH (1972) Effects of temperature on branchial sodium exchange and extrusion mechanisms in the seawater-adapted flounder *Platichthys flesus* L. J Exp Biol 56:565–585
14. Toneys ML, Coble DW (1980) Mortality, hematocrit, osmolality, electrolyte regulation, and fat depletion of young-of-the-year freshwater fishes under simulated winter conditions. Can J Fish Aquat Sci 37:225–232
15. Hochachka PW (1988) Channels and pumps—determinants of metabolic cold adaptation. Comp Biochem Physiol B 90:515–519
16. Buhariwalla HEC, Osmond EM, Barnes KR, Cozzi RRF, Robertson GN, Marshall WS (2012) Control of ion transport by mitochondrion-rich chloride cells of eurythermic teleost fish: cold shock vs. cold acclimation. Comp Biochem Physiol A 162:234–244
17. Allanson BR, Bok A, van Wyk NI (1971) The influence of exposure to low temperature on *Tilapia mossambica* Peters (Cichlidae). II. Changes in serum osmolarity, sodium and chloride ion concentrations. J Fish Biol 3:181–185
18. Martin TJ (1988) Interaction of salinity and temperature as a mechanism for spatial separation of three co–existing species of Ambassidae (Cuvier) (Teleostei) in estuaries on the south-east coast of Africa. J Fish Biol A 33:9–15
19. Hurst TP, Conover DO (2002) Effects of temperature and salinity on survival of young-of-the-year Hudson River striped bass (*Morone saxatilis*): implications for optimal overwintering habitats. Can J Fish Aquat Sci 59:787–795
20. Barans CA (1972) Spotted hake, *Urophycis regius*, of the York River and lower Chesapeake Bay. Chesapeake Sci 13:59–62
21. Burr BM, Schwartz FJ (1986) Occurrence, growth, and food habits of the spotted hake, *Urophycis regia*, in the Cape Fear Estuary and adjacent Atlantic Ocean, North Carolina. Northeast Gulf Sci 8:115–127
22. Freeland HJ (1990) Sea surface temperatures along the coast of British Columbia: regional evidence for a warming trend. Can J Fish Aquat Sci 47:346–350
23. Caputi N, De Lestang S, Feng M, Pearce A (2009) Seasonal variation in the long-term warming trend in water temperature off the Western Australian coast. Mar Freshwater Res 60:129–139
24. Ishiguro NB, Miya M, Nishida M (2003) Basal euteleostean relationships: a mitogenomic perspective on the phylogenetic reality of the "Protacanthopterygii". Mol Phylogenet Evol 27:476–488
25. Dodson JJ, Laroche J, Frederic L (2009) Contrasting evolutionary pathways of anadromy in euteleostean fishes. Am Fish Soc Symp 69:63–77
26. Inoue JG, Miya M, Miller MJ, Sado T, Hanel R, Hatooka K, Aoyama J, Minegishi Y, Nishida M, Tsukamoto K (2010) Deep-ocean origin of the freshwater eels. Biol Lett 6:363–366
27. Murata S, Takasaki N, Saitoh M, Okada N (1993) Determination of the phylogenetic relationships among Pacific salmonids by using short interspersed elements (SINEs) as temporal landmarks of evolution. Proc Natl Acad Sci USA 90:6995–6999
28. Saunders RL, Muise BC, Henderson EB (1975) Mortality of salmonids cultured at low temperature in sea water. Aquaculture 5:243–252
29. McCormick SD (1994) Ontogeny and evolution of salinity tolerance in anadromous salmonids: hormones and heterochrony. Estuaries 17:26–33.

# CHAPTER 3

# Freshwater Fish as Sentinel Organisms: From the Molecular to the Population Level, a Review

**Jacinto Elías Sedeño-Díaz[1] and Eugenia López-López[2]**

[1] *Polytechnic Coordination for Sustainability, National Polytechnic Institute. Mexico, Distrito Federal, México*

[2] *Ichtiology and Limnology Laboratory, National School of Biological Sciences, México, Distrito Federal, México*

## 1. INTRODUCTION

Fish around the world are found occupying almost any aquatic habitat. In particular, freshwater fish are severely threatened as the freshwater ecosystems are considered the most endangered of the world [1]. The ultimate destination of most contaminants is water; rivers, lakes, aquifers, or sea, are receptors of wastewaters with a complex mixture of xenobiotics. The variety of contaminants and their mixtures that daily reach the water bodies coupled with a multitude of irresponsible water management practices and destructive land uses, are currently threatening freshwater ecosystems [2], such is the case of discharge of municipal and industrial wastewaters, deforestation, increase of land crops, and water extraction from water bodies to human consumption and other uses. The impact of contaminants in an aquatic ecosystem is complex, therefore has increased the need for determining the ambient status in order to provide an indication of changes induced by anthropogenic activities and their influence on aquatic organisms. As physicochemical analyses shed no light on the biological status of ecosystems, a biological approach is needed to evaluate environmental health; moreover, the biological effects of contaminant interactions cannot be expressed by physicochemical investigations [3].The aquatic ecosystem health is often reflected by the health of organisms that reside in that system. Fish in their natural environments are typically exposed to numerous stressors including unfavorable or fluctuating temperatures, high water velocities and sediment loads, low dissolved oxygen concentrations, limited food availability, and among other types of natural episodic variables. In addition, anthropogenic stressors such as contaminant loading can add to the insults that fish may already experience in many systems. All these factors, individually or together, can impose considerable stress on physiological systems of fish and impair their health [4, 5]. Environmental contaminants are known to induce measurable biochemical changes in exposed aquatic organisms [6]. Likewise, stressors can load or limit physiological systems, reduce growth, impair reproduction, predispose fish to disease, and reduce the capacity of fish to tolerate additional stressors. Many species of fish, in particular those species near the top of the food chain, are generally regarded as integrators of environmental conditions and may reflect, therefore, the health of aquatic ecosystems [4] and therefore, they are excellent indicators of the relative health of aquatic ecosystems and their surrounding watersheds [7]. Thus, effects of contaminants on aquatic organisms may be manifested at all levels of biological organization (in a hierarchical scale that can be at cellular level, organisms, populations, communities, and ecosystems). In this way, the measuring of a suite of indicators across such levels of organization is often necessary to assess ecological integrity; these indicators also should include molecular, biochemical, physiological, population, community, and ecosystem responses.

The indicators allow us to isolate key aspects of the environment from an overwhelming array of signals [8]. Ecological indicators have been defined as measurable characteristics of the structure (e.g. genetic, population, habitat, and landscape pattern), composition (e.g., genes, species, populations, communities, and landscape types), or function (e.g., genetic, demographic/life history, ecosystem, and landscape disturbance processes) of ecological systems [9]. On the other hand, other authors [10], established that bioindicators are organisms or communities of organisms, which reactions are observed representatively to evaluate a situation, giving clues for the condition of the whole ecosystem; Gerhardt also indicate that bioindicators are species reacting to anthropogenical effects on the environment, concluding that a biological indicator would be: a species or group of species that readily reflects the abiotic or biotic state of an environment, represents the impact of environmental change on a habitat, community or ecosystem or is indicative of the diversity of a subset of taxa or the whole diversity within an area. In this sense, the primary role of ecological indicators is to measure the response of the ecosystem to anthropogenic disturbances [9]. A sentinel species can be defined as any domestic or wild microorganism, plant or animal, that can be used as an indicator of exposure to and toxicity of a xenobiotic that can be used in assessing the impact on human and/or environmental health because of the organism's sensitivity, position in a community, likelihood of exposure, geographic and ecological distribution or abundance [11].

The specific objective of this review is to provide a short framework of effects of xenobiotics on the responses of freshwater fish across molecular to population level when have been exposed to environmental stressors. Likewise, the present review considers the use of fish as sentinel organisms to assess the anthropogenic impacts over the freshwater ecosystems. The review asks whether fish can be able to reflect the environmental damage from molecular to population levels. Also, the present review offers a selection of examples of studies employing fish as sentinel organisms in ecological, toxicological and environmental risk assessments.

## 2. SUBORGANISMAL RESPONSES

When an organism is exposed to stressors like contaminants or a mixture of them, energy is demanded to deal with that stress [4]; stressors tend to impact ecosystems at lower levels of organization first [12]. One of the methods to quantify the exposure to xenobiotics and its potential impact on living organisms is the monitoring by the use of the so-called biomarkers [13]. Biomarkers have been defined by several authors, all of them, in reference to biological responses to contaminants exposure, as a) measurements in body fluids, cells or tissues indicating biochemical or cellular modifications due to the presence and magnitude of toxicants, or of host response [8]; b) a change in a biological response (ranging from molecular through cellular and physiological responses to behavioral changes) which can be related to exposure to or toxic effects of environmental chemicals [14]; c) any biological response to an environmental chemical at the subindividual level, measured inside an organism or in its products (urine, faeces, hair, feathers, etc.), indicating a deviation from the normal status that cannot be detected in the intact organism [15]; d) a xenobiotically induced variation in cellular or biochemical components or processes, structures, or functions that is measurable in a biological system or samples [16]; e) contaminant-induced physiological, biochemical, or histological response of an organism, and f) as functional measures of exposure to stressors expressed at the sub-organismal, physiological or behavioural level. Considering these definitions of biomarkers, we could adopt our own definition: "*any biological measurable response from an organism, induced by the exposure to a xenobiotic or complex mixture of them*". Biomarkers can provide valuable information in field or semifield testing and be used to measure a wide range of physiological responses to chemicals at the biochemical, cellular, or tissular level [17].

In concordance with [18] and other authors [19, 20], biomarkers have been classified in three different categories: a) biomarkers of exposure, which represent responses such as induction or inhibition of specific enzymes involved in biotransformation and detoxification as a consequence of chemical exposure [21], b) biomarkers of effect, are any changes in a biological system that reflects qualitative or quantitative impairment resulting from exposure [20], including responses measurable at level biochemical, physiological or some other alterations within tissues or body fluids of an organism that can be recognized as associated with an established or possible health impairment or disease [19], and c) biomarkers of susceptibility, which serve as indicators of a particular sensitivity of individuals to respond to the challenge of exposure to a effect of a xenobiotic or to the effects of a group of such compounds, in this case, individual changes included genetic factors and changes in receptors which alter the susceptibility of an organism to that exposure [19]. However, other authors have been subdivided the biomarkers in exposure biomarkers, effects biomarkers and predictive biomarkers.

Responses of fish at suborganismal level to xenobiotic exposure are complex and varied and depending of type of contaminant and time of exposure. The most general effect of xenobiotics on fish is oxidative stress, which is experienced when antioxidant defenses are overcome by prooxidant compounds. Oxidative stress include a variety of oxidative reactions, usually started by free radicals and propagated by molecular oxygen, which results in the oxidation of lipids, proteins, and nucleic acids [22]. Free radicals are atoms, molecules, or ions with unpaired electrons on an otherwise open shell configuration. These unpaired electrons are usually highly reactive due to which radicals are likely to take part in chemical reactions. Very often free radicals are confused with reactive oxygen species (ROS) such as molecular and singlet oxygen, superoxide anion, hydroxyl radical and some their derivatives; however, hydrogen peroxide is not a radical, but it is a reactive species because has higher activity than molecular oxygen [23]. Hydroxyl radical is the most important free radical of biological importance, because of its potent oxidative potential and indiscriminate reactivity with cellular components of enzymes and DNA [24, 25]; likewise, being oxidant, all ROS are agents which at high concentrations are toxic to cells. Oxidative stress is a risky condition in which increases in free radical production, and/or decreases in antioxidant levels can lead to potential damage. The antioxidant system in aerobic organisms includes several biochemical safety mechanisms such as antioxidant enzymes and other compounds like vitamins, glutathione, matallothioneins, and others.

Antioxidant defense enzymes are induced by various environmental pollutants under pro-oxydant conditions among these enzymes we can found superoxide dismutase (SOD), Catalase (Cat) and Glutation Peroxidase (GPx), Glutation reductase and Glutation S Transferase (GST, catalyze the nucleophilic conjugation of different biologically and potentially carcinogenic compounds [20]). The role of SOD is to catalyze the reaction of superoxide radical ($O_2^-$) to peroxide ($H_2O_2$). CAT detoxifies $H_2O_2$ to $H_2O$ and $O_2$. GPx detoxifies mainly organic peroxides. CAT is an enzyme with high biological relevance because reduce the concentration of peroxide, a precursor of OH-, which is a highly reactive toxic form of ROS [26]. Cytochrome P450 monooxygenases (CYPs) are a multi-gene family of enzymes that play a key role in the biotransformation of pollutants, such as dioxins, pesticides, polychlorinated biphenyls (PCBs) and polycyclic aromatic hydrocarbons (PAHs). One of the most common and highly conserved is the CYP1A subfamily. The CYP1A biomarker is widely used as a biomarker of effect both in vertebrates and invertebrates for environmental biomonitoring, especially in marine bivalves and fish. The induction of CYP1A is triggered via the cytosolic aryl hydrocarbon (Ah) receptor due to exposure to pollutants, such as PCBs, dioxins, and numerous PAHs. CYP1A activity is typically measured using the substrate ethoxyresorufin, which is o-deethylated by ethoxyresorufin-O-deethylase (EROD) to a fluorescent product, resorufin, which can be easily measured. Because EROD activities are generally measured using liver homogenates that also tend to accumulate numerous CYP1A substrates, activity may be inhibited by residual substrates or metals [20].

# 3. INDIVIDUAL RESPONSES

Morphological alterations is one of the most individual level parameters that are measure to identified damage in sentinel organisms [27, 28]. In reference [28] identified severe histological damage in gills and liver of *Goodea atripinnis* a goodeid fish from Central Mexico, after the chronic exposure to Yerbimat, an herbicide with glyphosate. After 75 days of exposure to pesticide, they found lamellar hypertrophy and leukocyte infiltration in gills, and hepatocytes with vacuolization in the cytoplasm and piknotic nuclei in liver, concluding that Yerbimat induces histological alterations in the gills and liver that might impair normal organ functioning that could lead to health damage in fish because of the important physiological roll of these organs.

Also, haematological parameters, such as erythrocyte and leucocyte count, erythrocyte indices and thrombocyte number vis-a-vis coagulation of blood has been considered bioindicators of toxicosis in fish following exposure to xenobiotics [29]. Any alteration in normal cellular components (morphology and number), and level of fluids of blood is named dyscrasia. Some authors [29] carry out an extensive review over dyscrasia in fish. They show several examples over alterations in morphological changes in blood cells, total count, haemoglobin content, thrombocytes and clotting time, concluding that haematological parameters are not specific to faced acute and chronic exposure of fish to xenobiotics, mainly organochlorinated, organophosphonate, pyrethroid and carbamate pesticides.

# 4. POPULATION RESPONSES

The status of a fish population is a reflection of the overall condition of the aquatic environment in which that population resides. As such, fish population characteristics can be used as indicators of environmental health. Although changes in population structure may act as a sensitive indicator of changing environmental conditions, the timing, degree and nature of the feedback response to altered conditions will vary with the intensity, identity and the number of stressors, as well as the availability of energy [30].

Bioindicators are responses to environmental effects that occur at higher levels of biological organization than sub-organism (biomarkers); This kind of responses can be measured at different high levels of biological organization, from individual, through population (reproductive success, mortality, size distribution, reduction in abundance and biomass), community (primary production, disruption of the nutrient cycle) to ecosystem levels [31, 32], whose main characteristic is that the measure change with exposure to negative environmental factors. Organosomatic indices are common approaches for assessing fish health [4]. In this review we consider three organosomatic indices: Condition factor (CF), hepatosomatic index (HSI) and gonadosomatic index (GSI).

Since the nutritional status of fish can change according to different factors, such as season, food availability, and among they the exposure to xenobiotics, is important to have a measure of corporal condition of fish. The chronic exposure to contaminants may cause changes in feeding behavior, leading to a deterioration of the health. The CF is a frequently used index for fish biology study, as it furnishes important information related to fish physiological state [33]. It is a measure of corporal condition since measuring the body mass associated with the body length. A fish that is heavier for a given length (higher CF) is considered to be a healthier fish, because extra weight means extra energy reserves. While a lighter fish lack energy reserves and therefore, tend to be more susceptible to environmental stressors. A low body condition may also suggest muscle wasting (proteolysis) indicating a starvation response [34]. It has also been suggested that females with a lower body condition reduce reproductive investment yet still have an increased risk of mortality.

The liver plays a major role in the metabolism of xenobiotic compounds with biochemical alterations occurring under some toxic conditions; likewise, the liver is a primary detoxification organ in fish [35]. Therefore, this strong activity can lead to an increase in liver size, from hypertrophy (an increase in size) to hyperplasia (an increase in number) of hepatocytes [35], or both. Studies evaluating the relative liver size of fishes from contaminated and reference sites often utilize the Hepatosomatic Index (HIS), which expresses the ratio of liver weight to body weight as a percentage.

Gonadosomatic index (GSI) is also a percentage relationship between the gonad weight and fish weight. Depending on the severity of exposure to xenobiotics, the sublethal effects can be to limit physiological capacity, reduce growth, and impair reproduction, therefore GSI is a convenient organosomaitc index [36].

## 5. SPECIFIC CASE STUDIES ABOUT FISH AS SENTINEL ORGANISMS

In México some studies have analyzed the use of freshwater fish as sentinels. These studies include the use of biomarkers and sentinel fish and the use of the whole fish population as indicators of environmental change. Furthermore, these studies are in areas of contrasting environmental conditions, in the case of biomarkers studies are in: the course of a river in the Atlantic slope (Río Champotón), a Lake in the Central Plateau (Yuriria Lake), a spring and a reservoir in the upper portion of a river of the Pacific slope (Ameca river). These studies make evident the utility of freshwater fish as sentinels and are briefly exposed.

## 5.1. Case Of Study Of *Astyanax Aeneus* In The Champotón River

The Champotón river, located in the humid subtropics of southeastern Mexico in terrain with a high content of karstic material, is the main surface stream in the Yucatán Peninsula; is a coastal river with 48 km in length to its outlet with a drainage basin surface area of 650 km2. The fish studied inhabit the fresh water zone of the river with salinity up to 1.2 practical salinity units [37]. The climatic regime is hot subhumid with summer rain (June to September) and occasional winter precipitation as a result of the windy (northerly) and hurricane seasons. The main anthropogenic activity in the basin is agriculture and livestock raising [38]. During the study period (2007-2008) the region was affected by several hurricanes (mostly from August to October), that caused the river overflow. This study assessed the effects of the environmental conditions along the freshwater portion of the Champotón river on the native fish *A. aeneus*, analyzing responses between lower and higher levels of organization, and linking spatial and seasonal fish responses with water quality features. A Water Quality Index (WQI) was employed as an indicator of environmental conditions, a set of sub-organismal biomarkers in *A. aeneus* lipid peroxidation (LPO), GST, EROD, and lactate dehydrogenase (LDH)) was monitored to determine the Integrated biomarker response (IBR), and organosomatic indices: GSI, HSI and CF were characterized. Three study sites were analyzed: San Juan Carpizo (SJC) in the upper portion of the river, San Antonio del Río (SAR) in the middle portion of the river (where there are a rustic swimming spot lacking of sanitary facilities), and downstream Ulumal (U); the study periods were: April, July, and November 2007 and February 2008.

The WQI scores exhibited spatial and temporal variations (from 53.21 to 78.49) on a scale of 0 to 100. The lowest value was recorded in July at site SAR, where the lack of sanitary facilities provoke fecal materials are swept away by runoff during the rainy season (July), increasing coliform numbers and lowering the WQI. In addition, this river flows through a region in which calcareous substrates predominate; with a high content of calcium carbonate and increase in conductivity both provoked decreases WQI scores, particularly during the drought. However, WQI scores were higher in November and February, following the hurricane season that brought large amounts of precipitation and increased river flow favoring dilution. As a result, the values of several WQI parameters (including hardness and conductivity) decreased while the river was in flood. Several studies reported similar WQI fluctuation patterns. WQI scores in the Lerma-Chapala Basin, Mexico, indicated severe degradation of the basin, particularly during the dry season, when its rating ranged from contaminated to highly contaminated; however, WQI improved during the wet season, [39, 40].

Although the WQI scores indicated that the Champotón river had acceptable water quality, some pollutants (residues of persistent organic compounds (POCs), such as PCBs, hexachlorocyclohexanes, aldrin-related pesticides, heptachlor, and dichlorodiphenyltrichloroethane) have been detected in the Champotón river [41]. In reference [42] reported that sediments from several Champotón river sites contained two or more of the 16 PAHs considered by the Environmental Protection Agency United States as priority pollutants that represent a potential threat to exposed organisms. Seasonal variations in POCs were found by [41]; PCBs and hexachlorocyclohexanes reached their highest values during the rainy season, while dichlorodiphenyltrichloroethane, drines, and heptachlor peaked during the dry season. Additionally, high episodic loadings of contaminants have been detected in aquatic ecosystems following flooding events [43].

Regarding biomarkers in the sentinel fish *A. aeneus*, the highest LPO values were detected in November and February (post-hurricane and windy seasons), while the lowest values were detected in July at all study sites. GST activity was highest in November at site U and lowest in July and April at all study sites. EROD activity in general was highest in April, while the lowest means occurred in July at all sites. Spatial analysis revealed that SAR recorded the highest mean EROD activity, while the lowest occurred at sites SJC and U. LDH values peaked in November and February following the hurricane and windy seasons, while minimum LDH levels occurred in April during the dry season. Spatial variability occurred; sites in the middle reaches (U, SAR) had the highest LDH values, while the lowest were recorded at headwater site SJC [44].

The increased LPO values in the post-hurricane season (November) and in the windy season (February) was associated with the hurricane season in the Champotón river which provokes flooding of adjacent areas where field crops are treated with agrochemicals that, along with the POCs and PAHs detected in sediments by [41], may be incorporated into the aquatic system. Similar results were found by [45] that detected the mobilization of agriculture-related xenobiotics, during the flooding of the river Elbe. Oxidative stress in fish after extensive flooding was also detected by [43] in the Pamlico Sound estuary. The observed increase in LPO levels in *A. aeneus* may be related to climate-induced stress and by exposure to the mixture of xenobiotics that may be mobilized during that period.

Produced ROS are detoxified by antioxidant defense mechanisms, which are essential for protection of cellular systems against xenobiotic-induced oxidative stress [46]. GST is particularly interesting, since it is involved in elimination of reactive compounds and is the transport system for glutathione [22]. GST activity in *A. aeneus* peaked during the same period that LPO increased, following the hurricane season, which may reflect a defense of the fish against oxidative stress, as reported by [47] in *Prochilodus lineatus* in Argentina and by [48] in three sub-species of *Salmo truta* in Turkey. The liver usually contains high levels of antioxidant enzyme activity, which may be due to high rates of free radical generation in this tissue [49].

*A. aeneus* exhibited seasonal fluctuations in EROD activity, peaking in April at the SAR site. Several authors [43, 50] point out that POCs (including PAHs) typically cause elevated levels of EROD activity, and fossil fuel spills and infiltrations are a major source of PAH input [51]. In *A. aeneus* EROD activity was highest at the SAR site, a recreational spot reached only by unpaved road where motor vehicles park on the riverbank. EROD induction may thus be due to surface runoff carrying PAHs.

LDHs are cytoplasmic enzymes that catalyze the reversible reduction of pyruvate to lactate [52], an important step in the energy processes of many animal groups. Its use as a biomarker is based on the assumption that organisms subjected to chemical stress must obtain additional energy rapidly, thus increasing anaerobic glycolysis. The LDH response is apparently time-dependent and may vary with the pollutant and organism involved [41]. LDH in *A. aeneus* increased at all sites in November and February, coinciding with the period of maximum LPO as well as the post-hurricane season and highest river flow.

The IBR enables evaluation of the global variations of biomarkers, taking into account the contributions and variations in the biomarkers assessed [17]. IBR data revealed seasonal fluctuations, the maximum total IBR values occurred in April and November (April, 15.73; November, 14.73) and the minimum values in February (2.53). Response in April and November suggested that *A. aeneus* was exposed to greater stress during this period, one at the end of the dry season and the other coinciding with the post-hurricane season that also led to high LPO values (oxidative stress), high GST levels (antioxidant responses), and high levels of LDH (high energetic need). Multiple stressors are involved during flooding, such as altered habitat, changes in hydrological regime, and mobilization of pollutants. Responses of biota to environmental stressors are the integrated result of natural and anthropogenic stressors that can be ultimately manifested in biotic changes at several levels of organization [53]. IBR differences among the study periods may reflect a compensatory mechanism by which the fish regains homeostasis following the period of highest stress. Fish display a large variety of physiological stress responses that manifest as an increase in certain biomarkers after a stressful event [54]. These responses are considered adaptations of the fish to adjust itself to the disturbance and regain homeostasis [55]. Fish have also been found to exhibit recovery responses dependent on stress duration and magnitude. If the stressors are too severe or persistent and the fish is unable to regain homeostasis, the responses may become maladaptive and may pose a risk to the health and wellbeing of the fish [54].

Regarding somatic indices, GSI displayed the reproductive period of *A. aeneus* occurred from April (end of the dry season) to July (early rainy season), with the reproductive peak in July. HSI was highest in *A. aeneus* during periods of reproductive inactivity, which is interpreted as an increase in liver-stored reserve materials for later use during gametogenesis. Furthermore, low HSI values prior to and during reproduction may result from the transfer of liver energy reserves toward gonadal maturation and the reproductive event, with consequent depletion of these reserves and a decrease in HSI values [56]. CF remained constant among sites and study periods, with maximum values during July. CF reflects the interactions between abiotic and biotic factors in the physiological conditions of the fish [57]. Observations demonstrate that *A. aeneus* maintains a stable, robust condition. CF and HSI trends may reflect physiological conditions in *A. aeneus* that enable oocyte maturation and release, suggesting that oocyte production relies on energy stored in the liver and not on energy stored in the musculature [58, 59]. GSI, HSI, and CF revealed that the reproductive success of the sentinel species had not been affected, since the GSI values concurred with those reported for other species in the genus *Astyanax*. HSI values showed the transfer of energy during the reproductive period, and CF remained stable throughout the study, evidencing no effects on the general condition of the fish during periods of higher stress and indicating that the stress was temporal and the fish were able to compensate for it.

Despite WQI scores suggest that the Champotón river water is not highly polluted, the set of *A. aeneus* biomarkers constitutes a more sensitive and effective tool for identifying periods of environmental conditions adverse to fish health. Markers of oxidative damage (LPO), energy processes (LDH), detoxification (EROD), and antioxidant activity (GST) suggested that two stress periods affected the health condition of *A. aeneus* in different ways. These biomarkers can be used as early warning signals of environmental change prior to the onset of irreversible damage at the population level. Indeed, the IBR values highlighted the two periods of high biomarker response. Overall, this study provides evidence supporting the use of a biomarker set in assessing the health of aquatic systems, corroborating the suitability of *Astyanas aeneus* as a sentinel species.

## 5.2. Study Case Of *Chirostoma Jordani* In Yurira Lake

*Chirostoma jordani* is an Atherinopsid fish endemic of Central México. Data presented are from a population living in Yuriria Lake, one of the most important lakes in México, located in the Central Plateau (hydrologic region Lerma-Chapala-Santiago 20 ° 20'24" -19 ° 04'48" N and 101 ° 55'48" -100 ° 48 '36" W). It is an artificial, small and shallow lake (area= 66 $km^2$ and maximum depth of 3.2 m) feed by a diversion of Lerma river (their main tributary), that carries wastes from mining activities, livestock, industrial, urban and rural areas [60]. In the western end the lake receives water from two small an intermittent tributaries. Yuriria Lake supplies water for the surrounding farming areas, harbors migratory bird populations, and supports fisheries and tourism of several human settlements on the littoral zone.

This study shows an assessment of water quality by means of a WQI and a battery of oxidative stress: LPO, and the activity of antioxidant enzymes, SOD, CAT, and GPx along with somatic indices, such as the GSI, HSI, and CF were analyzed to assess the health condition of *C. jordani* in Yuriria Lake. The study was carried on in a period with scarce pluvial precipitation and prolonged drought (May, August, November 2009 and February and May 2010).

Yuriria Lake is characterized by a high deterioration in their water quality, WQI values ranged from 55 to 70, with a global mean of 65.85. The lake has spatial differences in water quality, the limnetic zone has higher scores (63 to 70), and the tributaries have the lower quality (55 to 58). Yuriria Lake being located in the Central Plateau, one of the most highly populated areas in México, displays the general problem of water quality of the basin (the Lerma-Chapala basin), where urban and industrial wastewater discharges, and leachates of agrochemicals are the main pollutants that diminish the water quality [61]. Particularly the middle Lerma (where Yuriria Lake is located) is recognized as the most affected area, with WQI scores between 41.1 and 54.2 in 1999 [40, 61, 62]. Furthermore, previous studies have recognized the entry of pollutants in Yuriria Lake [61]; however, the effect of the mixtures of these pollutants on the aquatic biota inhabiting the lake had not been analyzed.

The biomarker assessment suggests that the lake conditions exert stress on the fish *C. jordani.* The biomarker response showed pronounced seasonal variations. The gills presented higher values of LPO. May 2009 displayed the highest levels and November the lowest. In the liver the higher levels of LPO were detected during November and February and May 2010 (the end of the rainy season and the dry season). In muscle, the highest level of LPO was observed during February and May 2010. Gills being the first organ of contact with water are exposed directly to any xenobiotic in the aquatic environment and their biomarker responses are the result of the exposure to stressors. In addition, toxics can also enter via the intake of water and food and be absorbed and transported by the portal system to the liver before entering the general circulatory system; in consequence, the liver is one of the most sensitive organs to environmental stressors [63].

The activity of the antioxidant enzymes in liver and gills also showed a marked seasonal variation. SOD and GPx significantly increased during November, mainly in the gills, compared to the rest of the seasons. CAT also showed higher values in activity during November; however, its highest value was found in gills during February. In general, the activity of antioxidant enzymes decreased from November to May 2010.

Exposure to various xenobiotics, such as metals and organic compounds that enter water bodies, can promote the formation of ROS and induce oxidative stress [12]. The increase in the level of LPO in liver observed in November 2009 and February and May 2010, suggests the existence of pro-oxidant agents in Yuriria Lake and indicates increased oxidative stress in these seasons. Seasonal variation in LPO values could be related to the rainy and the dry seasons; the rainy season can promote dilution of xenobiotics that induce less stress during this season. Rainfall also increases leaching and runoff that enhance the entry of xenobiotics (chemicals), in this study results show that the damage generated in the fish liver became evident from November (the end of the rainy season) until February and May 2010 (dry season), when the processes of evaporation and consequent concentration of xenobiotics could be higher. Furthermore, in May 2009, there were higher levels of LPO in gills, which may indicate that water in the lake at the beginning of the rainy season provoke oxidative stress in gills.

An increase in LPO levels in fish can trigger an antioxidant response as a defense mechanism to prevent cell damage caused by pro-oxidant agents [19, 64] and could be expressed as increased or depleted CAT, SOD, and GPx activities [12]; in both cases, the result is damage to the antioxidant system. In *C. jordani* inhabiting Yuriria Lake, both responses were detected: 1) stimulation of the activity, when LPO levels in liver and muscle were higher (November and February) and antioxidant activity showed the highest values; and 2) depletion in antioxidant activity during May 2009 with high levels of LPO in the gills, and May 2010 with increased LPO in liver and muscle. These highly variable responses of the antioxidant system depend on the type and concentration of contaminant to which fish are exposed, as well as on the intensity and duration of exposure [65]. The clear decrease in the activities of CAT, SOD, and GPx in *C. jordani* is extremely important because it suggests severe damage to the antioxidant system of fish. These damages have been previously documented by [66], who recorded a collapse of the antioxidant defense system of *Liza aurata* in the Ria de

Aveiro, Portugal, with exposure to mercury. In addition, [67] reported damage to the antioxidant system of *Oreochromis niloticus* from acute and chronic exposure to Cd, Cr, Cu, Zn, and Fe.

At the population level the assessment can reveal changes in the fish biology and ecology resulting from natural fluctuations and/or ecosystem changes caused by environmental degradation. The standard length of *C. jordani* ranges from 21.16 to 77.61 mm. Three size classes were determined: Class I with a mean size of 26 mm, represented only in the month of May 2010; Class II with a mean size of 56 mm; and class III with a mean size of 62 mm. There was a gap in the size frequencies in the class of 32–50 mm from May to August 2009, whose abundances were not sufficient to form this cohort. The lack of smaller sizes from May to November 2009 can be interpreted as a possible reproductive failure during that year, resulting in low recruitment and consequently precluding estimation of the cohort [68]. These gaps or missing cohorts have been previously documented for other fish as a result of overfishing and/or environmental degradation [69]. Several causes could explain this event. According to [70], the hydrological cycle plays an important role in the development of different biological attributes such as gonadal maturation, migration, spawning, larval development, growth, and feeding. Prolonged periods of drought are also associated with failures in the recruitment and subsequently reduced adult stocks and serious effects on fisheries.

The somatic indices revealed that K displayed small variation between sites and between periods. HSI showed significant differences between sites in August, and between periods in May, values were significantly lower than those of other periods. The GSI showed the greatest variation, with a clear reproductive peak during May 2010. There was a positive correlation between the GSI and HSI. By size class, only the GSI showed variations between seasons. Class I was significantly lower than the rest of the classes. Classes II and III the GSI showed a reproductive peak in May 2010.

In fish, the cost of reproduction may be considerable; thus, fish can express different patterns of energy storage and depletion in relation to reproductive cycles, with an alternation in energy storage (56). The comparison of K and the HSI with the GSI could therefore be useful for estimating the possible balances or energy transfer between the reproductive period and nutritional status [71]. *C. jordani* in Yuriria Lake show that K was maintained at stable levels. Furthermore, there was no alternation of energy storage between the liver and gonads; GSI and HSI correlated positively, indicating that the reproductive period did not compromise energy reserves or the liver or the soma.

According to [72], fish living in waters contaminated with domestic sewage exhibit higher K and GSI, these authors suggest it is because these sites have more available food for fish, enabling them to compensate for the environmental impairments. This scenario is likely occurring in Lake Yuriria, where *C. jordani*, having enough food, can complete reproductive cycles and reach larger sizes despite the presence of stressors in the lake.

According to the results critical periods in the health of *C. jordani* occur from November, February to May 2010 (end of the rainy season and the dry season) because in this period, the higher LPO, the lowest antioxidant response, and the lowest K were observed in contrast to the higher WQI scores detected in the same period. This result indicates that fish health assessed by biological indicators as oxidative stress biomarkers and the lack of a cohort, are highly sensitive to environmental conditions imposed by the dry season. The drought has been recognized as one of the critical periods in fish health because during this period, the dilution capacity of aquatic ecosystems is low, which increases the risk of exposure to high concentrations of pollutants [73].

The findings suggest that *C. jordani* faces oxidative stress resulting from the presence of pro-oxidant agents in Yuriria Lake. At the population level, *C. jordani* has adapted to the conditions in Yuriria Lake, with mean values of HSI and GSI greater than those in other sites and K values stable throughout the year. Changes were observed in recruitment and reproductive success associated with low water levels in the lake in 2009 which shows that the fish population is highly dependent on water levels in the lake and climate changes. Previous studies indicate that Yuriria Lake receives various xenobiotics with levels that vary both spatially and temporally. Biomarkers of oxidative stress, somatic indices, and monitoring of the size classes in the fish *C. jordani* are appropriate indicators of Yuriria Lake conditions.

## 5.3. Study Case Of *Ameca Splendens* And *Goodea Atripinnis* In Ameca River

The Ameca River is located in the western slope of Mexico which drains to the Pacific Ocean. It is a river characterized by their great fish biodiversity, however, the upper portion of Ameca River, is affected by several environmental disturbers: the construction of a reservoir, the inputs of wastewater from a sugar-processing facility and water extraction, which have resulted in a drastic reduction in fish biodiversity [74, 75]. In addition, some endemic fish species such as *Ameca splendens*, have suffered a reduction in their range, and have become more prone to extinction (NOM-059, 2002) than those with a broad distribution (like *Goodea atripinnis*).

The authors of this paper [32] analyze biomarkers and bioindicators of two viviparous fish species, *A. splendens* and *G. atripinnis* living in a reservoir of the upper portion of the Ameca River, which receives wastewater, and in a spring of the same river that is free from such polluting water. In this study a comparison of the biomarker responses and bioindicators in two fish species were assessed, according to the main objectives: a) to assess water quality of a spring (ER reference site) and De LaVega reservoir (LV impacted site) where *G. atripinnis* and *A. splendens* coexist; b) to examine the health of both fish species in the reference and impacted sites by means of a battery of biomarkers; and c) to analyze physiological condition indices and population level assessment by mean of bioindicators (population measurements). A WQI was assessed. The set of biomarkers was composed by: enzyme activities of gamma-glutamyl-transpeptidase (γ-GTP), acetylcholinesterase (AChE), EROD and the LPO were determined. Additionally, somatic index were analyzed: CF, HSI and GSI.

Data of WQI scores showed spatial and temporal variations, the spring ER achieved the highest scores in all months over the course of the study and the highest values occurred in March in both sites.

Regarding biomarkers highest values of LPO were found in September (rainy season); the highest values were detected for female livers and gills of *A. splendens* in the impacted site LV and for female livers and male gills of *G. atripinnis* in LV. The LPO activity in the upper Ameca River displayed the stress to which organisms are subjected, since in the reservoir LPO showed the highest values, being *A. splendens* the most affected species. Many environmental pollutants and their metabolites have shown to exert toxic effects associated to oxidative stress, producing free radicals that initiate the LPO and cause damage to membrane proteins [76].

Results of γ-GTP showed less marked seasonal differences than LPO. In this study, in most comparisons between sites, γ-GTP activity was slightly higher at the spring; the inhibitory effect of this activity at LV could indicate a diminution in the amount of membrane proteins caused by LPO [77]. On the other hand, an increase of γ-GTP activity towards March in LV could be to prevent increases in LPO [76]; this increase coincides with the rainy season. Several authors have found seasonal variations in the response of this enzyme as a result of exposition to alkylphenols, the final degradation products of pesticides, detergents and other formulated products [78].

There was a seasonal variation in AchE activity. Organophosphates and carbamates, as well as PAHs, have been widely recognized for causing AchE inhibition, through their reaction with the serine at active site of the enzyme [79]. In nervous tissue AchE is responsible for the breakdown of acetylcholine (Ach) during transmission of an impulse; if the enzyme is inhibited, Ach is accumulated and thus a prolonged transmission of impulses could result in tetani and often in respiratory failure and death. In September in *A. splendens* living at LV, AchE exhibited lower values related to reference site that could have resulted from fish exposure to diverse pesticides used in the adjacent agricultural lands that run off in the rainy season. In March during the dry season, in *G. atripinnis*, living at LV, AchE also exhibited lower values related to ER, may be the effect of evaporation and consequently the concentration of total solid dissolved at LV, including pesticides and other xenobiotics.

Regarding EROD, the hydrocarbons discharged by the sugar-processing facility and other effluents into LV from December to June, could be responsible for activating EROD detoxification mechanism in fishes living there [80]. This mechanism is considered as the main measure of the CYP1A activity, which in turn constitutes a part of the enzyme complex of the Mixed-Function Oxidase (MFO). Since MFO facilitates the excretion of aromatic contaminants from the body induction of this complex is an effective biomarker of exposure [81, 82]; there are several studies reporting elevated levels of MFO activity in liver fishes as a result of exposure to organic contaminants, such as PAHs, dioxins, PCBs and agricultural and urban wastewater [83]. Also, there are other factors, such as UV radiation, that causes increase in MFO activity [84], moreover, damaged livers, like those of LV organisms, are less capable to MFO induction [85].

Responses to environmental stress also were reflected in bioindicators in both species studied. The major HSI values in LV concurs with [86] who reported major HSI values, related with higher EROD induction, at contaminated sites in comparison with a reference site. Moreover, in our study, higher LSI in LV concurs with higher IBR values. High values of HSI could have resulted from exposition to hydrocarbons which cause hypertrophia in liver [86].

Species may differ in the nature of their physiological response and reproductive consequences to stressors [87]. Tolerant species to environmental stress, like *G. atripinnis*, are more abundant in more disturbed environments, like LV [74, 88]; on the other hand, *A. splendens* is more abundant in a more stable environment, like ER. In LV females/male ratio for *A. splendens* could be affected by environmental estrogens, like pesticides [89]; these could act by merging receptor binding properties of estradiol, alteration of estradiol/testosterone ratios or estrogen receptor levels [12].

The higher SL, weight and CF values in LV are in concordance with [71], that found higher CF and GSI in fishes living in waters polluted with untreated domestic sewage; they suggested that fishes in these sites could find abundant food availability, and they are able to compensate for environmental changes caused by untreated domestic sewage discharges. Fishes could have major GSI values, higher fecundity and lower maturity age even under conditions of high pH [89]; in the present study, *A splendens* in LV presented these features; only in *G. atripinnis* organisms GSI was higher in the ER. [90] found a decrease in GSI as a result of the exposition of chubs to effluents carrying out organic pollutants and metals. [71] revealed a negative relation between CF and HSI with GSI, but this relation was observed only in *G. atripinnis*. The larger size, higher growth, longevity and reproductive success of organisms living at LV suggest a tactic to compensate for the stress to which the populations of both species studied are subjected; however, offspring is smaller and has a lower weight.

Throughout this study, water quality was higher in ER than in LV due to human activities; but in both sites there were different spatial and temporal factors that produced stress on fishes living there. Therefore, fishes had responses at biomarker and population levels of biological organization. Every biomarker and IBR in this study showed seasonal variation and they were useful environmental tools to demonstrate that, as consequence of pollution, LV is a more stressing place to organisms living them in comparison with ER. In general, organisms in LV presented oxidative stress by the LPO levels, and then neurotoxic impacts by the AchE and some detoxification mechanisms were evident by the γ-GTP and EROD activities. Bioindicators showed evidences of physiological changes due to contaminants exposure and make evident the plasticity of the organisms to survive in this site, in turn the responses should be considered as tactics to survive under stress condition. Moreover, both biomarkers and bioindicators revealed that *A. splendens* is a less tolerant species than *G. atripinnis* to environmental stress. Differences in biological response could be attributed to different physiological status of each fish species during the wet and dry season as well as to differences in the type and quantity of the xenobiotics that input at LV due to the period of maximum and minimal activity of the sugar industry; the lixiviation of the agrochemicals from the adjacent lands to the water bodies and also to the complexity of the mixtures of pollutants that are conform at LV that provoke several biological responses.

# 6. HOLISTIC APPROACH

Environmental stressors can cause several and different damages over aquatic organisms. These damages could be from molecular to population levels, likewise community and ecosystem levels. Through the biomarkers such as defined in this document (any biological measurable response from an organism, induced by the exposure to a xenobiotic or complex mixture of them) we can determine only some of possible causal relationships. Therefore, it is necessary always, measure a set of biomarkers to identify different stressors or damage on sentinel organisms. Several indices have been proposed to try to integrate the multi responses of different biomarkers in a single number that is indicative of the severity of the damage or stress. Such is the case of the IBR proposed by [17]. In this index, the biomarker data must first be normalized and standardized; then the score is represented by the area of a star plot. IBR considers the responses of activation or inhibition of the biomarkers assessed. IBR is an exploratory tool and should be appropriate only if an a priori justification exists for each biomarker used and if the physiological significance of the changes to each biomarker is well known [17].

An other case of index based on a battery of biomarkers was proposed by [91]. This biomarker index was obtained by summing the biomarker values expressed in term of classes. Classes were determined by a distribution-free approach derived from the theory of rough sets. No synergistic or antagonistic assumptions were incorporated into this index.

In [4] the authors proposed a quantitative health assessment index for rapid evaluation of fish condition in the field named Health Assessment Index (HAI). This index is not based on a battery of biomarkers; however, it is a quantitative index that allows statistical comparisons of fish health among data sets. Index variables are assigned numerical values based on the degree of severity or damage incurred by an organ or tissue from environmental stressors.

The Bioeffect Assessment Index (BAI), is based on the integration of several pathological endpoints measured in the liver of fish [21]. The BAI represents a modification of the HAI since it includes solely validated biomarkers reflecting toxically induced alterations at different levels of biological organisation in order to quantify the effects of environmental pollution. BAI is able to reflect deleterious effects of several classes of xenobiotics such as heavy metals, organochlorines, pesticides, PAHs, and therefore is also considered as an integrative index of health in aquatic ecosystems.

When we use sentinel organisms, a key point is the study of baseline or natural variation of responses of the sentinel organism selected, or characterizing the response of the same sentinel organism in reference sites.

The use of fish as sentinel organism is feasible for pollution monitoring in aquatic systems; however, the survey should consider the application of a suite of measurable responses (biomarkers and bioindicators) to identify potential sources of stress and damage to which organisms are exposed, as shown in the case studies presented above. The set of biomarkers or bioindicators should also, consider several levels of biological organization in order to identify effects of environmental stressors, spatio temporal trends in environmental conditions and to identify early warning signals to prevent that damage continue from low biological organizing levels to higher levels of organization.

## REFERENCES


1. Dudgeon D, Arthington AH, Gessner O, Kawabata ZI, Knowler DJ, Lévêque C, Naiman RJ, Prieur-Richard AH, Soto D, Stiassny MLJ, and Sullivan CA. Freshwater biodiversity: importance, threats, status and conservation challenges. Biological Reviews 2005; 81: 163–182.
2. Strobl RO, Robillard PD. Network design for water quality monitoring of surface freshwaters: A review. Journal of Environmental Management 2008; 87: 639–648.
3. Vasseur P, Cossu Leguille C. Biomarkers and community indices as complementary tools for environmental safety. Environment International 2003; 28: 711 – 717.
4. Adams SM, Brown AM, Goede RW. A Quantitative Health Assessment Index for Rapid Evaluation of Fish Condition in the Field. Transactions of the American Fisheries Society 1993; 122: 63-73.
5. Wedemeyer GA, McLeay DJ, Goodyear CP. Assessing the tolerance of fish and fish populations to environmental stress: the problems and methods of monitoring. In: Cairns P, Hodson V, Nriagu JO. (ed) Contaminated effects on fisheries. Wiley, New York; 1984. p163-195.
6. Eufemia NA, Collier TK, Stein JE, Watson DE, and Di Giulio RT. Biochemical responses to sediment-associated contaminants in brown bullhead (Ameriurus nebulosus) from the Niagara River ecosystem. Ecotoxicology 1997; 6: 13–34.
7. Fausch KD, Lyons J, Karr JR and Angermeier PL. Fish Communities as Indicators of Environmental Degradation. American Fisheries Society Symposium 1990; 8:123-144.
8. NRC: Committee on Biological Markers of the National Research Council. Biological markers in environmental health research. Environmental Health Perspectives 1987; 74: 3-9.
9. Niemi GJ, McDonald ME. Application of Ecological Indicators. Annual Review of Ecology, Evolution, and Systematics 2004; 35: 89–111.
10. Gerhardt A. Bioindicator species and their use in biomonitoring. Environmental Monitoring I. EOLSS.
11. US Goverment Printing Office. http://www.gpo.gov/fdsys/pkg/CZIC-z5863-m65-p33-1995/xml/CZIC-z5863-m65-p33-1995.xml (accesed 12 july 2012).
12. Sherry JP. The role of biomarkers in the health assessment of aquatic ecosystems. Aquatic Ecosystems Health and Management 2003; 6: 443-440.
13. Livingstone d. Biotechnology and pollution monitoring: use of molecular biomarkers in the aquaitc environment. Journal of Chemistry, Technology and Biotechnology 1993; 57: 195-211.
14. Peakall DB, Walker CH. The role of biomarkers in environmental risk assessment. 3: Vertebrates. Ecotoxicology 1994; 3: 173-179.
15. Van Gestel CAM, Van Brummelen TC. Incorporation of the biomarker concept in ecotoxicology calls for a redefinition of terms. Ecotoxicology 1996; 5: 217-225.
16. Shugart LR, McCarthy JF, Halbrook RS. Biological markers of environmental and ecological contamination: An overview. Risk Analysis 1992; 12: 353–360.
17. Beliaeff B, Burgeot T. Integrated biomarker response: a useful tool for ecological risk assessment. Environmental Toxicology and Chemistry 2002; 21: 1316-1322.
18. WHO International Programme on Chemical Safety (IPCS). Biomarkers and risk assessment: concepts and principles. Environmental Health Criteria 155, World Health Organization, Geneva 1993.

19. van der Oost R, Beyer J, Vermeulen NPE. Fish bioaccumulation and biomarkers in environmental risk. Environmental Toxicology and Pharmacology 2003; 13: 57-149.
20. Rojas-García AE, Medina-Díaz IM, Robledo-Marenco ML, Barrón-Vivanco BS, Pérez-Herrera N. Pesticide Biomarkers. In Margarita Stoytcheva (ed.) Pesticides in the modern world-Pest Control and pesticides exposure and toxicity assessment. InTech; 161-190.
21. Broeg K, Westernhagen HV, Zander S, Körting W, Koehler A. The "bioeffect assessment index" (BAI) A concept for the quantification of effects of marine pollution by an integrated biomarker approach. Marine Pollution Bulletin 2005; 50: 495–503.
22. Turrens JF. Topical Review Mitochondrial formation of reactive oxygen species. Journal of Physiology 2003; 552(2): 335-344.
23. Lushchak VI. Environmentally induced oxidative stress in aquatic animals. Aquatic Toxicology 2011; 101: 13-30.
24. Valavanidis A, Vlahogianni T, Dassenakis M, Scoullos M. Molecular biomarkers of oxidative stress in aquatic organisms in relation to toxic environmental pollutants. Ecotoxicology Environmental Safety 2006; 64: 178-189.
25. Stadtman ER, Levine RL. Protein oxidation. Annals of the New York Academy of Sciences 2000; 899: 191–208.
26. Vlahogianni T, Dassenakis M, Scoullos MJ, Valavanidis A. Integrated use of biomarkers (superoxide dismutase, catalase and lipid peroxidation) in mussels Mytilus galloprovincialis for assessing heavy metals' pollution in coastal areas from the Saronikos Gulf of Greece. Marine Pollution Bulletin 2007; 54: 1361 - 1371.
27. Prieto AI, Jos A, Pichardo S, Moreno I, Cameán AM. Differential oxidative stress responses to microcystins LR and RR in intraperitoneally exposed tilapia fish (Oreochromis sp.). Aquatic Toxicology 2006; 77: 314-321.
28. Ortiz-Ordoñez E, Uría-Galicia E, Ruiz-Picos RA, Sánchez Duran AG, Hernández Trejo Y, Sedeño-Díaz JE, López-López E. Effect of Yerbimat Herbicide on Lipid Peroxidation, Catalase Activity, and Histological Damage in Gills and Liver of the Freshwater Fish Goodea Atripinnis. Archives of Environmental Contamination and Toxicology 2011; 61: 443–452.
29. Nath Singh N, Srivastava AK. Haematological parameters as bioindicators of insecticide exposure in teleosts. Ecotoxicology 2010; 19: 838–854.
30. Munkittrick KR, Dixon DG. A holistic approach to ecosystem health assessment using fish population characteristics. Hydrobiologia 1989; 188-189: 123-135.
31. Oertel N, Salánki J. Biomonitoring and bioindicators in aquatic ecosystems. In: Ambasht RS, Ambasht NK (ed) Modern trends in applied aquatic ecology. Nueva York, USA: Kluwer Academic 2003 p219–246.
32. Tejeda-Vera R, López-López E, Sedeño-Díaz JE. Biomarkers and bioindicators of the health condition of Ameca splendens and Goodea atripinnis (Pisces: Goodeaidae) in the Ameca River, Mexico. Environment International 2007; 33: 521–531.
33. Lima-Junior SE, Braz Cardone I, Goitein R. Determination of a method for calculation of Allometric Condition Factor of fish. Maringá 2002; 24(2): 397-400.
34. Barton BA, Morgan JD, Vijayan MM. Physiological and condition-related indicators of environmental stress in fish. In Biological Indicators of Aquatic Ecosystem Stress, ed. S. Marshall Adams, Maryland: American Fisheries Society. 2002; 111–148.
35. Hinton DF, Laurén DJ. Liver structural alterations accompanying chronic toxicity in fishes: potential biomarkers of exposure. In: McCarthy JF, Shugart LR (eds) Biomarkers of environmental contamination. Lewis, Boca Raton, 1990. pp17–57
36. Courtney J, Klinkmann T, Courtney A, Torano J, Courtney M. Relative Condition Factors of Fish as Bioindicators One Year after the Deepwater Horizon Oil Spill. www.arXiv.org > q-bio > arXiv:1208.5095.
37. López-López E, Sedeño-Díaz JE, López-Romero F, Trujillo-Jiménez P. Spatial and seasonal distribution patterns of fish assemblages in the Río Champotón, southeastern México. Reviews in Fish Biology and Fisheries 2009; 21: 127-142.

38. CONABIO (Comisión Nacional para el Conocimiento y Uso de la Biodiversidad). Uso del suelo y vegetación modificado por CONABIO. Escala 1:1,000,000. Comisión Nacional para el Conocimiento y Uso de la Biodiversidad. 1999. Ciudad de México, México.
39. Mestre, R.J.E., 2002. Integrated approach to River Basin management: Lerma-Chapala case study- attribution and experiences in water management in Mexico. Water Int. 22, 140-152.
40. Sedeño-Díaz JE, López-López E. Water quality in the Río Lerma, Mexico: An overview of the Last Quarter of the Twentieth Century. Water Resources Management 2007; 21: 1797-1812.
41. Rendón von Osten J. Uso de biomarcadores en ecosistemas acuáticos. In: Botello AV, Rendon von Osten J, Gold-Bouchot G, Agraz-Hernández C. (Eds.) Golfo de México contaminación e impacto ambiental: Diagnóstico y tendencias. México: Univ Autón Campeche, Univ Nal México, Instituto Nacional de Ecología. 2005 pp121-140.
42. Quetz L, Memije M, Benítez J, Rendón von Osten J. Hidrocarburos aromáticos policíclicos en sedimentos del río y costa de Champotón, Campeche. Jaina 2009; 20, 27-34.
43. Adams SM, Greeley Jr MS, McHugh JL, Noga EJ, Zelikoff JT. Application of multiple sublethal stress indicators to assess the health of fish in Pamlico Sound following extensive flooding. Estuaries 2003; 26, 1365-1382.
44. Trujillo-Jiménez P, Sedeño-Díaz JE, Camargo JA, López-López E. Assessing environmental conditions of the Río Champotón (México) using diverse indices and biomarkers in the fish Astyanax aeneus (Günther, 1860). Ecological Indicators 2011; 636-1646.
45. Einsporn S, Broeg K, Koehler A. The Elbe flood 2002—toxic effects of transported contaminants in flatfish and mussels of the Wadden Sea. Marine. Pollution Bulletin 2005; 50: 423–429.
46. Ochoa DM, González JF. Estrés oxidativo en peces inducido por contaminantes ambientales. Rev. Med. Vet. Zoo. 2008; 55: 115-126.
47. Cazenave J, Bacchetta C, Parma JM, Scarabotti AP, Wunderlin AD. Multiple biomarkers responses in Prochilodus lineatus allowed assessing changes in the water quality of Salado River basin (Santa Fe, Argentina). Environmental Pollution 2009; 157, 3025–3033.
48. Aras NM, Bayir A, Sirkecioglu AN, Bayir M, Aksakal E, Haliloglu HI. Seasonal changes in antioxidant defence system of liver and gills of Salmo trutta caspius, Salmo trutta labraz and Salmo trutta macrostigma. Journal of Fish Biology 2009; 74, 842-856.
49. Mohammed SAA, Al-Qahtani A. The distribution of glutathione and glutathione-S-transferase activity in the oprgans of dhub (The agamid Lizard; Uromastyx aegyptius). Journal of Biological Sciences 2007; 7: 558-561.
50. Havelkova M, Blahova J, Kroupova H, Randak T, Slatinska I, Leontovycova D, Grabic R, Pospisil R, Svobodova Z. Biomarkers of contaminant exposure in Chub (Leuciscus cephalus L.) – Biomonitoring of major rivers in the Czech Republic. Sensors. 2008; 8: 2589–2603.
51. Hartl MGL. Benthic fish as sentinel organisms of estuarine sediment toxicity, in: Bright M, Dworschak PC, Stachowitsch M. (Eds.) The Vienna School of Marine Biology: A Tribute to Jörg Ott. Facultas Universitätsverlag, Wien. 2002 pp89–100.
52. López JE, Marcano Torres M, López Salazar JE, López Salazar Y, Fasanella H, Michelangelli L. Lactato deshidrogenasa total y sus isoenzimas LDH1, LDH2,LDH3, LDH4, LDH5, en el diagnóstico diferencial del síndrome ascítico. Gaceta Médica de Caracas 1996; 104(4): 345-364.
53. Adams SM. Assessing cause and effect of multiple stressors on marine systems. Marine Pollution Bulletin 2005; 51, 649-657.
54. Barton AB. Stress in fishes: A diversity of responses with particular reference to changes in circulating corticosteroids. Integrative and Comparative Biology 2002; 42: 513-525.

55. Wendelaar BSE. The stress response in fish. Physiological Reviews 1997; 77: 591-625.
56. Dethloff MG, Christopher J, Schmitt JC. Condition factor and organosomatic indices, in: Schmitt CJ, Dethloff MG (ed) Biomonitoring of Environmental Status and Trends (BEST) Program: selected methods for monitoring chemical contaminants and their effects in aquatic ecosystems. U.S. Geological Survey, Biological Resources Division, Columbia, (MO): Information and Technology Report USGS/BRD-2000--2005. 2000; pp13-18.
57. Lizama MAP, Ambrósio AM. Condition factor in nine species of fish of the Characidae family in the upper Paraná river floodplain, Brazil. Brazilian Journal of Biology 2002; 62: 113-124.
58. Zimmerman M. Maturity and fecundity of arrow tooth flounder, Atheresthes stomias, from the Gulf of Alaska. Fishery Bulletin 1997; 95: 598-611.
59. Yoneda M, Tokomura M, Fujita H, Takeshita N, Takeshita K, Matsuyama M, Matsuura S. Reproductive cycle and sexual maturity of the anglerfish Lophiomus setigerus in the East China Sea with a note on specialized spermatogenesis. Journal of Fish Biology 1998; 53: 164-178.
60. López-Hernández M, Ramos-Espinosa MG, Carranza-Fraser J. Análisis multimétrico para evaluar contaminación en el Río Lerma y Lago de Chapala, México. Hidrobiológica 2007; 17: 17-30.
61. Jiménez-Cisneros BE. Sustentabilidad, un debate a fondo: Información y calidad del agua en México. Trayectorias 2007; 9: 45-56.
62. Torres de la O, M. Salud de veinte embalses de la cuenca Lerma-Chapala. Instituto Mexicano de Tecnología del Agua. 2007; México.
63. Peña CE, Carter DE, Ayala-Fierro F. Toxicología ambiental: Evaluación de riesgos y restauración ambiental. Southwest hazardous waste program. University of Arizona. USA. 2007.
64. Cheung CCC, Zheng GJ, Li AMY, Richardson BJ, Lam PKS. Relationships between issue concentrations of polycyclic aromatic hydrocarbons and antioxidative responses of marine mussels, Perna viridis. Aquat. Toxicol. 2001; 52: 189-203.
65. Mieriro CM, Ahmad I, Pereira ME, Duarte AC, Pacheco M. Antioxidant system breakdown in brain of feral Golden grey mullet (Liza aurata) as an effect of mercury exposure. Ecotoxicology 2010; 19: 1034-1045.
66. Atili G, Canli M. Response of antioxidant system of freshwater fish Oreochromis niloticus to acute and chronic metal (Cd, Cu, Cr, Zn, Fe) exposures. Ecotoxicology and Environmental Safety 2010; 73: 1884-1889.
67. Houde ED. Recruitment variability, in: Jacobsen, T., Fogarty, M., Megrey, B., Mocksness, E. (Eds), Fish reproductive biology, Implications for assessment and management, Wiley-Blackwell, USA, 2009 pp91-171.
68. Secor DH. Longevity and resilience of Chesapeake Bay striped bass. ICES J. Mar. Sci. 2000; 57: 808-815.
69. Agostinho A, Gomes LC, Verıssimo S, Okada EK. Flood regime, dam regulation and fish in the Upper Paraná River: effects on assemblage attributes, reproduction and recruitment. Rev. Fish. Biol. Fish. 2004; 14: 11-19.
70. González P, Oyarzun C. Variabilidad de índices biológicos en Pinguipes chilensis valenciennes 1833 (Perciformes, Pinguipedidae): ¿están realmente correlacionados?. Gayana (Concep.). 2002; 66, 249-253.
71. Alberto A, Camargo A,Verani JR, Costa OFT, Fernándes MN. Health variables and gill morphology in the tropical fish Astyanax fasciatus from a sewage-contaminated river. Ecotoxicol. Environ. 2005; 61: 247-255.
72. Adedeji OB. Acute Efects of Diazinon on Blood Paramters in the african Catfish (Clarias Gariepinus). Journal of Clinical Medicine Research 2010; 2: 1-6. http://www.academicjournals.org/jcmr/PDF/PDF2010/Jan/Adedeji.pdf. Accessed 7 November 2011
73. López-López E, Paulo-Maya J. Changes in the fish assemblages in the upper Río Ameca, México. Journal of Freshwater Ecology 2001; 16: 179–87.

74. López-López E, Paulo-Maya J, Carvajal AL, Ortiz-Ordóñez E, Uría-Galicia E, Mendoza-Reynosa E. Populations of the butterfly goodeid (Ameca splendens) in the upper Rio Ameca Basin, Mexico. Journal of Freshwater Ecology 2004; 19(4): 575–80.
75. Oruç EÖ, Üner. Combined effects of 2,4-D and azinphosmethyl on antioxidant enzymes and lipid peroxidation in liver of Oreochromis niloticus. Comp Biochem Physiol C 2000; 127: 291–6.
76. Gutteridge JMC. Lipid peroxidation and antioxidants as biomarkers of tissue damage. Clin Chem 1995; 41(12): 1819–28.
77. Kinnberg K, Korsgaard B, Bjerregaard P, Jespersen Å. Effects of nonylphenol and 17β-estradiol on vitellogenin síntesis and testis morphology in male platyfish Xiphophorus maculatus. J Exp Biol 2000; 203: 171–81.
78. Bocquené G, Chantereau S, Clérendeau C, Beausir E, Ménard D, Raffin B, et al. Biological effects of the "Erika" oil spill on the common mussel (Mytilus edulis). Aquat Living Resour 2004; 17: 309–16.
79. Arinç E, Sen A, Bozcaarmutlu A. Cytochrome P4501A and associated mixed function oxidase induction in fish as a biomarker for toxic carcinogenic pollutants in the aquatic environment. Pure Appl Chem 2000; 72(6): 985–94.
80. Cash KJ, Gibbons WN, Munkittrick KR, Brown SB, John Carey J. Fish health in the Peace, Athabasca and Slave river systems. J Aquat Ecosyst Stress Recovery 2000; 8: 77–86.
81. Fent K. Ecotoxicological effects at contaminated sites. Toxicology 2004; 205: 223–40.
82. Sarkar A, Ray D, Shrivastava AN, Sarker S. Molecular biomarkers: their significance and application in marine pollution monitoring. Ecotoxicology 2006; 15: 333–40.
83. Whyte JJ, Jung RE, Schmitt DJ, Tillin DE. Ethoxyresorufin-O-deethylase (EROD) activity in fish as a biomarker of chemical exposure. Crit Rev Toxicol 2000; 30: 348–495.
84. Webb D, Gagnon MM. Biomarkers of exposure in fish inhabiting the Swan-Canning Estuary, Western Australia: a preliminary study. J Aquat Ecosyst Stress Recovery 2002; 9: 259–69.
85. Corsi I, Mariottini M, Sensini C, Lancini L, Focardi S. Fish as bioindicators of brackish ecosystem health: integrating biomarker responses and target pollutant concentrations. Oceanol Acta 2003; 26: 129–38.
86. Schreck CB, Contreras-Sanchez W, Fitzpatrick MS. Effects of stress on fish reproduction, gamete quality, and progeny. Aquaculture 2001; 197 (1–4): 3–24.
87. Varela-Romero A, Ruiz-Campos G, Yépiz-Velázquez LM, Alaníz-García J. Distribution, habitat and conservation status of desert pupfish (Cyprinodon macularius) in the Lower Colorado River Basin, Mexico. Rev Fish Biol Fish 2002; 12: 157–65.
88. Toppari J, Larsen JC, Christiansen P, Giwercman A, Grandjean P, Guillette Jr LJ, et al. Male reproductive health and environmental xenoestrogens. Environmental Health Perspectives Supplement l 1996; 104(S4): 741–803.
89. Heibo E, Vøllestad LA. Life-history variation in perch (Perca fluviatilis L.) in five neighbouring Norwegian lakes. Ecol Freshw Fish 2002; 11: 270–80.
90. Flammarion P, Devaux A, Nehls S, Migeon B, Noury P, Garric J. Multibiomarker responses in fish from the Moselle River (France). Ecotoxicol Environ Saf 2002; 51: 145–53.
91. Chèvre N, Gagné F, Blaise C. Development of a biomarker-based index for assessing the ecotoxic potential of aquatic sites. Biomarkers 2003; 8: 287–98.

# CHAPTER 4

# Intracohort Cannibalism and Methods for Its Mitigation in Cultured Freshwater Fish

**Karolina Naumowicz. Joanna Pajdak.Elżbieta Terech-Majewska.Józef Szarek**

*Department of Pathophysiology, Forensic Veterinary Medicine and Administration, Faculty of Veterinary Medicine, University of Warmia and Mazury in Olsztyn, Olsztyn, Poland*

*Department of Epizootiology, Faculty of Veterinary Medicine, University of Warmia and Mazury in Olsztyn, Olsztyn, Poland*

## ABSTRACT

Cannibalism is a common phenomenon that occurs in over 30 fish families. Most of the species are valuable objects of aquaculture and cannibalism is highly undesirable. Cannibalism is an aggressive behaviour, which can be caused by stress induced by various population and environmental factors and can occur in two main forms: early larval and late juvenile (or adult). Each of these forms can lead to severe losses of stock. Losses are caused not only by fish eating each other, but also by an increase in the number of infections (the result of body surface damage). It is important to understand the phenomenon of cannibalism in order to properly prevent, diagnose and reduce it. The main aim of this study was to present and examine some of the well-examined methods of cannibalism mitigation. Most of the methods consist of creating rearing conditions similar to natural conditions and to reduce stress. Moreover, because of large interspecies differences, similar methods can result in different effects depending on the species. Additional studies are needed to better understand the biology of each cultured species with particular emphasis on the impact of domestication, as well as to improve fish rearing conditions.

**Keywords:** Cannibalism Cultured fish Aquaculture Fish rearing conditions

# INTRODUCTION

Cannibalism is a predatory feeding strategy that involves killing and eating individuals of the same species. In nature, this phenomenon can occur between parents and offspring, in a given age group between siblings or between individuals from separate populations (regardless of the kinship) (Smith and Reay 1991; Baras and Jobling 2002; Qin et al. 2004 and see Table 1). This behaviour represents a specific protective measure against an excessive density of a given population and an alternative source of food when other food resources become scarce (van den Bosch et al. 1988).

**Table 1** Types of cannibalism occurring in the natural environment (from Smith and Reay 1991)

| Criterion | Term | Definition |
|---|---|---|
| Developmental stage of prey | Egg cannibalism | Cannibalism of unhatched fertilized ova |
| | Larval, juvenile and adult cannibalism (post-hatching cannibalism) | Cannibalism of the post-hatching stages |
| Genetic relationship of cannibal to prey | Filial cannibalism | Cannibalism by parents of offspring |
| | Sibling cannibalism | Cannibalism of full siblings |
| | Non-kin cannibalism | Cannibalism of presumably unrelated individuals |
| Age relationship of cannibal and prey | Intracohort cannibalism | Cannibalism of members of the same age cohort |
| | Intercohort cannibalism | Cannibalism of members of a younger age cohort |

Cannibalism is reported in both wild and farmed fish, particularly in aquaculture of predatory fish. This behaviour may reflect disrupted welfare, poorly adapted rearing conditions (especially with regard to feeding schedules or modes of food distribution), or unmet behavioural needs in recently domesticated species (Kaleta 2013). Another possible cause of cannibalism may be the establishment of the group hierarchy among old larvae and subsequent stages of ontogeny (Smith and Reay 1991). This phenomenon is associated not only with a dramatic decline in fish survival, but also in injuries in situations when the prey is too large to be totally consumed by a cannibal. Furthermore, cannibalism generates substantial economic losses (Baras and Jobling 2002; Kestemont et al. 2003; Qin et al. 2004; Król and Zieliński 2015).

The objective of this paper was to review the current knowledge of cannibalism in freshwater fish reared in aquaculture, including clinical aspects for veterinarians involved in the examination of fish. We also review the methods applied to mitigate the prevalence of cannibalism.

## Causes, occurrence and development

Cannibalism in cultured fish occurs in different age groups and its type and development depend mostly on the species and production technology. It represents a major problem in the commercial production of many fish species (especially predatory species) and ranges from 15% to over 90% of individuals (Hecht and Appelbaum 1988). This phenomenon is mainly observed in fish from the following families: airbreathing catfish (*Clariidae*), pikes (*Esocidae*), percids (*Percidae*), characids (*Characidae*), latids (*Latidae*), gadids (*Gadidae*) and in over 30 other families, including cyprinids (*Cyprinidae*) and salmonids (*Salmonidae*), (Smith and Reay 1991; Hecht and Pienaar 1993; Qin et al. 2004).

Under rearing conditions, the intensity of cannibalism is directly influenced by population and abiotic factors, along with genetic predispositions (Fig. 1). However, the extent to which cannibalistic behaviour is genetically controlled in fish remains controversial, and it is still unclear if cannibals in many species are "natural born killers" or just "lottery winners" (Baras and Jobling 2002; Król et al. 2013; Yang et al. 2015). Some studies have demonstrated that this phenomenon is to some extent genetically determined (Thibault 1974; Ferguson and Noakes 1983; van Damme et al. 1989; Amundsen et al. 1999; Yang et al. 2015). However, the interactions between genotype and environment play a major role in the expression of cannibalism (Baras and Jobling 2002; Yang et al. 2015). Intracohort cannibalism may also be sex-based, especially in species that show sexual growth dimorphism during early ontogeny, e.g. Vundu catfish (*Heterobranchus longifilis*) (Baras 1999a), or in species with diet dimorphism, e.g. old hake females (family *Gadidae*) (Garrison and Link 2000). Moreover, the influence of the genotype of a maternal parent on cannibalism in progeny is higher than the paternal one (Mousseau 1998; Yang et al. 2015).

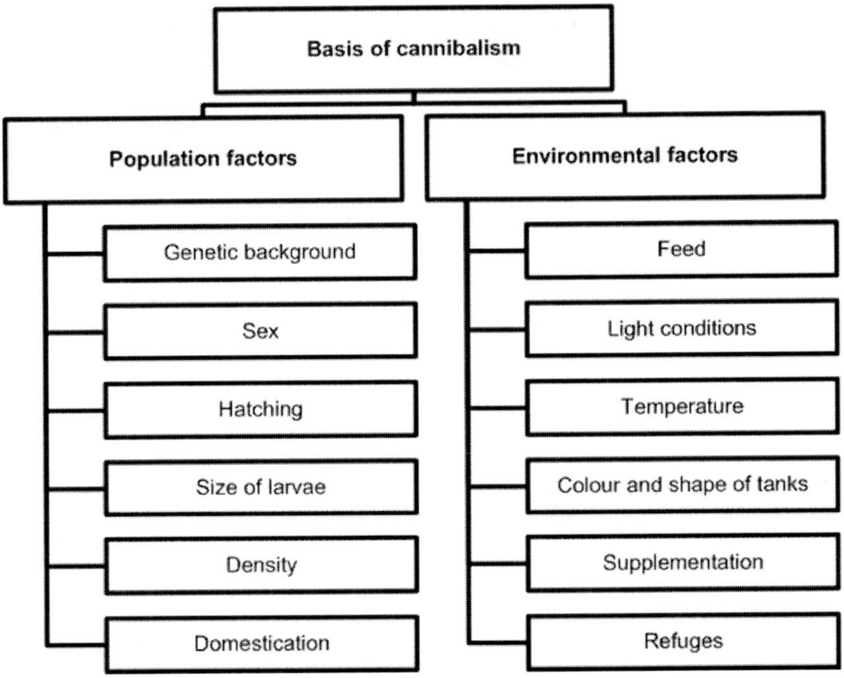


**Figure. 1**Population and environmental factors influencing cannibalism

Population factors that influence cannibalism include the size and density of larvae, with both being positively correlated with the incidence of cannibalistic behaviour. Abiotic factors mainly encompass the availability and energy value of feed, diet supplementation, feeding schedule, temperature and clarity of water, light intensity, photoperiod, shape and colour of internal walls of rearing tanks or the lack of refuges for small-sized individuals (Smith and Reay 1991; Britz and Pienaar 1992; Hecht and Pienaar 1993; Folkvord and Otterå 1993; Baras et al. 2000a; Baras and Jobling 2002; Kestemont et al. 2003; Szczepkowski 2009; Król and Zieliński 2015). As all factors interact with each other, they should be approached holistically (Hecht and Pienaar 1993; Kestemont et al. 2003) in order to mitigate cannibalism. These factors are discussed further in "Methods used to mitigate cannibalistic behaviour" section.

Intracohort cannibalism is typical of farming conditions and is divided into two types: type I, the so-called "early" type, which occurs in the larval phase and is independent of a diversity in fish sizes, where the victim is not completely ingested or consumed; and type II, a later form associated with heterogeneous growth, when the victim is consumed whole (Hecht and Appelbaum 1988; Smith and Reay 1991; Baras and Jobling 2002). A potential type I cannibal mostly attacks its prey from the tail end, eats only a damaged part of the tail and disgorges the rest of the larva as it is unable to swallow it due to the lack of difference in size. This does not affect some species, i.e. those with very large mouth dimensions or sharp oral teeth in the early ontogenetic stage—in such situations, cannibals can attack siblings' heads or trunks (Hecht and Appelbaum 1988; Baras 1999b; Baras and Jobling 2002). Type II cannibalism is determined by the mouth dimensions of a potential cannibal in relation to a potential prey, particularly to its body depth (Hecht and Appelbaum 1988; Smith and Reay 1991; Sogard and Olla 1994). The gape size limitations differ between species and ontogenetic stages, owing to the allometric growth of mouthparts (Baras and Jobling 2002). For type II cannibalism, prey is attacked from the head end or from the side and usually swallowed whole (Baras and Jobling 2002; Kestemont et al. 2003). However, because of a large predator-to-prey ratio, in type II cannibalism the mode of ingestion may be of little importance (Baras and Jobling 2002). An exception for this is Acanthopterygians, where prey is eaten head first as a cannibal then avoids the spiny fin rays of the dorsal or pectoral fins which threaten a predator. Otherwise, the erected rays may cause the death of the cannibal and its prey (Brabrand 1995; Qin et al. 2004). In most fish species, type I cannibalism is replaced by type II together with increasing gape size limitations, heterogeneity of a stock and growing chances of swallowing a potential prey. Type I cannibalism may trigger growth heterogeneity because digestible nutrients are often higher in fish than in plankton prey, which leads to the growth advantage of cannibals and, in effect, to the occurrence of type II cannibalism (Baras and Jobling 2002; Król and Zieliński 2015).

The outcomes of these two types of cannibalism differ in the level of mortality. In the case of type I, it ranges between 1.5 and 12.0% of the initial stock of European perch (*Perca fluviatilis*) larvae (Baras et al. 2003; Babiak et al. 2004; Król and Zieliński 2015) and from 10 to 17% in the stock of pikeperch (*Sander lucioperca*) larvae (Zakęś and Demska-Zakęś 1996; Zakęś 2012; Król and Zakęś 2016) or even 40% in dorada (*Brycon moorei*) (Baras et al. 2000b). Type II usually generates significantly higher losses and is mainly reported in a stock characterised by a large heterogeneity in size (Król and Zieliński 2015).

## Hormonal background

Intracohort aggression has often been linked to the mortality of aquaculture organisms (Manley et al. 2015). It is hormonally regulated via different modulators (Table 2), and is the basis of cannibalistic behaviour

(Hecht and Appelbaum 1988; Folkvord 1997; Yang et al. 2015; Vallon et al. 2016).

**Table 2** The impact of neuromodulators on aggressive behaviour in fish (own compilation based on Kania et al. 2012)

| Modulator | Physiological role | Impact on aggressive behaviour |
|---|---|---|
| Vasotocin | Ionoregulation<br>Osmoregulation | Gregarious fish: enhancement of aggressive behaviour; territorial fish: attenuation of aggressive behaviour |
| Isotocin<br>IT peptide | A role in reproductive physiology (the exact impact still unknown) | Gregarious and monogamous fish: enhanced aggressive behaviour (directed at individuals that disrupt social relations) |
| Serotonin | Immunomodulation; regulation of glucose metabolism | Mitigation of aggressive behaviour |
| Dopamine | A role in reproductive physiology | Enhancement of aggressive behaviour |
| Testosterone | Regulation of spermatogenesis; enhancement of the secondary sexual traits | Enhancement of aggressive behaviour (aiming at protecting the territory) |
| Oestrogens | Regulation of oogenesis; enhancement of the secondary sexual traits; stimulation of vitellogenin synthesis in the liver | Enhancement of aggressive behaviour |
| Glucocorticoids | Regulation of metabolism; contribution in adaptive responses to stressors | A high, persistent level in subordinates—an increase in serotonergic transmission in the brain resulting in a reduction of aggressive behaviour; reduced motor activity |
| CRH (corticotrophin-releasing hormone) | Regulation of the anterior pituitary lobe; regulation of TSH secretion | Mitigation of aggressive behaviour; increased motor activity |
| Neuropeptide Y | Regulation of blood pressure; circadian rhythm; food ingestion; pituitary hormone secretion | Mitigation of aggressive behaviour |

For more information on the direct influence of serotonin on cannibalism, please also see "Balanced nutrition and supplementation" section, where the role of serotonin precursor, i.e. tryptophan, is described (Hseu et al. 2003a; Hoglund et al. 2005)

In some species or life stages, an organism also may be aggressive or cannibalistic toward a conspecific when defending its own space (Hecht and Pienaar 1993; Baras and Jobling 2002), fighting for food during low feeding frequency and periods of low food availability, when conspecifics offer the only type of alternative prey (Hecht and Pienaar 1993; Baras and Jobling 2002; Kestemont et al. 2003). However, there are also exceptions, for example, in Vundu catfish, type II cannibalism requires no preliminary aggression or chase, or even contact with the prey, suggesting that cannibalism can be independent of aggressive behaviour (Baras 1999b).

## DIAGNOSTICS

Due to the complex causes of cannibalism, diagnostics should always start with case history—gathering information from the owner of the fish farm, which should be as extensive as possible. The questions should include the exact percentage of losses with a particular interest in the age of fish and the diversity in growth. It is important to consider how the losses were spread among a given age group and individual tanks. In order to rule out welfare issues, water parameters should be analysed in the preceding period (at least 2 weeks). Such a procedure enables detection of possible situations, which may trigger stress and behavioural changes, for example, alterations in temperature, oxygen or pH level, power outages, bio-filter faults or excessive waste. The owner should also be asked about situations that may have disrupted the established hierarchy, namely, transfer or sorting fish and tank cleaning. Data on nutrition is equally important; they refer to both feed types and feeding strategies.

The next step is a physical examination which includes a thorough inspection of fish with special attention paid to any injuries (inspection with a magnifying glass, a microscopic examination under a small magnification) and their localization. This should help to exclude the injuries caused by handling, tank design or coexisting diseases. In the case of tail injuries, type I cannibalism is suspected while the wounds to the head and body sides may indicate type II cannibalism. In addition, the gastrointestinal content of a potential cannibal may be examined. In most species, fish of different sizes are differentially sensitive to handling, but handling generally has no significant effect on fish survival (Baras et al. 2011a). However, special attention is needed when very small and fragile Percids larvae are examined, as they are by nature very sensitive to any handling and parameter alterations, which results in deaths (Król and Zieliński 2015) and may be mistaken for an outcome of cannibalism.

## Methods used to mitigate cannibalistic behaviour

Different approaches may be implemented to reduce the incidence of cannibalistic behaviour and resulting mortality. Previous research suggests that applying different methods simultaneously will produce a synergistic effect (Qin et al. 2004) and that methods should be adjusted according to a given species and its biology.

## Abiding by density standards

In nature, cannibalism is determined by the population density and usually occurs in years with high numbers of larvae (Baras and Jobling 2002). In aquaculture, excessive density may reduce or increase the incidence of cannibalistic behaviour, depending on the biology of a given species. With limited availability of food or improper feeding techniques (limited access, bullying of weaker individuals), an increased number of larvae is associated with lower resources per larva, which results in using a rescue source of food, i.e. siblings. Under rearing conditions, intensified cannibalistic events due to excessive stock density (despite the correct amount of food) may result from a higher chance of a cannibal-prey interaction (especially in species demonstrating "meet-and-eat" behaviour) (Sogard and Olla 1994; Baras and Jobling 2002).

On the other hand, increasing stock density may reduce aggression, especially that associated with territorial behaviour and setting the dominance hierarchy (Smith and Reay 1991; Sloman and Armstrong 2002). It is suggested that with high densities, a limited surface results in a lower number of aggressive individuals (in reference to the whole population) protecting the territory and demonstrating cannibalistic attitudes (Kucharczyk et al. 1998; Baras and Jobling 2002). In addition, at a certain stock density, fish start forming shoals and cannibals become perceptively confused and cannot select a prey (as the number of potential prey is too high) (Baras and Jobling 2002).

The impact of different stock density levels has been studied in many cultured fish species. A proportional increase in cannibalistic behaviour resulting from rising stock density and an increase in mortality, as a result, has been observed, for instance, in pike (*Esox lucius*) (Kucharczyk et al. 1998), pikeperch (Szkudlarek and Zakęś 2007), European perch (Baras et al. 2003) and African sharptooth catfish (*Clarias gariepinus*) larvae (Solomon and Udoji 2011). For these species, it is suggested to rear larvae initially at higher stock densities and then to reduce the stock proportionally (Kestemont et al. 2003). An opposite relation is seen in such species as Atlantic cod (*Gadus morhua*) (Baskerville-Bridges and Kling 2000), Vundu catfish (Imorou Toko et al. 2008) or European chub (*Leuciscus cephalus*) (Żarski et al. 2008), in which there is a reduction in cannibalistic behaviour at high stock densities.

## Size-sorting

In commercial farming of many fish species, size-sorting is carried out in order to ensure good husbandry practice, which is economically justified as it allows a more accurate estimation of feed consumption and weight gains in fish (Jobling et al. 1993). On the other hand, size-sorting would also have the effect of disrupting established dominance hierarchies, and it is widely believed that size-sorting results in greater gain in biomass because the suppressive effect of larger fish on the growth of smaller individuals is removed at regular intervals (Jobling et al. 1993). Dominance hierarchies often re-establish soon after fish have been rearranged into new groups, especially when group size is low and the fish are homogenous in size. The establishment of a dominance hierarchy is generally accompanied by heterogeneous food intake and growth, generally to the detriment of the smallest fish, thereby resulting in an overall slower growth rate, greater size dispersal and increased skewness (Azaza et al. 2010).

Size-sorting regarding type I cannibalism is connected mainly with maintaining the ontogenetically homogenous stock, because sorting does not reduce type I cannibalism, as this form is independent of larval sizes. At early stages, intraspecific predation is not greatly limited by size heterogeneity, as cannibals ingest conspecific prey (Ribeiro and Qin 2016). Synchronization of hatching and initial diversity of larval sizes play an important role among many fish species in maintaining a homogenously-sized stock. Both of them are associated with the size of a female and the number of her reproductive seasons. Very small females and those used reproductively for the first time demonstrate a tendency for producing a significantly smaller brood (Król and Zieliński 2015). That is why mixing of progenies with contrasted hatching times (with exception of fish at the same ontogenetic stage and homogenous in size) or progenies from females of different sizes, should be avoided -female size, egg size and size at hatching may be correlated (Wootton 1989; Baras and Jobling 2002; Steenfeldt et al. 2011; Trabelsi et al. 2013; Król and Zieliński 2015; Trabelsi et al. 2016). Therefore, it is recommended to select proper sized females at the stage of spawn collection. This is of particular importance for such species as Atlantic cod (Folkvord and Otterå 1993), African catfish

(Hecht and Appelbaum 1988), rainbow trout (*Oncorhynchus mykiss*) (Baras and Jobling 2002), largemouth bass (*Micropterus salmoides*) (DeAnglis et al. 1980), walleye (*Sander vitreus*) (Loadman and Moodie 1986), European seabass (*Dicentrarchus labrax*) (Katavić et al. 1989) and dorada (Baras et al. 2000a).

In walleye (Cuff 1980) and dorada (Baras et al. 2000b), initial size heterogeneity among embryos presumably has a limited effect on the impact of type I cannibalism. In contrast, the emergence of cannibalism was dependent strictly on the development of oral teeth at the embryonic stage (Baras et al. 2000a, b), suggesting that fish hatching slightly later, or developing at a slightly slower rate than others, incur a major penalty for this very short delay. In particular, this is likely to apply to fish species with oral teeth that facilitate the exercise of incomplete cannibalism (which does not require that the cannibal be larger than its victim), as individuals with a slightly faster growth than others during the first days of exogenous feeding would not be safe from cannibalism (Baras et al. 2014). The size advantage gained through the exercise of type I cannibalism at the embryonic and larval stages, later enables cannibals to exert type II cannibalism over smaller siblings, resulting in high mortality throughout the larval and early juvenile stages (Hecht and Appelbaum 1988; Baras 1999b). The same principle applies to deformed fish, which may constitute easily accessible prey to cannibals, contribute to their faster growth and further predation on normal siblings.

Because of size-dependence of type II cannibalism, substantial research has been done on morphometric assessments of the prevalence of cannibalism depending on the ratio of the mouth-dimensions of cannibals and the body dimensions of victims (Folkvord and Otterå 1993; Mélard et al. 1996; Hseu et al. 2003b; Fessehaye et al. 2005; Mandiki et al. 2007; Ribeiro and Qin 2013). Due to species–dependent morphological differences, the variability of dimensions during ontogeny and behaviour typical of a given species, diverse mathematical models are applied. The most commonly used models include those based on direct measurements of total length (TL) of cannibals and their prey or models based on such parameters as mouth width (MW), mouth gape (MG), body depth (BD) of a prey, head depth (HD) of a prey and head width (HW) of a prey (Folkvord and Otterå 1993; Baras and Dalmeida 2001; Hseu 2002; Kailasam et al. 2002; Kestemont et al. 2003; Hseu et al. 2003b; Fessehaye et al. 2005; Wallat et al. 2005; Hseu et al. 2007; Policar et al. 2013; Baras et al. 2014; Hseu and Huang 2014; Ribeiro and Qin 2015).

Selecting size of prey has been well-described for orange-spotted grouper (*Ephinephelus coioides*) which can consume prey almost as large as the cannibal. However, they rarely do so and seemingly prefer much smaller prey. Studies have shown that when a cannibal is offered prey of three sizes, orange-spotted grouper cannibals do prefer the smaller prey to the larger ones (Hseu and Huang 2014). Similar situations have been found in other cannibalistic fish such as yellow perch (*Perca flavescens*) (Post and Evans 1989), Arctic charr (*Salvelinus alpinus*) (Amundsen et al. 1995), Atlantic cod (Blom and Folkvord 1997), pike (Nilsson 1999), dorada (Baras et al. 2000b), *Pangasius djambal* (Baras et al. 2010) and Amazonian catfish *Pseudoplatystoma punctifer* (Baras et al. 2011b).

The size-sorting method based on comparing the length is the simplest and most practical in use, but is applicable only for a few species with direct ontogeny. However, there is some information provided, for example, in barramundi (*Lates calcarifer*), it is believed that a difference in length of 50% should be regarded as the absolute threshold at which cannibals should be removed (Ribeiro and Qin 2013). In other species, the threshold is 88% for pike (Szczepkowski 2009), 75% for Atlantic cod (Folkvord and Otterå 1993) and 55% for Arctic char (*S. alpinus*) (Jobling et al. 1993). In the case of Nile tilapia (*Oreochromis niloticus*), it is recommended to keep the largest individual bigger than the smallest one by a maximum of 13 times (Fessehaye et al. 2005).

In pikeperch, it has been demonstrated that larvae sorting has a beneficial impact on an increase in their survival through reduced cannibalistic behaviour (Szczepkowski et al. 2011). A similar effect of sorting has been reported for turbot (*Scophthalmus maximus*) (Sunde et al. 1998), African catfish (Biu et al. 2015), and yellow perch (Wallat et al. 2005). However, for the European perch, sorting has a negative impact on cannibalism prevalence and weight gains (Mélard et al. 1996). Arctic charr reared in size-sorted groups also suffer from a growth disadvantage when compared with groups containing fish of a wider range of sizes (Baardvik and Jobling 1990). In these species, removal of the largest individuals is most likely associated with establishing a new hierarchy in a stock, which triggers a reoccurrence of type I cannibalism (Mandiki et al. 2007). In this species, it is worth considering removing the smallest larvae that are the target of cannibals, as such an approach allows levelling the sizes and maintaining the current hierarchy in a stock at the same time (Baras and Jobling 2002).

## Balanced nutrition and supplementation

Correct selection of feed, in terms of its physical and chemical characteristics (e.g. size and texture of pellets, drowning speed, shape, colour, hardness, digestibility and beneficial value of FCR), is an important element of the preventive strategy against cannibalistic behaviour (Zakęś and Demska-Zakęś 1996; Altaff and Janakiraman 2013; Król and Zieliński 2015; Ljubobratović et al. 2015). However, rearing period and feeding schedule, specific for each species, are equally important (such as daily feed ration, feeding frequency, feed access surface and transition from live feed to commercial feed) (Jobling et al. 1993; Folkvord and Otterå 1993; Francis and Bengtson 1999; Baras et al. 2000a; Kestemont et al. 2003; Szczepkowski 2009; Conceição et al. 2010; Altaff and Janakiraman 2013; Piotrowska et al. 2013; Król and Zieliński 2015). This is important, because any type of feeding restriction not only promotes forage activities by hunger motivation, encounter rates of conspecifics and prey vulnerability (Hecht and Pienaar 1993; Baras et al. 2013; Ribeiro and Qin 2016), but can also lead to food competition, variability in food intake and establishment of a dominant hierarchy. Every situation which leads to inequities in food intake, results in growth depensation and size heterogeneity (Hecht and Pienaar 1993; Baras and Jobling 2002; Kestemont et al. 2003; Baras et al. 2013; Ribeiro 2015). The combination of all of the above mentioned factors, including the biology of a given species, should satisfy nutritional requirements, support homogenous growth of larvae and eliminate competition for food.

Live feed is usually preferred by fish larvae (Conceição et al. 2010; Ljubobratović et al. 2015). Importantly, it should be adjusted to the requirements and capacities of a given species. Some authors recommend mixed feeding to mitigate cannibalism, for example, in the case of Koi carp (*Cyprinus carpio koi*), the best performance results (growth and survival ratio) and the least expressed cannibalistic behaviour are reported with feeding rotifer larvae (*Brachionus plicatilis*) and copepods (*Apocyclops dengizicus*) (Altaff and Janakiraman 2013). For Atlantic sturgeon larvae (*Acipenser oxyrhynchus oxyrhynchus*), most often fed with *Artemia* sp. nauplii, it has been shown that mixing the nauplii with micronematodes of the *Panagrellus* sp. family is beneficial, as this reduces mortality and incidence of cannibalistic behaviour and is favourable from an economic standpoint (Piotrowska et al. 2013). However, in some cases, mixed feeding may also cause different individuals to feed on different types of food with different nutritional values, which may result in growing size heterogeneity (Baras et al. 2014).

In fish fed with commercial feed from the beginning of rearing period, the addition of live feed may generate rational effects which reduce the prevalence of cannibalism, e.g. *Daphnia* in the feeding of African catfish and rainbow trout (Hecht and Pienaar 1993). Decreased territorial aggression and cannibalistic behaviour when live feed is supplemented for African catfish and Vundu catfish is most likely caused by redirecting food aggression onto another, actively swimming source of food (Solomon and Udoji 2011). In addition, growth and survival parameters are better in fish fed with mixed feed in comparison to fish fed only with commercial feed, which is related to the differences in the protein-fat ratio (Akbary et al. 2010) and the secretion of additional proteolytic enzymes by the nauplii, which enhances food absorption in larvae (Bengtson et al. 1991). As a result, larvae fed with mixed feed are more efficient in feed conversion and better tolerate the transition to commercial feed.

The weaning from live food to commercial feed is a time point of particular importance for many fish species (Strand et al. 2007). This also applies to feeding perch larvae. When feed is changed, fish should be transitioned throughout a certain period (when both live food and commercial feed are used) long enough to allow larvae to become familiar with a new type of food. Individuals that fail are weaker and smaller and thus become easy prey for potential cannibals (Hecht and Pienaar 1993; Strand et al. 2007). The duration of the individual feeding stages also plays a role. For example, different experiments suggest that the initial phase during which only live food is used, actually slows the growth rate and increases the incidence of deformities in pikeperch (Zakęś and Demska-Zakęś 1996). In addition, excessively long feeding with live food makes fish become used to this type of feed and makes them unwilling to accept a new form of feed, thus triggering stress and a resultant increase in cannibalistic behaviour (Conceição et al. 2010).

Experiments are currently underway that are attempting to determine which feed additives may inhibit the intensity of cannibalism. L-tryptophan is one of them; this amino acid is a serotonin precursor and serotonin exhibits an inhibitory effect on aggressive behaviour via the increased serotoninergic activity of the brain (Winberg et al. 2001). Stressors such as social subordination (a low position in the dominance hierarchy), handling and exposure to predators trigger a rapid activation of the serotoninergic system in the brain (Lepage et al. 2003). Diet supplementation with L-tryptophan impacts the course of stress response in rainbow trout by lowering the level of cortisol in serum. However, with long-term administration, the regulatory mechanisms which limit this effect are switched on (Lepage et al. 2003; Mandiki et al. 2016). The level of absorbed L-tryptophan is positively correlated with the amount of ingested food: the highest doses are taken by the strongest individuals with the ultimate positions in the social hierarchy. Hence, aggressive and cannibalistic behaviours are most effectively suppressed in such animals. Short-term supplementation with L-tryptophan in the diet leads to weakening of the hierarchic structure, lowers the level of social stress, reduces the negative impact of stress on the immunity, levels the sizes of animals and, as a result, reduces the severity of cannibalism (Winberg et al. 2001).

Enrichment of live feed with probiotic preparations is another measure resulting in reduction of cannibalism. The use of probiotics in aquaculture undoubtedly brings about benefits such as growth promotion, increased survival and immunomodulation (Gatesoupe 1999; de Azevedo and Braga 2012). It has been proven that they contribute to a reduction in type II cannibalistic behaviour in the rearing of European perch. This effect was generated by supplementing live feed (*Artemia* nauplii) with probiotic *Bacillus licheniformis*, *Bacillus subtilis* and *Bacillus pumilus* strains (Mandiki et al. 2011).

## Temperature regulation

Environmental temperature has an enormous impact on the metabolic rate in fish, which rises with increasing water temperature (Baras and Jobling 2002; Barron et al. 2012; Frisk et al. 2012). For most fish species, this results in an increased demand for food, faster depletion of food resources, marked heterogeneity of larvae and, as a result, in potential cannibalistic behaviour (Sunde et al. 1998; Baras and Jobling 2002; Kestemont et al. 2003; Zakęś 2012). Rearing temperature should fall within the thermal comfort zone, which is specific for a given species and population. Fish raised at different temperatures generally have different growth-feed ratios, thereby the comparisons of the survival rates between fish of identical ages is not informative, because they will be at a different level of ontogenetic development (Baras et al. 2011a). However, due to the interval nature of this parameter, it is possible to rear fish closer to a lower or higher end of the comfort zone and, in this way, manipulate the metabolism rate.

Rearing at lower temperatures can exert a beneficial impact on retarding cannibalism and is connected with a stunted larvae growth rate (Kestemont et al. 2003). Pikeperch larvae behave similarly (Zakęś 2012), although in burbot (*Lota lota*) the impact of higher temperature on an increase in cannibalistic incidents has only been reported for larval and juvenile developmental stages and the relationship with temperature has not been examined (Barron et al. 2012). Pike larvae behave the opposite; the lowest losses due to cannibalistic behaviour have been observed at the highest temperatures investigated (28 °C) (Szczepkowski 2009). However, the influence of ontogenetic differences cannot be excluded in the cases mentioned here.

In facilities with the possibility of water temperature regulation, i.e. recirculating aquaculture systems, in order to optimize the outcome of rearing under periodic feeding conditions, the following scheme may be applied: after the last daily feeding, the water temperature may be slowly decreased (safe gradient up to 5 °C) (Baras and Jobling 2002). Such an approach should ensure optimal growth and reduce the severity of cannibalistic behaviour. Unfortunately, this method cannot be applied in species which demand constant feeding 24 h a day or in many outdoor rearing systems. However, such temperature fluctuations exist naturally in these type of rearing systems (lower temperatures at night) but without the possibility of regulation (Azaza et al. 2010).

## Proper light conditions and water clarity

Light intensity, length of lighting period and water clarity impact the regulation of circadian rhythms and, therefore, influence the prevalence of cannibalism. The spectral bandwidth of light available for vision is narrower under water than on land due to the spectrally selective absorption of natural water. The colour of natural water is highly variable (Kusmic and Gualtieri 2000). In general, freshwater bodies are subject to greater variations in light than marine waters owing to the higher concentrations of suspended organic and mineral matter and organic compounds in a solution, causing a shift in the peak transmission towards the red wavelength in varying amounts, within the range of 480– 590 nm (Knowles and Dartnall 1977).

Larvae of many fish species, which rely mainly on sight in food seeking (Baras and Jobling 2002), demonstrate positive phototaxy and are short-sighted (they see at a distance approximately 1/3 of their body length), which influences feed seeking, since to ingest food, they have to see it. This is favoured by clear lighting, the colour of tank walls contrasting with feed colour and a longer lighting period, which prolongs the feed-seeking period and shortens nocturnal fasting period (which prompts cannibalism) (Baras et al. 2000a; Tamazouzt et al. 2000; Kestemont et al. 2003; Qin et al. 2004). In these species, a prolonged period of darkness is associated with reduced activity and staying near the bottom of a tank, which leads to locally increased stock density and more severe aggressive behaviour (Baras et al. 2000a). Therefore, for such species as pike (Szczepkowski 2009), pikeperch (Król and Zakęś 2016) or European perch (Tamazouzt et al. 2000; Król and Zieliński 2015), the best rearing outcomes are obtained when fish are kept under a prolonged lighting period and at a proper light intensity for a given species.

Nevertheless, long periods of good visibility translate into long periods for contact between a potential cannibal and its prey, which may contribute to an increased frequency of cannibalistic behaviour and higher heterogeneity in a stock, representing a positive feedback mechanism (Baras and Jobling 2002). Moreover, excessive light intensity may be a source of stress and intensify aggressive behaviour, as reported in Nile tilapia and Barramundi perch (Baras et al. 2000a; Jesu Arockiaraj and Appelbaum 2011). In these species, reduced light intensity and water clarity have a beneficial effect on survival and performance.

Species that prefer darkness (such as African catfish—which finds food mainly with chemoreceptors and sensor organs) react to excessively intense light with intensified behaviour associated with chasing resting individuals, searching for refuges, gathering at the bottom of a tank and increased territorial aggression associated with locally increased stock density (Britz and Pienaar 1992; Baras and Jobling 2002). In addition, chemical substances are released from wounds (injuries incurred when attacked) into the environment which are a strong stimulus triggering cannibalistic behaviour (Mukai et al. 2013). Hence, turbid water, limiting the inflow of light and prolonging the darkness period for a darkness-preferring species, prompts a reduction in aggression and cannibalism as well as prolonged feeding and more homogenous weight gains (Solomon and Udoji 2011).

In farming facilities with the possibility of light wavelength regulation, a spectrum with added light intensity could be used to gain better rearing results. Sallehudin and Mukai (2014) demonstrated that the juveniles of African catfish had low biting activity under light intensities of 0.0017 and 0.021 $\mu mol/m^2/s$ for white, blue, and yellow light wavelengths, and 0.19 and 1.40 $\mu mol/m^2/s$ for blue and red light wavelengths. Only the blue light wavelength showed low biting activity for all levels of light intensity. Similar results were obtained in beluga (Banan et al. 2011), where blue light (by establishing comfort) resulted in increased growth performance and had a positive effect on physiological conditions. Contrary to this, red light had a negative impact on beluga growth performance.

## Providing refuges for weaker and poorly developing individuals

It is important to design refuges for smaller and weaker individuals that may become potential prey for cannibals. The efficacy of refuges in preventing cannibalism depends to a certain extent on lighting conditions. This aspect has been studied in barramundi perch (Qin et al. 2004) and African catfish (Hecht and Appelbaum 1988; Pienaar 1990). In these species, a significant reduction in cannibalistic behaviour and territorial aggression has been demonstrated to occur after introduction of refuges (Solomon and Udoji 2011). The amount of aggressive behaviour associated with chasing and biting was also reduced.

The size, design and number of refuges should be adapted to the biology of a given species. With too large refuges (letting in both small and large individuals), the opposite effect, i.e. an increased level of cannibalism, may occur (Qin et al. 2004). Incorrect design will not bring the expected effects if hiding requires swimming at depths unnatural for a given species (Francis and Bengtson 1999). In addition, the number of refuges should be sufficiently high to satisfy the needs of all individuals; if the number of refuges is too low, territorial aggression and fighting for refuges may rise (Britz and Pienaar 1992). Moreover, special care should be taken for darkness-preferring species such as European chub. If an opaque, dark hiding place is designed and too intense light is applied, fish could gather in such an opaque hiding place due to their marked preference for darkened locations. With incorrect lighting, fish would be less eager to leave refuges for feeding and ultimately they could die of starvation or oxygen deficiency.

Major limitations in the implementation of this method are the constant need for modification of refuge dimensions according to the growth ratio, especially during periods of quick changes in development, as well as the fact that refuges hinder many routine operations and can cause sanitary problems. For these reasons many consider using refuges laborious and inconvenient for culture.

## Colour and shape of tanks

Light perception ability varies in fish due to morphological differences in eye anatomy (Kusmic and Gualtieri 2000). Colour of inside walls and shape of tanks should be adjusted for each fish species, in regards to the spectrum of used light and water turbidity as these factors influence visibility and, thus, feed ingestion. This is of particular importance for species that rely on sight in feed seeking (Sogard and Olla 1994; Baras and Jobling 2002; Üstündağ and Rad 2015). Depending on the fish species, light intensity or developmental stage, the same tank colour may exert a different impact (Tamazouzt et al. 2000). For example, for rainbow trout especially under low light intensities, beige coloured tanks are more suitable for rearing (Üstündağ and Rad 2015). Imanpoor and Abdollahi (2011) demonstrated that colour can have large impact on feeding rate, hematocrit, final body lipid content and all body colour parameters in juvenile Caspian kutum (*Rutilus kutum*). Survival was also influenced by the tank wall colour with light grey showing the best result (17%). In beluga (*Huso huso*), most of growth parameters suggested a positive impact of rearing in black tanks (Banan et al. 2011). In Eurasian perch, the greatest growth in weight and length was observed in tanks with light grey and white walls, which were strongly illuminated, while the lowest growth was recorded in a tank with black walls and 250 lx illumination (Tamazouzt et al. 2000). Moreover, in Eurasian perch, light walls seems to induce a strong contrast between brown-yellow food particles and the tank walls and, consequently, improve food perception as observed in some other fish species, such as seabream *Sparus auratus* (Ounais-Guschemann 1989) and walleye (Bristow and Summerfelt 1994).

The shape and size of a tank influences the movement of water, as it circulates differently in tanks with a round, square or rectangular bottom (Rożyński and Zakęś 2015). Depending on the shape and size of a tank, fish will have different access to its edges that may serve as refuges; smaller tanks have more of them (Baras and Jobling 2002). On the other hand, cannibals may more easily spot prey in a small tank (Sogard and Olla 1994). Large tanks allow establishment of more complex relations between individuals. In addition, tanks may differ in depth, and depending on the design the smaller the depth, the faster fish may reach the water surface to ingest feed or breathe in air. This is important in fish species that rely on sight, as the time needed to select a potential prey shortens as the prey swims closer to the water surface, which reduces the chances for cannibalistic behaviour (Baras and Jobling 2002). These factors influence the movement of fish in tanks and indirectly the weight gains and occurrence of cannibalism. In pike rearing, the lowest incidence of cannibalistic behaviour was found in tanks with a square bottom (Szczepkowski 2009) whereas for pikeperch, tanks with square bottoms were more efficient for older stages and larvae had better survival parameters in tanks with round bottoms, although in this experiment the impact of different lagoon depths cannot be excluded (Kozłowski et al. 2009).

## Degree of domestication

The impact of domestication on the physiology and behaviour of fish is a new but widely discussed topic in the scientific world. However, further research is warranted. Only two fish species have been completely domesticated: carp (*Cyprinus carpio carpio*) and goldfish (*Carassius auratus auratus*) (Balon 2004). Other species still have reproductive cells acquired from wild populations, which may influence both their health and behaviour. With progressive domestication, susceptibility to stress decreases (Douxfils et al. 2011) and wild or partially domesticated fish demonstrate a higher tendency for aggressive behaviour than fish with a greater degree of domestication (Hedenskog et al. 2002). On the other hand, cannibalistic strains may arise during domestication, because the characteristics deemed by aquaculturists such as fast growth, may exacerbate cannibalism through increased size heterogeneity (Hecht and Pienaar 1993; Baras and Jobling 2002).

## Health of stock and use of pharmaceuticals

The health condition of fish stocks is important regarding the possibility of the occurrence of cannibalism. Health problems and diseases can impair the growth of ill individuals, and increase the potential for cannibalism. In such cases, it is recommended to use therapeutics. One can choose between antibiotics for direct treatment or disinfectants as preventives. For food safety, the use of antibiotics should be limited as much as possible and they should not be overused. Subagja et al. (1999) demonstrated that use of disinfectants such as chloramine T and formalin in Asian catfish can significantly increase the survival rate and health status of fish stock.

Another approach to pharmaceutical treatment is the use of sedatives to reduce the level of stress and, thereby, cannibalism. However, this would only be treating the symptoms and not the root cause of the stress. Moreover, such an approach may be economically unjustified and may raise doubts as to the quality and safety of animal origin food products, as it would require determining the residue level standards (Gomułka and Antychowicz 1999; Neiffer and Stamper 2009).

## SUMMARY

Cannibalism represents a problem under intensive commercial production of different fish species and is impossible to manage without understanding the factors impacting its occurrence and degree of severity. It is thus necessary to constantly expand the knowledge of biology and natural habitats of reared species and species which have been newly introduced to aquaculture. This is becoming even more important as aquaculture is continuously developing and an increasing number of species are being adapted to intensive rearing and breeding. The key is to include consideration of the phenomenon of cannibalism in selection and breeding programmes.

Each case of cannibalism should be considered in the context of biology and specificity of a given species. There is not a single, universal preventive measure that would be equally effective in all species and for all types of cannibalism. In addition, a reduction in cannibalistic behaviour is not necessarily correlated with the best outcomes in production and, hence, the parameters should be carefully selected. Furthermore, there may need to be decisions made between fish survival and the final fish biomass.

# REFERENCES


1. Akbary P, Hosseini SA, Imanpoor M et al (2010) Comparison between live food and artificial diet on survival rate, growth and body chemical composition of *Oncorhynchus mykiss* larvae. Iran J Fish Sci 9:19–32
2. Altaff K, Janakiraman A (2013) Impact of different feeds on cannibalism in early larval stages of Koi carp (*Cyprinus carpio*). J Appl Biol Biotechnol 1:35–39
3. Amundsen P-A, Damsgård B, Arnesen AM et al (1995) Experimental evidence of cannibalism and prey specialization in Arctic charr, *Salvelinus alpinus*. Environ Biol Fishes 43:285–293. doi:10.1007/BF00005860
4. Amundsen PA, Svenning MA, Slikavuoplo SI (1999) An experimental comparison of cannibalistic response in different Arctic charr (*Salvelinus alpinus* (L.)) stocks. Ecol Freshw Fish 8:43–48. doi:10.1111/j.1600-0633.1999.tb00051.x
5. Azaza MS, Legendre M, Kraiem MM, Baras É (2010) Size-dependent effects of daily thermal fluctuations on the growth and size heterogeneity of Nile tilapia *Oreochromis niloticus*. J Fish Biol 76:669–683. doi:10.1111/j.1095-8649.2009.02524.x
6. Baardvik BM, Jobling M (1990) Effect of size-sorting on biomass gain and individual growth rates in Arctic charr, *Salvelinus alpinus* L. Aquaculture 90:11–16. doi:10.1016/0044-8486(90)90278-U
7. Babiak I, Mandiki SNM, Ratsinjomanana K, Kestemont P (2004) Initial weight and its variation in post-larval Eurasian perch affect quantitative characteristics of juvenile cohorts under controlled conditions. Aquaculture 234:263–276. doi:10.1016/j.aquaculture.2003.12.010
8. Balon EK (2004) About the oldest domesticates among fishes. J Fish Biol 65:1–27. doi:10.1111/j.1095-8649.2004.00563.x
9. Banan A, Kalbassi MR, Bahmani M, Sadati MAY (2011) Effects of colored light and tank color on growth indices and some physiological parameters of juvenile beluga (*Huso huso*). J Appl Ichthyol 27:565–570. doi:10.1111/j.1439-0426.2011.01682.x
10. Baras E (1999a) Functional implications of early sexual growth dimorphism in vundu. J Fish Biol 54:119–124. doi:10.1111/j.1095-8649.1999.tb00616.x
11. Baras É (1999b) Sibling cannibalism among juvenile vundu under controlled conditions. I. Cannibalistic behaviour, prey selection and prey size selectivity. J Fish Biol 54:82–105. doi:10.1006/jfbi.1998.0847
12. Baras É, Dalmeida AF (2001) Size heterogeneity prevails over kinship in shaping cannibalism among larvae of sharptooth catfish *Clarias gariepinus*. Aquat Living Resour 14:251–256. doi:10.1016/S0990-7440(01)01118-4
13. Baras E, Jobling M (2002) Dynamics of intracohort cannibalism in cultured fish. Aquac Res 33:461–479. doi:10.1046/j.1365-2109.2002.00732.x
14. Baras E, Maxi MYJ, Ndao M, Mélard C (2000a) Sibling cannibalism in dorada under experimental conditions. II. Effect of initial size heterogeneity, diet and light regime on early cannibalism. J Fish Biol 57:1021–1036. doi:10.1006/jfbi.2000.1365
15. Baras E, Ndao M, Maxi MYJ et al (2000b) Sibling cannibalism in dorada under experimental conditions. I. Ontogeny, dynamics, bioenergetics of cannibalism and prey size selectivity. J Fish Biol 57:1001–1020. doi:10.1006/jfbi.2000.1366
16. Baras E, Kestemont P, Mélard C (2003) Effect of stocking density on the dynamics of cannibalism in sibling larvae of *Perca fluviatilis* under controlled conditions. Aquaculture 219:241–255. doi:10.1016/S0044-8486(02)00349-6
17. Baras É, Hafsaridewi R, Slembrouck J et al (2010) Why is cannibalism so rare among cultured larvae and juveniles of *Pangasius djambal*? Morphological, behavioural and energetic answers. Aquaculture 305:42–51. doi:10.1016/j.aquaculture.2010.04.004
18. Baras É, Raynaud T, Slembrouck J et al (2011a) Interactions between temperature and size on the growth, size heterogeneity, mortality and cannibalism in cultured

larvae and juveniles of the Asian catfish, *Pangasianodon hypophthalmus* (Sauvage). Aquac Res 42:260–276. doi:10.1111/j.1365-2109.2010.02619.x

19. Baras É, Silva del Aguila DV, Montalvan Naranjos GV et al (2011b) How many meals a day to minimize cannibalism when rearing larvae of the Amazonian catfish *Pseudoplatystoma punctifer*? The cannibal's point of view. Aquat Living Resour 24:379–390. doi:10.1051/alr/2011141
20. Baras É, Hafsaridewi R, Slembrouck J et al (2013) Do cannibalistic fish possess an intrinsic higher growth capacity than others? A case study in the Asian redtail catfish *Hemibagrus nemurus* (Valenciennes, 1840). Aquac Res 45:68–79. doi:10.1111/j.1365-2109.2012.03205.x
21. Baras É, Dugué R, Legendre M (2014) Do cannibalistic fish forage optimally? An experimental study of prey size preference, bioenergetics of cannibalism and their ontogenetic variations in the African catfish *Heterobranchus longifilis*. Aquat Living Resour 27:51–62. doi:10.1051/alr/2014006
22. Barron JM, Jensen NR, Anders PJ et al (2012) Effects of temperature on the intensive culture performance of larval and juvenile North American Burbot (*Lota lota maculosa*). Aquaculture 364:67–73
23. Baskerville-Bridges B, Kling LJ (2000) Larval culture of Atlantic cod (*Gadus morhua*) at high stocking densities. Aquaculture 181:61–69. doi:10.1016/S0044-8486(99)00220-3
24. Bengtson DA, Léger P, Sorgeloos P (1991) Use of Artemia as a food source for aquaculture. In: Browne RA, Sorgeloos P, Trotman CNA (eds) Artemia biology. CRC Press, Boca Raton, pp 255–285
25. Biu AA, Gwana AM, Bukar-Kolo YM et al (2015) Effects of sorting on cannibalism in Catfish (*Clarias gariepinus*) rose in concrete tanks in maiduguri, North-Eastern Nigeria. Anim Vet Sci 3:67–73. doi:10.11648/j.avs.20150302.16
26. Blom G, Folkvord A (1997) A snapshot of cannibalism in 0-group Atlantic cod (*Gadus morhua*) in a marine pond. J Appl Ichthyol 13:177–181. doi:10.1111/j.1439-0426.1997.tb00118.x
27. Brabrand Å (1995) Intra-cohort cannibalism among larval stages of perch (*Perca fluviatilis*). Ecol Freshw Fish 4:70–76. doi:10.1111/j.1600-0633.1995.tb00119.x
28. Bristow BT, Summerfelt RC (1994) Performance of larval walleye cultured intensively in lear and turbid water. J World Aquac Soc 25:454–464. doi:10.1111/j.1749-7345.1994.tb00230.x
29. Britz PJ, Pienaar AG (1992) Laboratory experiments on the effect of light and cover on the behaviour and growth of African catfish, *Clarias gariepinus* (Pisces: Clariidae). J Zool 227:43–62. doi:10.1111/j.1469-7998.1992.tb04343.x
30. Conceição LEC, Yúfera M, Makridis P et al (2010) Live feeds for early stages of fish rearing. Aquac Res 41:613–640. doi:10.1111/j.1365-2109.2009.02242.x
31. Cuff WR (1980) Behavioral aspects of cannibalism in larval walleye, *Stizostedion vitreum*. Can J Zool 58:1504–1507
32. de Azevedo RV, Braga LGT (2012) Use of Probiotics in aquaculture. In: Rigobelo E (ed) Probiotic in animals. InTech, pp 103–118. doi: 10.5772/50056
33. DeAnglis DL, Hackney PA, Webb JC (1980) A partial differential equation model of changing sizes and numbers in a cohort of juvenile fish. Environ Biol Fishes 5:261–266. doi:10.1007/BF00005361
34. Douxfils J, Mandiki SNM, Marotte G et al (2011) Does domestication process affect stress response in juvenile Eurasian perch *Perca fluviatilis*? Comp Biochem Physiol Part A Mol Integr Physiol 159:92–99. doi:10.1016/j.cbpa.2011.01.021
35. Ferguson MM, Noakes DLG (1983) Movers and stayers: genetic analysis of mobility and positioning in hybrids of lake charr, *Salvelinus namaycush*, and brook charr, *S. fontinalis* (Pisces, Salmonidae). Behav Genet 13:213–222. doi:10.1007/ BF01065 670
36. Fessehaye Y, Rezk M, Bovenhuis H, Komen H (2005) Size dependent cannibalism in juvenile Nile Tilapia (*Oreochromis niloticus*). In: Bolivar RB, Mair GC, Fitzsimmons

K (eds) 6th international symposium on Tilapia in Aquaculture Philippine International Convention Center Roxas Boulevard. Manila, Philippines, pp 230–238

37. Folkvord A (1997) Ontogeny of cannibalism in larval and juvenile fishes with special emphasis on Atlantic cod. In: Chambers RC, Trippel EA (eds) Early life history and recruitment in fish population. Springer Netherlands, Dordrecht, pp 251–278. doi:10.1007/978-94-009-1439-1
38. Folkvord A, Otterå H (1993) Effects of initial size distribution, day length, and feeding frequency on growth, survival, and cannibalism in juvenile Atlantic cod (*Gadus morhua* L.). Aquaculture 114:243–260. doi:10.1016/0044-8486(93)90300-N
39. Francis AW Jr, Bengtson DA (1999) Partitioning of fish and diet selection as methods for the reduction of cannibalism in *Paralichthys dentatus* larviculture. J World Aquac Soc 30:302–310. doi:10.1111/j.1749-7345.1999.tb00680.x
40. Frisk M, Skov PV, Steffensen JF (2012) Thermal optimum for pikeperch (*Sander lucioperca*) and the use of ventilation frequency as a predictor of metabolic rate. Aquaculture 324–325:151–157. doi:10.1016/j.aquaculture.2011.10.024
41. Garrison LP, Link JS (2000) Diets of five hake species in the northeast United States continental shelf ecosystem. Mar Ecol Prog Ser 204:243–255. doi:10.3354/meps204243
42. Gatesoupe FJ (1999) The use of probiotics in aquaculture. Aquaculture 180:147–165. doi:10.1016/S0044-8486(99)00187-8
43. Gomułka P, Antychowicz J (1999) Anaesthetics used in fish. Med Weter 55:89–93
44. Hecht T, Appelbaum S (1988) Observations on intraspecific aggression and coeval sibling cannibalism by larval and juvenile *Clarias gariepinus* (Clariidae: Pisces) under controlled conditions. J Zool 214:21–44. doi:10.1111/j.1469-7998.1988.tb04984.x
45. Hecht T, Pienaar AG (1993) A review of cannibalism and its implications in fish larviculture. J World Aquac Soc 24:246–261. doi:10.1111/j.1749-7345.1993.tb00014.x
46. Hedenskog M, Petersson E, Järvi T (2002) Agonistic behavior and growth in Newly emerged brown trout (*Salmo trutta* L.) of sea-ranched and wild origin. Aggress Behav 28:145–153. doi:10.1002/ab.90016
47. Hoglund E, Bakke MJ, Overli O et al (2005) Suppression of aggressive behaviour in juvenile Atlantic cod (*Gadus morhua*) by L-tryptophan supplementation. Aquaculture 249:525–531. doi:10.1016/j.aquaculture.2005.04.028
48. Hseu JR (2002) Effects of size difference and stocking density on cannibalism rate of juvenile grouper *Epinephelus coioides*. Fish Sci 68:1384–1386. doi:10.1046/j.1444-2906.2002.00578.x
49. Hseu JR, Huang WB (2014) Application of logistic regression analysis to predict cannibalism in orange-spotted grouper *Epinephelus coioides* (Hamilton) fry. Aquac Res 45:868–873. doi:10.1111/are.12031
50. Hseu JR, Lu FI, Su HM et al (2003a) Effect of exogenous tryptophan on cannibalism, survival and growth in juvenile grouper, *Epinephelus coioides*. Aquaculture 218:251–263. doi:10.1016/S0044-8486(02)00503-3
51. Hseu JR, Chang HF, Ting YY (2003b) Morphometric prediction of cannibalism in larviculture of orange-spotted grouper, *Epinephelus coioides*. Aquaculture 218:203–207. doi:10.1016/S0044-8486(02)00315-0
52. Hseu JR, Shen PS, Bin Huang W, Hwang PP (2007) Logistic regression analysis applied to cannibalism in the giant grouper *Epinephelus lanceolatus* fry: short paper. Fish Sci 73:472–474. doi:10.1111/j.1444-2906.2007.01358.x
53. Imanpoor MR, Abdollahi M (2011) Effects of tank color on growth, stress response and skin color of Juvenile caspian kutum *Rutilus frisii Kutum*. Glob Vet 6:118–125
54. Imorou Toko I, Fiogbé ED, Kestemont P (2008) Determination of appropriate age and stocking density of vundu larvae, *Heterobranchus longifilis* (Valenciennes 1840), at the weaning time. Aquac Res 39:24–32. doi:10.1111/j.1365-2109.2007.01857.x

55. Jesu Arockiaraj A, Appelbaum S (2011) Sibling cannibalism in juvenile barramundi, *Lates calcarifer* (Actinopterygii: Perciformes: Centropomidae), reared under different light conditions. Acta Ichthyol Piscat 41:7–11. doi:10.3750/AIP2011.41.1.02
56. Jobling M, Jørgensen EH, Arnesen AM, Ringø E (1993) Feeding, growth and environmental requirements of Arctic charr: a review of aquaculture potential. Aquac Int 1:20–46. doi:10.1007/BF00692662
57. Kailasam M, Thirunavukkarasu AR, Chandra PK, Subburaj R (2002) Influence of size variation and feeding on cannibalism of Asian sea bass *Lates calcarifer* (Bloch) during hatchery rearing. Indian J Fish 49:107–113
58. Kaleta T (2013) Behavior of captive lower vertebrates as an indicator of animal welfare. Życie Weter 88:860–866
59. Kania BF, Zawadzka E, Dębski B (2012) Neurohormonal basis of aggression in fish. Med Weter 68:195–198
60. Katavić I, Jug-Dujaković J, Glamuzina B (1989) Cannibalism as a factor affecting the survival of intensively cultured sea bass (*Dicentrarchus labrax*) fingerlings. Aquaculture 77:135–143. doi:10.1016/0044-8486(89)90197-X
61. Kestemont P, Jourdan S, Houbart M et al (2003) Size heterogeneity, cannibalism and competition in cultured predatory fish larvae: biotic and abiotic influences. Aquaculture 227:333–356. doi:10.1016/S0044-8486(03)00513-1
62. Knowles A, Dartnall HJA (1977) Habitat, habit and visual pigments. In: Davson H (ed) The Eye: Photobiology of Vision, Vol 2B, 2nd edn. Academic Press, NewYork, pp 581–648
63. Kozłowski M, Szczepkowski M, Wunderlich K et al (2009) Results of rearing juvenile pikeperch (*Sander lucioperca*) in different types of tanks. Komun Rybackie 6:6–10
64. Król J, Zakęś Z (2016) Effect of dietary L-tryptophan on cannibalism, survival and growth in pikeperch *Sander lucioperca* (L.) post-larvae. Aquac Int 24:441–451. doi:10.1007/s10499-015-9936-1
65. Król J, Zieliński E (2015) Effects of stocking density and weaning age on cannibalism, survival and growth in European perch *Perca fluviatilis* larvae. Pol J Nat Sci 30:403–415
66. Król J, Zakęś Z, Kestemont P (2013) Emergence of cannibalism in European percid fish—size heterogeneity or natural born killers consequence? In: Larvi 2013: 6th fish & shellfish larviculture symposium, pp 231–232
67. Kucharczyk D, Mamcarz A, Kujawa R, Skrzypczak A (1998) Development of cannibalism in larval northern pike, *Esox lucius* (Esocidae). Ital J Zool 65:261–263. doi:10.1080/11250009809386829
68. Kusmic C, Gualtieri P (2000) Morphology and spectral sensitivities of retinal and extraretinal photoreceptors in freshwater teleosts. Micron 31:183–200. doi:10.1016/S0968-4328(99)00081-5
69. Lepage O, Vílchez IM, Pottinger TG, Svante W (2003) Time-course of the effect of dietary L-tryptophan on plasma cortisol levels in rainbow trout *Oncorhynchus mykiss*. J Exp Biol 206:3589–3599. doi:10.1242/jeb.00614
70. Ljubobratović U, Kucska B, Feledi T et al (2015) Effect of weaning strategies on growth and survival of Pikeperch, *Sander lucioperca*, larvae. Turkish J Fish Aquat Sci 15:325–331. doi:10.4194/1303-2712-v15_2_15
71. Loadman NL, Moodie GEE (1986) Significance of cannibalism in larval Walleye (*Stizostedion vitreum*). Can J Fish Aquat Sci 43:613–618
72. Mandiki SNM, Babiak I, Król J et al (2007) How initial predator-prey ratio affects intracohort cannibalism and growth in Eurasian perch *Perca fluviatilis* L. larvae and juveniles under controlled conditions. Aquaculture 268:149–155. doi:10.1016/j.aquaculture.2007.04.036
73. Mandiki SNM, Milla S, Wang N et al (2011) Effects of probiotic bacteria on growth parameters and immune defence in Eurasian perch *Perca fluviatilis* L. larvae under

intensive culture conditions. Aquac Res 42:693–703. doi:10.1111/j.1365-2109.2010.02744.x

74. Mandiki SNM, Redivo B, Baekelandt S et al (2016) Long-term tryptophan supplementation decreased the welfare and innate immune status of pikeperch juveniles. Fish Shellfish Immunol 53:113–114. doi:10.1016/j.fsi.2016.04.090
75. Manley CB, Rakocinski CF, Lee PG, Blaylock RB (2015) Feeding frequency mediates aggression and cannibalism in larval hatchery-reared spotted seatrout, *Cynoscion nebulosus*. Aquaculture 437:155–160. doi:10.1016/j.aquaculture.2014.11.012
76. Mélard C, Baras E, Mary L, Kestemont P (1996) Relationships between stocking density, growth, cannibalism and survival rate in intensively cultured larvae and juveniles of perch (*Perca fluviatilis*). Ann Zool Fennici 33:643–651. doi:10.1016/j.infsof.2008.09.005
77. Mousseau T (1998) The adaptive significance of maternal effects. Trends Ecol Evol 13:403–407. doi:10.1016/S0169-5347(98)01472-4
78. Mukai Y, Sanudin N, Firdaus RF, Saad S (2013) Reduced cannibalistic behavior of African catfish, *Clarias gariepinus*, larvae under dark and dim conditions. Zoolog Sci 30:421–424. doi:10.2108/zsj.30.421
79. Neiffer DL, Stamper MA (2009) Fish sedation, analgesia, anesthesia, and euthanasia: considerations, methods, and types of drugs. ILAR J 50:343–360. doi:10.1093/ilar.50.4.343
80. Nilsson PA (1999) Foraging among cannibals and kleptoparasites: effects of prey size on pike behavior. Behav Ecol 10:557–566. doi:10.1093/beheco/10.5.557
81. Ounais-Guschemann N (1989) Définition d'un modèle d'élevage larvaire intensif pour la daurade *Sparus auratus*. Dissertation, Université d'Aix-Marseille II
82. Pienaar AG (1990) A study of coeval sibling cannibalism in larval and juvenile fishes and its control under culture conditions. Rhodes University, Grahams Town
83. Piotrowska I, Szczepkowska B, Wunderlich K, Szczepkowski M (2013) Results of the larviculture of Atlantic sturgeon (*Acipenser oxyrinchus*) fed different types of diets. Arch Pol Fish 21:53–61. doi:10.2478/aopf-2013-0006
84. Policar T, Stejskal V, Kristan J et al (2013) The effect of fish size and stocking density on the weaning success of pond-cultured pikeperch *Sander lucioperca* L. juveniles. Aquac Int 21:869–882. doi:10.1007/s10499-012-9563-z
85. Post JR, Evans DO (1989) Experimental evidence of size-dependent predation mortality in juvenile yellow perch. Can J Zool 67:521–523. doi:10.1139/z89-076
86. Qin JG, Mittiga L, Ottolenghi F (2004) Cannibalism reduction in juvenile barramundi *Lates calcarifer* by providing refuges and low light. J World Aquac Soc 35:113–118. doi:10.1111/j.1749-7345.2004.tb01067.x
87. Ribeiro FF (2015) Cannibalism in Barramundi *Lates calcarifer*: understanding functional mechanisms and implication to aquaculture. Flinders University, Bedford Park
88. Ribeiro FF, Qin JG (2013) Modelling size-dependent cannibalism in barramundi *Lates calcarifer*: cannibalistic polyphenism and its implication to aquaculture. PLoS ONE 8:1–8. doi:10.1371/journal.pone.0082488
89. Ribeiro FF, Qin JG (2015) Prey size selection and cannibalistic behaviour of juvenile barramundi *Lates calcarifer*. J Fish Biol 86:1549–1566. doi:10.1111/jfb.12658
90. Ribeiro FF, Qin JG (2016) Bioenergetics of cannibalism in juvenile barramundi *Lates calcarifer* (Bloch): exploring growth advantage of fish fed live prey and formulated diet. Aquac Res 47:2324–2333. doi:10.1111/are.12685
91. Rożyński M, Zakęś Z (2015) Wpływ parametrów technicznych basenów i właściwości fizycznych pasz komponowanych na efekty podchowu ryb. Komun Rybackie 3:12–44
92. Sallehudin F, Mukai Y (2014) Cannibalistic behaviour of African catfish juveniles, *Clarias gariepinus* under different light wavelengths and intensities. In: Nejadkoorki F (ed) 3rd international conference on applied life sciences (ICALS2014). ISALS Publishing, Malaysia, pp 51–55

93. Sloman KA, Armstrong JD (2002) Physiological effects of dominance hierarchies: laboratory artefacts or natural phenomena? J Fish Biol 61:1–23. doi:10.1111/j.1095-8649.2002.tb01733.x
94. Smith C, Reay P (1991) Cannibalism in teleost fish. Rev Fish Biol Fish 1:41–64. doi:10.1007/BF00042661
95. Sogard SM, Olla BL (1994) The potential for intracohort cannibalism in age-0 walleye pollock, *Theragra chalcogramma*, as determined under laboratory conditions. Environ Biol Fishes 39:183–190. doi:10.1007/BF00004936
96. Solomon RJ, Udoji FC (2011) Canibalism among cultured African catfishes (*Heterobranchus longifillis* and *Clarias gariepinus*). Nat Sci 9(9):1–13
97. Steenfeldt S, Lund I, Höglund E (2011) Is batch variability in hatching time related to size heterogeneity and cannibalism in pikeperch (*Sander lucioperca*)? Aquac Res 42:727–732. doi:10.1111/j.1365-2109.2010.02681.x
98. Strand Å, Alanärä A, Staffan F, Magnhagen C (2007) Effects of tank colour and light intensity on feed intake, growth rate and energy expenditure of juvenile Eurasian perch, *Perca fluviatilis* L. Aquaculture 272:312–318. doi:10.1016/ j.aquaculture. 2007.08.052
99. Subagja J, Slembrouck J, Hung LT, Legendre M (1999) Larval rearing of an Asian catfish *Pangasius hypophthalmus* (Siluroidei, Pangasiidae): Analysis of precocious mortality and proposition of appropriate treatments. Aquat Living Resour 12(1):37–44. doi:10.1016/S0990-7440(99)80013-8
100. Sunde LM, Imsland AK, Folkvord A, Stefansson SO (1998) Effects of size grading on growth and survival of Juvenile turbot at two temperatures. Aquac Int 6:19–32. doi:10.1023/A:1009265602388
101. Szczepkowski M (2009) Impact of selected abiotic and biotic factors on the results of rearing juvenile stages of northern pike *Esox lucius* L. in recirculating systems. Arch Pol Fish 48:107–147. doi:10.2478/v10086-009-0010-3
102. Szczepkowski M, Zakęś Z, Szczepkowska B, Piotrowska I (2011) Effect of size sorting on the survival, growth and cannibalism in pikeperch (*Sander lucioperca* L.) larvae during intensive culture in RAS. Czech J Anim Sci 56:483–489
103. Szkudlarek M, Zakęś Z (2007) Effect of stocking density on survival and growth performance of pikeperch, *Sander lucioperca* (L.), larvae under controlled conditions. Aquac Int 15:67–81. doi:10.1007/s10499-006-9069-7
104. Tamazouzt L, Chatain B, Fontaine P (2000) Tank wall colour and light level affect growth and survival of *Eurasian perch* larvae (Perca fluviatilis L.). Aquaculture 182:85–90. doi:10.1016/S0044-8486(99)00244-6
105. Thibault RE (1974) Genetics of cannibalism in a viviparous fish and its relationship to population density. Nature 251:138–140. doi:10.1038/251138a0
106. Trabelsi A, Gardeur JN, Teletchea F et al (2013) Hatching time effect on the intra-spawning larval morphology and growth in Northern pike (*Esox lucius* L.). Aquac Res 44:657–666. doi:10.1111/j.1365-2109.2011.03070.x
107. Trabelsi A, Jaworski A, Kamler E et al (2016) The effect of hatching time on the bioenergetics of northern pike (*Esox lucius*) larvae from a single egg batch during the endogenous feeding period. Fish Physiol Biochem 42:593–606. doi:10.1007 /s10695-015-0162-6
108. Üstündağ M, Rad F (2015) Effect of different tank colors on growth performance of Rainbow trout juvenile (*Oncorhynchus mykiss* Walbaum, 1792). J Agric Sci 21:144–151
109. Vallon M, Grom C, Kalb N et al (2016) You eat what you are: personality-dependent filial cannibalism in a fish with paternal care. Ecol Evol 6:1340–1352. doi:10.1002/ece3.1966
110. van Damme P, Appelbaum S, Hecht T (1989) Sibling cannibalism in Koi carp, *Cyprinus carpio* L., larvae and juveniles reared under controlled conditions. J Fish Biol 34:855–863. doi:10.1111/j.1095-8649.1989.tb03369.x
111. van den Bosch F, de Roos AM, Gabriel W (1988) Cannibalism as a life boat mechanism. J Math Biol 26:619–633. doi:10.1007/BF00276144

112. Wallat GK, Tiu LG, Wang HP et al (2005) The effects of size grading on production efficiency and growth performance of yellow perch in earthen ponds. N Am J Aquac 67:34–41. doi:10.1577/FA04-003.1
113. Winberg S, Øverli Ø, Lepage O (2001) Suppression of aggression in rainbow trout (*Oncorhynchus mykiss*) by dietary L-tryptophan. J Exp Biol 204:3867–3876
114. Wootton RJ (1989) Ecology of teleost fishes. Springer, Netherlands
115. Yang S, Yang K, Liu C et al (2015) To what extent is cannibalism genetically controlled in fish? A case study in juvenile hybrid catfish *Silurus meridionalis-asotus* and the progenitors. Aquaculture 437:208–214. doi:10.1016/ j.aquaculture. 2014.12.005
116. Zakęś Z (2012) The effect of body size and water temperature on the results of intensive rearing of pike-perch, *Stizostedion lucioperca* (L.) fry under controlled conditions. Arch Pol Fish 20:165–172. doi:10.2478/v10086-012-0020-4
117. Zakęś Z, Demska-Zakęś K (1996) Effect of diets on growth and reproductive development of juvenile pikeperch, *Stizostedion lucioperca* (L.), reared under intensive culture conditions. Aquac Res 27:841–845. doi:10.1046/j.1365-2109.1996.00801.x
118. Żarski D, Kucharczyk D, Kwiatkowski M et al (2008) The effect of stocking density on the growth and survival of larval asp, *Aspius aspius* (L.), and European chub, *Leuciscus cephalus* (L.), during rearing under controlled conditions. Arch Pol Fish 16:371–381. doi:10.2478/s10086-008-0025-1.

# CHAPTER 5

# Estuarine Fisheries Community-Level Response to Freshwater Inflows

**James M. Tolan[1]**

*[1] Texas Parks and Wildlife Department, Coastal Fisheries Division, Natural Resource Center 2501, Unit 5846, Corpus Christi, TX, USA*

## 1. INTRODUCTION

Every estuary needs freshwater inflow (FWI) to maintain proper salinity regimes, nutrient loading, and sediment inputs to support its geographically unique levels of biological productivity [1-4]. Watershed elevations and soil types determine surface and groundwater flows into estuaries, and these flows have source, timing, and velocity components that can be significantly affected by anthropogenic alterations at the landscape level. It is estimated that approximately 60% of the global storage of freshwater is now contained behind reservoirs and dams [5] and 77% of the total water discharge from 139 of the largest river systems in the northern hemisphere are either strongly or moderately affected by dams, interbasin transfers, and surface water withdrawals [6]. Hydrologic modifications of estuarine watersheds influence wetland and open-water salinity patterns, nutrients, sediment fertility, bottom topography, dissolved oxygen, and concentrations of xenobiotics [7]. Because demand for freshwater is only expected to increase as population continues to grow [8], it is incumbent upon resource managers to examine the environmental effects and biological consequences of hydrologic alterations within coastal ecosystems [9-11].

Resource-based approaches seek to link freshwater inflows to a number of fishery species generally considered valuable by society [12]. The optimization model utilized by Powell and Matsumoto [13] uses a series of relationships between monthly inflows and the catch of a number of commercially and recreationally important finfish (red drum *Sciaenops ocellatus*, black drum *Pogonias cromis*, spotted seatrout *Cynoscion nebulosus*, southern flounder *Paralichthys lethostigma*), crustaceans (blue crab *Callinectes sapidus*, white shrimp *Litopenaeus setiferus*, brown shrimp *Farfantepenaeus aztecus*, pink shrimp *F. duorarum*) and mollusks (eastern oyster *Crassostrea virginica*) to arrive at a set of targeted monthly freshwater inflows to maintain healthy ecological conditions in estuaries. The goal of this method, which jointly considers the salinity tolerance range of each of the target organisms and limits the inflow volume solution by imposing numerous process constraints (such as fishery biomass and harvest ratios; monthly, bi-monthly, and yearly freshwater volumes; upper and lower bounds for salinity; nutrient and sediment loading; see [13]), is to estimate the minimum amounts of FWI needed to maintain historical fisheries production. Although the inflow-harvest equations were originally based on fishery-dependent commercial catch records, recent modeling efforts have incorporated a greater proportion of fishery-independent data sources [14].

A problem with the resource-based approach is that it focuses on adults, which are harvestable. Although these are estuarine dependent species, the adults are widely distributed along salinity gradients [15]. A number of transient taxa (sensu, facultative estuarine-dependent, see [16]) which recruit from offshore spawning areas are known to have size-specific use patterns within shallow habitats of the oligohaline-to-freshwater portions of estuaries [17-20]. Inflows, especially those large pulses associated with flooding events, can displace seaward the boundary between the brackish and freshwater interface, on a kilometers-to-estuary wide scale. Taxa with specific nursery habitat requirements could therefore be restricted from ingress into portions of the estuary, potentially altering an important habitat for juvenile nekton.

The goal of this study was to expand the focus of interest of the resource-based approach beyond the limited number of fisheries target species and to include juvenile stages of fisheries species to examine the functional role of FWI in shaping the total nekton assemblage structure in estuaries. The approach was to perform an analysis of a long-term, state agency, bag seine monitoring program. Bag seine samples are fishery independent and contain juvenile stages of fishery species. There are three FWI gradients examined; within estuaries from river to sea, among estuaries along a climatic gradient, and over time as changes in freshwater inflows, in the form of flood pulses and drought events, dramatically alter the salinity structure of the estuary.

# 2. MATERIALS AND METHODS

## 2.1. Study Area

Texas' coastline extends along 600 km of open Gulf of Mexico shoreline and contains 3,420 kilometers of bay-estuary-lagoon shoreline. This is a biologically rich and ecologically diverse region of the state, supporting more than 247,576 hectares of fresh, brackish, and salt marshes. Within the state, over 305,600 km of rivers and streams coalesce into 15 major river systems, and these rivers empty into seven major estuaries (Figure 1). All seven estuaries have similar geomorphic structure and physiography, yet each is quite diverse hydrologically. This is primarily due to a climatic gradient influencing freshwater inflows. This gradient of decreasing rainfall from northeast to southwest is one of the most distinctive features of the coastline (Table 1). Along this gradient, rainfall decreases by a factor of two, yet inflow decreases by almost two orders of magnitude. The Laguna Madre, a hypersaline lagoon, has a negative inflow balance because this estuary lacks any major riverine inflow and evaporation normally exceeds precipitation. The net effect is a gradient of estuaries with similar physical characteristics but greatly differing salinity regimes.

From the Louisiana border to the Trinity-San Jacinto estuary the coastline is characterized by marshy plains with low, narrow beach ridges, and from there to the border of Mexico the coastline is characterized by long barrier islands and large, shallow lagoons. Barrier islands are parallel to the mainland along the coast, and between the barrier islands and the mainland are lagoons. These lagoons are interrupted with drowned river valleys that form the bays and estuaries. Inlets through the barrier island connect the Gulf of Mexico to these lagoons, with each lagoon opening into a large primary bay. There is typically a constriction between the primary and secondary (and in some cases tertiary) bays. While ungauged coastal watershed runoff can locally influence estuarine salinity, most inflow into each bay is supplied primarily by just one or two gauged rivers draining hydrologically isolated watersheds.

## 2.2. Bag Seine Community Structure Data

Starting in January 1992 and continuing through the present, 20 replicate bag seine samples have been collected each month within each major estuary system along the Texas coast. Sampling locations are randomly selected from a grid system of one minute latitude and one minute

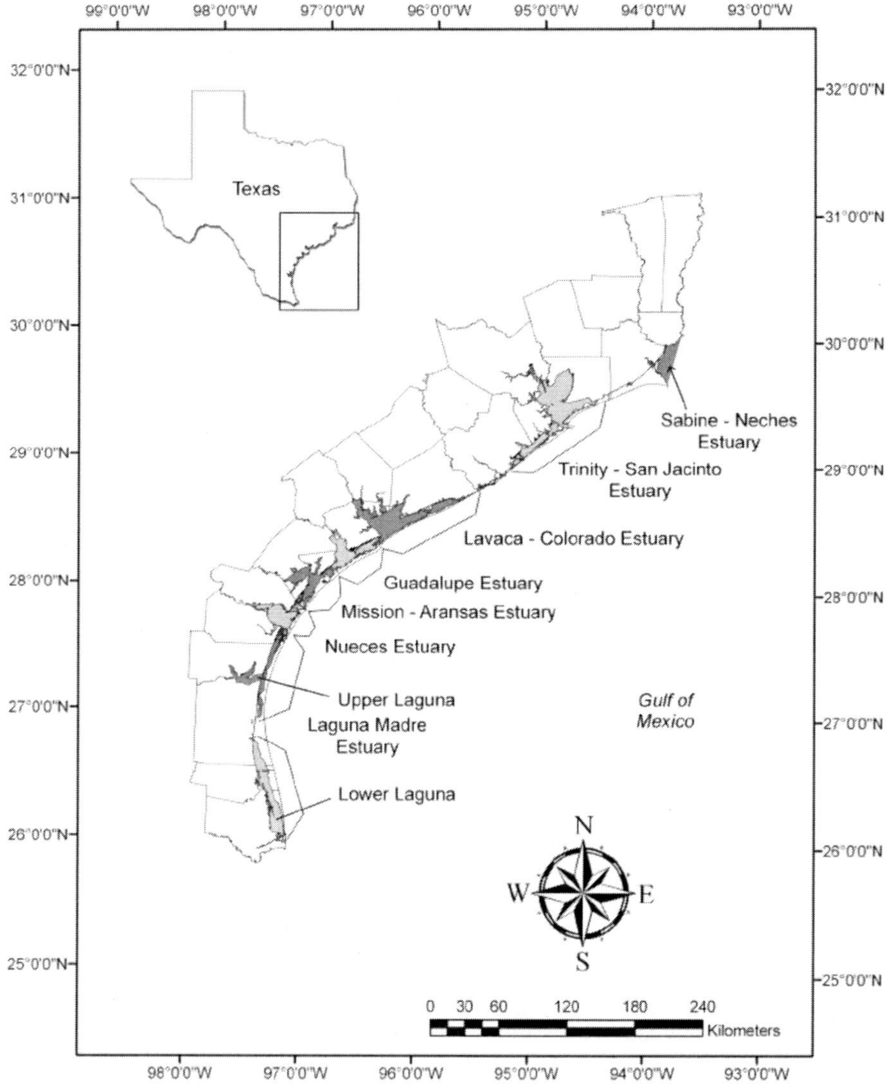


**FIGURE 1.** Map of Texas showing the location along the coast of each of the major estuarine systems.

**TABLE 1.** Climatic gradient in Texas estuaries, listed from north to south. Drainage basin size (Drainage), bay surface area (Area), and Habitat (SAV = submerged aquatic vegetation) characteristics from USEPA (1999). Average annual Inflow Balance, Rainfall, and Salinity characteristics (1941-1999) from the Texas Water Development Board; http://www.twdb.state.tx.us/data/ bays_estuaries/bays_estuary_toc.asp

| | | | | | | Habitat | |
|---|---|---|---|---|---|---|---|
| **Estuary** | **Drainage ($km^2$)** | **Area ($km^2$)** | **Rainfall ($cm\ y^{-1}$)** | Inflow ($10^6 m^3\ y^{-1}$) | **Salinity (PSU)** | **Wetlands ($km^2$)** | **SAV ($km^2$)** |
| Sabine-Neches | 45 705 | 243 | 142 | 16 897 | 8 | 967 | - |
| Trinity-San Jacinto | 57 900 | 1399 | 112 | 14 000 | 16 | 1594 | 73 |
| Lavaca-Colorado | 111 890 | 1158 | 102 | 3801 | 18 | 348 | 28 |
| Guadalupe | 26 330 | 551 | 91 | 2664 | 16 | 271 | 65 |
| Mission-Aransas | 7860 | 453 | 81 | 265 | 19 | 393 | 85 |
| Nueces | 43 350 | 433 | 76 | 298 | 29 | 121 | 53 |
| Laguna Madre | 29 695 | 3658 | 69 | -893 | 36 | 1825 | 773 |

longitude, with no selected grid sampled more than once per calendar month. For each sample, a bag seine (18.3 m X 1.8 m, 1.9 cm stretch nylon multifilament; central bag, 1.8 m wide, 1.3 cm stretch mesh) is pulled parallel to the shore for 15.2 m [21]. The surface area sampled is estimated using the distance pulled and length of extension of the bag seine. All fish and invertebrates collected in each sample are identified, enumerated, and measured. Total catch of each taxon is standardized and expressed as catch per hectare. Prior to each bag seine collection, surface salinity in Practical Salinity Scale were measured with either handheld Hydrolab or YSI multiprobes calibrated to the manufacturers specifications.

## 2.3. Data Analysis

The experimental unit defined for this study was each bay system as a whole, as each of the major estuaries along the Texas coast can be defined by the underlying hydrologic gradient. To assemble the bag seine collections into a time series, catch data for each taxa from the monthly replicate samples were summed across estuaries individually, and reported as total catch per month. Salinity records were similarly transformed into a time series, although the replicate values were first averaged across each estuary and reported as mean salinity per month. For each estuary, a categorical 'Inflow Condition' variable was defined by evaluating the average salinity time series. Salinities above the $85^{th}$percentile were deemed indicative of 'Drought' conditions and values below the $15^{th}$ percentile representative of high flow or 'Flood' conditions. Values between these two extremes were identified as 'Normal' flows. The Laguna Madre Estuary was further sub-divided into an Upper and Lower components, providing for eight estuaries under investigation (see Figure 1). This sub-division is based on a natural sand sheet or land bridge (the Land Cut) connecting Padre Island with the mainland [21]. The Gulf Intracoastal Waterway bisects this extensive sand flat, thus connecting the two lagoons via the Land Cut.

Community analyses were performed using Primer-E (Version 6.0) software [22]. A matrix of Bray-Curtis Distance similarities between each total catch per month sample was created. Catch data was initially transformed [$Log_{10}$ + 1] to down-weight the most abundant taxa. Significant differences in rank similarities between groups of samples were then tested by Analysis of Similarity (ANOSIM). In the ANOSIM procedure, the probability of *a priori* groupings of samples is estimated by repeated permutations of the original data matrix. Values of the $R$ statistic can range from -1 to 1, although $R$ will usually fall between 0 and 1 with $R$ values > 0.4 indicating higher degrees of discrimination among groups. The *a priori* factors tested with the ANOSIM procedure were the external factors associated with each sample (e.g., season of collection and inflow condition) within a common estuary. The entire collection was then merged across estuaries, and then tested for differences in community structure among estuaries. The SIMPER (SIMilarity PERcentages) routine was used to examine the contribution of individual species to the community structure seen among the *a priori* factors. Similarities among the samples are graphically represented with non-metric multidimensional scaling (MDS) ordinations [23]. Although outcomes of the ANOSIM are not dependant on MDS ordinations, the ordinations are presented here as they are a helpful way of visualizing patterns in the data. Stress values indicate how well the two-dimensional plot represents relationships among samples in the multidimensional space. Stress values < 0.15 indicate a good fit. MDS ordinations may be arbitrarily rotated so axes are not labeled.

## 3. RESULTS

From 1992 through 2006, bag seine sampling resulted in 28,786 individual collections from the eight major estuarine systems. This time series of 180 months revealed dramatically fluctuating mean salinities throughout the study period (Table 2). Despite dramatic differences in total inflows across the coastal hydrologic gradient, temporal inflow patterns were generally similar across the coast. The timing of extended flooding conditions (i.e., on the order of 10 months during 1992 and again in 1997) or droughts (the majority of the calendar years of 2000 and 2001) were similar within each estuary (Figure 2).

**TABLE 2.** Salinity summary statistics by estuary for the study period January 1992 through December 2006.

| | *n* | Mean | Std. Dev. | Min. | Max. |
|---|---|---|---|---|---|
| Estuary | | | | | |
| Sabine-Neches | 3599 | 6.73 | 6.23 | 0.0 | 32.0 |
| Trinity-San Jacinto | 3597 | 16.52 | 9.21 | 0.0 | 41.0 |
| Lavaca-Colorado | 3599 | 18.47 | 9.57 | 0.0 | 40.0 |
| Guadalupe | 3594 | 16.62 | 11.36 | 0.0 | 45.0 |
| Mission-Aransas | 3600 | 18.13 | 9.70 | 0.0 | 41.0 |
| Nueces | 3600 | 28.66 | 7.42 | 0.0 | 59.0 |
| Upper Laguna Madre | 3599 | 35.44 | 10.64 | 0.0 | 78.0 |
| Lower Laguna Madre | 3598 | 32.04 | 7.98 | 0.0 | 64.0 |

The bag seines recorded 3,583,061 individuals from 387 unique taxa. Analysis of Similarity of the entire collection showed that in each estuary, community structure was significantly different across seasons (Table 3), and these seasonal differences were repeated annually across all inflow categories. The greatest disparity in community composition involved comparisons across opposite seasons (e.g., winter vs. summer, spring vs. fall), with significant pairwise comparison R values ranging from 0.609 – 0.971 (see Table 3). While seasonal differences in communities were quite evident, there appears to be little correspondence between community structure and synoptic-scale inflow events (Figure 3). This general disconnect between shallow water nekton assemblages and inflows was evident in every estuary along the Texas coast (Figure 4). The only estuaries to display significant community-level differences across the different inflow conditions were from opposite ends of the salinity spectrum. The Sabine-Neches estuary (mean salinity approximately 7) had significantly different community compositions during a drought relative to flood conditions ($R = 0.520$, $p < 0.001$). Greater abundances of white shrimp (7 fold increase), brown shrimp (17 fold increase), pinfish *Lagodon rhomboides* (6 fold increase), white mullet *Mugil curema* (12 fold increase), spotted seatrout (17 fold increase), and sheepshead minnow *Cyprinodon variegatus* (9 fold increase) were recorded during the periods of elevated

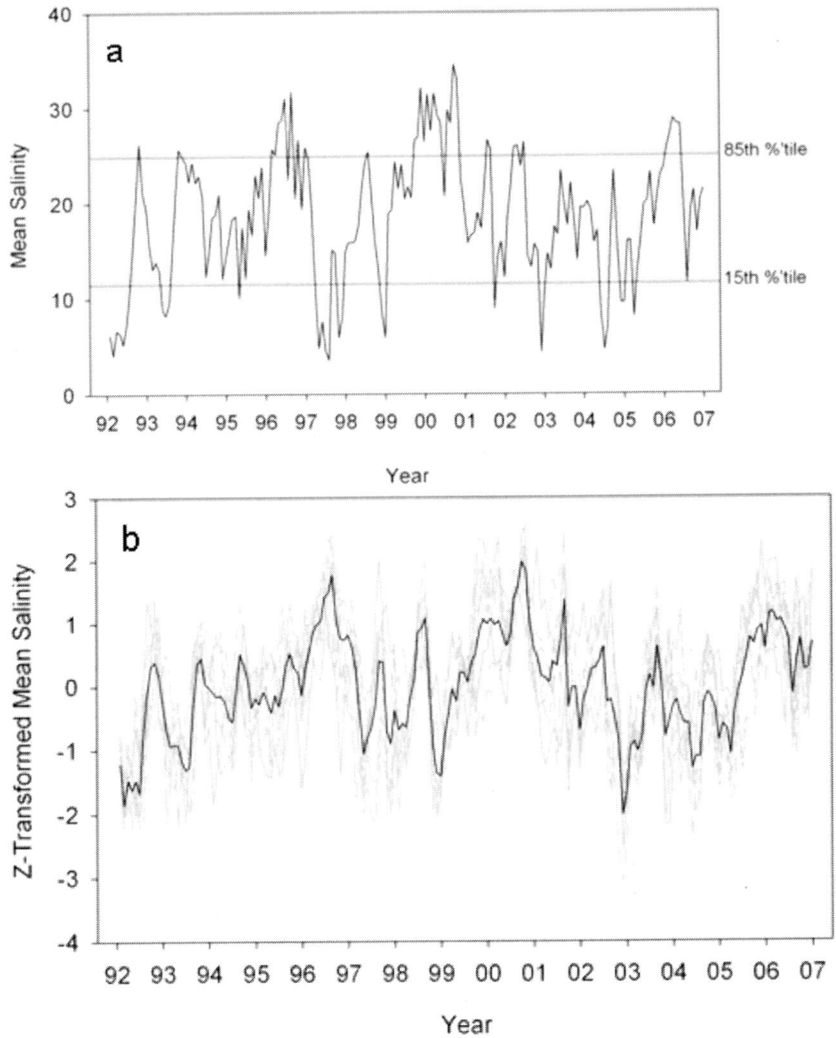


**FIGURE 2.** Estuarine-wide mean salinity time series during the study period of 1992 through 2006: (a) from a representative estuary (Lavaca-Colorado), and (b) from all eight major estuaries. Salinities in (b) are Z-transformed (not labeled individually for clarity), and a mean line added to aid in interpretation.

**TABLE 3.** Analysis of Similarity results of community structure within each estuary across seasons. Global R by Estuary, *** = $p < 0.001$, pairwise comparison R values by season (significant pairwise R values > than the Global R in bold).

| Estuary | Global R | | Seasonal Pairwise Comparison R | | |
|---|---|---|---|---|---|
| | | | Winter | Spring | Summer |
| | | Spring | 0.464 | - | |
| Sabine-Neches | 0.669*** | Summer | **0.908** | 0.566 | - |
| | | Fall | **0.836** | **0.834** | 0.501 |
| | | Spring | 0.628 | - | |
| Trinity-San Jacinto | 0.649*** | Summer | **0.900** | 0.553 | - |
| | | Fall | **0.815** | **0.786** | 0.362 |
| | | Spring | 0.576 | - | |
| Lavaca-Colorado | 0.695*** | Summer | **0.971** | **0.697** | - |
| | | Fall | **0.820** | **0.796** | 0.391 |
| | | Spring | 0.573 | - | |
| Guadalupe | 0.661*** | Summer | **0.897** | 0.660 | - |
| | | Fall | **0.756** | **0.786** | 0.411 |
| | | Spring | 0.582 | - | |
| Mission-Aransas | 0.677*** | Summer | **0.848** | 0.652 | - |
| | | Fall | **0.763** | **0.845** | 0.502 |
| | | Spring | 0.565 | - | |
| Nueces | 0.647*** | Summer | **0.802** | 0.627 | - |
| | | Fall | **0.700** | **0.873** | 0.450 |
| | | Spring | 0.493 | - | |
| Upper Laguna Madre | 0.516*** | Summer | **0.651** | 0.420 | - |
| | | Fall | 0.515 | **0.659** | 0.381 |
| | | Spring | 0.382 | - | |
| Lower Laguna Madre | 0.532*** | Summer | **0.609** | **0.572** | - |
| | | Fall | **0.629** | **0.764** | 0.302 |

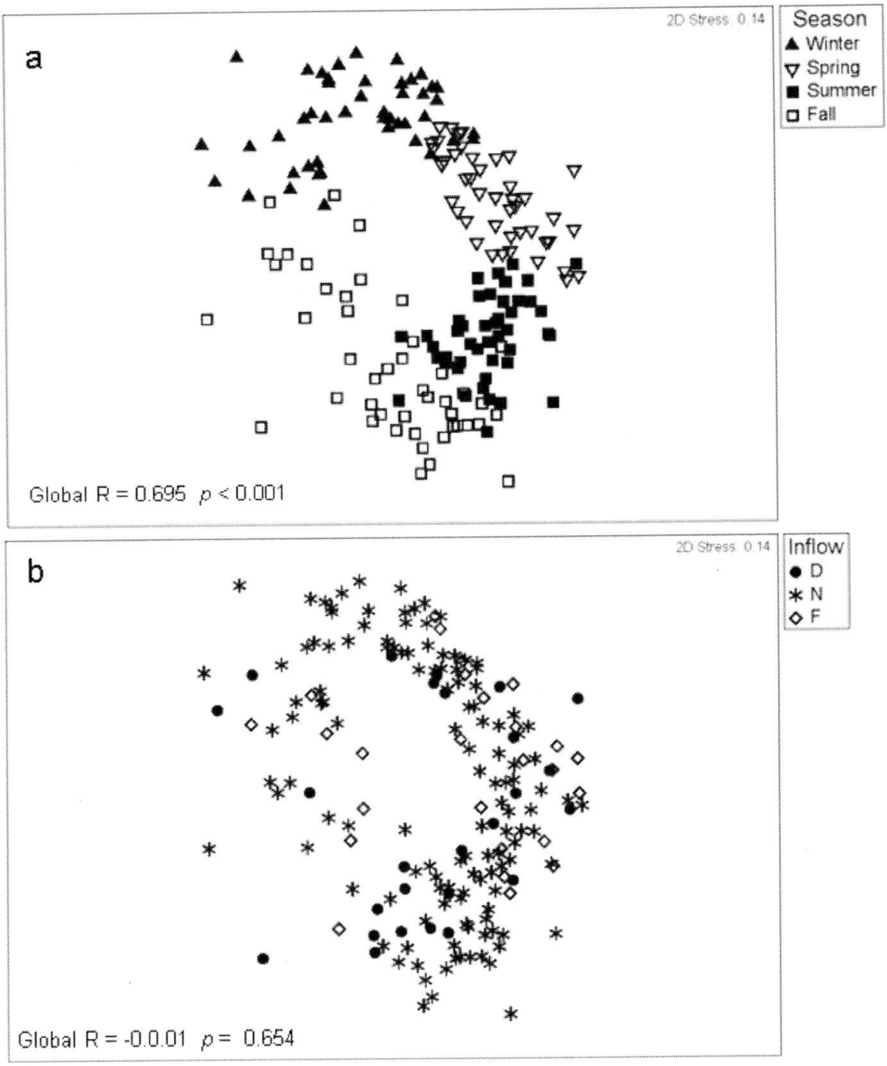


**FIGURE 3.** Multidimensional scaling (2D) configuration of bag seine community structure from a representative estuary (Lavaca-Colorado) overlaid with (a) Season, and (b) Inflow Condition. Season of collection defined as: Winter (Dec, Jan, Feb); Spring (Mar, Apr, May); Summer (Jun, Jul, Aug); and Fall (Sep, Oct, Nov). Inflow Condition designations; D = Drought, N = Normal, F = Flood. Global R values for each Analysis of Similarity test included.

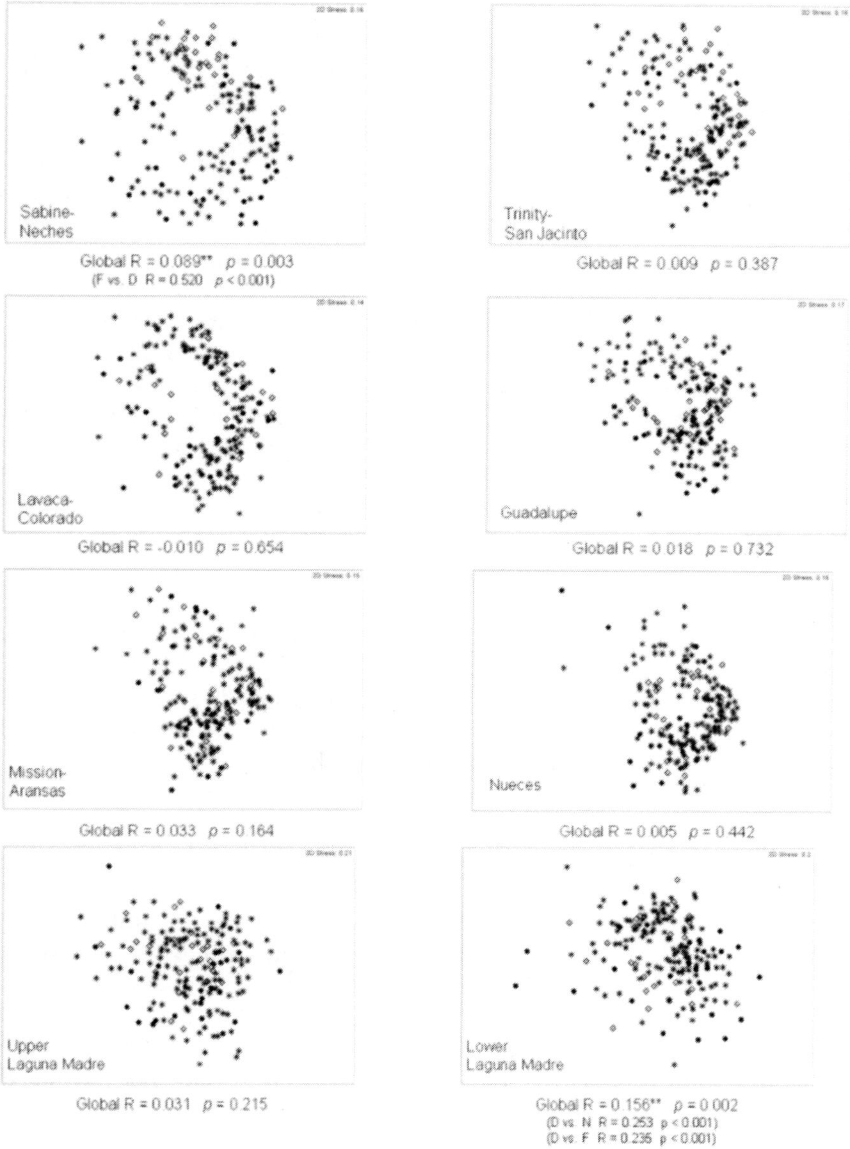


**FIGURE 4.** Multidimensional scaling (2D) configurations of bag seine derived community structure for each Texas estuary identifying the categorical Inflow Condition, with the Global R values for the seasonal Analysis of Similarity test. Inflow Condition symbols follow Figure 3.

salinities. The Lower Laguna Madre (mean salinity 32) also had significantly different community compositions during drought conditions (Drought vs. Normal comparison, R = 0.253, $p < 0.001$; Drought vs. Flood comparison, R = 0.235, $p < 0.001$), although the elevated salinities in this estuary during drought conditions (mean salinity > 40) led to lower abundances of some of these same taxa. Substantial decreases in brown shrimp (5 fold), white shrimp (15 fold), and Atlantic croaker *Micropogonias undulatus* (7 fold), as well as lower abundances of rainwater killifish *Lucania parva* (4 fold decrease) and red drum (2 fold decrease) were noted during extended low inflow conditions.While estuary-specific assemblages do not appear to be responding to synoptic inflow events, the inclusive role of salinity on overall community structure can be seen in Table 4. Across the estuaries, patterns of community structure roughly matched the NE-SW salinity gradient present on the Texas coast, with the freshest estuaries on the upper coast having significantly different communities than the more saline estuaries found on the lower coast. The middle coast estuaries (Lavaca-Colorado, Guadalupe, and Mission-Aransas) showed the greatest degree of overlap in their community structure. Of the hundreds of taxa that constitute the nekton recorded with the bag seines, differences among the estuaries could be explained by examining only a fraction of this total. Abundance levels of 36 taxa accounted for the vast majority of the individuals found in each estuary, ranging from a low of 99.5 % in the Lavaca-Colorado system to a high of 100% in the Nueces Estuary (Table 5). Of the commercially and recreationally important species currently used for TxEMP modeling, only blue crab, white shrimp, and brown shrimp contributed substantially to nekton community structure patterns. Other taxa used for TxEMP either contributed little to the overall community (e.g., red drum ranked no higher than $12^{th}$ from any estuary; spotted seatrout no higher than $14^{th}$) or were identified as a characteristic component from only a single estuary (southern flounder contributed to at least 1% of the community structure only in the Sabine-Neches estuary). Black drum were not identified as a significant component from any estuary. Community structure across the salinity gradient of estuaries appears to be driven by the relative proportion of only a few oligohaline (e.g., Atlantic croaker, bay anchovy *Anchoa mitchilli*, and Gulf menhaden *Brevoortia patronus*) and polyhaline to metahaline taxa (pinfish, Gulf killifish *Fundulus grandis*, sheepshead minnow, longnose killifish *F. similis*, and grass shrimp *Palaemonetes* spp.). Ubiquitous euryhaline taxa that were identified at equivalent ranks across the coastal salinity gradient included blue crab, striped mullet *Mugil cephalus*, spot *Leiostomus xanthurus*, brown shrimp, white shrimp, and silversides *Menidia* spp.

**TABLE 4.** Matrix of pairwise comparison R values for the Analysis of Similarity test of community structure among estuaries. Global R = 0.455, $p < 0.001$; significant pairwise R values > than the Global R in bold.

| Estuary | Sabine-Neches | Trinity-San Jacinto | Lavaca-Colorado | Guadalupe | Mission-Aransas | Nueces | Upper Laguna Madre |
|---|---|---|---|---|---|---|---|
| Trinity-San Jacinto | 0.299 | - | | | | | |
| Lavaca-Colorado | 0.324 | 0.081 | - | | | | |
| Guadalupe | **0.686** | 0.423 | 0.347 | - | | | |
| Mission-Aransas | **0.555** | 0.283 | 0.213 | 0.295 | - | | |
| Nueces | **0.710** | 0.371 | 0.335 | 0.418 | 0.116 | - | |
| Upper Laguna Madre | **0.875** | **0.806** | **0.727** | 0.403 | **0.579** | **0.679** | - |
| Lower Laguna Madre | **0.784** | **0.578** | **0.527** | 0.324 | 0.381 | 0.336 | **0.485** |

**TABLE 5.** Rank order of the nekton taxa contributing to the top 90% of average similarity within each estuarine assemblage. Species identified by an asterisk (*) represent commercially or recreationally important target species currently used in TxEMP modeling. Blank entries represent taxa recorded from each estuary, but their overall contribution to the community in that estuary is less than 1%.

| Species | Sabine-Neches | Trinity-San Jacinto | Lavaca-Colorado | Guadalupe | Mission-Aransas | Nueces | Upper Laguna Madre | Lower Laguna Madre |
|---|---|---|---|---|---|---|---|---|
| *Micropogonias undulatus* | 1 | 2 | 4 | 12 | 13 | 17 | | 14 |
| *Callinectes sapidus** | 2 | 1 | 2 | 5 | 3 | 2 | 7 | 5 |
| *Anchoa mitchilli* | 3 | 5 | 10 | 11 | 11 | 13 | 8 | 17 |
| *Brevoortia patronus* | 4 | 3 | 11 | 16 | | | | |

| | | | | | | | | |
|---|---|---|---|---|---|---|---|---|
| Menidia beryllina/ peninsulae | 5 | 4 | 3 | 2 | 2 | 5 | 2 | 8 |
| *Mugil cephalus* | 6 | 6 | | 8 | 10 | 10 | 10 | 9 |
| *Litopenaeus setiferus** | 7 | 10 | 6 | 15 | 12 | 14 | | 13 |
| *Palaemonetes spp.* | 8 | 7 | 1 | 1 | 1 | 6 | 3 | 12 |
| *Leiostomus xanthurus* | 9 | 8 | 9 | 9 | 8 | 7 | 11 | 6 |
| *Farfantepenaeus aztecus** | 10 | 9 | 5 | 6 | 7 | 8 | 9 | 3 |
| *Lagodon rhomboides* | 11 | 12 | 8 | 3 | 4 | 3 | 6 | 1 |
| *Sciaenops ocellatus** | 12 | 16 | 15 | 14 | 14 | 16 | 16 | 18 |
| *Paralichthys lethostigma** | 13 | | | | | | | |
| *Cynoscion arenarius* | 14 | 17 | 18 | | | | | |
| *Fundulus similis* | | 13 | 7 | 7 | 6 | 4 | 4 | 4 |
| *Fundulus grandis* | | 14 | 13 | 10 | 9 | 9 | 5 | 15 |
| *Cynoscion nebulosus** | | | | 17 | 17 | 18 | 14 | |
| *Gobiosoma bosc* | | | | 18 | | 19 | | |
| *Eucinostomus argenteus* | | | | 19 | 18 | | | 16 |
| *Cyprinodon variegatus* | | 11 | 12 | 4 | 5 | 1 | 1 | 2 |
| *Mugil curema* | | 15 | 14 | 13 | 15 | 15 | 15 | 11 |
| *Citharichthys spilopterus* | | 18 | 19 | | | | | |
| *Callinectes similis* | | 19 | | | | 12 | | 10 |
| *Menticirrhus americanus* | | 20 | 16 | | | | | |

| | | | | | | | | |
|---|---|---|---|---|---|---|---|---|
| *Arius felis* | | 21 | 17 | | | | | |
| *Lucania parva* | | | | | | | 12 | |
| *Farfantepenaeus duorarum* | | | | | 16 | 11 | | 7 |
| *Syngnathus scovelli* | | | | | | | 13 | |
| *Percent Total Abundance* | 99.7 | 99.7 | 99.6 | 99.5 | 99.8 | 100 | 99.8 | 99.7 |

# 4. DISCUSSION

Long-term data sets are fundamental to an understanding of factors that regulate system level processes, because the inherent complexity and variability of open natural systems make it difficult to establish causal relationships between and among the important components. These data are needed to ensure that the environmental conditions which potentially can lead to dramatic fluctuations in observed nekton abundance levels are recorded at least once and preferably several times [24]. Decadal-scale continuous records of biological data utilizing uniform sampling strategies are the exception rather than the rule for most estuarine and coastal realms [25]. Many of the estuarine studies that do take into account the spatial and temporal aspects of the physical environment often utilize commercial catch and effort records [26-30], and these catch per unit effort (CPUE) indices of abundance are not without problems. Technological advances and external economic factors which can directly affect actual effort are either poorly documented, or often entirely dismissed. Circumventing some of the inherent problems associated with fishery-dependent CPUE indices, TPWD utilizes fisheries-independent sampling methodologies to assemble long-term data sets of estuarine biotic and abiotic structure [21]. Besides providing resource managers with uniform information that is reliably documented and collected under standardized sampling designs and techniques, these long-term data sets offer the antithesis to short-term management solutions dictated by monetary constraints that emphasize research and monitoring projects of limited temporal and spatial duration [31].

The current TxEMP methodology which uses salinity as a proxy for FWI to establish inflow–species spatial relationships has demonstrated varying levels of correspondence between abundance and salinity gradients [14, 32-33]. In estuaries receiving substantial inflows that facilitate defined salinity zonations (e.g., Trinity-San Jacinto and Guadalupe), peak densities of many target species were spatially correlated with specific salinity zones. Conversely, in estuaries receiving lower amounts of inflows (e.g., Nueces and Laguna Madre), well defined salinity zonations were either dramatically compressed into the upper-most reaches of the estuary, or absent altogether, and consequently these same target species were far less associated with their recognized salinity preferences. Expanding the spatial scale beyond individual estuaries and using the bag seine information to encompass the entire nekton community, the present analysis shows that a much lower degree of correspondence exists between the synoptic-scale FWI signal and community assemblage. While this general lack of correspondence between the motile nekton and FWI may seem contradictory to the reported positive flow effects on fisheries abundance [29-30, 34-35], similar neutral responses by fisheries to FWI have been reported in other studies conducted at equivalent spatial scales as was used for this study. For example, in East Bay, Florida, Livingston et al. [2] found that river flow and primary production were associated mainly with changes in the communities at the lower trophic levels (herbivores and omnivore), whereas the carnivores (e.g., spotted seatrout, southern flounder, and red drum) were associated primarily with other animal trophic interactions. Their study showed that salinity changes were only indirectly involved in biological interactions at the highest trophic levels. Similarly, Griffiths [36] showed that yellowfin bream *Acanthopagrus australis* (the functional equivalent of pinfish used for this study) and striped mullet were generally resilient to salinity perturbations in Shellharbour Lagoon, Australia. Neutral responses to fluctuating salinities is not exclusive to finfish, as both Kimmerer [37] reporting on California bay shrimp *Crangon franciscorum* (functional equivalent of white shrimp) in the northern San Francisco Estuary, California, and Rozas et al. [11] working with brown shrimp in Breton Sound, Louisiana, each showed a general de-coupling of abundance levels and hydrologic conditions. Similarly, increases in temporal scales have also revealed a general de-coupling between abundance levels and FWI, as both spot and Atlantic croaker did not correlate with year-to-year variation of river discharge in Apalachicola Bay, Florida [38]. Weinstein et al. [39] also showed that shallow-water fish assemblages in the Cape Fear River estuary were not affected by annual differences in river discharge.

The repeatability of species assemblage composition and abundances from year to year across the salinity spectrum in the estuarine systems along the Texas coast is one of the most prominent features of this study. From Figure 4, it is clear that an orderly seasonal succession in abundance and species composition of the dominant components confirms the many published accounts of annually repeating community structure from a variety of locations [40-43]. A common theme found within these studies is that the identification of quite specific arrival times, or dates of first occurrence within each season of recruitment, can be shown for a number of taxa, regardless of the hydrologic conditions within an estuary at the time of recruitment. Interannual variations in these dates of first occurrence are typically small, suggesting that temporal stability of assemblage structure may be more closely related to temperature [40] or seasonal photoperiods than to salinity. The current analysis shows that the greatest disparity in community composition, regardless of any underlying salinity level difference, involved comparisons across opposite seasons. These seasonal differences were steadfastly replicated year after year, in spite of the dramatically different levels of freshwater inflows producing temporally unpredictable flood and drought conditions. During these environmental extremes, no wholesale changes in community composition were noted; only changes in the relative abundances within a set of common taxa.

Absent from the current analysis is a recognition of the role of physical habitat in structuring nekton community compositions (reviewed in [44]). From Table 1, it is quite clear that major differences in the areal extent of fringing wetlands and submerged aquatic vegetation exists among the eight estuaries under investigation, and despite these obvious differences, the major nekton components of each community assemblage are, for the most part, the same limited suite of taxa (Table 5). Many estuarine organisms have increased (sometimes dramatically) abundances in areas closest to the freshwater source, and these same oligo- and mesohaline areas are noted for supporting much of the wetland habitats cited in Table 1. Even though the direct effect that FWI has on wetlands and the species that use them has not been definitely demonstrated [12], there is very good evidence that these relationships exist, at least for some size-groups or life history stages [7, 18, 36, 45]. Transient groups of young-of-the-year clupeiforms (Gulf menhaden, bay anchovy), perciforms (Atlantic croaker, spot, red drum, pinfish, spotted seatrout, and both species of mullets) and pleuronectiforms (southern flounder, bay whiff) that are spawned in deeper estuarine, nearshore, or offshore waters have all been shown to enter the shallow portions of estuaries and occur in very high densities in their recruitment and residence periods [17-19, 34,47]. Those nekton communities from the fresher, upper coast estuaries with large amounts of surrounding wetlands supported greater proportions of Atlantic croaker, bay anchovy, Gulf menhaden, and striped mullet than did the more southern estuaries with less fringing marsh. In the more saline estuaries, where seagrasses generally replace fringing marsh systems as one of the dominant structured habitats, the nekton communities identified by the bag seines were characterized by increases in cyprinidontiforms (sheepshead minnows and longnose killifish), grass shrimp, pinfish. Except for pinfish, all these taxa are estuarine-residents that do not recruit from nearshore or offshore spawning grounds, therefore the intermediate linkage between freshwater inflows and physical habitats may not be as important for these populations to be successful.

Many investigations have suggested that variability in estuarine production can be attributed either directly or indirectly to the fertilizing effects of freshwater input [1, 10, 48-49]. This estuarine 'agricultural model' is based on a mechanistic link between nutrient loading and increased phytoplankton production, ultimately leading to increased fisheries yields. Although Sutcliffe's [48] arguments have been disputed on interpretation and statistical grounds [46, 50], the concept persists and is fundamental to the implementation of the TxEMP methodology. Relating the flow effects to animal populations requires trophic transfer up the food web, and numerous studies have focused on the relative importance of 'top-down' vs. 'bottom-up' control of aquatic food webs [51-54]. A distinct dichotomy in the response of FWI controlling factors within estuarine systems appears to be that the herbivores and omnivores are more directly linked to physical and chemical controls associated implicitly with primary production ('bottom-up' regulation), whereas the carnivores (primary, secondary, and tertiary) are more closely associated with 'top-down' biological factors [2, 38]. Examination of the target taxa used for the TxEMP inflow-species relationships (Table 5) shows that all the vertebrates fall into the tertiary carnivore class, and the epibenthic macroinvertebrates are classified by as primary and secondary carnivores [2]. Omnivorous taxa that are more likely to have a more direct trophic linkage to the effects of FWI included pinfish, spot, striped mullet, white mullet, hardhead catfish *Arius felis*, and Gulf pipefish *Syngnathus scovelli*. Spot, a bottom-feeding perciform characteristic of the assemblage structure in every estuary, are potentially an important linkage between inflows and production because of their ability to regulate benthic invertebrates [55]. The connection between the benthos and FWI associated production appears show a much stronger mechanistic link [38, 56-57], although the benthic environment is represented only by eastern oyster within the current modeling paradigm. Thus, evaluating the biological effects of FWI within Texas estuaries is currently dependent upon taxa that empirically have been shown to display the weakest mechanistic couplings.

The time steps involved in TxEMP FWI modeling are on the order of a calendar month, whereas the nekton appear to be operating on the order of months (the seasonal signal was clearly evident in every estuary across the coast) to a calendar year (the repeating pattern of seasonal compositions common to each estuary resulted in the circular configuration of the samples seen in Figure 3a). Conversely, the drought and flood FWI signal common to the entire Texas coast is due by climate-level drivers operating at fundamental frequencies of approximately 11, 5, and 3.5 years [58]. While all of the commercially and recreationally important finfish used to quantify optimal FWI are long lived species and can contribute a number of different year-classes to the nekton community, the macroinvertebrates used by TxEMP all have life spans less than even the shortest frequency inflow signal driver. The shrimp species are all considered annual species [15], with maximum life spans from 18 months to 2 years, whereas blue crabs are reported to have a life span approaching 3 years. For the shrimp species abundant throughout the Texas coast, physical timing of their recruitment periods appear to be more in synchrony with species-specific temperature ranges instead of estuarine salinity requirements. Brown shrimp recruit to the estuary from February through May, while white shrimp typically show up from June through October. Once recruited from offshore spawning areas, white shrimp juveniles can migrate farther into the less saline waters of the upper estuary because they are more tolerant of lower salinities than the other shrimp species. This pattern is evidenced by the higher rank abundance values for white shrimp seen in the less saline upper coast estuaries, whereas rank values for brown shrimp were higher in the more saline lower coast estuaries. While the interaction of available habitat and salinity tolerance levels can therefore aid in successful recruitment, the annual frequency in shrimp spawning does not appear to be closely tied to the multiyear to decadal frequencies of the inflows. These observations conform to the conclusions of Allen and Barker [19], in that 'responses of populations to major changes in the estuarine environment are more strongly expressed as alterations in the magnitude than in the timing of habitat utilization'.

Because estuarine fishes have evolved to exploit one of the more physiologically challenging environments, it should not too surprising that they do not appear to be dramatically responding to the synoptic-scale inflow events that are currently used to quantify 'ecological health'. Even when utilizing species ranks to adjust for any gross differences in relative abundance among estuaries (e.g., biomass), the present analysis reveals that in each estuary, the contributions of a very limited number of taxa were strikingly similar. To evaluate the functional role of freshwater inflow into estuaries and determine estuarine FWI needs for the future, incorporating more sensitive 'measuring stick' organisms are recommended. One way this could be accomplished is to incorporate a greater range of trophic structure, utilizing some of the lower trophic level taxa that constituted a majority portion of the community assemblage (e.g., Gulf menhaden, bay anchovy, Gulf killifish, striped mullet, sheepshead minnow), or base the FWI modeling on taxa that appear to show a definite salinity response (e.g., Atlantic croaker, longnose killifish, white mullet, pinfish). Still another more challenging option would be to move down the trophic food web and index measures of 'estuarine health' to benthic taxa that show more direct mechanistic linkages to FWI.

# ACKNOWLEDGEMENTS

I am sincerely indebted to the more than twenty years of field staff and technicians at the Coastal Fisheries Division of Texas Parks and Wildlife Department for their diligent collection of the biotic and abiotic parameters used for this study. This project was never explicitly funded by research grants, but I gratefully acknowledge the continued support of Sportfish Restoration Funds, without which the time for data synthesis and interpretation would not have been possible.

# REFERENCES

1. Flint, R.W. Long-term estuarine variability and associated biological response. Estuaries 1985;8 158-169.
2. Livingston, R.J., N. Xufeng, F.G. Lewis, III, and G.C. Woodsum. Freshwater input to a Gulf estuary: long-term control of trophic organization. Ecological Applications 1997;7 277-299.
3. Logeragan, N.R., and S.E. Bunn. River flows and estuarine ecosystems: implications for coastal fisheries from a review and case study of the Logan River, southeast Australia. Australian Journal of Ecology 1999;24 431-440.
4. Alber, M. A conceptual model of estuarine freshwater inflow management. Estuaries 2002;25 1246-1261.
5. Vörösmarty, C.J., and D. Sahagian. Anthropogenic disturbance of the terrestrial water cycle. BioScience 2000;50 753-765.
6. Dynesius, M., and C. Nilsson. Fragmentation and flow regulation of river systems in the northern third of the world. Science 1994;266 753-762.
7. Sklar, F.H., and J.A. Browder. Coastal environmental impacts brought about by alterations to freshwater flow in the Gulf of Mexico. Environmental Management 1998;22 547-562.
8. Rijsberman, F.R. Water scarcity: Fact or fiction? Agricultural Water Management 2006;80 5-22.
9. Skreslet, S. Freshwater outflow in relation to space and time dimensions of complex ecological interactions in coastal waters. In: S. Skreslet (ed.) The role of freshwater outflow in coastal marine ecosystems. Berlin, Germany: Springer-Verlag; 1986. p3-12.
10. Mallin, M.A., H.W Paerl, J. Rudek, and P.W. Bates. Regulation of estuarine primary productivity by watershed rainfall and river flow. Marine Ecology Progress Series 1993;93 199-203.
11. Rozas, L.P., T.J. Minello, I. Munuera-Fernandez, B. Fry, and B. Wissel. 2005. Macrofaunal distributions and habitat change following winter-spring releases of freshwater into the Breton Sound estuary, Louisiana (USA). Estuarine, Coastal and Shelf Science 65:319-336.
12. Longley, W.L., editor. Freshwater inflows to Texas bays and estuaries: Ecological relationships and methods for determination of needs. Texas Water Development Board and Texas Parks and Wildlife Department, Austin, Texas. 1994.
13. Powell, G.L., and J. Matsumoto. Texas estuarine mathematical programming model: a tool for freshwater inflow management. In: K.R. Dyer and R.J. Orth (eds.) Changes and fluxes in estuaries. Fredensborg, Denmark: Olsen and Olsen; 1994. p401-406.

14. Pulich, W. Jr, J. Tolan, W.Y. Lee, and W. Alvis. Freshwater inflow recommendation for the Nueces Estuary. http://www.tpwd.state.tx.us/l andwater/water /conservation/freshwater_inflow/nueces/ (accessed 20 Feb 2010).
15. Patillo, M.E., T.E. Czapla, D.M. Nelson, and M.E. Monaco. Distribution and abundance of fishes and invertebrates in Gulf of Mexico estuaries, Volume II: Species life history summaries. ELMR Report Number 11. NOAA/NOS Strategic Assessment Division, Silver Spring, MD. 1997.
16. Able, K.W. A re-examination of fish estuarine dependence: Evidence for connectivity between estuarine and ocean habitats. Estuarine, Coastal and Shelf Science 2005;64 5-17.
17. Weinstein, M.P., and H.A. Brooks. Comparative ecology of nekton residing in a tidal creek and adjacent seagrass meadow: community composition and structure. Marine Ecology Progress Series 1983;12 15-27.
18. Rogers, S.G., T.E Targett, and S.B. Van Sant. Fish-nursery use in Georgia salt-marsh estuaries: the influence of springtime freshwater conditions. Transactions of the American Fisheries Society 1984;113 595-606.
19. Allen, D.M., and D.L. Barker. Interannual variation in larval fish recruitment to estuarine epibenthic habitats. Marine Ecology Progress Series 1990;63 113-125.
20. Hare, J.A., and K.W. Able. Mechanistic links between climate and fisheries along the east coast of the United States: explaining population outbursts of Atlantic croaker (Micropogonias undulatus). Fisheries Oceanography 2007;16 31-45.
21. Martinez-Andrade, F., P. Campbell, and B. Fuls. Trends in relative abundance and size of selected finfishes and shellfishes along the Texas Coast: November 1975-December 2003. Management Data Series No. 232, Texas Parks and Wildlife Department, Coastal Fisheries Division. Austin, Texas. 2005.
22. McKee, D.A. Fishes of the Laguna Madre: A guide for anglers and naturalists. College Station, Texas: Texas A&M University Press. 2008.
23. Clarke, K.R., and R.M. Warwick. Change in marine communities: an approach to statistical analysis and interpretation, 2nd edition. Plymouth : PRIMER-E. 2001.
24. Kruskal, J.B. Multidimensional scaling by optimizing goodness of fit to a non-metric hypothesis. Psychometrika 1964;29 1-27.
25. Rose, K.A., and J.K. Summers. Relationships among long-term fisheries abundances, hydrographic variables, and gross pollution indicators in northeastern U.S. estuaries. Fisheries Oceanography 1992;1 281-293.
26. Wolfe, D.A., M.A. Champ, D.A Flemer, and A.J. Mearns. Long-term biological data sets: their role in research, monitoring, and management of estuarine and coastal marine systems. Estuaries 1987;10 181-193.
27. Summers, J.K., T.T. Polgar, J.A. Tarr, K.A. Rose, D.G. Heimbuch, J. McCurley, R.A. Cummins, G.F. Johnson, K.T. Yetman, and G.T. DiNardo. Reconstruction of long-term time series for commercial fisheries abundance and estuarine pollution loadings. Estuaries 1985;8 114-124.
28. Pearson, T.H., and P.R.O. Barnett. Long-term changes in benthic populations in some west European coastal areas. Estuaries 1987;10 220-226.
29. Houde, E.D., and E.S. Rutherford. Recent trends in estuaries fisheries-predictions of fish production and yield. Estuaries 1993;16 161-176.
30. Jassby, A.D., W.J. Kimmerer, S.G. Monismith, C. Armor, J.E. Cloern, T.M. Powell, J.R. Schubel, and T.J. Vendlinski. Isohaline position as a habitat indicator for estuarine populations. Ecological Applications 1995;5 272-289.
31. Diop, H., W.R. Keithly, Jr., R.F. Kazmierczak, Jr., and R.F. Shaw. Predicting the abundance of white shrimp (Litopenaeus setiferus) from environmental parameters and previous life stage. Fisheries Research 2007;86 31-41.
32. Champ, M.A. Monitoring: Painting a moving train. Sea Technology 1986;27 73.
33. Pulich, W. Jr, J. Tolan, W.Y. Lee, and W. Alvis. Freshwater inflow recommendation for the Nueces Estuary. http://www.tpwd.state.tx.us/landwater/ water/conserv ation/freshwater_inflow/nueces/ (accessed 20 Feb 2010).

34. Tolan, J.M., W.Y. Lee, G. Chen, and D. Buzan. Freshwater inflow recommendation for the Laguna Madre Estuary system. Texas Parks and Wildlife Department. Austin, Texas. 2004.
35. Copeland, B.J. Effects of decreased river flow on estuarine ecology. Journal of the Water Pollution Control Federation 1966;38 1831-1839.
36. Weinstein, M.P., and M.P. Walters. Growth, survival and production in young-of-year populations of Leiostomous xanthurus Lacépède residing in tidal creeks. Estuaries 1981;4 185-197.
37. Griffiths, S.P. Factors influencing fish composition in an Australian intermittently open estuary. Is stability salinity-dependent? Estuarine, Coastal and Shelf Science 2002;52 739-751.
38. Kimmerer, W.J. Effects of freshwater flow on abundance of estuarine organisms: physical effects or trophic linkages? Marine Ecology Progress Series 2002;243 39-55.
39. Kobylinski, G.J., and P.F. Sheridan. Distribution, abundance, feeding and long-term fluctuations of spot, Leiostomus xanthurus, and croaker, Micropogonias undulates, in Apalachicola Bay, Florida, 1972-1977. Contributions in Marine Science 1971;22 149-161.
40. Weinstein, M.P., S.L. Weiss, and W.F. Walters. Multiple determinants of community structure in shallow marsh habitats, Cape Fear River estuary, North Carolina, USA. Marine Biology 1980;58 227-243
41. McGovern, J.C., and C.A. Wenner. Seasonal recruitment of larval and juvenile fishes into impounded and non-impounded marshes. Wetlands 1990;10 203-221.
42. Witting, D.A., K.W. Able, and M.P. Fahay. Larval fishes of a Middle Atlantic Bight estuary: assemblage structure and temporal stability. Canadian Journal of Fisheries and Aquatic Sciences 1999;56 222-230.
43. Hagan, S.M., and K.W. Able. Seasonal changes of the pelagic fish assemblage in a temperate estuary. Estuarine, Coastal and Shelf Science 2003;56 15-29.
44. Tolan, J.M. Larval fish assemblage response to freshwater inflow: a synthesis of five years of ichthyoplankton monitoring within Nueces Bay, Texas. Bulletin of Marine Science 2008;82 275-296.
45. Minello, T.J. Nekton densities in shallow estuarine habitats of Texas and Louisiana and the identification of essential fish habitat. In L. Benaka (ed.) Fish habitat: Essential fish habitat and habitat restoration. American Fisheries Society, Symposium 22. Bethesda, MD. 1999. p43-75.
46. Sinclair, M., G.L. Bugden, C.L. Tang, J.C. Therriault, and P.A. Yeats. Assessment of effects of freshwater runoff variability on fisheries production in coastal waters. In S. Skreslet (ed.) The role of freshwater outflow in coastal marine ecosystems. Berlin, Germany: Springer-Verlag. 1986. p139-160.
47. Martino, E.J., and K.W. Able. Fish assemblage across the marine to low salinity transition zone of a temperate estuary. Estuarine, Coastal and Shelf Science 2003;56 969-987.
48. Sutcliffe, W.H., Jr. Some relation of land drainage, particulate matter, and fish catch in to eastern Canadian bays. Journal of the Fisheries Research Board of Canada 1972;29 357-362.
49. Cloern, J.E., A.E. Alpine, B.E. Cole, R.L.J Wong, J.F. Arthur, and M.D. Ball. 1983. River discharge controls phytoplankton dynamics in the northern San Francisco Bay estuary. Estuarine, Coastal and Shelf Science 1983;16 415-429.
50. Drinkwater, K.F., and R.A. Myers. Testing predictions of marine fish and shellfish landings from environmental variables. Canadian Journal of Fisheries and Aquatic Sciences 1987;44 1568-1573.
51. McQueen, D.J., R.S. Johannes, J.R. Post, T.J. Stewart, and D.R.S. Lean. Bottom-up and top-down impacts on freshwater pelagic community structure. Ecological Monographs 1989;59 289-309.
52. Menge, B.A. Community regulation: under what conditions are bottom-up factors important on rocky shores. Ecology 1992;73 755-765.

53. Flinkman, J., E. Aro, I. Vuorinen, and M. Viitasalo. Changes in northern Baltic zooplankton and herring nutrition from 1980s to 1990s: top-down and bottom-up processes at work. Marine Ecology Progress Series 1998;165 127-136.
54. Micheli, F. Eutrophication, fisheries, and consumer-resource dynamics in marine pelagic ecosystems. Science 1999;285 1396-1398.
55. Killam, K.A., R.J. Hochberg, and E.C. Rzemiem. Synthesis of basic life histories of Tampa Bay species. Tampa Bay National Estuary Program, Technical Publication Number 10-92. 1992.
56. Nilsson, P., B. Jönsson, I.L. Swanberg, and K. Sundbäck. Response of a marine shallow-water sediment system to an increased load of inorganic nutrients. Marine Ecology Progress Series 1991;71 275-290.
57. Montagna, P.A., and R.D. Kalke. The effect of freshwater inflow on meiofaunal and macrofaunal populations in the Guadalupe and Nueces Estuaries. Estuaries 1992;15 307-326.
58. Tolan, J.M. El Nino-Southern Oscillation impacts translated to the watershed scale: salinity patterns along the Texas Gulf Coast, 1982 to 2004. Estuarine, Coastal and Shelf Science 2007;72 247-260.

# CHAPTER 6

# Patterns and Processes in The Drift of Early Developmental Stages of Fish in Rivers: A Review

**Aaron Lechner . Hubert Keckeis . Paul Humphries**

*Department of Limnology and Oceanography, Faculty of Life Sciences, University of Vienna, Althanstrasse 14, 1090 Vienna, Austria*

*School of Environmental Sciences, Charles Sturt University, PO Box 789, Albury, NSW 2640, Australia*

## ABSTRACT

Current-mediated downstream dispersal by the early developmental stages of fish in rivers is a common phenomenon. Knowledge of patterns and processes in the dispersal, or 'drift', of young fishes provides important information on spawning location and spawning success, habitat use, movement paths and flow-ecology relationships more generally, all of which are critical for effective river conservation and management. But despite the importance of such information, our understanding of the patterns and processes of the drift of the early life stages of riverine fishes is limited. Furthermore, riverine fish drift research has tended to occur in isolation from movement studies of other organisms, limiting its integration with higher level concepts and theory. This manuscript reviews the literature on the dispersal of young fishes in running waters. Relevant studies from all climatic zones and geographical regions are investigated, with particular attention given to the types and life history stages of fishes that drift and the seasonal and diel patterns of drifting. We then consider how fish enter the drift and their mode of drifting, attempting to reconcile a long-running discussion, under what we call the 'active–passive conundrum'. We argue that, aside from eggs, the early stages of fish are not exclusively either passive or active drifters, but usually a mixture of the two, which we term 'actipassive' drift. Finally, we evaluate existing knowledge in the context of a general conceptual framework for movement ecology, identifying gaps in our understanding of the roles of internal state, navigation capacity, motion capacity, external factors and internal factors in influencing the dispersal process.

**Keywords:** Fish larvae Free embryos Drift Dispersal Rivers

## INTRODUCTION

Placing a fine-mesh net in a river in spring or early summer will usually catch the eggs, free embryos and larvae of fishes moving downstream, sometimes in vast numbers. These early developmental stages of fishes are participating in the first mass movement of their lives, dispersing from spawning or nursery sites to what are presumed appropriate rearing habitats (Pavlov 1994). This group includes individuals that seem deliberately to use the current to help them move (Braaten et al. 2012), and others that are apparently accidentally entrained (Harvey 1987; Wolter and Sukhodolov 2008). Whatever the mechanism, current-mediated dispersal—or 'drift'—of the young stages of fish is an integral part of the life cycle of many riverine species (Muth and Schmulbach 1984; Bujold et al. 2004; Humphries 2005). It enhances habitat connectivity and gene flow, promotes community stability and extends the range of populations (Janac et al. 2013; Roberts et al. 2013).

While drifting, eggs, free embryos and larvae are relatively simple to collect, and these collections can provide much information on the dynamics of early life. This information can elucidate patterns and processes associated with spawning and larval production (Zitek et al. 2004a; King et al. 2005; Braaten et al. 2010; Borcherding et al. 2014), it may allow forecasting of future year-class-strength (Johnston et al. 1995) and provide estimates of stock size of spawning adults (Usvyatsov et al. 2013). It also assists in understanding how fishes and their environment—especially the flow environment–interact. This is vital for effective river management, because the presence of drifting fishes provides evidence of in situ reproduction of native and alien species, common and rare species, and species important for recreational and commercial fisheries (Humphries and Lake 2000; Jiang et al. 2010).

This sort of information is also critical if we are to assess, and perhaps counter, the impact of flow alteration and other anthropogenic disturbances (Pavlov et al. 2008; Bracken and Lucas 2013; Lechner et al. 2014a). But despite the importance of such information, our understanding of the patterns and processes of the drift of the free embryos and larvae of riverine fishes is limited (Corbett and Powles 1986; Flecker et al. 1991; Schmutz and Jungwirth 1999). There are many studies from around the world that have described temporal and spatial patterns of drift of the larvae of riverine fishes (e.g. Gadomski and Barfoot 1998; Oesmann 2003; Zitek et al. 2004b). These include a range of temporal scales, from diel, daily to seasonal (e.g. de Graaf et al. 1999; Copp et al. 2002; Araujo-Lima and Oliveira 1998). But what catalyses fish larvae to drift in the first place, how far they drift, how they navigate to settlement sites, how they detect these sites, how they exit the current, how they avoid predation and other hazards *en route*, and how significant this dispersal is to population dynamics, are largely unknown (but see Johnston 1997; Robinson et al. 1998; Schludermann et al. 2012).

Here we review the literature on the patterns and processes associated with the dispersal of the free embryos larvae and juveniles of riverine fishes, with the overall aim of understanding why, how and where fish drift and how drifting fish interact with the riverine environment. We searched a range of scientific databases (Google Scholar, Web of Science, Scopus) and employed the search terms *fish larvae + drift + dispersal + rivers.*

The review is divided into three sections. In the first, we consider species–, stage–and location-specific characteristics of fish drift, and discuss drift patterns at different temporal scales. In the second section, we highlight processes associated with the downstream movement of young fish; i.e. the entry into the current and the subsequent behaviour in the flow. We discuss established classifications and introduce new approaches based upon recent insights. In the third section, the existing literature on fish drift is analysed in the context of the unifying organismal movement model of Nathan et al. (2008), by evaluating the state of knowledge relating to the *internal state*, *motion capacity* and *navigation capacity* of drifting fish. Knowledge gaps are identified and suggestions for future research, towards an integrated approach in drift studies, are made.

## WHAT TYPES OF FISHES DRIFT?

The intensity of drift, its governing mechanisms and the importance of the process in the life cycle of fishes is species-specific (Johnston et al. 1995; D'Amours et al. 2001). For example, Reichard et al. (2001) discriminated between taxa that continuously appeared in the flow of a Czech lowland river and others that were exclusively found during elevated discharge. Humphries and King (2004) classified species based on the relevance of drift to their life history. Species were categorized as *obligate-* (high abundance in drift, low abundance in other habitats), *facultative-* (occur in similar abundance in the drift and in low-flow habitats) or *non-drifters* (low abundance in the drift, drift entrance is likely accidental). The assignation to a specific group appears independent of velocity preference of the adults of particular species: some normally limnophilic species as adults (e.g. *Rhodeus sericeus,* Cyprinidae) drift as larvae and utilize the flow for dispersal (Jurajda 1998), whereas the larvae of some normally rheophilic species as adults (e.g. *Squalius cephalus, Gobio spp.*, Cyprinidae) avoid drifting (Brown and Armstrong 1985; Reichard et al. 2001).

Species-specific patterns in initial drifting patterns greatly depend on the spawning mode of the fish in question (Pavlov 1994). Pelagophilous fish release buoyant or semi-buoyant eggs that immediately enter the flow and, apart from the influence of buoyancy and shape, drift passively (Araujo-Lima and Oliveira 1998; Jiang et al. 2010; Widmer et al. 2012). Embryos and larvae successively develop in the water column during transport (Wootton 1998). By contrast, the early embryogenesis of salmonids (brood hiders) and many rheophilous cyprinids (open-substrate spawners) is characterized by an extended below-gravel period (Persat and Olivier 1995; Bardonnet 2001). Here, the first dispersers are free embryos or well-developed larvae that enter the current after the yolk sac is depleted. The influence of the spawning mode on subsequent drift events weakens as the fish grow and develop specific phenotypes and behaviours, resulting in variable drift patterns within the same reproductive guild (Pavlov 1994).

Other factors affecting patterns of drift are related to the migratory behaviour of the adults. In some amphidromous species, for example, larval drift is a continuous downstream movement, with a distinct spatial goal: the ocean (Iguchi and Mizuno 1990; Maeda and Tachihara 2010). Movement between freshwater and the sea is, in this case, an obligate part of the life history, and so it is unsurprising that drift is deliberate and controlled (Bell and Brown 1995). Wholly freshwater fishes, however, tend to show more diverse drift patterns (Pavlov et al. 1978). For example, whereas larvae of the anadromous smelt (*Osmerus eperlanus*, Osmeridae) use strong currents in the mid-channel of the River Elbe as a vehicle for swift downstream movements, potamodromous cyprinids instead drift in moderate currents along the shoreline (Oesmann 2003). Likewise, while larvae of some amphidromous sculpins (*Cottus aleuticus, Cottus asper*, Cottidae) in Smith and Van Duzen rivers, Northern California, expedite their dispersal to estuaries by drifting at elevated river flows, the potamodromous fish community here apparently has evolved life histories that minimize the risk of transport into saltwater (White and Harvey 2003).

However, comprehensive drift models for potamodromous fish are, with a few exceptions, largely absent or theoretical. Apparently, the larvae of long migratory, potamodromous sturgeon species feature a continuous drift over several days and hundreds of kilometres (see Online Resource). Drift distance and duration are highly variable among species, and may be genetically coded in order to reach distinct nursery habitats and to compensate for the long upstream migrations of the adults (D'Amours et al. 2001; Kynard and Horgan 2002; Kynard et al. 2007a, 2007b; Braaten et al. 2008, 2012). Instead of a continuous downstream movement, the early dispersal of other potamodromous species potentially includes a sequence of drifting phases (primarily during night), interspersed with periods of inshore settlement in nursery habitats (primarily during day) (Carter et al. 1986; Gadomski and Barfoot 1998). The cumulative distance covered during this pattern is not known and needs more research.

## AT WHAT LIFE HISTORY STAGE DO FISHES DRIFT?

Drifting typically occurs during defined intervals of ontogeny and at certain body lengths for particular species. A narrow size range of drifting individuals, for example, has been observed in two invasive benthic species (*Neogobius melanostomus,Proterorhinus semilunaris*, Gobiidae) in the River Dyje, Czech Republic (Janac et al. 2013), catfish alevins (*Ictalurus punctatus, Pylodictus olivaris*, Ictaluridae) in the Illinois River, Arkansas (Brown and Armstrong 1985), and particular native fishes (*Catostomus latipinnis, Catostomus discobolus*, Catostomidae; *Gila robusta, Rhinichthys osculus*, Cyprinidae) in the Colorado River, Colorado (Carter et al. 1986). The Murray cod (*Maccullochella peelii*, Percichthyidae) in the Murray River, Australia (Humphries 2005), and pumpkinseeds (*Lepomis gibbosus*, Centrarchidae) in the River Rhone, France (Copp and Cellot 1988), predominantly drift as free embryos. Cyprinids, on the other hand, mostly disperse during their early larval development (larval stages L1–L4; Copp et al. 2002; Zitek et al. 2004a; Sonny et al. 2006) or at the transition between the larval and the juvenile period (stages L6/J1; Reichard and Jurajda 2007; Lechner et al. 2014b). But the former generality, that downstream dispersal is most intensive during early embryogenesis and sharply decreases upon the achieving of the juvenile period (Pavlov 1994), does not always apply: e.g., the drift in two Czech rivers, Morava and Kyjovka, and in a bypass section of the River Rhone was dominated by juvenile cyprinids, percids and cobitids (Peňáz et al. 1992; Reichard et al. 2001).

Drifting by fishes at particular sizes and developmental stages may be an adaptation to coping with, or taking advantage of, riverine conditions.

Behavioural and physiological changes during development may enable certain stages to actively react to their environment and to temporally and spatially regulate their drift (Pavlov et al. 2008). Developmental stage, for instance, was found to correlate with change in swimming ability and habitat or diet shifts among seven drifting cyprinid species (Reichard and Jurajda 2007) and swimming ability of several Murray-Darling Basin fishes (Kopf et al. 2014). Alternatively, the size-structure of drifting cyprinids in the diel cycle might instead relate to an altered reaction to light during ontogeny. It may only be older, more-developed larvae that respond to the transition from day to night by moving into the current (Reichard et al. 2002a; Zitek et al. 2004b).

In fact, stage-specific drift patterns have also been observed in non-riverine species (*Rhodeus sericeus*, Cyprinidae; *L. gibbosus*, Centrarchidae) inhabiting running waters (Copp and Cellot 1988; Jurajda 1998), suggesting that stage-dependency could merely reflect susceptibility to entrainment in the current. Ontogenetic changes in morphology (i.e. filling of the swim bladder) and behaviour, as well as deficits (i.e. low visual acuity and swimming ability), or even improvements (i.e. enhanced swimming capacity and activity) in physical capabilities, may increase chance encounters with high currents, thereby inducing washouts and accidental drift (Harvey 1987; Persat and Olivier 1995). It must be borne in mind, however, that the stage and size distributions of larvae collected in a drift net may be an artefact of the spatial distribution of drifting larvae and how larvae are collected. For example, catch composition will be affected substantially by the spatial [e.g., distance from shore (Reichard et al. 2004), distance from hatching sites (Sonny et al. 2006) or vertical position in the water column (Pavlov et al. 1978)] and temporal [e.g., diel or seasonal (Carter and Reader 2000)] design of the sampling program or the used sampling gear (Tonkin et al. 2007).

## AT WHAT TIME OF THE YEAR DO FISHES DRIFT?

The timing of drift is directly linked to that of reproduction (Brown and Armstrong 1985). The peak in spawning- and hence drift activity is presumably adaptive and initiated by environmental cues that promise favourable conditions for the progeny (Somarakis et al. 2000). The most important factors, in this respect, are water temperature (Carter et al. 1986; Johnston 1997; D'Amours et al. 2001; Reichard et al. 2002b; Hay et al. 2008) and discharge (Johnston et al. 1995; Robinson et al. 1998; Araujo-Lima and Oliveira 1998; Auer and Baker 2002). Either because fish do not spawn or produce viable eggs and larvae (i.e. *reproductive effects*; Zitek et al. 2004a) or because larvae do not survive (i.e. *recruitment effects*;

Schiemer et al. 2001; Humphries et al. 2013), temperature and discharge not only govern the timing, but also the spawning success and hence the subsequent intensity of larval drift (Reichard et al. 2002b; Durham and Wilde 2008; Ellsworth et al. 2010b). Intra–and inter-annual variation in drift densities, for instance, may be attributed to fluctuations in water temperature and the related variability in egg mortality during incubation (Busch et al. 1975; Sonny et al. 2006). Rising discharge levels during certain periods may inundate breeding grounds, and ensure efficient oxygenation of the eggs, thereby increasing larval density and inducing seasonal drift peaks (Johnston et al. 1995; de Graaf et al. 1999; Martin and Paller 2008).

The time-density curves of larval drift are often location- or river-specific and can be informative about the composition of the resident fish assemblage (Humphries and Lake 2000; Reichard et al. 2002b). Bi-and multi-modal seasonal patterns are characteristic of taxa-rich waters and reflect time-staggered spawning and drifting of species (Jiang et al. 2010). Additionally, multiple spawning events throughout the season of frequently drifting species may contribute to these patterns (Zitek et al. 2004b; Janac et al. 2013). Unimodal drift patterns can occur when single species dominate in the drift, or peak drift densities of different species overlap in the seasonal course (Lechner et al. 2014b).

## AT WHAT TIME OF THE DAY DO FISHES DRIFT?

The pronounced circadian rhythm of drift, with peak densities during the night, is a unifying pattern that applies for virtually all fishes and fish-like organisms. For example, drifting lamprey ammocoetes (superclass: Agnatha) show a nocturnal increase in abundance (Johnston 1997; White and Harvey 2003; Bracken and Lucas 2013), as do some sturgeon species (superclass: Gnathostomata, infraclass: Chondrostei) (D'Amours et al. 2001; Kynard and Parker 2006) and many teleost fishes (superclass: Gnathostomata, infraclass: Teleostei) (Johnson and McKenna 2007; Durham and Wilde 2008; Jiang et al. 2010). Day-night ratios of drifting fish vary between 1:1.3 (Peňáz et al. 1992) and 1:190 (Reichard et al. 2004). The diurnal differences in downstream dispersal are most likely related to the light level and its influence on drift entrance (see next section). Different species and ontogenetic stages of fish respond to different levels of light, apparently resulting in time-staggered density peaks of species (dusk, night and dawn; Pavlov et al. 1978; Brown and Armstrong 1985, Janac et al. 2013), and sizes of larvae throughout the course of the night (Zitek et al. 2004b; Sonny et al. 2006). Similarly, varying light conditions not only induce intra-specific differences in drifting patterns among rivers

(Johnson and McKenna 2007), but also along the course of the same river (Iguchi and Mizuno 1991). Low illumination or water transparency during the day (e.g. high turbidity during flood events) can cause a cessation of diel periodicity in drifting (but see review by Reeves and Galat 2010). Nevertheless, the mechanisms driving diel patterns in drift are uncertain, and results far from unequivocal: for example, some drift studies have not detected temporal patterns (Muth and Schmulbach 1984; Robinson et al. 1998; Braaten et al. 2012; Borcherding et al. 2014) or the influence of water turbidity (Jurajda 1998).

## THE ACTIVE–PASSIVE-CONUNDRUM (APC)

There has been a long-standing debate about the active or passive nature of fish drift (Pavlov 1994; Reichard and Jurajda 2004; Copp et al. 2002; Schludermann et al. 2012). Entry into the current and the subsequent downstream movement may be deliberate, with the ultimate aim of efficient dispersal. Presumably, this behaviour occurs in response to particular environmental cues, and evolved to expedite the process of downstream dispersal (Robinson et al. 1998; Lechner et al. 2014b). If this is the case, drift should be considered a predominantly *active process*. The alternative explanation is that drift primarily occurs accidentally, and environmental factors, such as current and light, override the physical capabilities—swimming capacity and orientation—of small fish (Corbett and Powles 1986; Wolter and Sukhodolov 2008). In that case, drift should be considered a predominantly *passive process*.

In reality, knowledge about larval behaviour is still fragmentary (Reichard and Jurajda 2007; Pavlov et al. 2008), and many authors discuss both active and passive explanatory models; but typically they offer no resolution or integration (Gadomski and Barfoot 1998; Bardonnet 2001; Humphries and King 2004; Reichard et al. 2004; Sonny et al. 2006). An analysis of what we call the "Active–Passive-Conundrum" (APC), based on empirical studies, is outlined in the next sections. Specifically, we assess what we know about how fish larvae enter the drift and how they behave while drifting and contrast the internal and external factors involved. In our review, the work of Dmitry Pavlov is pivotal (Pavlov 1994, Pavlov et al. 1978, 1995, 2008, 2011). His concepts on the nature of fish drift are briefly summarized at the beginning of each of the following sections. We then offer a classification of drift entrance modes and suggest modifications to Pavlov´s model of active/passive drift, based upon our review of the drift literature.

## Drift entrance and its reasons

According to Pavlov (1994), drift entrance incorporates mechanisms of different orders. *First order* mechanisms enhance fish activity and distribute them in the open water. They constitute the prerequisite for drift entrance, and include behavioural responses that either can be related (*specific behavioural responses*) or unrelated (*non-specific responses*) to the current (Pavlov et al. 2008). *Second order* mechanisms inhibit rheoreaction, the fish's inherent behaviour to move against a current (Pavlov 1994). In the following, we use Pavlov's work, substantiated by other significant research in the field, to introduce three scenarios for how young fish can enter the current.

Entry into the drift can be: (a) accidental, caused by environmental conditions that exceed and inhibit the individual's ability to maintain position, and flushes them away from nursery or hatching sites (*passive drift entry*); (b) partly behavioural, when unrelated factors (e.g., feeding movements, flight responses) cause young fish in the open water to be exposed to strong currents and they are entrained (*coincidental drift entry*); or (c) deliberate, as an adaption to fluvial conditions, and relying on innate species- and stage-specific behavioural responses that are aimed at leaving or reaching specific habitats (*active drift entry*). Discriminating between these alternative explanations is largely speculative. We can only guess at what initiates, and the ultimate goal of, drifting behaviour. And mechanisms may change during ontogeny for the same species, where the probability of a passive entry could decrease with increasing physical and behavioural abilities (Pavlov 1994; Kynard et al. 2007b). In the following, we consider each mode of drift entry in turn.

## Passive drift entry

In passive drift entry, elimination of rheoreaction is triggered by the loss of visual cues at low illumination and in swift flows, or by the inability of fishes to resist overcritical current speeds (Pavlov 1994). In this respect, small fish are particularly vulnerable, because both swimming capacity and visual acuity increase with size, and are limited in embryos and larvae (Wanzenböck and Schiemer 1989; Flore et al. 2000).

The loss of visual orientation and the ensuing displacement from refuges is a common explanation for the increase in drift density of fish larvae with the onset of darkness and in highly turbid waters (Northcote 1962; Corbett and Powles 1986; Gadomski and Barfoot 1998; Peňáz et al. 1992; Reichard et al. 2001; Oesmann 2003; Reeves and Galat 2010). In this context, Bardonnet (1993) demonstrated that phosphorescent landmarks lessened the nocturnal drift peak of emerging brown trout (*Salmo trutta*, Salmonidae). Additionally, experiments by Pavlov et al. (1972) on various teleost species showed that the critical current velocities that young fish could resist were markedly higher in light than in darkness.

Other studies highlight the effect of high current velocities during flood events on influencing larvae to enter the drift (Reichard et al. 2001). The larvae of usually non-drifting species were frequently detected in the flow during rising water levels in two Australian rivers (Humphries and King 2004). The authors concluded that these individuals were flushed from backwaters and other still-water habitats. Harvey (1987) observed a taxon–and size-specific susceptibility of downstream displacement by flooding in Brier Creek, Oklahoma, where smaller fish (5–10 mm) were at greater risk of being entrained than larger ones (10–25 mm). Likewise, washouts of mainly eurytopic 0+ cyprinids during a summer flood in the river Oder, Germany, caused a drastic decline in fish densities in different mesohabitats (Bischoff and Wolter 2001). Drifting during flood events can lead to substantial mortality, likely caused by gill damage from suspended sediment (Mion et al. 1998). Notably, moderate variation in discharge does not appear to affect the abundance or taxonomic and size composition of drifting larvae (Corbett and Powles 1986; Robinson et al. 1998; Copp et al. 2002; Reichard and Jurajda 2004; Zitek et al. 2004b; Humphries 2005).

## Coincidental drift entry

Drift entry may be initiated by behaviours which are not targeted deliberately at dispersal, but expose young fish to strong currents (sensu *non-specific behavioural responses*; Pavlov et al. 2008). In that case, individuals inadvertently enter the flow while emerging from substrate, or during feeding activities and flight responses.

The directional movements of larvae in response to a light stimulus (=phototaxis) are considered crucial for the coincidental entrance of young fish into the water column (Pavlov 1994). Positive and negative photoreactions are species-specific, unconditioned reflexes to certain levels of illumination (Pavlov et al. 1978). They *inter alia* mediate the emergence of recently hatched larvae from the spawning substrate in order to fill the swim bladder and start exogenous feeding (Persat and Olivier 1995; Zitek et al. 2004b). Photoreactions are supposed to be important determinants of temporal drift patterns. Observations by Iguchi and Mizuno (1990), for example, indicate that diel drift patterns in the Japanese common goby (*Rhinogobius brunneus*, Gobiidae) depend on light preferences of the free embryos and changes in the ambient light conditions along a river course.

Armstrong and Brown (1983) and Brown and Armstrong (1985) concluded that coincidental drift entry relates to a higher risk of colliding with fast currents after initiation of feeding activities, because they detected only a few pro-larvae (with yolk sac), and mainly uniform-sized post-larvae (without yolk sac), in the Illinois River. Encounter rates with currents can also increase during exploratory behaviour and may depend on visual habitat heterogeneity. For example, in experiments with *Cichlasoma octofasciatum*, Cichlidae, Pavlov et al. (2008) discovered that motor activity during habituation in novel surroundings is higher when visual landmarks are missing.

## Active drift entry

Young fish commonly leave a specific area in order to escape unfavourable conditions or to reach the most advantageous habitats. Active drift entry, together with the utilization of the river current as means of transport, may have evolved to facilitate these movements (Gadomski and Barfoot 1998; Ellsworth et al. 2010a; Braaten et al. 2012). There are several reasons for relocations during early life-history. One of the most important may be the matching of larval requirements and environmental conditions at certain points of ontogeny. Mismatches in these two can arise immediately after hatching, when vulnerable embryos and early larvae face the swift-flowing, nutrient-poor spawning grounds chosen by their parents (Bardonnet 2001; Keckeis 2001). And they can occur later, when food or physico-chemical requirements change (Corbett and Powles 1986; Bardonnet 2001; Schiemer et al. 2003). Consequently, the initial drift entry at the spawning site, along with a rapid transport to suitable near-shore feeding areas and subsequent movements between these nurseries, are crucial for maximizing energy gain and survival (Jonsson 1991; Usvyatsov et al. 2013). By diluting the naturally high aggregation of individuals at spawning sites and nurseries, drift entry and dispersal also reduce the attraction of predators, the risk of cannibalism and competition for food and space (Copp et al. 2002; Humphries 2005). Population density effects, by way of example, are the main dispersal drivers in salmonid fry: downstream movement of territorial young brown trout (*Salmo trutta*, Salmonidae) ceased, independent of current speeds, after a constant final proportion of fish had been displaced (Daufresne et al. 2005). The smaller and lighter fry component of steelhead (*Oncorrhynchus mykiss*, Salmonidae) and Atlantic salmon (*Salmon salar*, Salmonidae) left their hatching sites, because they were out-competed by larger conspecifics in establishing and holding territories (Johnston 1997; Bujold et al. 2004; Johnson et al. 2013).

Although drift entry is often linked to illumination and current velocity, lack of correlation between these variables and abundance of drifting individuals may provide evidence for the deliberate nature of the process. For example, there was no relationship between light intensity and body length of drifting cyprinids in the River Morava (Reichard et al. 2002a). A negative correlation was expected if drift entrance was catastrophic and solely attributed to a size-dependent loss of visual orientation and neutralisation of rheoreaction. The conclusion that entry into the drift is a time-dependent behavioural decision is supported by other studies: Sonny et al. (2006) ascribed the observed differences in the diurnal size range of drifting chub (*Squalius cephalus*) and roach (*Rutilus rutilus*, Cyprinidae) in the River Meuse, Belgium, to diverging dispersal strategies of both species; Pavlov (1994) watched roach larvae actively entering the flow of a hydrodynamic channel during the night; in experiments with brown trout, Roussel and Bardonnet (1999) showed that upstream movements were predominantly nocturnal, indicating that darkness might instead of inhibiting a fish´s orientation, in fact allow it the freedom to disperse. Irrespective of why fish choose to enter the current, doing so during the night or at high levels of turbidity may be beneficial, and be a way of avoiding visual predators (Clark and Pearson 1980; Johnston 1997; Copp et al. 2002; Usvyatsov et al. 2013). Another hypothesis, that adaptive night-time drift entrance of young fish corresponds to the feeding on synchronously drifting invertebrates (Elouard and Leveque 1977), is rather implausible: capture success and consumption rate of larvae sharply decrease with increasing velocity (Flore and Keckeis 1998) and decreasing light (Blaxter 1986), making it difficult for drifting fish to locate and eat prey. Most larvae are caught with empty guts at night (Shepherd and Mills 1996).

Other results suggest that rising water levels and current speeds do not necessarily trigger a passive displacement, but an active movement of young fish. Increasing current speeds, during elevated discharge, neither affected abundance nor size and age of drifting young-of-the-year cyprinids (Reichard and Jurajda 2004). In a similar way to the effects of light, young fish may enter the drift only under particular hydraulic conditions (Johnston et al. 1995; Araujo-Lima and Oliveira 1998). Studies on Kootenai white sturgeon larvae (*Acipenser transmontanus*, Acipenseridae) in artificial streams concluded that there is a threshold velocity needed to trigger larval dispersal (Kynard et al. 2007b). Daufresne et al. (2005) deduced that drift entry is not only attributed to swimming abilities, because greater displacement rates of brown trout were observed at lower *and* higher velocities.

Other abiotic factors than light and current can provoke an active drift entry: Gale and Mohr (1978) interpreted the high densities of drifting larvae downstream of an acid mine drainage as escape from heavily polluted waters; and the negative correlation of drift density and suspended particular matter in the main channel of the Elbe has been attributed to active drift avoidance (Oesmann 2003).

## Drift control and orientation

Pavlov's classification of larval drift modes has served as a basis for almost all the work that has been carried out on this topic (but see his review papers: Pavlov 1994 and Pavlov et al. 2008). He cites the most common drift mode as constituting *passive drift*: non-oriented individuals are transported downstream at the same rate as the current speed. Passive drift mode likely corresponds to accidental drift entry, typical of the early larval stages and usually observed in the dark. The opposite, *active drift,* logically follows a deliberate entry into the current. Here, oriented individuals move faster downstream than the current, primarily during daytime. A hybrid form, *active–passivedrift*, is characterized by upstream oriented fish, moving downstream at lower rates than the mean current speed. This mode, originally attributed to impaired swimming ability at decreased water temperatures and during starvation (Pavlov et al. 2008), is now considered as one manifestation of negative rheoreaction; i.e. an oriented, active downstream movement (Pavlov et al. 2011).

Below, we review the evidence for active and passive drift in the literature. We suggest that recent work indicates that classification into discrete drift modes may not be the best model of what happens in the wild, and instead, we offer an alternative approach.

## Passive drift

A solely passive downstream transport of larvae is mostly ascribed to a combination of poor swimming abilities with highly variable flow conditions in rivers (Corbett and Powles 1986; Gadomski and Barfoot 1998; de Graaf et al. 1999). Average current speeds in rivers often exceed swimming speeds of recently hatched larvae (Wolter and Sukhodolov 2008). Additionally, turbulent flows make maintaining position or direction of swimming extremely difficult if not impossible for many larvae (Webb and Cotel 2011), increase the energetic costs of locomotion (Liao 2007) and reduce the critical swimming velocities young fish can maintain (Lupandin 2005).

If passive transport is the norm, vertical and lateral distributions of drifting fish larvae depend on the location of upstream spawning sites (Brown and Armstrong 1985; D'Amours et al. 2001) and a subsequent exposure to hydraulic forces (e.g. average current speeds, transverse flow circulations, rheogradients, turbulent mixing; Pavlov 1994; Pavlov et al. 2008). Therefore, passively drifting individuals should become redistributed according to their body shape and buoyancy (Copp et al. 2002). They should end up in depositional habitats (e.g. groyne fields; Lechner et al. 2014a) or be concentrated in dead zones, slackwater habitats (Wolter and Sukhodolov 2008; Kopf et al. 2014) and concave riverbanks (Pavlov et al. 2008). Depending on discharge levels, hydraulic forces may concentrate young fish in areas with high current speeds, such as the thalweg (Braaten et al. 2010; Ellsworth et al. 2010a). If larvae are kept in suspension under high discharges, this may considerably increase drift distance and influence longitudinal distribution patterns (Corbett and Powles 1986).

However, a growing number of studies, documenting behavioural, sensorial and physical abilities of young fish (Garner 1999; Hogan and Mora 2005; Stoll and Beeck. 2012), refute the model of a solely passive drift. Specifically, it has been shown that the spatio-temporal dispersal patterns of fish larvae differ from those of virtual (Schludermann et al. 2012) and physical (Lechner et al. 2014a) passive particles, and that passive transport models overestimate drift distance by far (Braaten et al. 2012).

## Active drift

The drifting free embryos and larvae of some species are well developed and are active swimmers, able to determine their location in the water column (Peňáz et al. 1992; Humphries 2005). Even at low light levels and overcritical currents, they may respond to hydraulic gradients and drift under certain conditions (Kaminskas 2011; Schludermann et al. 2012). Experiments on zebrafish larvae (*Danio rerio*, Cyprinidae), for instance, showed that rheoreaction is mediated by neuromasts of the lateral line, which enables young fish to sense water flows and orient in currents at night (Olszewski et al. 2012; Stewart et al. 2013).

If active transport is the norm, the spatial distribution of drifting larvae depends on deliberate swimming behaviour and the ability to locate preferred hydraulic habitat, or at least to orientate along gradients which lead to these (Robinson et al. 1998). Vertical positioning of drifting larvae may reflect an effort to avoid benthic predators and collisions with the substrate (Brown and Armstrong 1985; Gadomski and Barfoot 1998). Lateral positioning of drifting larvae, however, is thought to be a result of specific dispersal strategies; e.g., drifting near-shore to enhance the probability of being able to move into suitable habitat when coming close to it (Araujo-Lima and Oliveira 1998), or drifting offshore to accelerate transport and thereby minimize mortality (Reichard et al. 2004). The distance covered during active drifting will clearly vary depending if larvae choose to drift at low (Lechner et al. 2014b), intermediate (Schludermann et al. 2012) or high currents (Oesmann 2003) and whether they drift low or high in the water column (Kynard et al. 2007a).

## A continuum mode of dispersal: *actipassive* drift

We argue that, aside from eggs, the early developmental stages of fish are not exclusively either passive or active drifters. On the one hand, the hydraulic forces can exceed the swimming capabilities of free embryos and larvae (Wolter and Arlinghaus 2003). On the other hand, even tiny larvae display orientation in the dark at over-critical currents and can, to a degree, regulate their dispersal (Zens 2015). Therefore, a strict separation of drift modes, as applied in the majority of studies (*active*: Peňáz et al. 1992; Robinson et al. 1998; Humphries 2005; Braaten et al. 2012; *passive*: Armstrong and Brown 1983; Harvey 1987; de Graaf et al. 1999; Wolter and Sukhodolov 2008; Ellsworth et al. 2010a; Janac et al. 2013) appears inappropriate. Although, it may appear reasonable to evaluate the ratio of active and passive components for single dispersal events along a continuum (Fig. 1), this ratio must result from the strength of hydraulic forces and species—or stage-specific capabilities (behavioural, physical). A profound knowledge of larval skills (i.e. swimming performance, orientation ability, drift behaviour) and the limiting environmental conditions is required, if one wants to assess the proportion of each component. As a theoretical model we suggest instead, that the drift of larvae, free embryos and juveniles should be referred to as an *actipassive* process. Provided there is detailed information on the ratio, the terms *active–passive*—for a predominantly controlled mode of downstream movement—and *passive–active*—for a primary externally vectored transport–might be used.

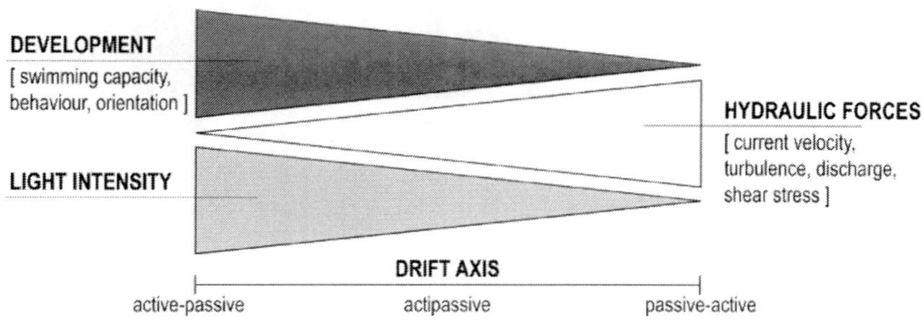


**Figure. 1**Illustration of the continuum drift mode. *Triangles* represent the range of parameters with low values for hydraulic forces and rudimentary developmental state at the apexes

## An integrated approach to the drift of early life stages of riverine fishes

The state of knowledge in research on the drift of early life stages of riverine fishes was reviewed, based on peer-reviewed publications between 1972 and 2014 (see Online Resource). Of the 60 publications, 89 % were field studies, and of these, 85 % used stationary drift nets (Fig. 2). The research was conducted in 54 rivers on 4 continents, but mostly in temperate floodplain—and upland rivers in North America and Europe (Fig. 2, Online Resource). Only 10 % of these studies used hydrodynamic or mathematical models to explain or predict drift patterns. We analysed all publications using a conceptual framework of organismal movement advocated by Nathan et al. (2008) (Fig. 3). According to Nathan et al. (2008), the movement path of a focal individual results from the dynamic interplay of four basic components:

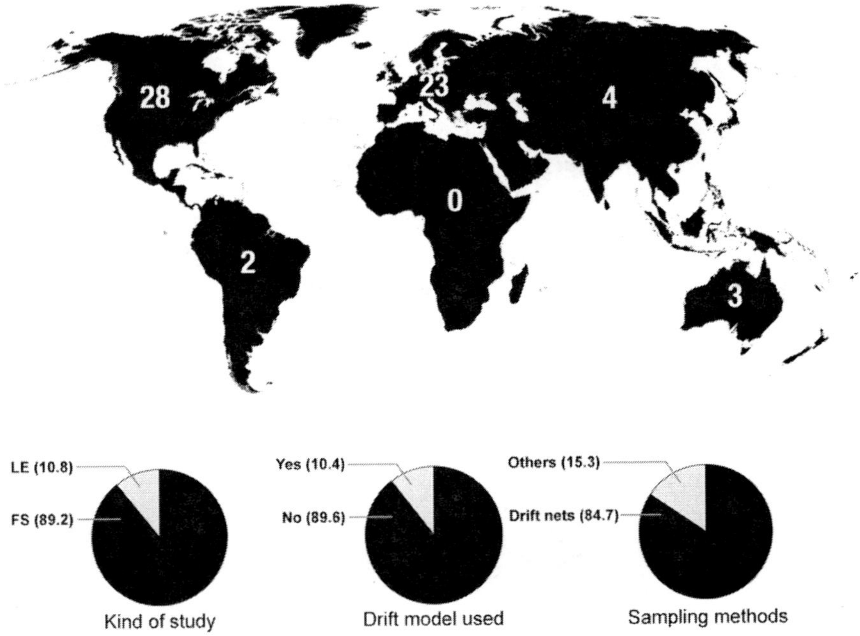


**Figure. 2** Overview of the geographical distribution of the 60 analysed drift studies. The percentage of field studies (FS) and laboratory experiments (LE) as well as the frequency of drift model- and driftnet-usage is given in the pie charts below

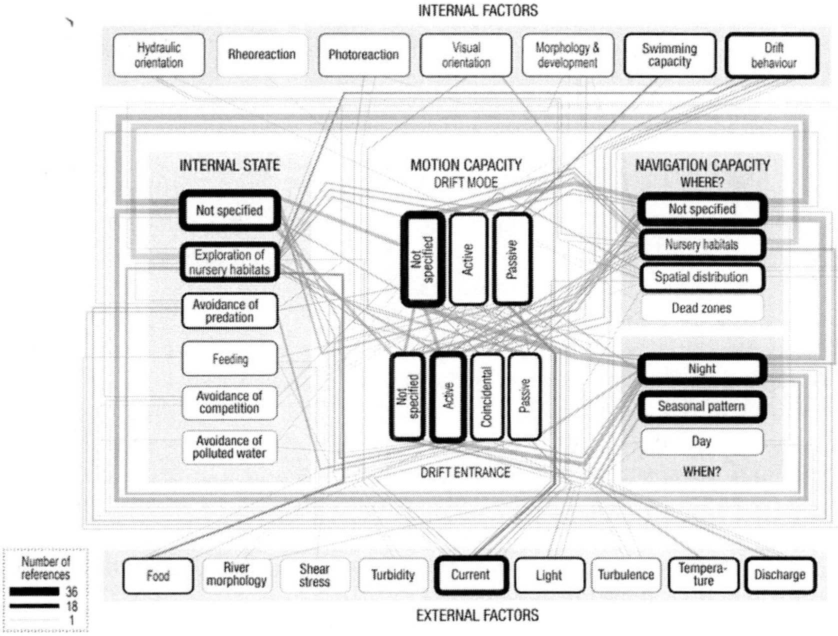


**Figure. 3** Characteristics of larval fish drift, as derived from the relevant literature, are displayed on the base of a general framework for movement ecology (*coloured blocks*; Nathan et al. 2008). Frame- and line thickness refer to the frequency of described cases and connections. For reasons of clearness, existing links between internal and external factors are not drawn

1. *internal state*—the different motivations to move or "why do the early life stages of riverine fishes drift?";
2. *navigation capacity*—the orientation ability of moving organisms or "where and when do the early life stages of riverine fishes drift?";
3. *motion capacity*—the different modes (active, passive) of movement or "how do the early life stages of riverine fishes drift?";
4. *external factors*—biotic and abiotic parameters that affect why, how, where and when to drift.
5. But because of the field of research, we introduced a fifth component:

6. *internal factors*—a set of intrinsic characteristics (e.g. physiology and morphology) that influence different components of drift movement.

Each publication was scrutinized for specific information associated with the above–outlined model. Information was categorized, where applicable, and displayed in framed white boxes within the five components (Fig. 3). The “not-specified” boxes refer to studies that provide no answers on why, how or where fish drift. Lines between boxes indicate co-occurrence or cause-and-effect relationships, as obtained from the literature. The thickness of frames and lines represent the relative frequency of references to a particular aspect. Both speculative and empirical conclusions–drawn by the authors themselves–on drift characteristics were included in this meta–analysis. A synopsis shows the most common categories (except “not specified”) and linkages (Fig. 4). This will serve as a roadmap to discuss the current state of knowledge in drift research, and identify requirements for future studies. Note, the data content was adapted to accord with Nathan’s framework. Therefore, drift entrance modes and temporal patterns of drift were excluded, and have already been considered in the above.

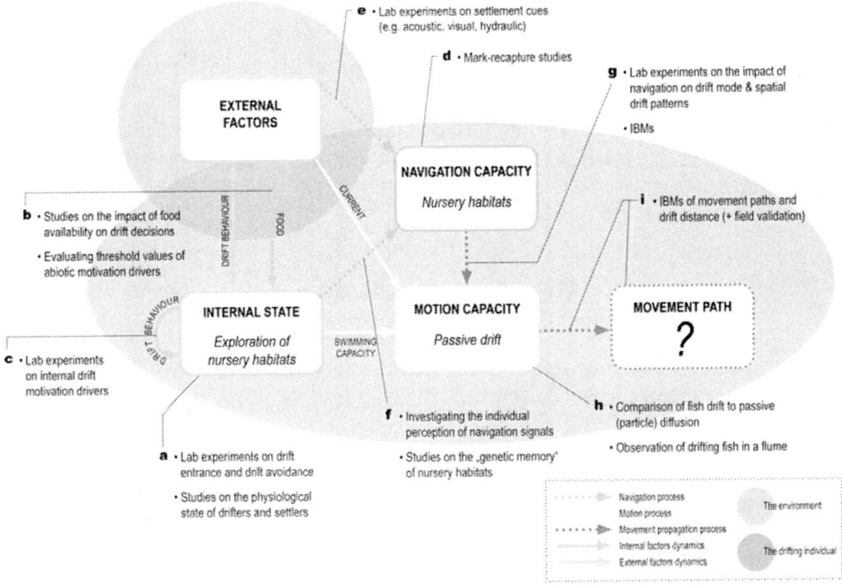


**Figure. 4** Conceptual model of fish drift inspired by Nathan et al. (2008), showing the processes and relationships (indicated by *arrows*) between the five basic components (*boxes*) of dispersal. *Dotted lines* show knowledge gaps in the reviewed literature (<5 references). *Black lines* (*a–i*) indicate suggested research needs for particular aspects of fish drift. Detailed information of the glossary is given in Nathan et al. (2008).

Briefly: the navigation process refers to the realized navigation capacity given the impact of the current location, internal state, and external factors on the fundamental navigation capacity of the focal individual; the motion process refers to the realized motion capacity given the impact of the current location, internal state, and external factors on the fundamental motion capacity of the focal individual; the movement propagation process refers to the realized movement produced by the motion process (optionally affected by the navigation process)

## *Internal State*: "why do the early life stages of riverine fishes drift?"

The internal state of drifting larvae is not specified in most studies reviewed (Fig. 3). Determining the motivation for a single drift event is a challenging task. It requires the assessment of the psychological (do larva have a psyche?) and physiological state of a small fish, at any given time, driving it to fulfil a particular goal by changing its position. This is compounded by complex phenotypic and genotypic behaviour-environment interactions. Perhaps, except for amphidromous species, where the objective is to reach the sea, it is always going to be uncertain whether drift is the result of a desire to reach a destination, or to avoid a threat, such as competition, predation or poor water quality, at its current location (Nathan et al. 2008).

The perception–perhaps anthropocentric–of larvae looking for the most advantageous feeding grounds underlies many drift studies (Fig. 4). Recently hatched fish are unlikely to be aware of suitable downstream areas (but see *Navigation capacity*). Rather than a 'motivation' per se, the 'exploration of nursery habitats' may be a by-product of an active drift entrance that potentially refers to ultimate evolutionary payoffs from moving (e.g. gaining energy, dampening population fluctuations by spreading larvae over heterogeneous environments etc.). Whilst this last is hard to prove, the motivation to leave disadvantageous areas can be tested in laboratory experiments on drift entrance by controlling conditions of water quality, predation or competition (Fig. 4a). In this respect, the impact of predator kairomones on young fishes' drift entrance dynamics, as shown for invertebrates (Winkelmann et al. 2008), could be of particular interest. Investigations of the physiological state of drifting (and settling) fish may clarify whether hunger is an important driver of drift.

Specific attention should be given to external factors that stimulate or discourage young fish to move. Food availability, for example, is mentioned a common reason to leave, or look for a specific habitat (Fig. 4). To the best of our knowledge, there is no study that actually investigated food quality or quantity at the starting- or end-point of a drift event. Evaluating the threshold values of other abiotic factors (e.g. current, light), which affect the fish's internal state, will provide deeper insights into why, when and where young fish enter the flow (Fig. 4b). Furthermore, a profound understanding of an individual's motivation to drift (i.e. drift behaviour) allows more informed conclusions to be drawn on the significance of environmental conditions encountered on the way (Fig. 4): larvae of potamodromous species, for example, that search the river shorelines for suitable habitats, potentially prefer lower currents than those of anadromous species that aim for the distant ocean.

The motivation for drifting is most likely self-regulated by internal factors (Fig. 4). However, detailed knowledge on how (ontogenetic) changes in morphology, swimming capacity, orientation ability or behaviour affect the readiness to move, or assist a specific drift strategy, is required. Again, laboratory experiments appear to be the appropriate method (Fig. 4c).

## *Navigation capacity*: "where and when do the early life stages of riverine fishes drift?"

The drift destination is not specified in most work on dispersing fish larvae in rivers (Fig. 3). Indeed, nursery areas are frequently assumed by authors to be where larvae stop drifting (Fig. 4) but there is no sure indication for this conclusion, as most studies do not observe settlement activities (Kennedy and Vinyard 1997; Schludermann et al. 2012; Lechner et al. 2014a). Hence, an increased application of mark-recapture experiments, focusing on drift exit and habitat choice along the shoreline, will provide valuable information (Fig. 4d). Using genetic and geochemical markers (see reviews by Hedgecock et al. 2007 and Thorrold et al. 2007) or trans-generational tagging approaches, offers the possibility to mark large numbers of fish larvae (Thorrold et al. 2006; Zite et al. 2014), and should be more commonly incorporated into riverine drift studies.

Almost nothing is known of the navigation ability of drifting fish and the influencing external and internal factors (Fig. 4). This kind of knowledge, however, is indispensable to understand how larvae detect and reach suitable inshore habitats and to predict spatial patterns of drift. Studies on marine species have shown that fish larvae navigate toward settlement areas by reacting to a variety of environmental stimuli (Dixson et al. 2011; Leis et al. 2011; Huijbers et al. 2012). Consequently, we encourage studies on the settlement cues for drifting fish in running waters (Fig. 4e). Those cues potentially encompass acoustic (e.g. the sound of the wave-wash) or visual (e.g. shading by vegetation) signals and hydraulic gradients (e.g. current speed, current direction, water depth) that indicate shore proximity. Given the limited observability of small fish in large rivers, the feasibility of such studies is mainly restricted to laboratory experiments.

The internal state of the individual (e.g. neurological and physiological) potentially influences the process of navigation. Studies are needed that examine the species—and stage-specific ability to sense and respond to navigation signals (Fig. 4f). Furthermore, the question arises whether fish larvae, as some birds do (Nathan et al. 2008), possess a genetically coded 'memory' guiding them towards suitable nursery areas. The long-time observation of downstream habitat use (i.e. 'nursery fidelity') by marked larvae from a known spawning ground could throw light on the matter (Fig. 4f).

In future research, the focus must shift to how individuals implement navigation decisions in the process of drifting (i.e. *Motion capacity*). Many studies ascribe the observed spatial distribution patterns (i.e. lateral and horizontal; Online Resource) of drifting fish to active navigation (i.e. active drift), but remain vague about the underlying mechanisms (Brown and Armstrong, 1985; Gadomski and Barfoot 1998; Robinson et al. 1998; Schludermann et al. 2012). Certainly, the observation of individuals in an artificial stream-channel is a potent method to correlate a variety of navigation-related cues to the drift mode under controlled conditions, thereby elucidating spatial patterns of transport (Fig. 4g). The ultimate significance of the suggestions above is debatable, as transferring laboratory results to the field has its drawbacks. A promising approach, in this respect, is the field validation of predicted spatial drift outcomes by biophysical models that incorporate larval navigation abilities recorded in the laboratory (Fig. 4g). These individual based models (IBM) are an emerging tool in oceanic research (Cowen et al. 2006; Christensen et al. 2007; Staaterman and Paris 2014), supporting the identification of potential nurseries and spawning grounds. In freshwater systems, this technique is in its infancy.

## *Motion capacity*: "how do the early life stages of riverine fishes drift?"

Generally, the mode of transport is not specified in drift studies (Fig. 3). This is at first attributable to the common sampling method, i.e. exposing stationary nets into the current to capture autochthonous fish fry. Missing information on where larvae had started drifting, and the environmental conditions along their route, make reliable statements on the drift mode impossible. The frequent reports of passive drift caused by strong currents (Fig. 4), thus often derive from a simple comparison between values of current speed (measured at the net opening) and larval swimming performance (Corbett and Powles 1986; Gadomski and Barfoot 1998; Wolter and Sukhodolov 2008).

The eulerian observation of larval movement (using drift nets) in relation to passive particles is a promising approach for a rough identification of the drift mode (Fig. 4h). The spatio-temporal drift patterns of marked and recaptured individuals can be either compared to those of physical floats, ideally featuring some larval attributes (e.g. shape and density), or to those of virtual particles, simulated using specific tracking software on the base of hydrodynamic models (Schludermann et al. 2012; Lechner et al. 2014a).

Flume experiments have a great potential to reveal specific peculiarities of the drift mode, however, they are rarely applied (Persat and Olivier 1995; Kynard et al. 2007a; Kaminskas 2011; Pavlov et al. 2011; Zens 2015). We suggest that future work should focus on the transport mode of different species (e.g. invasive, endemic, riverine, lentic, benthic), developmental stages, ecological guilds (e.g. reproduction and current) etc. in the laboratory, where abiotic factors (i.e. current, temperature, turbidity and turbulence) are changed while observing drifting fish (Fig. 4h).

## *Movement path*: "which way do drifting larvae take?"

So far, there is no existing knowledge on the exact route of drifting young fish in rivers (Fig. 4). The limiting factor is, again, the traditional method of collecting larva: drift net samples are point measurements, providing no information on the processes *between* two points. However, the instructive in situ tracking of larval movement, as demonstrated in marine studies (Huebert and Sponaugle 2009; Paris et al. 2013), is made problematic by swift currents, turbid water and the nocturnal drift activity in rivers (Kennedy and Vinyard 1997).

In our opinion, the most practicable and sophisticated way is to simulate larval movement paths with IBMs and validate the results in the field (Fig. 4i). This requires a specific understanding of *navigation capacity* and *motion capacity.* Furthermore, high resolution 3D-hydrodynamic river models are needed to predict a larva's reactions towards hydraulic gradients *en route.* The field validation should preferably be based upon mark-recapture studies at different spatial scales with well-understood model species (see Lechner et al. 2014a).

In order to set the spatial limits of a movement path, it might initially be useful to focus on the distance covered during a drift event. Mathematical models have been used to estimate drift distances of some sturgeon species by integrating flow velocity and empirical data on drift duration, drift mode and swimming height (Kynard et al. 2007a; Braaten et al. 2008, 2012), but these sorts of studies are rare. There is an urgent need for research regarding the drift distance of other fishes (Fig. 4i).

Data on the spatial characteristics (i.e. route and distance) of fish drift are important to evaluate the connectivity of key habitats (spawning-rearing) and will have implications for river management and restoration projects. Information on the average drift distance, for example, could influence the designation of protected areas downstream of known reproduction sites.

## CONCLUSION

This review clearly shows that we are far from a general understanding of the processes and patterns of fish drift. As stated previously, the most common technique in the investigation of the drift of the young stages of fish is the deployment of stationary drift nets. This technique is unsuitable for finding answers to the pressing questions in the field that we have identified. We believe that laboratory observations and experiments are essential to elucidate the sensorial, physical and behavioural capabilities of fish larvae, which will in turn provide information on the processes involved in *navigation capacity*, *motion capacity* and *internal state*. The information gained will improve riverine drift models. These are a rarely-used, but powerful tool for understanding dispersal, and will shed light on movement path and/or settlement behaviour.

Increasing our understanding of the nature of the dispersal ecology of the early life stages of riverine fishes is urgently needed. The increasing human impact on rivers worldwide impinges on all aspects of the ecology of fishes, but young fish are particularly vulnerable. Implications of river channelization and flow regulation: (1) disturb localized settlement and movement in inshore areas (Braaten et al. 2012); (2) disrupt natural dispersal signals and affect drift distance (Ellsworth et al. 2010b; Usvyatsov et al. 2013); (3) increase washout-effects and catastrophic drift entrance during high water levels and navigation-induced wave wash (Bischoff and Wolter 2001; Kucera-Hirzinger et al. 2009; Schludermann et al. 2013); and (4) jeopardize drifting fish by entrainment into water abstraction sites (Carter and Reader 2000; Pavlov et al. 2008; Bracken and Lucas 2013). An integrated approach to the ecology of drifting fish in rivers will greatly contribute to effective conservation and management of riverine fish populations.

## ACKNOWLEDGMENTS

Open access funding provided by University of Vienna. This work was financed by Austrian Science Fund (FWF; project MODI P22631-

B17; http://www.fwf.ac.at;). J. Waldhör, A. Pivo and A. Seidel helped with the figures. Fruitful discussions with our colleagues, especially during the 39th Larval Fish Conference in Vienna, improved the manuscript enormously.

**Online Resource:** A tabulated analysis of 60 relevant publications on the dispersal of the early developmental stages of freshwater fish. Glossary: Types of study were field study (FS), laboratory experiment (LE) and review (R); Developmental stages under investigation (DUI) were alevins (ALE), ammocetes (AMO), eggs (EGG), embryos (EMB), fry (FRY), juveniles (JUV), larvae (LAR) and young-of-the-year (YOY). "Not specified" (n.s.) marks a lack of relevant information in the reference. Species names were adopted from the reviewed papers and have not been adjusted to possible taxonomic shifts. (PDF 45 kb)

## REFERENCES

1. Araujo-Lima CARM, Oliveira EC (1998) Transport of larval fish in the Amazon. J Fish Biol 53:297–306. doi:10.1006/jfbi.1998.0815
2. Armstrong ML, Brown AV (1983) Diel drift and feeding of channel Catfish Alevins in the Illinois River, Arkansas. Trans Am Fish Soc 112:302–307. doi:10.1577/1548-8659(1983)112<302:Ddafoc>2.0.Co;2
3. Auer NA, Baker EA (2002) Duration and drift of larval lake sturgeon in the Sturgeon River, Michigan. J Appl Ichthyol 18:557–564. doi:10.1046/j.1439-0426.2002.00393.x
4. Bardonnet A (1993) Use of visual landmarks by young trout (Salmo-Trutta) during their diel downstream postemergence displacement in experimental channels. J Fish Biol 43:375–384. doi:10.1111/j.1095-8649.1993.tb00572.x
5. Bardonnet A (2001) Spawning in swift currents: implications for eggs and larvae. Archiv für Hydrobiologie Supplement 135(2-4):271–291. doi:10.1127/lr/12/2001/271
6. Bell KNI, Brown JA (1995) Active salinity choice and enhanced swimming endurance in 0-D-Old to 8-D-old larvae of diadromous gobies, including sicydium-punctatum (Pisces), in Dominica, West-Indies. Mar Biol 121:409–417. doi:10.1007/Bf00349450
7. Bischoff A, Wolter C (2001) The flood of the century on the river oder: effects on the 0+ fish community and implications for floodplain restoration. Regul Rivers-Res Manag 17:171–190. doi:10.1002/Rrr.612
8. Blaxter JHS (1986) Development of sense-organs and behavior of teleost larvae with special reference to feeding and predator avoidance. Trans Am Fish Soc 115:98–114. doi:10.1577/1548-8659(1986)115<98:NLFCDO>2.0.CO;2
9. Borcherding J, Breukelaar AW, Winter HV, Konig U (2014) Spawning migration and larval drift of anadromous North Sea houting (Coregonus oxyrinchus) in the River IJssel, the Netherlands. Ecol Freshw Fish 23:161–170. doi:10.1111/eff.12058
10. Braaten PJ, Fuller DB, Holte LD et al (2008) Drift dynamics of larval pallid sturgeon and shovelnose sturgeon in a natural side channel of the upper Missouri River, Montana. North Am J Fish Manag 28:808–826. doi:10.1577/M06-285.1
11. Braaten PJ, Fuller DB, Lott RD, Ruggles MP, Holm RJ (2010) Spatial distribution of drifting pallid sturgeon larvae in the missouri river inferred from two net designs and multiple sampling locations. North Am J Fish Manag 30:1062–1074. doi:10.1577/M09-149.1

12. Braaten PJ, Fuller DB, Lott RD et al (2012) An experimental test and models of drift and dispersal processes of pallid sturgeon (Scaphirhynchus albus) free embryos in the Missouri River. Environ Biol Fishes 93:377–392. doi:10.1007/s10641-011-9925-9
13. Bracken FSA, Lucas MC (2013) Potential impacts of small-scale hydroelectric power generation on downstream moving lampreys. River Res Appl 29:1073–1081. doi:10.1002/Rra.2596
14. Brown AV, Armstrong ML (1985) Propensity to drift downstream among various species of fish. J Freshw Ecol 3:3–17. doi:10.1080/02705060.1985.9665087
15. Bujold V, Cunjak RA, Dietrich JP, Courtemanche DA (2004) Drifters versus residents: assessing size and age differences in Atlantic salmon (Salmo salar) fry. Can J Fish Aquat Sci 61:273–282. doi:10.1139/F03-162
16. Busch WDN, Scholl RL, Hartman WL (1975) Environmental factors affecting strength of walleye (Stizostedion-Vitreum Vitreum) Year-Classes in Western Lake Erie, 1960-70. J Fish Res Board Can 32:1733–1743. doi:10.1139/f75-207
17. Carter KL, Reader JP (2000) Patterns of drift and power station entrainment of 0+ fish in the River Trent, England. Fish Manag Ecol 7:447–464. doi:10.1046/j.1365-2400.2000.00224.x
18. Carter JG, Lamarra VA, Ryel RJ (1986) Drift of larval fishes in the Upper Colorado River. J Freshw Ecol 3:567–577. doi:10.1080/02705060.1986.9665151
19. Christensen A, Daewel U, Jensen H, Mosegaard H, St John M, Schrum C (2007) Hydrodynamic backtracking of fish larvae by individual-based modelling. Mar Ecol Prog Ser 347:221–232. doi:10.3354/Meps06980
20. Clark AL, Pearson WD (1980) Diurnal variations in ichthyoplankton densities at Ohio River Mile 571. Trans of the Ky Acad Sci 41:116–121
21. Copp GH, Cellot B (1988) Structure and function of ecosystems of the french Upper-Rhone River. 51. Drift of embryonic and Larval Fishes, especially Lepomis-Gibbosus (L), in the Upper-Rhone River. J Freshw Ecol 4:419–424. doi:10.1080/02705060.1988.9665193
22. Copp GH, Faulkner H, Doherty S, Watkins MS, Majecki J (2002) Diel drift behaviour of fish eggs and larvae, in particular barbel, Barbus barbus (L.), in an English chalk stream. Fish Manag Ecol 9:95–103. doi:10.1046/j.1365-2400.2002.00286.x
23. Corbett BW, Powles PM (1986) Spawning and Larva drift of sympatric walleyes and white suckers in an Ontario Stream. Trans Am Fish Soc 115:41–46. doi:10.1577/1548-8659(1986)115<41:Saldos>2.0.Co;2
24. Cowen RK, Paris CB, Srinivasan A (2006) Scaling of connectivity in marine populations. Science 311:522–527. doi:10.1126/science.1122039
25. D'Amours J, Thibodeau S, Fortin R (2001) Comparison of lake sturgeon (Acipenser fulvescens), Stizostedion spp., Catostomus spp., Moxostoma spp., quillback (Carpiodes cyprinus), and mooneye (Hiodon tergisus) larval drift in Des Prairies River, Quebec. Can J Zool-Revue Canadienne De Zoologie 79:1472–1489. doi:10.1139/cjz-79-8-1472
26. Daufresne M, Capra H, Gaudin P (2005) Downstream displacement of post-emergent brown trout: effects of development stage and water velocity. J Fish Biol 67:599–614. doi:10.1111/j.1095-8649.2005.00759.x
27. de Graaf GJ, Born AF, Uddin AMK, Huda S (1999) Larval fish movement in the River Lohajang, Tangail, Bangladesh. Fish Manag Ecol 6:109–120. doi:10.1046/j.1365-2400.1999.00124.x
28. Dixson DL, Jones GP, Munday PL et al (2011) Terrestrial chemical cues help coral reef fish larvae locate settlement habitat surrounding islands. Ecol Evol. doi:10.1002/Ece3.53
29. Durham BW, Wilde GR (2008) Composition and abundance of drifting fish larvae in the Canadian River, Texas. J Freshw Ecol 23:273–280. doi:10.1080/02705060.2008.9664199
30. Ellsworth CM, Tyler TJ, VanderKooi SP (2010a) Using spatial, seasonal, and diel drift patterns of larval Lost River suckers Deltistes luxatus (Cypriniformes:

catostomidae) and shortnose suckers Chasmistes brevirostris (Cypriniformes: Catostomidae) to help identify a site for a water withdrawal structure on the Williamson River, Oregon. Environ Biol Fishes 89:47–57. doi:10.1111/j.1095-8649.2010.02701.x

31. Ellsworth CM, Belk MC, Keleher CJ (2010b) Residence time and drift patterns of larval June sucker Chasmistes liorus in the lower Provo River as determined by otolith microstructure. J Fish Biol 77:526–537. doi:10.1007/s10641-010-9688-8

32. Elouard JM, Leveque C (1977) Nycthemeral Periodicity in Drift of Insects and Fishes in Ivory-Coast Rivers. Cah Orstom Hydrobiol 11:179–183

33. Flecker AS, Taphorn DC, Lovell JA, Feifarek BP (1991) Drift of Characin Larvae, Bryconamericus-Deuterodonoides, during the dry season from Andean Piedmont Streams. Environ Biol Fishes 31:197–202. doi:10.1007/Bf00001020

34. Flore L, Keckeis H (1998) The effect of water current on foraging behaviour of the rheophilic cyprinid Chondrostoma nasus (L.) during ontogeny: evidence of a trade-off between energetic gain and swimming costs. Regul Rivers-Res Manag 14:141–154. doi:10.1002/(Sici)1099-1646(199801/02)14:1<141:Aid-Rrr492>3.3.Co;2-W

35. Flore L, Reckendorfer W, Keckeis H (2000) Reaction field, capture field, and search volume of 0 + nase (Chondrostoma nasus): effects of body size and water velocity. Can J Fish Aquat Sci 57:342–350. doi:10.1139/cjfas-57-2-342

36. Gadomski DM, Barfoot CA (1998) Diel and distributional abundance patterns of fish embryos and larvae in the lower Columbia and Deschutes rivers. Environ Biol Fishes 51:353–368. doi:10.1023/A:1007485015830

37. Gale WF, Mohr HW (1978) Larval fish drift in a large river with a comparison of sampling methods. Trans Am Fish Soc 107:46–55. doi:10.1577/1548-8659(1978)107<46:Lfdial>2.0.Co;2

38. Garner P (1999) Swimming ability and differential use of velocity patches by 0+ cyprinids. Ecol Freshw Fish 8:55–58. doi:10.1111/j.1600-0633.1999.tb00053.x

39. Harvey BC (1987) Susceptibility of young-of-the-year fishes to downstream displacement by flooding. Trans Am Fish Soc 116:851–855. doi:10.1577/1548-8659(1987)116<851:Soyftd>2.0.Co;2

40. Hay CH, Franti TG, Marx DB, Peters EJ, Hesse LW (2008) Relationships between larval fish drift and abiotic factors in the missouri river. J Freshwater Ecol 23:589–597

41. Hedgecock D, Barber PH, Edmands S (2007) Genetic approaches to measuring connectivity. Oceanography 20:70–79. doi:10.5670/oceanog.2007.30

42. Hogan JD, Mora C (2005) Experimental analysis of the contribution of swimming and drifting to the displacement of reef fish larvae. Mar Biol 147:1213–1220. doi:10.1007/s00227-005-0006-5

43. Huebert KB, Sponaugle S (2009) Observed and simulated swimming trajectories of late-stage coral reef fish larvae off the Florida Keys. Aquat Biol 7:207–216. doi:10.3354/Ab00200

44. Huijbers CM, Nagelkerken I, Lossbroek PAC et al (2012) A test of the senses: fish select novel habitats by responding to multiple cues. Ecology 93:46–55. doi:10.1890/10-2236.1PubMed

45. Humphries P (2005) Spawning time and early life history of Murray cod, Maccullochella peelii (Mitchell) in an Australian river. Environ Biol Fishes 72:393–407. doi:10.1007/s10641-004-2596-z

46. Humphries P, King AJ (2004) Drifting fish larvae in Murray-Darling Basin rivers: composition, spatial and temporal patterns and distance drifted. Canberra Workshop June 2003, p 3–4

47. Humphries P, Lake PS (2000) Fish larvae and the management of regulated rivers. Regul Rivers-Res Manag 16:421–432. doi:10.1002/1099-1646(200009/10)16:5<421:Aid-Rrr594>3.0.Co;2-4

48. Humphries P, Richardson A, Wilson G, Ellison T (2013) River regulation and recruitment in a protracted-spawning riverine fish. Ecol Appl 23:208–225

49. Iguchi K, Mizuno N (1990) Diel Changes of Larval drift among amphidromous gobies in Japan, especially rhinogobius-brunneus. J Fish Biol 37:255–264. doi:10.1111/j.1095-8649.1990.tb05857.x
50. Iguchi KI, Mizuno N (1991) Mechanisms of embryonic drift in the amphidromous goby, Rhinogobius-brunneus. Environ Biol Fish 31:295–300
51. Janac M, Slapansky L, Valova Z, Jurajda P (2013) Downstream drift of round goby (Neogobius melanostomus) and tubenose goby (Proterorhinus semilunaris) in their non-native area. Ecol Freshw Fish 22:430–438. doi:10.1111/Eff.12037
52. Jiang W, Liu HZ, Duan ZH, Cao WX (2010) Seasonal variation in drifting eggs and larvae in the Upper Yangtze, China. Zool Sci 27:402–409. doi:10.2108/Zsj.27.402
53. Johnson JH, McKenna JE (2007) Diel periodicity of drift of larval fishes in tributaries of Lake Ontario. J Freshw Ecol 22:347–350. doi:10.1080/02705060.2007.9665057
54. Johnson JH, McKenna JE, Douglass KA (2013) Movement and feeding ecology of recently emerged steelhead in Lake Ontario tributaries. J Appl Ichthyol 29:221–225. doi:10.1111/Jai.12032
55. Johnston TA (1997) Downstream movements of young-of-the-year fishes in Catamaran Brook and the Little Southwest Miramichi River, New Brunswick. J Fish Biol 51:1047–1062. doi:10.1111/j.1095-8649.1997.tb01543.x
56. Johnston TA, Gaboury MN, Janusz RA, Janusz LR (1995) Larval fish drift in the Valley River, Manitoba: influence of abiotic and biotic factors, and relationships with future year class strengths. Can J Fish Aquat Sci 52:2423–2431. doi:10.1139/F95-833
57. Jonsson N (1991) Influence of water flow, water temperature and light on fish migrations in rivers. Nord J Freshw Res 66:20–35
58. Jurajda P (1998) Drift of larval and juvenile fishes, especially Rhodeus sericeus and Rutilus rutilus, in the river Morava (Danube basin). Archiv Für Hydrobiol 141:231–241
59. Kaminskas T (2011) Downstream movement of Murray cod free-embryos and larva. Charles Sturt University, Bathurst
60. Keckeis H (2001) Influence of river morphology and current velocity conditions on spawning site selection of Chondrostoma nasus (L.). Arch Hydrobiol Suppl 135:341–356. doi:10.1127/lr/12/2001/341
61. Kennedy TB, Vinyard GL (1997) Drift ecology of western catostomid larvae with emphasis on Warner suckers, Catostomus warnerensis (Teleostei). Environ Biol Fishes 49:187–195. doi:10.1023/A:1007329010680
62. King A, Crook DA, Koster WM, Rylah A (2005) Comparison of larval fish drift in the lower Goulburn and mid-Murray Rivers. Ecol Manag Restor 6:136–138. doi:10.1111/j.1442-8903.2005.230-3.x
63. Kopf SM, Humphries P, Watts RJ (2014) Ontogeny of critical and prolonged swimming performance for the larvae of six Australian freshwater fish species. J Fish Biol 84:1820–1841. doi:10.1111/jfb.12399
64. Kucera-Hirzinger V, Schludermann E, Zornig H, Weissenbacher A, Schabuss M, Schiemer F (2009) Potential effects of navigation-induced wave wash on the early life history stages of riverine fish. Aquat Sci 71:94–102. doi:10.1007/s00027-008-8110-5
65. Kynard B, Horgan M (2002) Ontogenetic behavior and migration of Atlantic sturgeon, Acipenser oxyrinchus oxyrinchus, and shortnose sturgeon, A-brevirostrum, with notes on social behavior. Environ Biol Fishes 63:137–150. doi:10.1023/A:1014270129729
66. Kynard B, Parker E (2006) Ontogenetic behaviour and dispersal of Kootenai River white sturgeon early life intervals: a laboratory study. Final Report to Idaho Department of Fish and Game, Boise
67. Kynard B, Parker E, Pugh D, Parker T (2007a) Use of laboratory studies to develop a dispersal model for Missouri River pallid sturgeon early life intervals. J Appl Ichthyol 23:365–374. doi:10.1111/j.1439-0426.2007.00908.x

68. Kynard B, Parker E, Kynard B, Parker T (2007b) Dispersal characteristics, drift distance, and wintering behaviour of young Kootenai River White Sturgeon: a laboratory study. Final Report to Idaho Department of Fish and Game, Boise
69. Lechner A, Keckeis H, Schludermann E et al (2014a) Shoreline configurations affect dispersal patterns of fish larvae in a large river. ICES J Mar Sci 71:930–942. doi:10.1093/icesjms/fst139
70. Lechner A, Keckeis H, Schludermann E, Humphries P, McCasker N, Tritthart M (2014b) Hydraulic forces impact larval fish drift in the free flowing section of a large European river. Ecohydrology 7:648–658. doi:10.1002/Eco.1386
71. Leis JM, Siebeck U, Dixson DL (2011) How nemo finds home: the neuroecology of dispersal and of population connectivity in larvae of marine fishes. Integr Comp Biol 51:826–843. doi:10.1093/Icb/Icr004
72. Liao JC (2007) A review of fish swimming mechanics and behaviour in altered flows. Philos Trans R Soc B-Biol Sci 362:1973–1993. doi:10.1098/rstb.2007.2082
73. Lupandin AI (2005) Effect of flow turbulence on swimming speed of fish. Biol Bull 32:461–466. doi:10.1007/s10525-005-0125-z
74. Maeda K, Tachihara K (2010) Diel and seasonal occurrence patterns of drifting fish larvae in the Teima Stream, Okinawa Island. Pac Sci 64:161–176
75. Martin FD, Paller MH (2008) Ichthyoplankton transport in relation to floodplain width and inundation and tributary creek discharge in the lower Savannah River of Georgia and South Carolina. Hydrobiologia 598:139–148. doi:10.1007/s10750-007-9146-6
76. Mion JB, Stein RA, Marschall EA (1998) River discharge drives survival of larval walleye. Ecol Appl 8:88–103. doi:10.2307/2641313
77. Muth RT, Schmulbach JC (1984) Downstream transport of fish larvae in a shallow Prairie River. Trans Am Fish Soc 113:224–230. doi:10.1577/1548-8659(1984)113<224:Dtofli>2.0.Co;2
78. Nathan R, Getz WM, Revilla E et al (2008) A movement ecology paradigm for unifying organismal movement research. Proc Natl Acad Sci USA 105:19052–19059. doi:10.1073/pnas.0800375105
79. Northcote TG (1962) Migratory behaviour of juvenile rainbow trout, Salmo-Gairdneri, in outlet and inlet streams of Loon Lake, British-Columbia. J Fish Res Board Can 19:201–270
80. Oesmann S (2003) Vertical, lateral and diurnal drift patterns of fish larvae in a large lowland river, the Elbe. J Appl Ichthyol 19:284–293. doi:10.1046/j.1439-0426.2003.00503.x
81. Olszewski J, Haehnel M, Taguchi M, Liao JC (2012) Zebrafish larvae exhibit rheotaxis and can escape a continuous suction source using their lateral line. Plos One. doi:10.1371/journal.pone.0036661
82. Paris CB, Atema J, Irisson JO, Kingsford M, Gerlach G, Guigand CM (2013) Reef odor: a wake up call for navigation in reef fish larvae. Plos One. doi:10.1371/journal.pone.0072808
83. Pavlov DS (1994) The downstream migration of young fishes in rivers—mechanisms and distribution. Folia Zool 43:193–208
84. Pavlov DS, Sbikin YN, Vashchinnikov AE, Mochek AD (1972) The effect of light intensity and water temperature on current velocities critical to fish. Vopr Ikhtiologii 12:769–778
85. Pavlov DS, Kuragina GN, Nezdoliy VK, Nekrasova NP, Brodskiy DA, Ersler AL (1978) Some features of the downstream migrations of Juvenile Fishes in the Volga and Kuban Rivers. J Ichthyol 17:63–374
86. Pavlov DS, Lupandin AI, Degtyareva NG, Dedov SM (1995) The role of turbulence in the formation of fish larvae distribution patterns in wide and narrow water streams. Dokl Akad Nauk SSSR 341:842–845
87. Pavlov DS, Mikheev VN, Lupandin AI, Skorobogatov MA (2008) Ecological and behavioural influences on juvenile fish migrations in regulated rivers: a review of

experimental and field studies. Hydrobiologia 609:125–138. doi:10.1007/s10750-008-9396-y
88. Pavlov DS, Kostin VV, Zvezdin AO (2011) Patterns of movements of fish with negative type of rheoreaction. J Ichthyol 51:536–542. doi:10.1134/S0032945211040138
89. Peňáz M, Roux AL, Jurajda P, Olivier JM (1992) Drift of larval and juvenile fishes in a bypassed floodplain of the upper river Rhone, France. Folia Zool 41:281–288
90. Persat H, Olivier JM (1995) The first displacements in the early stages of Chondrostoma nasus under experimental conditions. Folia Zool 44:43–50
91. Reeves KS, Galat DL (2010) Do larval fishes exhibit diel drift patterns in a large, turbid river? J Appl Ichthyol 26:571–577. doi:10.1111/j.1439-0426.2010.01481.x
92. Reichard M, Jurajda P (2004) The effects of elevated river discharge on the downstream drift of young-of-the-year cyprinid fishes. J Freshw Ecol 19:465–471. doi:10.1080/02705060.2004.9664921
93. Reichard M, Jurajda P (2007) Seasonal dynamics and age structure of drifting cyprinid fishes: an interspecific comparison. Ecol Freshw Fish 16:482–492. doi:10.1111/j.1600-0633.2007.00229.x
94. Reichard M, Jurajda P, Vaclavik R (2001) Drift of larval and juvenile fishes: a comparison between small and large lowland rivers. Archiv Für Hydrobiol–Suppl 135(2-4):373–389. doi:10.1127/lr/12/2001/373
95. Reichard M, Jurajda P, Ondrackova M (2002a) The effect of light intensity on the drift of young-of-the-year cyprinid fishes. J Fish Biol 61:1063–1066. doi:10.1006/jfbi.2002.2116
96. Reichard M, Jurajda P, Ondrackova M (2002b) Interannual variability in seasonal dynamics and species composition of drifting young-of-the-year fishes in two European lowland rivers. J Fish Biol 60:87–101. doi:10.1006/jfbi.2001.1813
97. Reichard M, Jurajda P, Smith C (2004) Spatial distribution of drifting cyprinid fishes in a shallow lowland river. Archiv Für Hydrobiologie 159:395–407. doi:10.1127/0003-9136/2004/0159-0395
98. Roberts JH, Angermeier PL, Hallerman EM (2013) Distance, dams and drift: what structures populations of an endangered, benthic stream fish? Freshw Biol 58:2050–2064. doi:10.1111/Fwb.12190
99. Robinson AT, Clarkson RW, Forrest RE (1998) Dispersal of larval fishes in a regulated river tributary. Trans Am Fish Soc 127:772–786. doi:10.1577/1548-8659(1998)127<0772:Dolfia>2.0.Co;2
100. Roussel JM, Bardonnet A (1999) Ontogeny of diel pattern of stream-margin habitat use by emerging brown trout, Salmo trutta, in experimental channels: influence of food and predator presence. Environ Biol Fishes 56:253–262. doi:10.1023/A:1007504402613
101. Schiemer F, Keckeis H, Reckendorfer W, Winkler G (2001) The "inshore retention concept" and its significance for large rivers. Arch Hydrobiol (Suppl.) (Large Rivers) 135:509–516
102. Schiemer F, Keckeis H, Kamler E (2003) The early life history stages of riverine fish: ecophysiological and environmental bottlenecks. Comp Biochem Physiol Part A 133:439–449. doi:10.1016/S1095-6433(02)00246-5
103. Schludermann E, Tritthart M, Humphries P, Keckeis H (2012) Dispersal and retention of larval fish in a potential nursery habitat of a large temperate river: an experimental study. Can J Fish Aquat Sci 69:1302–1315. doi:10.1139/F2012-061
104. Schludermann E, Liedermann M, Hoyer H, Tritthart M, Habersack H, Keckeis H (2013) Effects of vessel-induced waves on the YOY-fish assemblage at two different habitat types in the main stem of a large river (Danube, Austria). Hydrobiologia 729:3–15. doi:10.1007/s10750-013-1680-9
105. Schmutz S, Jungwirth M (1999) Fish as indicators of large river connectivity: the Danube and its tributaries. Archiv Für Hydrobiol 115(3):329–348. doi:10.1127/lr/11/1999/329

106. Shepherd WC, Mills EL (1996) Diel feeding, daily food intake, and Daphnia consumption by age-0 gizzard shad in Oneida Lake, New York. Trans Am Fish Soc 125:411–421. doi:10.1577/1548-8659(1996)125<0411:Dfdfia>2.3.Co;2
107. Somarakis S, Maraveya E, Tsimenides N (2000) Multispecies ichthyoplankton associations in epipelagic species: is there any intrinsic adaptive function? Belg J Zool 130:129–133
108. Sonny D, Jorry S, Watriez X, Philippart JC (2006) Inter-annual and diel patterns of the drift of cyprinid fishes in a small tributary of the Meuse River, Belgium. Folia Zool 55:75–85
109. Staaterman E, Paris CB (2014) Modelling larval fish navigation: the way forward. ICES J Mar Sci 71:918–924. doi:10.1093/icesjms/fst103
110. Stewart WJ, Cardenas GS, McHenry MJ (2013) Zebrafish larvae evade predators by sensing water flow. J Exp Biol 216:388–398. doi:10.1242/Jeb.072751
111. Stoll S, Beeck P (2012) Larval fish in troubled waters - is the behavioural response of larval fish to hydrodynamic impacts active or passive? Can J Fish Aquat Sci 69:1576–1584. doi:10.1139/F2012-086
112. Thorrold SR, Jones GP, Planes S, Hare JA (2006) Transgenerational marking of embryonic otoliths in marine fishes using barium stable isotopes. Can J Fish Aquat Sci 63:1193–1197. doi:10.1139/F06-048
113. Thorrold SR, Zacherl DC, Levin LA (2007) Population connectivity and larval dispersal using geochemical signatures in calcified structures. Oceanography 20:80–89. doi:10.5670/oceanog.2007.31
114. Tonkin Z, King A, Mahoney J (2007) Testing a modification to a standard passive drift net to capture drifting ichthyofauna. Fish Manag Ecol 14:299–301. doi:10.1111/j.1365-2400.2007.00558.x
115. Usvyatsov S, Picka J, Taylor A, Watmough J, Litvak MK (2013) Timing and extent of drift of shortnose sturgeon larvae in the Saint John River, New Brunswick, Canada. Trans Am Fish Soc 142:717–730. doi:10.1080/00028487.2012.760484
116. Wanzenböck J, Schiemer F (1989) Prey detection in cyprinids during early development. Can J Fish Aquat Sci 46:995–1001. doi:10.1139/f89-129
117. Webb PW, Cotel AJ (2011) Assessing possible effects of fish-culture systems on fish swimming: the role of stability in turbulent flows. Fish Physiol Biochem 37:297–305. doi:10.1007/s10695-011-9497-9
118. White JL, Harvey BC (2003) Basin-scale patterns in the drift of embryonic and larval fishes and lamprey ammocoetes in two coastal rivers. Environ Biol Fishes 67:369–378. doi:10.1023/A:1025891216345
119. Widmer AM, Fluder JJ, Kehmeier JW, Medley CN, Valdez RA (2012) Drift and retention of pelagic spawning minnow eggs in a regulated river. River Res Appl 28:192–203. doi:10.1002/Rra.1454
120. Winkelmann C, Petzoldt T, Koop JHE, Matthaei CD, Benndorf J (2008) Benthivorous fish reduce stream invertebrate drift in a large-scale field experiment. Aquat Ecol 42:483–493. doi:10.1007/s10452-007-9101-7
121. Wolter C, Arlinghaus R (2003) Navigation impacts on freshwater fish assemblages: the ecological relevance of swimming performance. Rev Fish Biol Fish 13:63–89. doi:10.1023/A:1026350223459
122. Wolter C, Sukhodolov A (2008) Random displacement versus habitat choice of fish larvae in rivers. River Res Appl 24:661–672. doi:10.1002/Rra.1146
123. Wootton RJ (1998) Ecology of teleost fishes (fish and fisheries), vol 24. Kluwer Academic Publishers, Dordrecht
124. Zens B (2015) Linking behaviour and hydraulics: rheoreaction and movement patternsof early stages of C. nasus under different flow conditions. MSc, University of Vienna
125. Zite A, Irrgeher J, Cervicek M et al (2014) Individual-specific transgenerational marking of common carp, Cyprinus carpio L., using Sr-86/Sr-84 double spikes. Mar Freshw Res 65:978–986. doi:10.1071/Mf13235

126. Zitek A, Schmutz S, Unfer G, Ploner A (2004a) Fish drift in a Danube sidearm-system: i. site-, inter- and intraspecific patterns. J Fish Biol 65:1319–1338. doi:10.1111/j.1095-8649.2004.00533.x

127. Zitek A, Schmutz S, Ploner A (2004b) Fish drift in a Danube sidearm-system: II. seasonal and diurnal patterns. J Fish Biol 65:1339–1357. doi:10.1111/j.1095-8649.2004.00534.x

# CHAPTER 7

# Taxonomy, Distribution and Evolution of Trisopterine Gadidae by Means of Otoliths and Other Characteristics

**Pieter A. M. Gaemers**

*Joost van den Vondelstraat 30, 7103 XW Winterswijk, The Netherlands*

## ABSTRACT

In a greater study of the recent fossil Gadidae, the object of this paper is to better define the trisopterine species and their relationships. The taxonomy of the four recent species usually included in the genus Trisopterus is further elaborated by means of published and new data on their otoliths, by published data on general external features and meristics of the fishes, and their genetics. Fossil otoliths, from the beginning of the Oligocene up to the present, reveal much of their evolution and throw more light on their relationships. Several succeeding and partly overlapping lineages representing different genera are recognized during this time interval. The genus Neocolliolus Gaemers, 1976, for Trisopterus esmarkii (Nilsson, 1855), is more firmly based. A new genus, Allotrisopterus, is introduced for Trisopterus minutus (Linnaeus, 1758). The similarity with Trisopterus capelanus (Lacepède, 1800) is an example of convergent evolution. The tribe Trisopterini Endo (2002) should only contain Trisopterus, Allotrisopterus and Neocolliolus as recent genera. Correct identification of otoliths from fisheries research and from sea bottom samples extends the knowledge of the present day geographical distribution of T. capelanus and T. luscus (Linnaeus, 1758). T. capelanus is also living along the Atlantic coast of Portugal and at least up to and including the Ría de Arosa, Galicia, Spain. There it can easily be mistaken for A. minutus that is also living there. Otoliths of T. luscus have been identified from the Evvoïkós Channel between Euboia and the mainland of Greece, thus it must live also in the Aegean Sea. Otoliths prove to be a powerful tool in taxonomy, biogeography and evolution of teleosts.

**Keywords**: Trisopterus; Neocolliolus; new genus Allotrisopterus; Trisopterini; new data on systematics; evolution; and biogeography

# 1. INTRODUCTION

From Linnaeus onwards the cod family Gadidae included a large and often variable cluster of different gadiform fishes (see for instance [1,2,3,4,5,6]). Progressive insight of many researchers has rightly curtailed the number of species that should be included in this family. In the literature, this group is usually ranked as the subfamily Gadinae (see [2,3,4,5,6,7,8]). Following Howes [9] and Fahay [10], I define the Gadidae family here as comprising only the species with three dorsal fins. Twelve genera are usually distinguished among the living representatives of this family, comprising 22 recent species. Raniceps raninus (Linnaeus, 1758), a primitive gadiform species with two dorsal fins of which the first is tiny and the second very long, should not be included in the Gadidae as indicated by Eschmeyer [11], but should be placed in the separate family Ranicipitidae, erected by T.P. Gill [12], see [8].

The Gadidae live in cold to temperate seas, where there is much less food available in the winter. Therefore, the cods are generalists, opportunistic feeding on a large variety of prey species. Closely related species are often too much competitors of one another because of a similar position in the food web. This explains the poor cladogenesis in this family: five genera, Arctogadus, Boreogadus, Melanogrammus, Merlangius and Theragra are monotypic (see revision of Arctogadus in Jordan et al. [13] and of Theragra in Byrkjedal et al. [14] (Theragra finnmarchica Koefoed, 1956 is supposed to be synonymous with Theragra chalcogramma (Pallas, 1811), see [10]); five genera, Eleginus, Gadiculus, Microgadus, Micromesistius and Pollachius, comprise two species, the genus Gadus comprises three species (but never more than two in the same area), and the genus Trisopterus, as it is usually defined up to now, comprises four species, of which often three are living in the same area (in this study it will be shown that Trisopterus capelanus (Lacepède, 1800) is a separate species and not a subspecies). For the following four genera it is certain that at present never more than one species of the genus occurs in an area, because of geographic separation: Eleginus, Gadiculus, Microgadus, and Micromesistius. It should be noted that Gadiculus argenteus Guichenot, 1850 and Gadiculus thori Schmidt, 1913, are not subspecies as it has been assumed since Svetovidov [2,15,16], but are actually two different species based on otolith morphology and ontogeny, larval pigmentation and external features of adults (see also Schmidt [17]). Trisopterus thus seems to be an exception, with a more extensive cladogenesis and more species living together in the same area than any other gadid.

The taxonomy and evolution of the trisopterine Gadidae have been the subject of several specific studies since the beginning of the 21st century, stimulated by new genetic data within this group [18,19,20,21,22], and by new genetic data of a larger group of Gadiformes including some trisopterine species [7,8]. The profound investigations by Delling et al. [21] moreover are indispensable because they have straightened up the taxonomic mess in the trisopterines that started already in the beginning of scientific taxonomy before Linnaeus and continued up to their publication. In a wise manner they designated neotypes for A. minutus and T. capelanus, and a lectotype for T. luscus, as well as neotypes for several synonyms of these species.

The aim of the present study is to further update the taxonomy and evolution of the trisopterine Gadidae based on recent and fossil otoliths, with the inclusion of other data from the literature. Evidence will be given that cladogenesis in trisopterine Gadidae is not more extensive than in other gadids. With the genera Neocolliolus and Allotrisopterus included, the number of genera in the Gadidae amounts to fourteen.

In order to clearly distinguish between recent and fossil taxa a dagger (†) is placed before a genus or species that is extinct.

# 2. RESULTS, INCLUDING A SURVEY OF CHARACTERISTICS OF RECENT TRISOPTERINES

## 2.1. Factors Influencing Otolith Shape

1. Shape changes during ontogeny. It is important to shortly describe the changes that occur in otoliths during ontogeny. Otoliths usually start to grow before hatching as a tiny sphere. In the larval stage, at first, length increases more rapidly than the other dimensions. Thereafter, the height of the otolith increases more rapidly than thickness. Therefore, a more specific shape starts forming. Soon also a beginning of ornamentation develops. The number of knobs and furrows increases during early growth and also the relief becomes more pronounced. Usually in the late youth or early

adulthood of the fish the relief is highest and the number of knobs stops augmenting. When becoming older the relief flattens more and more. In the course of growth the relative width and depth of the sulcus also increase in Gadidae and other families. Finally, the length-height and length-thickness ratios of otoliths are not only changing in the beginning of the life of the fish, but remain in a state of change during the whole life of the fish. In the first part of the life of a fish, otoliths become more slender and thinner. Some species show this tendency in their otoliths during their whole life, but many show a reversal trend in later life, when otoliths gradually become higher and thicker again [3]. These allometrical developments can be weak or strong, depending on the species and/or the phylogenetic stage. If this allometry is strong, it is possible that early and late ontogenetic stages of the same species are wrongly interpreted as different species.

2. In any species, there is variability among its individuals. Otoliths are no exception in this respect. Shape differences in otoliths may be the result of genetic differences, but also environmental factors may influence otolith shape. Sometimes aberrant shapes occur and occasionally the left and right otolith are remarkably different. It is not yet known what causes such phenomena.

3. Within an evolutionary lineage of a gadid species the shape of the otoliths gradually change according to a fixed pattern that probably is genetically controlled [3]. Every lineage suddenly starts without a gradual change, originating from an older lineage that usually has a big overlap in geological time with the younger lineage. The otoliths of the new lineage have many characteristics in common with the old lineage, but the new lineage also shows some new features, so that they usually can be distinguished easily by an experienced otolith specialist. In its first phylogenetic stage the otoliths are relatively short and thick, becoming more slender and thinner in its second stage. In the third stage, the otoliths are most slender and thin; afterwards, they become again shorter and thicker in the last, fourth, stage. Every gadid species evolves in this way and there is no fifth stage, so every species becomes extinct at a certain moment in time. The otoliths of many gadid species gradually evolve from smaller to larger in the course of hundreds of thousands or millions of years [3], but there is now much evidence that the otoliths of still a considerable number of species evolve to smaller sizes. Rarely the size of gadid otoliths remains the same during the whole evolution of the species. It is logical to suppose that the size of the fishes in an evolutionary lineage changes in the same direction as the otoliths do. However, complete fossil gadid fish skeletons are rare, thus this assumption usually cannot be proven by fossils.

4. Sexual dimorphism can occur in otoliths. This phenomenon was discovered for the first time in recent small cichlid species from Lake

Victoria, Tanzania. It occurs only in species that have large size differences between males and females [23]. Individuals of the larger sex do not only get larger otoliths, but also get more slender otoliths with more ornamentation. Later it was also found in several extinct Gadidae species, which have (much) smaller otoliths than those of Gadiculus argenteus Guichenot 1850, the smallest cod species living today [24]. It is logical to assume that the maximum size of these extinct species was (distinctly) smaller than 15 cm, the maximum total length of G. argenteus. The smallest fossil Gadidae species probably did not become larger than 5 cm.

Recent gadid females grow always larger and become older than males, and there is no reason to suppose that this was different for extinct species and genera in this family. Therefore the larger, more slender and ornamented otoliths within a population of otoliths of a fossil gadid are considered to be female otoliths. Why larger gadid species with size difference between the sexes do not—or only slightly—show sexual dimorphism in their otoliths, is not known.

5. There is also evidence that temperature may have some influence on the shape of otoliths. Near the northern (coldest) or southern (warmest) limits in several cases the otoliths of a species have different OL/OH and OL/OT ratios (OL: otolith length in mm; OH: otolith height (width) in mm; OT: otolith thickness in mm). In the coldest areas otoliths may be more slender and thinner, in the warmest areas they may be shorter (higher) and thicker. Examples of more slender and thinner otoliths are those of Limanda limanda (Linnaeus, 1758) [25]: Plate 91, figures A–F, in Härkönen (1986) compared to those from the same species in the southern North Sea. The same applies to the otoliths of Microstomus kitt (Walbaum,1792) [25]: Plate 93, figures C–F, compared to Plate 93, figures A,B; the latter figures represent higher otolith shapes that are known from the southern North Sea. The same may be true for the more slender otoliths of Phrynorhombus norvegicus (Günther, 1862) [25]: see Plate 88, figures C–F, Trachinus draco (Linnaeus, 1758) [25]: see Plate 72, figures A–D, and Taurulus bubalis (Euphrasen, 1786) [25]: Plate 50, figures A–F, compared to more southern otoliths in my own collection of recent otoliths. Furthermore, I observed that otoliths of Micromesistius poutassou (Risso, 1826) from the Mediterranean are higher than those in the Atlantic, and also the otoliths of Merlangius merlangus (Linnaeus, 1758) from the Adriatic Sea are higher than those in the Atlantic. I never found the reverse situation, that the otoliths of a certain species in warmer areas are more slender, so it seems to be a rule, that if the otolith shape changes under different temperature conditions, the change is consistently always in the same direction.

From the above-mentioned factors it is clear that one needs to study a large amount of otoliths in order to have a complete overview of the sizes and shapes of the otoliths of a (sub)species, and so ensuring good identification of closely related (sub)species. If this is done for one or a few (sub)species, it is not necessary to do it equally extensively for all (sub)species because overall variability can then be estimated also for other related (sub)species.

It is very important to picture otoliths of closely related species at the same magnifications in order to be able to notice differences between species and different phylogenetic stages of species at a given size of the otoliths. Otherwise the changes in shape during ontogeny and phylogeny often cannot be appreciated in the correct way, and can easily lead to misidentifications. Unfortunately, different otolith taxonomists use different magnifications and some taxonomists even use various magnifications in one publication. This hinders easy comparison. I would highly recommend to use standard magnifications (for instance 5×, 10×, 15×, 20×), depending on the maximum size of the otoliths of a species.

## 2.2. Characteristics of Gadidae Otoliths

In order to understand the taxonomic importance of otoliths, firstly it is necessary to describe and illustrate different primary characteristics of the external morphology of Gadidae otoliths (Figure 1). When using the term otolith, always the sagitta is meant, which is the largest otolith or ear stone in most bony fish families. The most important element of an otolith, functionally as well as taxonomically, is the sulcus acusticus or hearing groove, where numerous offshoots of the hearing nerve end. The sulcus acusticus, also shortly named sulcus, is situated on the inner (medial) surface of the otolith. The sulcus runs lengthwise on the otoliths, parallel to the length of the whole fish. The sulcus can be divided in two or three different portions. The anterior portion is called the ostium, the posterior portion is the cauda. In between, a collum can be present or absent. In the Gadidae the outline of the sulcus is always homosulcoid: ostium and cauda are clearly distinguishable and have about the same shape, so that the sulcus is more or less symmetrical [26]; the type of sulcus opening in the Gadidae usually is pseudobiostial: the sulcus ends indistinctly very near to the anterior and posterior margins, so that both ostium and cauda are closed toward the margins; sometimes the sulcus opening in the Gadidae is biostial: both ostium and cauda are open at the margins [26]. Ostium and cauda are usually filled partly or nearly entirely with cushions, called ostial respectively caudal colliculum. The colliculi do not extend to the margins. In otoliths of some Gadidae species a pseudocolliculum (a narrow ridge) occurs at the ventral part of the collum. Nolf and Steurbaut [27] use the term central collicular crest for it. The sulcus is often surrounded by narrow ledges or kinks, the crista superior and the crista inferior. The outline of this surface shows a broad anterior part in Gadidae otoliths, called the rostral margin, and a more or less pointed posterior end. Various angular portions may occur along the outline. In Gadidae otoliths pre- and postdorsal angles are usually present. Furthermore, ornamentation of lower or higher rounded knobs may occur along the margins. On the reverse side, the outer surface, usually rounded knobs occur, separated by sharp furrows. The ventral portion of a Gadidae otolith is usually thicker than the dorsal portion; sometimes ventral and dorsal portions are equally thick. The ventral and dorsal views show the thickness of the otolith and if the otolith is flat or bent lengthwise; they also show details of the curvature of the inner and outer surfaces.

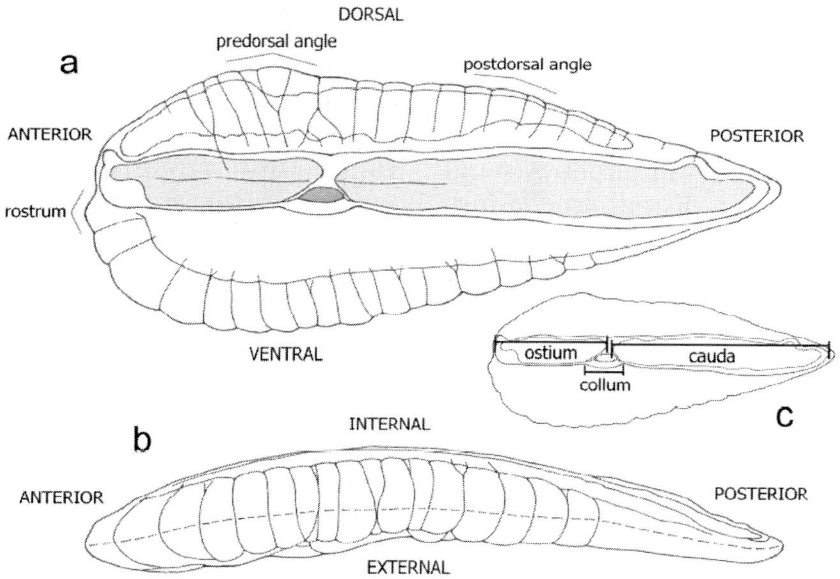


**Figure 1**. Left otolith of † Colliolus mistensis Gaemers, 1990. (a) Inner surface showing length and height (or width) of the otolith; the colliculi are light grey, the medially situated pseudocolliculum is dark grey; the ventral margin is regularly bent, the dorsal margin has a distinct predorsal angle and a weak postdorsal angle; (b) Ventral view showing length, thickness and lengthwise curvature of the otolith; (c) Overview with ostium, cauda and collum.

## 2.3. Otolith Characteristics of Recent Trisopterine Gadidae

Recent trisopterine otoliths can easily be distinguished from otoliths of other gadids:

- Pear shaped (also in Gadiculus-like and extinct Colliolus-like otoliths)

- Thick, especially the adult otoliths
- Sulcus (very) wide
- Colliculi long and wide, largely filling sulcus and touching one another
- Collum only at ventral part of sulcus
- Pseudocolliculum absent

## 2.4. Other Characteristics of Trisopterine Gadidae

A synopsis of several biological characteristics of trisopterine Gadidae summarized from the literature is given in Table 1 and Table 2. Trisopterine fishes are small gadids, most of which usually do not become larger than 25 cm in total length, whereas their maximum size rarely exceeds 30 cm; only Trisopterus luscus (Linnaeus, 1758) normally can grow up to 46 cm (Table 1). Their life span is short, generally between 3 and 5 years, and they mature mostly in their second year (Table 1). Otoliths, of course, become larger when the fishes grow, but the largest fishes do not necessarily have the largest otoliths (compare fish size and otolith size in Table 1). This is caused by a greater variability in the OL/TL ratio in larger fish (OL: otolith length in mm; TL: total fish length in mm).

**Table 1**. Synopsis of ages and sizes of recent trisopterine Gadidae.

| Species | Age of Females When 50% Are Mature | Age of Males When 50% Are Mature | Maximum Known Age (Normal Max. Age) | TL of Females When 50% Are Mature (in mm) | TL of Males When 50% Are Mature (in mm) | Normal Maximum TL (in mm) | Extreme Maximum TL (in mm) | Maximum Observed OL (in mm) [a] |
|---|---|---|---|---|---|---|---|---|
| *Neocolliolus esmarkii* | 1.5 years [b] | 1.2 years [b] | 7 [c] (3 [d]–4 [e]) | 140 [e] | 150 [e] | 235 | 350 [f] | 9.07 (TL 205 mm) |
| *Allotrisopterus minutus* | 1.7 years [g] | 2.7 years [g] | 7 [g] (4–5) | 160–190 [h] | 140–200 [h] | 245 | 280 [g] | 11.68 (TL 269 mm) |
| *Trisopterus luscus* | 1.7 years [i] | 1.8 years [i] | 9 [j] (5) | 220 [j,k]<br>192 [k] | 234 [j] | 460 [j] | 490 [l] | 14.55 (TL 415 mm) |
| *Trisopterus capelanus* | end 1st year [m] | end 1st year [m] | 6 [m,n] (4) | 133 [o]<br>138.2 [p] | 125 [o]<br>123.9 [p] | 250 [g] | 320 [q] | 12.82 (TL 288 mm) |

[a] My own observations and measurements; [b] [28]; [c] [29], exceptional age; [d] [30], North Sea; [e] [31], Iceland; [f] [32]; [g] [33]; [h] [34]; [i] [35]; [j] [36]; [k] [37], Average length at maturity established at 220 mm macroscopically and at 192 mm histologically (microscopically); [l] [38], caught by B. Katgerman (Dordrecht, The Netherlands) on 5 May 1995; [m] [39]; [n] [40], Adriatic Sea; [o] [41]; [p] [42]; [q] [43], Adriatic Sea. TL: Total fish length in mm, OL: otolith length in mm.

**Table 2.** Synopsis of depth distribution and schooling characteristics of recent trisopterine Gadidae.

| Species | Preferred Depth | Common Depth | Depth in Bays and Estuaries | Minimum Depth in Open Sea | Maximum Depth | Shoaling |
|---|---|---|---|---|---|---|
| *Neocolliolus esmarkii* | 100–200 m [a] Highest abundance I-group and older: 200 m [b] | 50–250 m (80–200 m [c]) | Few meters [d] | 40 m [c] | 560 m [e] | Very large shoals |
| *Allotrisopterus minutus* | 40–160 m [f] | 25–300 m [c] | 0.5 m [g] | 15 m [h] | 440 m [i] | Small shoals |
| *Trisopterus luscus* | 30–100 m [j] | 3–300 m [c] | Few meters [d] | 3 m [c] | 650 m<br>300 m [n] | Small to large shoals |
| *Trisopterus capelanus* | 50–120 m [k] | 40–250 m [l] | | 0 m [m] | 310 m [o]<br>450 m [m] | Small shoals |

[a] [31]; [b] [44]; [c] [45]; [d] [46]; [e] [47]; [f] [33]; [g] [48], mainly at night; [h] [49]; [i] [50]; [j] [16]; [k] [42]; [l] [51]; [m] [52]; [n] [53]; [o] [54].

Trisopterines inhabit the continental shelf and the upper part of the continental slope. The smallest and youngest fishes live in the shallowest waters; adult fishes can live in the deepest waters of their depth range (Table 2). Apart from young individuals, Neocolliolus esmarkii (Nilsson, 1855) lives in the deepest waters, usually in the lower part of the continental shelf.

All trisopterines are schooling to a lesser or greater extent (Table 2). Due to their relatively small size they are rather low in the food web and are common to abundant in European seas. This is also valid for the Tertiary and Quaternary, the more so as otoliths of ancestral lineages are known to grow to smaller maximum sizes than recent ones. Thus, fossil species must have remained smaller than the recent representatives. Trisopterine otoliths are common in marine sediments of northwestern Europe from the Oligocene up to the present day, so that it is well possible to reconstruct their phylogeny.

## 2.5. Neocolliolus: A Justifiable Taxon in the Gadidae Family

### 2.5.1. Introductory remarks

Based on a number of differences the new genus Neocolliolus for Trisopterus esmarkii was erected in a mainly paleontological publication [55]. Apparently, this taxonomic change has been overlooked by many ichthyologists up to now.

There is no doubt that Neocolliolus esmarkii is more closely related to the other trisopterines than to other gadids, because the mutual external characteristics of the fishes and of the otoliths resemble one another more in comparison with other species. All trisopterines are relatively small fishes. Full-grown individuals of N. esmarkii are small, usually not larger than 26 cm total fish length. In the Barents Sea individuals up to 35 cm have been caught [32].

The otoliths of N. esmarkii differ in various important respects from the other recent trisopterine species (Figure 2, Table 3): the ostium is only slightly shorter than the cauda; the ostial colliculum is relatively long and narrows above the collum as well as at its anterior part; the collum is long with a trapezoidal or ribbon like shape; and the frontal rim of the otolith is rounded. Because of growing insight it is necessary to change and enlarge the otolith diagnosis of the genus Neocolliolus (Gaemers, 1976) in several respects.

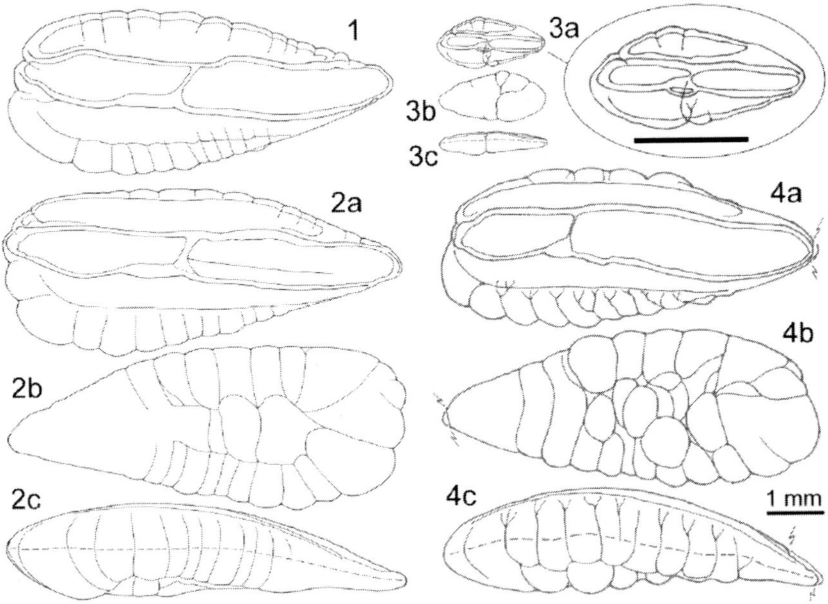


**Figure 2**. Differences between otolith morphology of Neocolliolus and Trisopterus. (1) Neocolliolus esmarkii, Recent, North Sea, TL 159, OL 6.85, coll. Mardik Leopold, number EU25; (2a–c) Neocolliolus esmarkii, Recent, North Sea, TL 157, OL 7.07, coll. Mardik Leopold, number EU22; (3a–c) Neocolliolus esmarkii, Recent, North Sea, TL 46, OL 1.715, mirror image, coll. Mardik Leopold, number CB1; (4a–c) New subspecies of Trisopterus capelanus, Early Miocene, Behrendorfian, Berchem Formation, Kiel Sands, borehole airfield Haamstede, The Netherlands, 132–133 m below surface, mirror image.

### 2.5.2. Amended otolith diagnosis of Neocolliolus

A small gadid genus with moderately small, rather slender and moderately thick otoliths up to a little more than 9 mm long. General shape oblongly pear-shaped, only slightly bent lengthwise. Inner surface slightly to moderately convex lengthwise and not twisted. Ventral portion of rostral end strongly rounded. Caudal end sharply pointed. Distinct, rounded predorsal angle at the dorsal portion of the rostral rim. Postdorsal angle usually not pronounced, but well visible especially in medium-sized otoliths. Parts between caudal point and postdorsal angle, and between postdorsal and predorsal angle usually straight. Sulcus acusticus wide and shallow, and nearly straight; it is nearly completely filled with the caudal and ostial colliculi; cauda only slightly bent towards the ventral rim. Ostial

**Table 3**. Differences and similarities between otoliths of Neocolliolus, Allotrisopterus and Trisopterus.

| *Neocolliolus esmarkii* Lineage | *Allotrisopterus minutus* Lineage | *Trisopterus luscus* and *T. capelanus* Lineages |
|---|---|---|
| Otoliths flat or slightly curved lengthwise | Otoliths slightly curved or flat lengthwise | Otoliths clearly curved lengthwise |
| Inner surface slightly to moderately convex lengthwise and not twisted | Inner surface usually moderately convex and twisted | Inner surface strongly convex lengthwise and twisted |
| Postdorsal angle usually distinct | Postdorsal angle usually distinct | Postdorsal angle usually absent |
| Strongest ornamentation: low relief with low to very low knobs and shallow grooves | Strongest ornamentation: very high relief with protruding knobs and deep grooves | Strongest ornamentation: very high relief with protruding knobs and deep grooves |
| Frontal rim rounded | Frontal rim ± straight, nearly perpendicular to long axis | Frontal rim ± straight, obliquely running with respect to the long axis |
| Collum long, trapezoidal or ribbon like | Collum short, about triangular | Collum short, about triangular |
| Ostium slightly shorter than cauda | Ostium clearly shorter than cauda | Ostium clearly shorter than cauda |
| Cauda (nearly) straight | Cauda with S-curve | Cauda with S-curve |
| Ostial colliculum relatively long, also above collum where it narrows; its anterior part narrows too | Ostial colliculum relatively short, regularly oval-shaped | Ostial colliculum relatively short, regularly oval-shaped |

colliculum only slightly shorter than caudal colliculum. Posterior part of ostial colliculum narrow above collum. Anterior part of ostial colliculum also narrowed, except in the largest otoliths. Rather big trapezoidal to ribbon like collum with a long base and only present on the ventral part of the sulcus acusticus. No pseudocolliculum in the collum of adults, but early juvenile otoliths may have a pseudocolliculum (Figure 2(3a)). Ornamentation on outer surface and along the rims weakly developed with low or even very low, rounded knobs, and shallow grooves on the outer surface and along the rims fading away in larger otoliths. Ventral part of otolith clearly thicker than dorsal part.

### 2.5.3. Other distinguishing characteristics of *Neocolliolus*

The special taxonomic position of this species cannot only be shown by the otoliths but also many characteristics of the fishes are different (Figure 3, Table 4). Several differences are connected with a different way of life and ecology of this species compared with the other trisopterines. The position of the mouth shows that N. esmarkii is equipped for foraging in mid water, and not on the sea-bottom like the other species. This is confirmed by the food that is regularly found in the stomachs of this species. It feeds mostly on planktonic crustaceans (copepods, euphausiids, shrimps and amphipods), but also on small fish and various eggs and larvae

[31,44,56,57]. The much higher number of gill rakers [16] shows that this species can filter smaller prey than the other trisopterines. In their diagnosis for the genus Cohen et al. [49] mention a well-developed chin barbel for all species assigned to Trisopterus, but this is not correct. N. esmarkii has a very small chin barbel. This feature and the shorter pelvic fin are in accordance with a mainly mesopelagic life as contrasted with the other trisopterines that live close to the sea-bottom and feed just above it or in it. The more streamlined body, the longer pectoral fin, the lower dorsal and anal fins, and the short interspaces between the dorsal fins probably all indicate that N. esmarkii is able to swim faster than the other trisopterine species. It is also recognized by a clearly counter shaded coloration with a dark, grey-brown dorsum, dull silvery sides, and a white belly [21,49].

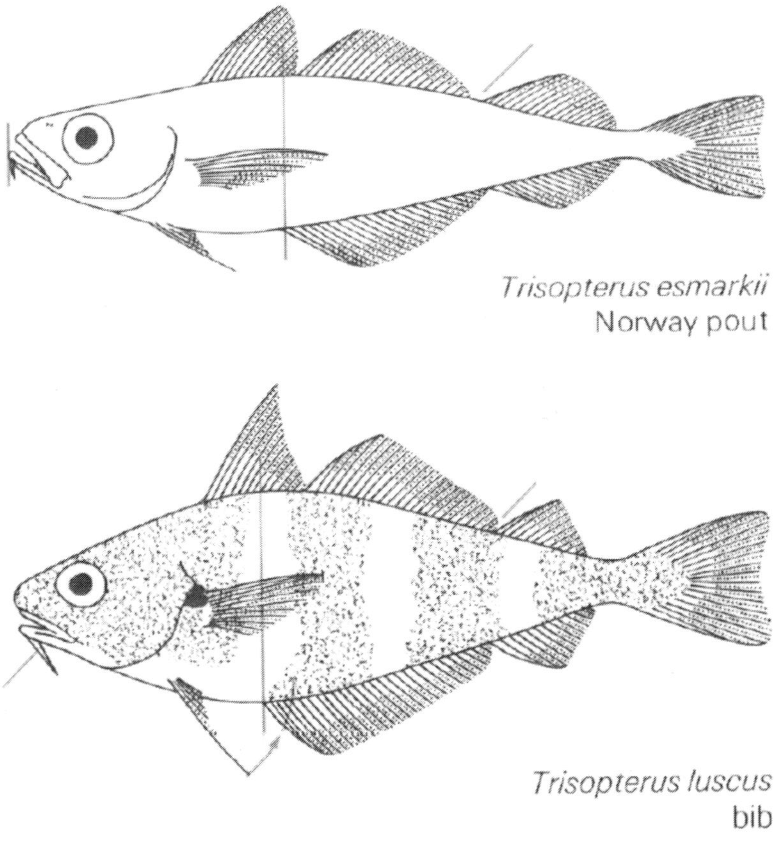


**Figure 3**. External features of Neocolliolus and Trisopterus showing important differences indicated with red lines. From Wheeler (1978) [45].

**Table 4**. Distinguishing external and ecological characteristics of Neocolliolus esmarkii with regard to the other recent trisopterine species.

| *Neocolliolus esmarkii* | Other Recent Trisopterine Species |
|---|---|
| Streamlined, slender body shape with a lower head profile and deepest body part more backwards | Less streamlined body shape with a higher head profile and deepest body part more forwards |
| Lower jaw slightly longer than upper jaw | Upper jaw clearly longer than lower jaw |
| Chin barbel short and thin | Chin barbel long and thick |
| Pelvic fin relatively short | Pelvic fin relatively long |
| Pectoral fin relatively long | Pectoral fin relatively short |
| Dorsal and anal fins lower and more rounded | Dorsal and anal fins higher and more pointed |
| Short interspaces between the dorsal fins | No interspaces between dorsal fins |
| Eye relatively large | Eye relatively smaller |
| High number of gill rakers (33–43) | Low number of gill rakers (14–22) |
| Benthopelagic to mesopelagic | Demersal |
| Food: mainly planktonic crustaceans | Food: various infaunal and epifaunal (in)vertebrates |

## 2.6. Trisopterus capelanus Is a Separate Species

The taxonomic status of Trisopterus capelanus (Lacepède, 1800) has been discussed much in the course of time. Originally described as a species, T. capelanus, it nowadays is regarded by most biologists as a subspecies of A. minutus, following Svetovidov's monograph [1,2] on the Gadiformes (published in English in 1962). This view is also hold by the authors of the last published big monograph of the Gadiformes [49]. Confusion was already created by Linnaeus [58] mentioning Gadus minutus only from the Mediterranean Sea, but he took this erroneously from Artedi [59] who described this species also from the Atlantic (Svetovidov 1973). It is, therefore, unclear which species Linnaeus meant under the name Gadus minutus. Müller [60] decided to give this name to fishes from Danish waters (see [2]), and from that time onwards ichthyologists always meant with Gadus minutus the so-called poor cod living in the Northeastern Atlantic and the North Sea. This explains why Müller by various taxonomists is mentioned as the author of G. minutus, but this is not correct. According to Lacepède [61], G. capelanus is living in the Mediterranean Sea as well as in the Atlantic Ocean. Thus, both species were said to occur in the same large area, adding more fuel to the confusion. Several taxonomists, especially in the past, didn't know the right authorship of G. capelanus, and ascribed this taxon erroneously to Rafinesque-Schmaltz [62] or Risso [63]. Otherwise, sometimes T. capelanus is considered a subspecies of T. luscus [64]. In Tuset et al. [65] the adult otolith from the western Mediterranean Sea named Trisopterus minutus on their figure 32(B3) belongs unmistakably to T. capelanus, whereas the other otoliths from the north-eastern Atlantic on their figure 32(B1, B2) are indeed A. minutus.

Svetovidov [1,2] was of the opinion that T. capelanus was very close to A. minutus in external form and osteology, although he knew the findings of

Fage [66], and Chaine and Duvergier [67]. Therefore he decided to place both taxa in one species. Svetovidov had only three specimens of T. capelanus at his disposal, and it is very likely that he has never seen a fresh fish of this species. However, if he would have read Fage's article carefully, he should have come to another conclusion. Svetovidov even mentions that Chaine and Duvergier [67] indicate differences in the external characteristics of the fishes as well as of the otoliths of T. capelanus, A. minutus and T. luscus; in the latter paper even illustrations of the otoliths have been published. Apparently Svetovidov was unaware of the possibilities of otoliths to discriminate between closely related species in most bony fish families, otherwise he probably would have come to another decision.

Although the fishes of A. minutus and T. capelanus resemble one another much at first glance, because externally both are slender and have about the same maximum size, there are much more important differences. A. minutus is more slender than T. capelanus; the upper part of its head is less steeply rising; its curvature of the lateral line is less strong, its first anal fin is shorter [66]. Its number of fin rays of the second and third dorsal fins and the second anal fin is distinctly larger, and the same applies to its number of gill-rakers and vertebrae [1,2].

Moreover, the otoliths of these species differ very much. Those of A. minutus are generally much shorter, higher and thicker than those of T. capelanus. There is only little overlap in the length-height ratios for otoliths with OL greater than 7 mm (A. minutus: TL 49–273; T. capelanus: TL 66–288) (Figure 4). There is somewhat more overlap in the length-thickness ratios of the otoliths of these species; the greatest differences occur at OL between 5 and 10 mm (Figure 5). The postdorsal angle is always well developed in A. minutus otoliths, and usually absent in T. capelanus. In medium-sized to large otoliths the margin between the postdorsal angle and the posterior point is nearly always concave in A. minutus, and nearly never concave in T. capelanus. The angle between the rostral margin and the greatest length of the otolith is greater in A. minutus, usually around 80°; in T. capelanus it is on an average about 60°. The mean OL-OH and OL-OT curves of both species that can be constructed show concave shapes in the measured interval, but the curves for T. capelanus are stronger bent and those for A. minutus are straighter, thus indicating that the otoliths of T. capelanus grow more allometrical.

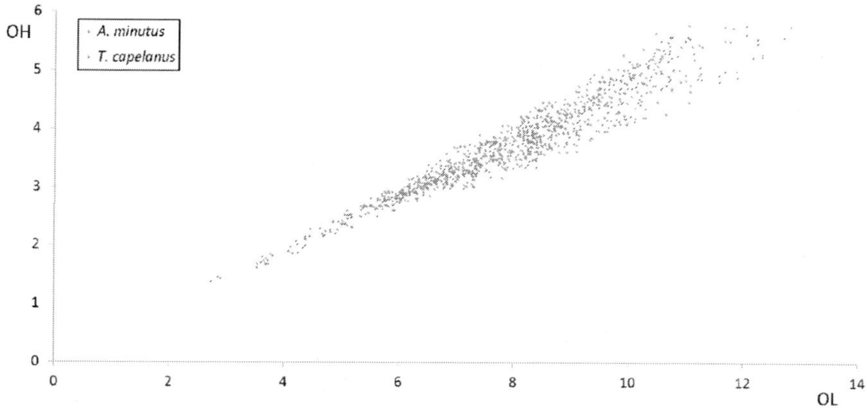


**Figure 4.** The ratios of otolith length (OL) and otolith height (OH) compared between Trisopterus capelanus ($n$ = 338) and Allotrisopterus minutus ($n$ = 764). OL and OH in mm.

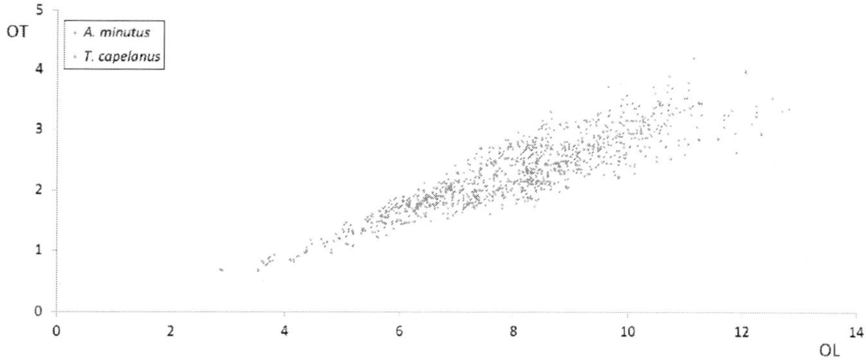


**Figure 5.** The ratios of otolith length (OL) and otolith thickness (OT) compared between Trisopterus capelanus ($n$ = 337) and Allotrisopterus minutus ($n$ = 741). OL and OH in mm.

The otoliths of T. capelanus more strongly resemble those of T. luscus in their general shape than those of A. minutus. The postdorsal angles are absent or weakly developed in both species. The inclination of their rostral rims is also similar, rather variable and largely overlapping in the two species, although that of T. capelanus on an average is a little bit less steep. The most obvious difference is that otoliths of T. capelanus are more slender than those of T. luscus, apparently reflecting the more slender external shape of the fishes of T. capelanus. The otoliths of the latter species are also larger with respect to the size of the fish (Figure 6). This is in agreement with the observed general tendency, that smaller species within a closely related group have relatively larger otoliths. This is in line with relatively larger eyes and various other organs in smaller species as well as in smaller individuals of a certain species. The fact is, that the maximum total fish length of T. capelanus (about 32 cm) remains much smaller than that of T. luscus (46 to 49 cm). Figure 6 also shows, that OL/TL ratio of both species is not linear but allometric: the convex curves that can be constructed for the mean values demonstrate that otolith lengths of both species become relatively smaller in larger fishes. Although the smallest specimens are not available, it is logical thinking that the curves must start at the zero point of the graph. For T. luscus Merayo and Villegas [68] however suggest that the OL/SL ratio (which is comparable with the OL/TL ratio) can be considered linear, but although they do not present a graph, their mean line clearly cannot start at the zero point because of the equations in the upper part of their Table 1. Their conclusion that the mean development of the OL/SL ratio follows a straight line is not true. In the last sentence of their section on otolith growth on page 117, they were aware of the negative allometry of this relationship because the power functions proved significantly different from 1.

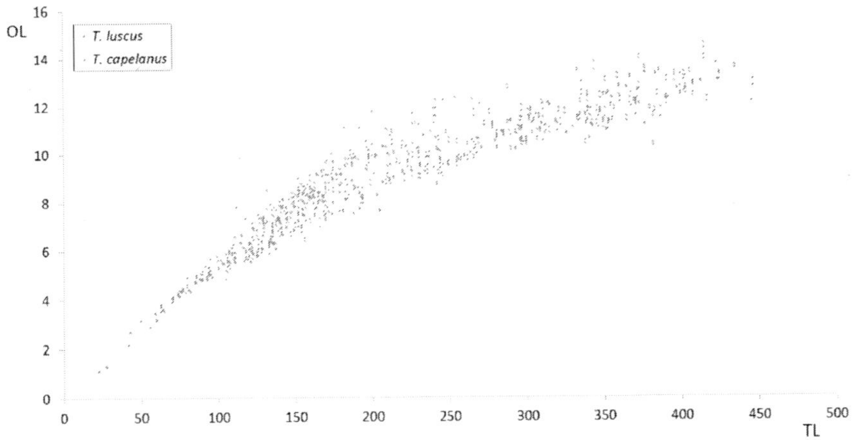


**Figure 6.** The ratios of otolith length (OL) and total fish length (TL) compared between Trisopterus capelanus ($n$ = 328) and Trisopterus luscus ($n$ = 767). OL and TL in mm.

The ornamentation of the outer surface of the otoliths of T. luscus culminates in a clearly stronger relief in medium-sized specimens: the knobs are then extremely high and the sharp grooves are very deep. The maximum expression of the ornamentation in T. capelanus is more normal and its relief smoothens earlier in later life. In T. luscus the length curvature of the outer surface is more regular, so that the thickest point of the otolith is not elevated above its surroundings. In larger otoliths of T. capelanus, in contrast, the thickest point is often clearly protruding forming an umbo. Although there might be some overlap in the variability of some characteristics of the otoliths of the two species, individual otoliths of these species usually can be distinguished rather easily. Based on many external characteristics of the fishes and of the otoliths it is obvious that T. luscus and T. capelanus are more related to one another than to the other species of the trisopterines, and they can be considered sister species. Genetic data support this close relationship very well: both the complete mitochondrial DNA of cytochrome b and the nuclear DNA of the rhodopsin genes point to a much closer relationship of T. luscus and T. capelanus than any other combination of two of the four trisopterine species [21,22]. This is a strong confirmation of earlier research in which was found that both species shared the highest number of loci with the same alleles extracted from muscle and liver tissue [18].

## 2.7. Allotrisopterus, a New Genus for Trisopterus Minutus

### 2.7.1. Allotrisopterus n. gen.

Type species: Gadus minutus Linnaeus, 1758;

Derivatio nominis: Allos (Greek): another one; tris (Greek): thrice; pteron (Greek): wing, fin. Diagnosis: **External characteristics**. A gadid genus of small trisopterine fishes (total fish length up to 26 cm) with a slender shape. Head profile low. Origin of first anal fin beneath the space between first and second dorsal fins or anterior to it. First anal fin only slightly longer than second anal fin.

**Meristic characteristics**. Numbers of gill rakers, vertebrae, pterygiophores of 3rd dorsal fin and pterygiophores of 2nd anal fin intermediate between N. esmarkii and T. luscus/T. capelanus. Number of 2nd dorsal fin pterygiophores close to N. esmarkii. Number of 1st dorsal fin pterygiophores close to T. luscus/T. capelanus.

**Otoliths**. Outline short and high. Flat or slightly curved lengthwise. Inner surface moderately convex lengthwise. Postdorsal angle usually distinct. Frontal margin nearly perpendicular to long axis.

### 2.7.2. Remarks of justification

There are a series of arguments for the erection of a new genus for Trisopterus minutus, based on external characteristics, meristics, otoliths and genetics.

Differences in **external characteristics** at first glance are not very obvious between A. minutus and T. capelanus. A. minutus is somewhat more slender than T. capelanus. The head profile in A. minutus is lower than in T. luscus and T. capelanus, and comes closer to that in N. esmarkii. In A. minutus the origin of the first anal fin is beneath the space between the first and second dorsal fins or anterior to it; about the same positioning is present in N. esmarkii [45]. In T. luscus and T. capelanus the origin of the first anal fin is halfway beneath the first dorsal fin. The first anal fin is only slightly longer than the second anal fin in A. minutus, whereas the first anal fin in T. luscus and T. capelanus is much longer than the second one.

The **meristic characteristics** show an intermediate position between N. esmarkii and the other trisopterines, especially the numbers of gill rakers, of vertebrae, of pterygiophores of the 3rd dorsal fin and of pterygiophores of the 2nd anal fin (Table 5). The number of 1st dorsal fin pterygiophores is more in the range of the T. luscus and T. capelanus, the number of 2nd dorsal fin pterygiophores is more in agreement with N. esmarkii.

**Table 5**. Meristics showing the intermediate position of Allotrisopterus minutus between Neocolliolus and Trisopterus (after reference [20]); numbers of vertebrae and gill rakers supplemented with data given in [1,2].

| Meristics | *N. esmarkii* | *A. minutus* | *T. capelanus* | *T. luscus* |
|---|---|---|---|---|
| Number of vertebrae | 51–55 | 47–51 | 45–48 | 46–49 |
| Number of gill rakers | 33–37 | 24–28 | 15–21 | 18–23 |
| Number of dorsal fin pterygiophores | | | | |
| 1st dorsal fin | 15–17 | 13–15 | 12–14 | 13–14 |
| 2nd dorsal fin | 26–28 | 24–28 | 18–23 | 22–25 |
| 3rd dorsal fin | 28–30 | 22–25 | 17–21 | 19–23 |
| 1st anal fin | 31–35 | 29–34 | 26–32 | 31–36 |
| 2nd anal fin | 27–30 | 23–25 | 17–22 | 19–22 |

The **otoliths** of A. minutus also show an intermediate position between Neocolliolus and Trisopterus (Table 3). The differences with Neocolliolus were already mentioned in the section on Neocolliolus as a justifiable taxon. The otoliths of A. minutus differ from those of T. luscus and T. capelanus by its distinct postdorsal angle, its steep frontal margin, its flatter side view, and its less convex inner surface. The shape of the otoliths of the sister species T. luscus and T. capelanus correlates very well with the shape of the fishes: both otoliths and fishes of T. capelanus are more slender than in T. luscus. If A. minutus were a sister species of the other two, it should have had even somewhat more slender otoliths than T. capelanus. However, A. minutus has the shortest and highest otoliths of all Recent trisopterine species.

Since the beginning of the 21st century a series of papers have been published about **genetic characteristics** of trisopterine species [18,19,20,21,22,69]. Allele frequencies at polymorphic loci obtained from muscle and liver tissues by Mattiangeli et al. [18] were not conclusive enough to unravel the closest relationship of A. minutus with one of the other trisopterines. They found that A. minutus and N. esmarkii share the most common allele at GPI-2*, while the most common allele at ADA-3* is shared by P. minutus and T. capelanus. Therefore, these data are contradictory. According to their calculated Nei's genetic distance Mattiangeli et al. [18] concluded that N. esmarkii would be more closely related to T. luscus and T. capelanus than to A. minutus. Their dendrogram of the intraspecific relationships among A. minutus from the various investigated areas (ibidem, Figure 3), however, shows an incorrect picture of the most related connections, giving proof of their indecisive phylogenetic data. Based on molecular data from the mitochondrial cytochrome b gene and the nuclear rhodopsin gene Delling et al. [21] and Gonzalez et al. [22] firmly concluded that A. minutus and N. esmarkii are more closely related to one another than to the sister species T. luscus and T. capelanus. In addition, Carr et al. [68] come to the conclusion that A. minutus is more closely related to N. esmarkii than to T. luscus based on portions of two mitochondrial DNA genes (cytochrome b and cytochrome oxidase I). The data in [21,22,70] are in full agreement with the data obtained from otolith characteristics.

All above-mentioned data taken together justify the erection of the new genus Allotrisopterus. Fossil trisopterine otoliths from the Tertiary affirm this (see section Evolution of trisopterines). The general resemblance between A. minutus and T. capelanus has to be attributed to convergence.

## 2.8. Otoliths from Bottom Samples of the Ría de Arosa (north-western Spain)

Among the mollusks and other marine animal remains in Recent and Holocene sediments Cadée [71] found otoliths in and off the Ría de Arosa in Galicia, north-western Spain at 19 stations (Figure 7). Ten bony fish species were recognized by means of otoliths: three gadids, one merluccciid, five gobiids and a sole (Table 6). Apart from several specimens of Lesueurigobius friesii the gobiids are restricted to depths between 7 and 10 m in the inner central bay and the marginal shallow zone. The one otolith of a sole has been found at 7 m depth in the marginal shallow zone. The European Atlantic hake is known from a juvenile otolith in one sample in the outer central bay. The gadid otoliths all belong to three species of trisopterines (Figure 8, Table 6). The most common species is A. minutus (Figure 8(5a–c,6a–c)) occurring at 9 stations in the outer central bay, the oceanic zone and the marginal deep zone bordering the outer central bay. T. capelanus has been found in the same environments, but is less common, occurring at 5 stations (Figure 8(1a–c,2a–c)). T. luscus has a wider distribution: it occurs in the same environments as the other two trisopterine species, but has also been found in the middle central bay and the marginal shallow zone (Figure 8(3a–c,4a–c)). Except for the shallow occurrence of all Trisopterus and Neocolliolus species in special environments like estuaries (Table 2), T. luscus is known to live in the shallowest waters in the north-eastern Atlantic and the North Sea. The presence of this species in shallower waters in the Ría de Arosa is in agreement with this. Despite its wider distribution, T. luscus has been found at no more than 6 stations.

**Table 6**. Distribution of otoliths in sea-bottom samples in the Ría de Arosa, Galicia, northwestern Spain.

| Ecological Zonation (Cadee, 1968) | Inner Central Bay | | Marginal Shallow Zone | | Middle Central Bay | | | Marginal Deep Zone | Outer Central Bay | | | | | | Oceanic Zone | | | | |
|---|---|---|---|---|---|---|---|---|---|---|---|---|---|---|---|---|---|---|---|
| Station | 87 | 88 | 47 | 116 | 13 | 51 | 172 | 246 | 186 | 475 | 479 | 579 | 721 | 931 | 771 | 775 | 841 | 900 | 901 |
| Depth in m | 7–8 | 10 | 7 | 12 | 34 | 24 | 8 | 15–35 | 18–20 | 60 | 35–38 | 35–40 | 50–60 | c. 62 | 68 | 50 | 40–55 | 85 | 60 |
| *Trisopterus capelanus* | | | | | | | | 1 | | | | | 1 | 1 | | | 2 | 5 | |
| *Trisopterus luscus* | | | | 1 | | | 1 | | | | 1 | | | | | | 1 | 6 | 1 |
| *Allotrisopterus minutus* | | | | | | | | 1 | 1 | 1 | | 1 | | | 2 | 1 | 2 | 24 | 2 |
| *Merluccius merluccius* | | | | | | | | | | | | | 1 | | | | | | |
| *Aphia minuta* | | | 1 | | | | | | | | | | | | | | | | |
| *Lesueurigobius friesii* | | 2 | | | 4 | 1 | | | | | | | | | | | | | |
| *Lesueurigobius sanzoi* | | 2 | | | | | | | | | | | | | | | | | |
| *Pomatoschistus minutus* | | 5 | 9 | | | | | | | | | | | | | | | | |
| *Thorogobius ephippiatus* | 1 | | | | | | | | | | | | | | | | | | |
| *Solea solea* | | | 1 | | | | | | | | | | | | | | | | |

At two stations in the oceanic zone at depths between 40 and 85 m, the three trisopterine species co-occur in the same samples (Table 6). The highly overlapping distribution of these species is additional proof that the three species are and were living together in this area, what was not known before. Thus, Lacepède [61] turns out to be right that G. capelanus is also living in the Atlantic Ocean.

## 2.9. Geographical and Depth Distribution

Trisopterus luscus (Figure 9). This species has a large geographical distribution. It is most common from the Nordfjord (62°N) in Norway to the Skagerrak, in the whole North Sea shallower than 100 m, around the entire United Kingdom and Ireland, south of the Faroe Islands, on Rockall Bank, in the English Channel, along the Atlantic coast of France and Spain up to the northern part of Portugal.

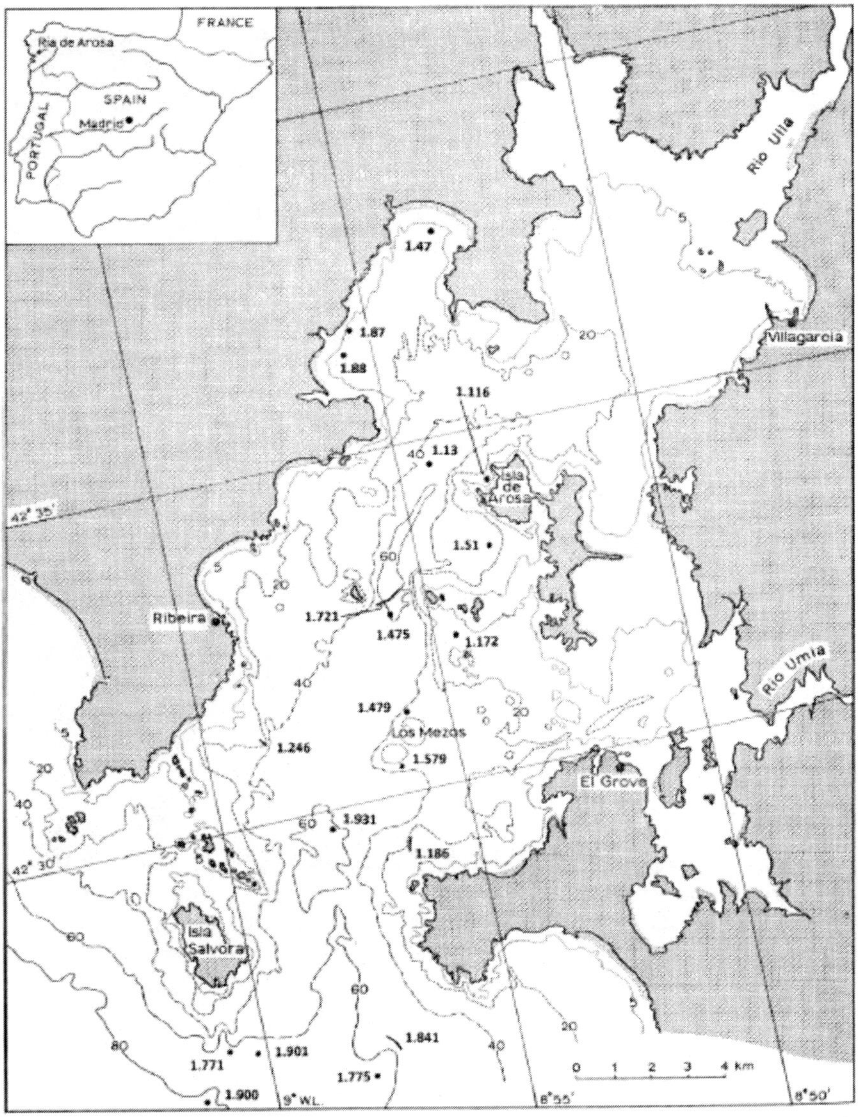


**Figure 7.** Map of the Ría de Arosa, Galicia, Northwest Spain with the stations at which otoliths were found in sediment samples. The stations are indicated by dots or lines. Modified after Cadeé [71].

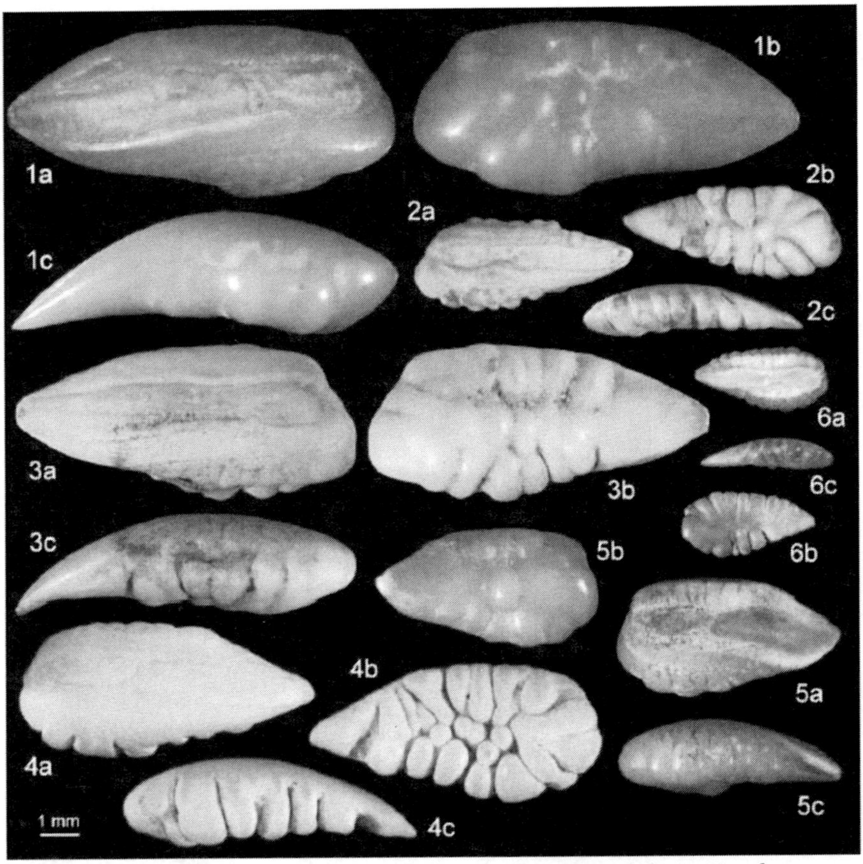


**Figure 8.** Trisopterine otoliths from the Ría de Arosa, Galicia, Northwestern Spain. (1a–c) Trisopterus capelanus, station 1.841, 40–55 m depth, MAB 001328; (2a–c) T. capelanus, station 1.721, 50–60 m depth, MAB 001327; (3a–c) T. luscus, station 1.246, 15–35 m depth, MAB 001321; (4a–c) T. luscus, station 1.901, 60 m depth, MAB 001325; (5a–c) A. minutus, station 1.841, 40–55 m depth, MAB 001313; (6a–c) A. minutus, station 1.246, 15–35 m depth, MAB 001308.

**Figure 9**. Present geographic distribution of Trisopterus luscus. Lighter orange in the North Sea means lesser abundance. Very light orange around Greece: probable distribution, but not yet proven.

The most northern location of this species is the Romsdalsfjord in Norway (62°44′N). It is less abundant in the Kattegat and the south-westernmost part of the Baltic Sea, in the deeper parts of the North Sea, from central Portugal to the south of Morocco in the Atlantic (to 25°N), around Madeira and east of Lanzarote and Fuerteventura (Canary Islands), around some of the Canary Islands, in the shallower depths of the western Mediterranean Sea up to the northern part of Tunisia and around Sicily, in the entire northern part of the Adriatic Sea shallower than 200 m [72]. It also occurs along the more southeastern coast of Italy, thus along the entire coast of Italy. According to the FAO aquatic species distribution map of T. luscus [73] this species continues its occurrence along the eastern coast of the Adriatic Sea (Kroatia, Montenegro, Albania) up to the Gulf of Corinth in Greece.

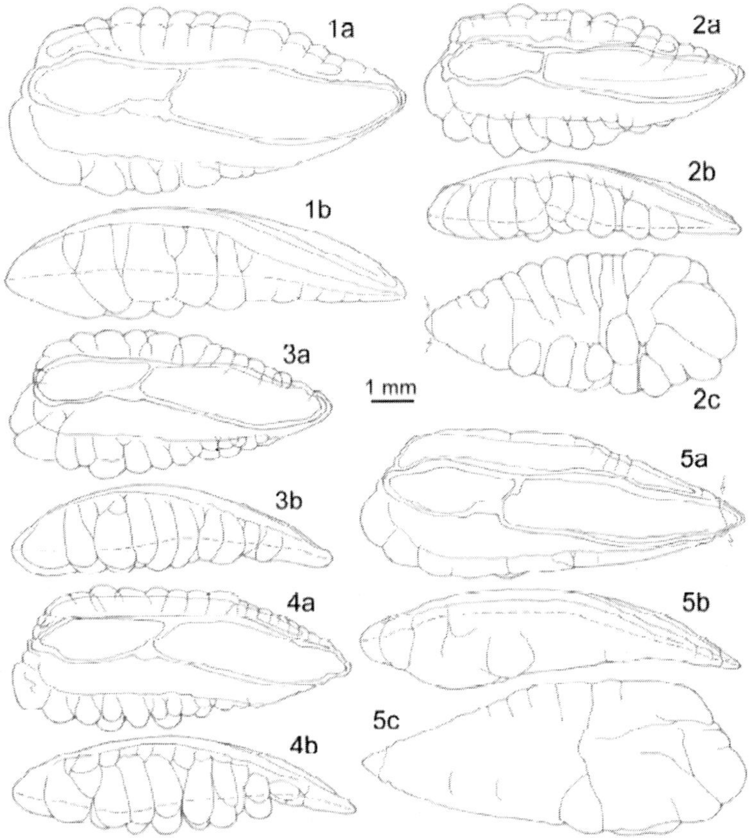


**Figure 10**. Trisopterus luscus otoliths from Greece and the North Sea, and (most likely) Trisopterus capelanus otolith, Faroe Islands. (1a,b,2a–c) T. luscus, Ovvoïkós Channel, east of mainland of Greece, below 100 m depth, mirror images; (1a,b) female, TL 245, OL 9.21; (2a–c) female, TL 149, OL 7.28; (3a,b,4a,b) T. luscus, North Sea, coll. M. Leopold, numbers AR 32 and AR 31, respectively; (3a,b) TL 166, OL 7.34; (4a,b) TL 163, OL 7.74; (5a–c) Most probably T. capelanus, Faroe Bank, Faroe Islands, cruise 03680025, number 5092, male, TL 181, TW 55, OL c. 8.45, mirror image.

And now it has also been recognized by means of otoliths in the Euvoikos Gulf, the Orei Channel and the Pagassitikos Gulf along the central eastern coast of Greece, which have an open connection with the Aegean Sea. Otoliths of 15 fishes of T. luscus appeared to be present among a collection of otoliths from 193 fishes identified as T. capelanus and obtained from Chrissy-Yanna Politou. Adult fishes of this species occur here mainly below

100 meters depth and younger fishes were not found shallower than 50 meters. The occurrence east of central Greece stimulates the supposition that it also occurs on the deeper parts of the shelf around southern Greece. There might be several reasons why T. luscus has not yet been recognized in this area: it is more rare there, it lives in deeper water where the local fishermen usually are not fishing, it was not expected there, and perhaps it lacks the characteristic distinct broad dark bars; and thus it has been confused there with T. capelanus. These Greek T. luscus otoliths have a relatively stocky shape like other Mediterranean specimens of this species (Figure 10(1a,b,2a–c)) and can easily be distinguished from the T. capelanus otoliths in the same area.

In the southern part of the North Sea T. luscus otoliths are fairly variable in shape; the same stocky shape as in the Mediterranean can be found there (Figure 10(3a,b)), but usually the otoliths are somewhat more slender (Figure 10(4a,b)).

Trisopterus capelanus (Figure 11). The main area of this species nowadays is the Mediterranean Sea. It is abundant in the entire western Mediterranean Sea occurring there along the Spanish, Italian, Moroccan and Algerian coasts, the northern coast of Tunisia, and around Corsica, Sardinia and Sicily. It is also abundant in the Adriatic Sea and the northern Aegean Sea. It is more rare in the southern part of the Aegean Sea and along the south coast of Turkey. According to Cohen et al. [49] it also occurs along eastern coasts of the eastern Mediterranean up to and including the Nile delta. It is probably absent along the coasts of Libya and Egypt west of the Nile delta. Svetovidov [1,2] mentions this species also from the Atlantic Ocean at the Strait of Gibraltar. It also occurs along the Moroccan Atlantic coast [16].

The occurrence in the Atlantic Ocean however is greater than published up till now. On 11th September, 1972, I collected otoliths from a fish with a total length of 250 mm in Salir do Porto, a small fishing-village belonging to the municipality of Caldas da Rainha, about 75 km north of Lisbon in Portugal, that clearly belong to this species (Figure 12(1a–c)) (collection P. Gaemers). Some of the fishes from Portugal, illustrated by Assis [74] and Andrade da Silva [75] as T. luscus, evidently belong to T. capelanus, because the otoliths show the typical slender shape of this species. It occurs even farther north, at least to 42°30′N, because I could identify without difficulty several Holocene to Recent otoliths as T. capelanus that were found from sea floor sediments in front of the Ría de Arosa and the outer part of this Ría off the coast of north-western Spain (Figure 8(1a–c,2a–c)). There it co-occurs with T. luscus and T. minutus, two times even in the same sediment samples (Table 6). Fresh fish of T. capelanus along the Portuguese and Spanish Atlantic may be more easily mistaken as T. minutus, because both species are slender and lack the dark vertical bars, which are usually present in subadult and smaller specimens of T. luscus.

**Figure 11**. Present geographic distribution of Trisopterus capelanus. The occurrence southwest of the Faroe islands is only based on one specimen. The most likely route to there is west of Great Britain or even west of Ireland, indicated with question marks.

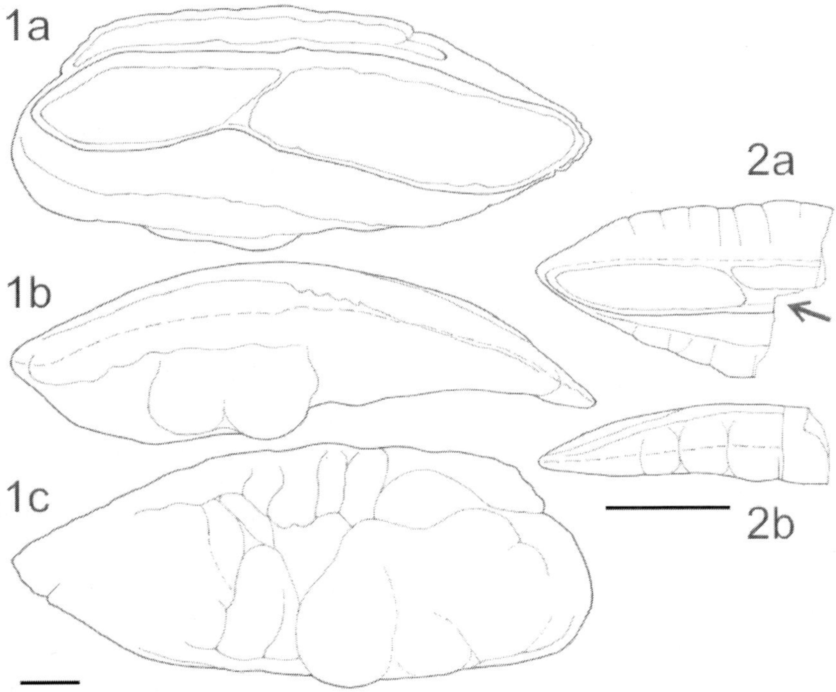


**Figure 12.** (1a–c) Trisopterus capelanus otolith from Salir do Porto, Portugal, TL 250 mm, coll. P. Gaemers; (2a,b) Incomplete fossil Neocolliolus esmarkii otolith, off southern Morocco (28°56′57″N, 12°30′56″W, depth 524 m), coll. P. Gaemers; (1a,2a) Inner surfaces, (1b,2b) Ventral views, (1c) outer surface; the arrow points to the partially preserved, ventrally situated collum. The bars represent one mm.

Furthermore, I discovered an otolith that resembles those of T. capelanus the most (Figure 10(5a–c)) among a large number of A. minutus otoliths from the Faroe Bank, southwest from the Faroe Islands [33,76]. The otolith is easily distinguishable because it is much more slender in shape than those of A. minutus, and can impossibly belong to this species. This otolith is also much too slender and too smooth for T. luscus as it is found in the North Sea (compare Figure 10(4a,b)). This otolith is even more slender than other known otoliths of T. capelanus of the same size. This is not surprising because otoliths of a species are often more slender towards their northern, colder, limits, as indicated in the section "Factors influencing otolith shape". It is, therefore, most likely that the otolith comes from T. capelanus, but it cannot yet be excluded completely, that it belongs to T. luscus. Hopefully more individuals of T. luscus or T. capelanus will be found on the Faroe

Bank in the near future. During the March fish survey of 2015 on the Faroe Bank scientist Peter Steingrund carefully looked for T. luscus among the 355 kg of A. minutus that were caught, but unfortunately he could not find any specimen of this species (personal communication Eiðfinn Magnussen). Both T. capelanus and T. luscus were not known from the Faroe Bank before (personal communication E. Magnussen). The haul in which this specimen was caught was made between 61°01′N 8°10′05W en 61°00′9N 8°16′W at depths between 120 and 114 m on 15 September 2003. The fish was a one-year-old male (TL 181 mm, FW (total fish weight in g) 55 g, OL 8.4 mm). It is very remarkable that this southern species has been found so far north. This very northern occurrence probably is quite recent and must be attributed to the global warming. In the warmer Miocene and Pliocene epochs T. capelanus was very common in the North Sea.

Allotrisopterus minutus (Figure 13). This species is known to occur abundantly from the Vikna Islands, Norway (65°N) to the Skagerrak, nearly the whole North Sea, around the Faroe Islands, on Rockall Bank, around Great Britain and Ireland, in the English Channel and in the Atlantic up to and including at least the northern part of Portugal [77]. The most northern occurrence known from the literature is at Saltfjorden near Bodö, Norway 67°15'N [78].

Less numerous it is in the Kattegat, the southern part of the Baltic Sea, and between the Faroe Islands and Iceland. The southernmost limit along the Portuguese coast is not exactly known, because in distribution maps A. minutus is taken together with T. capelanus, which latter species mostly is taken erroneously as a subspecies of A. minutus. However, from the data presented in Gomes et al. [79] it is clear that T. luscus and A. minutus are abundant up to the Tagus estuary (38°40′N) and are more rare south of it. From this it can be concluded that the southernmost limit of A. minutus will be at least at the Tagus estuary, but may be even more to the south.

**Neocolliolus esmarkii** (Figure 14). This species has the most northerly distribution of all four recent species. It is abundant from about 71°N off northern Norway to the Skagerrak, in the deeper, northern part of the North Sea, around Scotland, Ireland and the western part of England and Wales up to the northern part of Brittany and the western part of the English Channel, and also south and west of Iceland, at Rockall Bank, and south of the Faroe Islands. Less common is this species along the whole northern coast of Norway in the south-western part of the Barents Sea to off the southwestern part of Novaya Zemlya, south and southwest of Svalbard up to 80°N, in the Kattegat, around the Faroe Islands, north and east of Iceland, around Jan Mayen Island and Bear Island, central east of Greenland, and in the central and eastern part of the English Channel

[80,81]. The southernmost part of its certain distribution is off the Gironde estuary, France, in the Bay of Biscay at 45°40′N [82].

Interestingly, I have identified one broken, juvenile otolith of this species among a larger number of otoliths of various species of myctophids from a sediment sample from the present sea floor of the Atlantic Ocean off southern Morocco (28°56′57″N, 12°30′56″W, depth 524 m, collection P. Gaemers, from W. Sparreboom, The Netherlands) (Figure 12(2a,b)). This location is at the same latitude as the island Lanzarote (Canary Islands). The peculiar and unique shape of the collum of this otolith shows that it belongs to N. esmarkii. (The accompanying otolith fauna consists of the following species: Benthosema glaciale, Ceratoscopelus maderensis, Diaphus sp. 1, D.

**Figure 13**. Present geographic distribution of Allotrisopterus minutus.

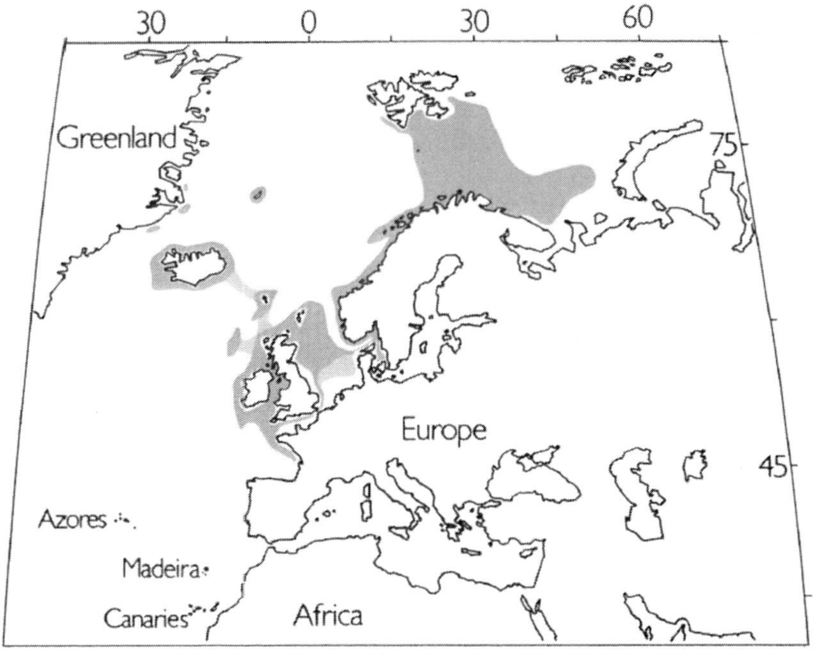


**Figure 14**. Present geographic distribution of Neocolliolus esmarkii. Lighter orange in the North Sea means lesser abundance. Lighter orange in the Atlantic Ocean means probable connection with more westerly occurrences.

sp. 2, Hygophum macrochir, Myctophum punctatum, Notoscopelus sp., Myctophidae indet. The geological age of this N. esmarkii otolith most likely is Weichselian (last glacial), because the southernmost occurrence of this species at present is the Atlantic in southern France [16], a long distance to the north and west from this location.

## 2.10. Evolution of Trisopterines

A large amount of new data is available on trisopterine fossil otoliths compared with my first paper on the evolution of cod otoliths in 1976 [3]. For a large part this material is still unpublished, thus a lot of data in this section are still provisionally presented and will be the subject of a paleontological paper in which the new taxa will be described. The new data show that the evolution of trisopterines is much more complex, comprising more lineages and much shorter lengths of lineages than earlier thought. This justifies the establishment of extinct fossil genera. During his whole career up to now Nolf did not want to recognize fossil genera based on otoliths only (see for instance [4] and [83]) and he refused to use such available genus names even when he must admit that he could not fit certain fossil otoliths in a recent genus, like Ensigadus ensiformis (Steurbaut and Herman, 1978) that is named "Gadida" ensiformis in Nolf [83] and "genus Gadidarum" ensiformis in earlier publications for example in Nolf and Steurbaut [4]. Moreover, describing new fossil species based on otoliths with the provisional genus names "genus Gadidarum" or "Gadida", is not allowed by the International Code for Zoological Nomenclature.

The study of fossil and recent Gadidae otoliths makes it possible to reconstruct many evolutionary lineages, of which many appear to be extinct. This knowledge of the Gadidae evolution can be used successfully for establishing many biozones by which the relative geological age of the sediments in which the otoliths are found, can be accurately determined [24,84,85,86,87]. With the many new data it will be possible to extend this biozonation considerably.

The earliest representatives of the trisopterines are known from the Early Oligocene (Rupelian), about 34 million years ago. Only one evolutionary stage, probably the last stage, is known from the oldest, still unnamed lineage (Figure 15(1a–c,2a–c)). The origin of this lineage must lie in the Late or Middle Eocene, but temperatures were too high for the Gadidae in the North Sea Basin before the Oligocene, so they must have lived more to the north from where no data are yet available. The small, thick and high otoliths show sexual dimorphism, and must belong to adult fishes. The pseudocolliculum is well developed and nearly completely fills the column. From this lineage the extinct genus Semeniolum originates by cladogenesis (Figure 15(3a–c,4a,b)), a lineage that is known from the base of the Early Oligocene (Rupelian, c. 34 million years ago) to the lower part of the Late Oligocene (Chattian A). Members of this lineage always have a pseudocolliculum in juvenile otoliths, but this feature is absent in adult ones. The otoliths become larger and somewhat more slender than those of the former lineage. The lineage succeeding Semeniolum is still unnamed and is only known from the Late Oligocene (Chattian, 28.5–23 million years ago). These otoliths again become larger and can be more slender than the ancestral lineage, and are strikingly thick. They show sexual dimorphism and do not have a pseudocolliculum (Figure 15(5a,b,6a,b). Another lineage with much thinner and again larger otoliths is also known from the base of the Late Oligocene, resembling the genus Trisopterus in many respects (Figure 15(7a–c)). Its origin may be connected with the former one, but could also directly be connected with the Semeniolum lineage by cladogenesis. The main differences with Trisopterus are, that the otoliths are less bent lengthwise, flatter and smaller. A pseudocolliculum is absent in this lineage. Undoubtedly this lineage is the forerunner lineage of Trisopterus, but it is still unknown from the Early Miocene.

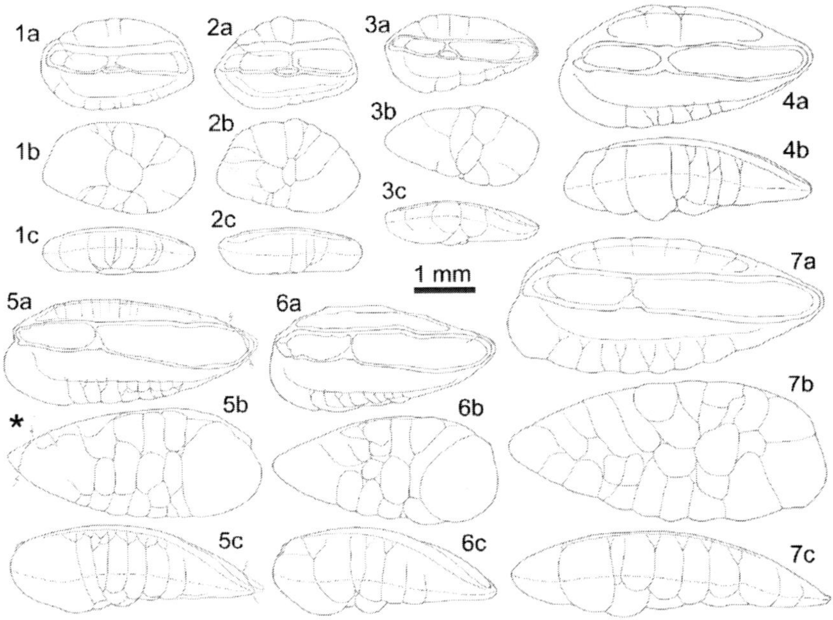


**Figure 15**. Examples of earlier evolution of trisopterines. (1a–c,2a–c) † Trisopterini, new genus and species 1, Early Oligocene, Rupelian, Kotten Member, borehole Tervoert, Winterswijk, The Netherlands; (1a–c) female, bed 28, 23.8–24.25 m below surface; (2a–c) male, bed 29—base bed 30, 22.3–22.8 m below surface; (3a–c) † Semeniolum rupelense Gaemers, 1984, Early Oligocene, Rupelian, Kotten Member, borehole Tervoert, Winterswijk, The Netherlands, top bed 23—base bed 24, 26.7–27.2 m below surface. (4a,b) † Semeniolum rupelense Gaemers, 1984, Late Oligocene, base of Chattian A, Söllingen, Germany. (5a–c,6a–c) † Trisopterini, new genus and species 2, Late Oligocene, Chattian A, (5a–c) female, mirror images, indicated with *, (6a–c) male; (7a–c) † "Trisopterus" kasselensis n. ssp., Late Oligocene, base of Chattian A.

Fossil otoliths show that the origin of the closely related T. luscus and T. capelanus lineages both are in the Early Miocene of the Northeast Atlantic and North Sea, at least 18 million years ago (Figure 16(1a–e,2a–c)). These sister species indeed form a monophyletic group, as correctly concluded in [21]. Furthermore, there are two extinct lineages that I provisionally include also in the genus Trisopterus, the Trisopterus spectabilis and Trisopterus arcuatus lineages. Interesting is, that the T. spectabilis lineage

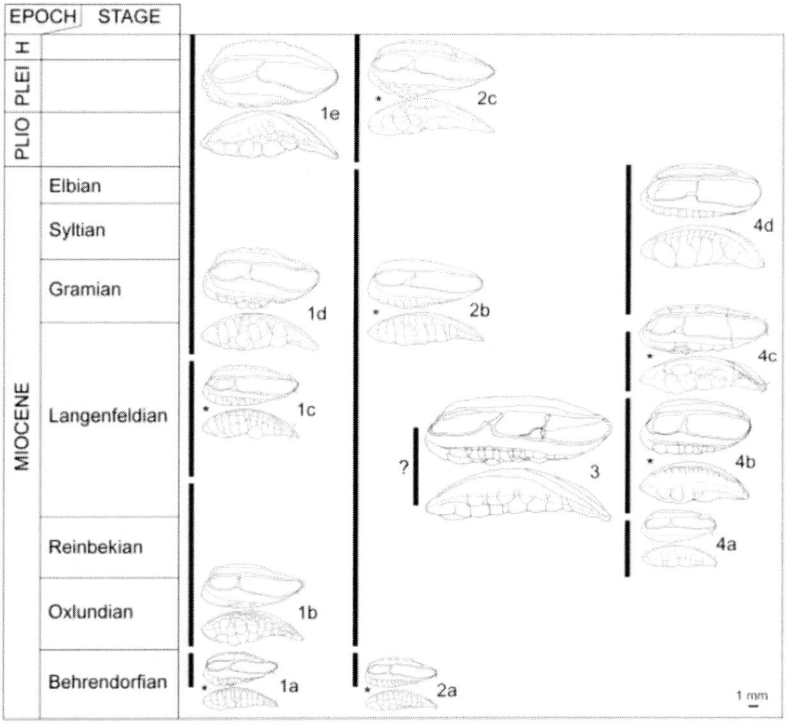


**Figure 16**. Evolutionary lineages of Trisopterus. Otoliths indicated with an asterisk are mirror images. (1a) T. luscus n. ssp. 1, Behrendorfian, Edegem Sands, bed 6, Slijkhoek, Antwerp, Belgium; (1b) T. luscus concavus (Gaemers, 1976), Oxlundian, Miste Bed, Miste, Winterswijk, The Netherlands. (1c). T. luscus n. ssp. 2, Langenfeldian, borehole Straat, 518.7–520.7 m below surface, near Roermond, The Netherlands; (1d) T. luscus luscus (Linnaeus, 1758), Gramian, borehole Straat, 494.7–495.7 m below surface, near Roermond, The Netherlands; (1e) T. luscus luscus (Linnaeus, 1758), Recent, North Sea, TL 388; (2a) T. capelanus n. ssp. 1, Early Miocene, Behrendorfian, Berchem Formation, Kiel Sands, borehole airfield Haamstede, The Netherlands, 132–133 m below surface; (2b) T. capelanus n. ssp. 2, Gramian, Gram, Denmark; (2c) T. capelanus capelanus (Lacepède, 1800), Recent, Aegean Sea, east of mainland of Greece, 50–100 m depth, female, TL 200; (3) † T. arcuatus (Radwanska, 1992), Badenian, Middle Miocene, Rybnica, Poland; (4a) † T. spectabilis n. ssp. 1, Reinbekian, borehole Straat, 636.5–638.5 m below surface, near Roermond, The Netherlands; (4b) † T. spectabilis spectabilis (Koken, 1891), Langenfeldian, Gross Pampau, Germany; (4c) † T. spectabilis n. ssp. 2, Langenfeldian, borehole Straat, 524.7–526.7 m below surface, near Roermond, The Netherlands; (4d) † T. spectabilis angustus (Gaemers and Schwarzhans, 1982), Syltian, Sylt, Germany.

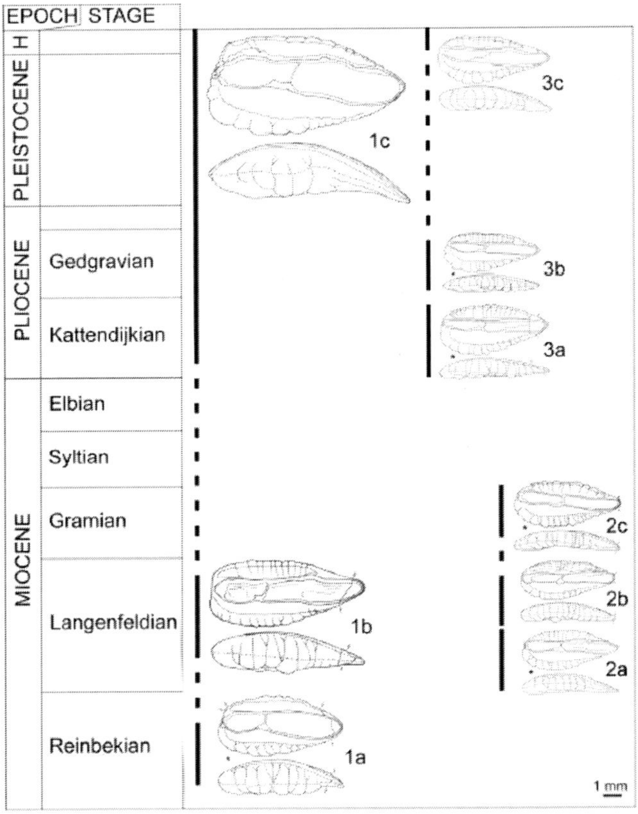


**Figure 17**. Evolutionary lineages of Allotrisopterus and Neocolliolus sensu lato. Otoliths indicated with an asterisk are mirror images. (1a) Allotrisopterus minutus n. ssp.1, Reinbekian, borehole Straat, 650.2–652.2 m below surface, near Roermond, The Netherlands; (1b) A. minutus incognitus (Gaemers, 1976), Langenfeldian, Hohen Woos, Germany; (1c) A. minutus minutus (Linnaeus, 1758), Recent, Faroe Bank, Faroe Islands, cruise 03680025, number 5091, female, TL 256, TW 155, OL 10.57; (2a) † "Neocolliolus" vikingensis n. ssp. 1, Langenfeldian, borehole Straat, 488.7–489.7 m below surface, near Roermond, The Netherlands; (2b) † "N." vikingensis n. ssp. 2, Gramian, borehole Straat, 430.5-431.5 m below surface, near Roermond, The Netherlands; (2c) † "N." vikingensis vikingensis Gaemers, 1987, Gramian, borehole J, Viking Graben, northern North Sea; (3a) Neocolliolus esmarkii pliocenicus (Gaemers and Schwarzhans, 1973), Kattendijkian, Early Pliocene, Kerkstraat, Antwerp, Belgium (3b) N. esmarkii n. ssp., Middle/Late Pliocene, Mill-Langenboom, The Netherlands; (3c) N. esmarkii esmarkii (Nilsson, 1855), Recent, North Sea, TL 140, OL 5.90.

goes more rapidly through its evolutionary cycle (Figure 16(4a–d)). It is well represented in the North Sea Basin and became extinct at the end of the Miocene, about 5.5 million years ago. Although it is not yet completely sure, it seems that this lineage started later (c. 15 or 16 million years ago) than those of T. luscus and T. capelanus. Furthermore, T. spectabilis displays a much stronger allometry in the ontogeny of the otoliths. These data suggest that the T. spectabilis lineage very likely is not a sister lineage of the other two. Finally, only one evolutionary stage of T. arcuatus (Radwańska, 1992) has been found in the Middle Miocene (Badenian) of southeastern Poland (Figure 16(3)). This lineage most likely became also extinct in the Late Miocene. The former sea in that area belonged to the Paratethys and—as far as is known—had no direct connection with the North Sea, but only via the Mediterranean Sea and the Atlantic Ocean. Precise chronostratigraphic correlation with the North Sea Basin is therefore difficult, indicated with a question mark in Figure 16(3).

The origin of the A. minutus lineage probably is in the Middle Miocene, c. 15 million years ago (Figure 17(1a–c)). The oldest known representatives of the N. esmarkii lineage are from the Early Pliocene, c. 5 million years ago (Figure 17(3a–c)), but its still unknown origin must be in the Late Miocene (probably a few million years earlier). Its forerunner lineage, "Neocolliolus" vikingensis (Gaemers, 1987) is partly known and ranges at least between c. 8 and 11 million years (Late Miocene) (Figure 17(2a–c)). From this it can be concluded that a direct common ancestor with the A. minutus lineage does not exist. The Neocolliolus lineage and its forerunner lineage both evolved more rapidly than the other trisopterine lineages.

It is striking that the Oligocene lineages have shorter evolutionary cycles than most younger ones. As a consequence the Oligocene lineages succeed one another more rapidly than the younger ones.

## 3. DISCUSSION AND CONCLUSIONS

As shown in this paper, fossil and recent otoliths are indispensable for a better understanding of the evolution, mutual relationships, systematics and geographic distribution of closely related species. If only data from recent species are available—even if they come from various sources (such as external features, meristics, otoliths and genetics) and have been elaborated with sophisticated mathematical methods—it may remain difficult to assign a species to the right genus and is it often hazardous to reconstruct the phylogeny and the biogeographic history. Cladistics, also called "phylogenetic systematics", introduced by Willy Hennig [88], originally is solely based upon recent species and their various characteristics, and is at best a serious attempt to a phylogenetic reconstruction. Cladograms constructed by this method are not evolutionary, phylogenetic trees because they do not show how ancestors are related to descendants. However, there are a lot of scientists who consider cladograms as a kind of phylogenetic trees. Without inclusion of fossil representatives a certain and reliable phylogeny can never be obtained with this method, even when genetic data are included. Fossil data are difficult to integrate in the cladistic system of Hennig, because time—and consequently also stratigraphic data—is not easy to fit into this system [89]. The uncertainty about phylogeny by means of cladistics can also be demonstrated by the proposal of several possible cladograms in many publications using cladistics, without having the ability to choose the right one. The term "phylogenetic systematics" for Hennig's method therefore is misleading and should be avoided if only recent species are involved, what is usually the case. Hennig's method should therefore only be called "cladistics", because this term does not cause confusion.

The much-used concept of vicariance in Hennig's method, for instance, is often not provable without fossils. A good definition of vicariance is: the geographical separation and isolation of two subpopulations of a species by some sort of barrier, resulting in two different populations leading to separate varieties and, in the end, to different species. This is a type of allopatric speciation inferred from Darwin's theory of evolution. Two related species in the same genus that nowadays live in two separate areas are therefore often called vicariant species. However, the present geographic distribution of two species in separate areas is not always guarantee that geographical separation and isolation is the original process of the speciation. In the past the geographical distribution of such species might have been very different. A good example is found in the species T. capelanus and A. minutus that were supposed to be subspecies of one species. The latter species is only living in the Atlantic, including the North Sea and Norwegian Sea. T. capelanus was thought to be living nowadays only as an endemic in the Mediterranean Sea. These species were seen as vicariant species for a long time. As demonstrated in this publication, fossil and recent otoliths, and also genetics, have shown convincingly that these species are not even sister species.

Gonzalez et al. [22] hypothesize that the most recent common ancestor of the real sister species T. luscus and T. capelanus had a large geographic distribution including the Atlantic as well as the Mediterranean. The time of origin of these species from a common ancestor at about 19.7 or 21 million years that these authors have calculated is a good estimate as the fossil otoliths show, but the final closure of the Tethys Ocean blocking the connection between the Mediterranean and the Indian Ocean around that time has nothing to do with the speciation of Trisopterus, because the oldest known ancestors in the T. luscus and T. capelanus lineages occur in the same Early Miocene deposits of the North Sea Basin and not in the Mediterranean. Thus, fossil otoliths of these lineages indicate that both originated in the northeastern Atlantic including the North Sea. There was, and still is, no barrier in that area, thus these species are an example of sympatric speciation. There is much evidence that T. luscus and T. capelanus lived together in the same area and are distinguishable by means of their otoliths during their whole known, long geological history, and there is no reason to suppose that it was different during a possible unknown part of their phylogeny [55,84,90,91,92] (see also Figure 16).

The Messinian salinity crisis (~6.0 to 5.3 mya) might have caused disappearance of many marine species from the Mediterranean including T. capelanus, as Gonzalez et al. [22] assume, but the only area where the T. capelanus lineage could have survived during that time was in the Atlantic where the separate T. luscus lineage continually was living. A post-glacial, very recent invasion of T. capelanus in the Mediterranean as considered by Tirard et al. [93] who synonymize this species with T. luscus, is certainly not realistic, because conditions for this species in the Mediterranean during the Pliocene and Pleistocene were good enough to live there in these periods.

N. esmarkii is the most aberrant species within the trisopterines. Many characteristics of the fishes themselves, their otoliths, their habitat and food justify its taxonomic position in the separate genus Neocolliolus.

Closer relationships between A. minutus and N. esmarkii, compared with the other trisopterines, are demonstrated by otolith characteristics and genetics. The greater taxonomic distance of A. minutus and N. esmarkii to T. luscus and T. capelanus is based upon various characteristics of the fishes, such as otolith morphology and genetic data [18,22], and also upon phylogenetic data inferable from fossil otoliths. These data together do not allow including A. minutus in the genus Trisopterus or Neocolliolus. All available data in combination justify the erection of the new genus Allotrisopterus for T. minutus. The similarities in fish shape and size of T. capelanus and A. minutus are therefore due to convergence, and not to close relationship.

The fossil otoliths demonstrate clearly that the trisopterines do not directly form a monophyletic group as assumed by most ichthyologists up to now, including Delling et al. [21]. The only real sister group relationship exists between T. luscus and T. capelanus, and therefore the genus name Trisopterus should be restricted to these taxa alone. Their common ancestry lies in the Early Miocene. The most recent common ancestor of the four recent trisopterines together is not yet known, but must have lived farther back in time, in the older part of the Late Oligocene or perhaps even in the younger part of the Early Oligocene.

The tribe Trisopterini introduced by Endo [5] must be redefined. Apart from fossil genera this tribe only contains the recent genera Trisopterus, Allotrisopterus and Neocolliolus. Micromesistius, Boreogadus and Arctogadus must be excluded from this tribe. Although the genus Micromesistius is not so distant from the trisopterines [21], it is more closely related to Gadiculus [7,94], and might be included in the tribe Gadiculini. The genus Boreogadus belongs to the tribe Gadini based on morphology and genetic data [7], and according to morphological characteristics Arctogadus should also be placed in the Gadini.

# 4. MATERIALS AND METHODS

## 4.1. Recent Otoliths

During a fish survey of the Rijksinstituut voor Visserijonderzoek (RIVO, IJmuiden, The Netherlands, now part of Wageningen IMARES, Institute for Marine Resources & Ecosystem Studies) in the North Sea, I collected the first otoliths of the three trisopterine species occurring in the North Sea in February 1969. I obtained many hundreds of recent otoliths of Trisopterus luscus (Linnaeus, 1758) by buying the fishes on the fish market in Leiden (The Netherlands) for our own consumption in the course of tens of years. I could study hundreds of otoliths of the three North Sea trisopterine species from the collection of Maarten Frederik (Mardik) Leopold (Texel, The Netherlands) from the North Sea. A series of otoliths of 194 recent fishes identified as Trisopterus capelanus appeared to include otoliths of 15 fishes of T. luscus off the eastern coast of Greece, and were kindly sent and given to me by Chrissy-Yanna Politou (Hellinikon, Greece). I could also study a large collection of recent otoliths of 265 fishes identified as Trisopterus minutus (Linnaeus, 1758) from the Faroe Bank around the Faroe Islands, that was used in the BSc Thesis of Marita Magnussen and kindly loaned to me by her supervisor Eyðfinn Magnussen (Torshavn, Faroe Islands). Two fishes of this collection appeared to belong to Neocolliolus esmarkii and one most likely to Trisopterus capelanus. On holiday trips and excursions in Spain, Portugal, Greece, France and Morocco I bought trisopterine fishes on fish markets and dissected the otoliths. Gerhard C. Cadeé (Texel, The Netherlands) found otoliths in bottom samples in and off the Ría de Arosa (Galicia, northwestern Spain) most of which belong to Recent and Holocene trisopterine species, collection Oertijdmuseum De Groene Poort (Museum Ammonietenhoeve Boxtel), Boxtel, Noord-Brabant, The Netherlands (collection numbers MAB 001307–001342). Smaller numbers of recent trisopterine otoliths were obtained from some other collectors. William N. Eschmeyer (California Academy of Sciences, San Francisco, CA, USA.)

allowed me to dissect the otoliths from some trisopterines (otoliths now in author's collection): 1 specimen, T. luscus (identified as Brachygadus minutus) from Venice (Italy), California Academy of Sciences no. 47126; 1 specimen, T. luscus from Palermo (Italy), California Academy of Sciences no. 47124; 1 specimen, T. capelanus from Naples (Italy), California Academy of Sciences no. 20894; 2 specimens, T. capelanus from Naples (Italy), Stanford University Natural History Museum no. 2461.

## 4.2. Fossil Otoliths

Many species and phylogenetic stages of fossil Gadidae otoliths are known in the literature from the North Sea Basin and the northeastern Atlantic from The Netherlands, Belgium, Germany, Denmark, England, France, Poland and the British part of the North Sea. There are also many trisopterine otoliths available that are still unpublished. I hope to publish this material in one or more other papers that are in preparation, because it is beyond the scope of this paper. A lot of important new otoliths of trisopterine Gadidae occurred in borehole Straat (numbered 58G-191 by the Geological Survey of The Netherlands, now part of TNO-NITG), in the middle of the province of Limburg, close to the city of Roermond, in The Netherlands. I collected large samples of the thick Miocene sequence in this borehole myself and I washed the large samples. The picking of the otoliths was done by Messrs André J. J. Jansen and Cor Karnekamp.

## 4.3. Picturing and Measuring

I have drawn the otoliths at magnifications of 15 × (12 × 1.25) and 31.25 × (25 × 1.25) under a Wild M5 binocular microscope with a drawing tube. The otoliths were measured with a sliding calliper with an accuracy of 0.05 mm; with a ten times magnifying achromatic hand-lens the measurements were estimated to a hundredth of an mm on the sliding calliper. Photographs were taken by the author at the Center of Biodiversity Naturalis The Netherlands, Leiden, using a digital multifocal camera (Leica MZ16A, Leica Microsystems, Wetzlar, Germany) with accompanying software coupled to a personal computer. All drawings of otoliths have been pictured as right sagittae in order to provide the easiest mutual comparison; thus all otoliths presented as mirror images are left sagittae, indicated by an asterisc.

**Acknowledgments:** I am thankful to Mardik F. Leopold (NIOZ Royal The Netherlands Institute for Sea Research, Texel, The Netherlands) and Eiðfinn Magnussen (University of the Faroe Island, Torshavn) for the loan of large numbers of trisopterine otoliths. I am grateful to Gerhard C. Cadeé (NIOZ Royal Netherlands Institute for Sea Research, Texel, The Netherlands) and Chrissi-Yianna Politou (National Centre of Marine Research in Hellinikon,

Greece) for the otoliths they gave me, as well as the late Juan Bauza-Rullan (Palma de Mallorca, Spain), William N. Eschmeyer (Institute of Biodiversity Science and Sustainability, California Academy of Sciences, San Francisco, CA, USA). Dirk Nolf (Royal Belgian Institute of Natural Sciences, Brussels, Belgium), I. Óskarsson, (Marine Research Institute, Reykjavík, Iceland) and Jouke van der Zee (Nijmegen, The Netherlands) for providing me with smaller numbers of otoliths. I thank Antoni Lombarte (Institut de Ciènces del Mar-CSIC, Barcelona, Spain) for measurements of a small series of Trisopterus otoliths. I thank André F.J. Jansen (Bovenkarspel, The Netherlands) and Cor Karnekamp (Diemen, The Netherlands) for picking the otoliths and other fish remains from the residues of borehole Straat near Roermond, The Netherlands). I thank Lars van den Hoek Ostende (Center of Biodiversity Naturalis Netherlands, Leiden) for access to the photographic equipment and Jaap J. Oppedijk (Leiden, The Netherlands) for assembling and digitalizing the figures and plates. No funding or grants were received for this study. Finally, I thank the two anonymous reviewers for their comments by which the text could be improved.

## ABBREVIATIONS

The following abbreviations are used in this publication:

| | |
|---|---|
| TL | Total fish length in mm |
| FW | Total fish weight in g |
| OL | Otolith length in mm |
| OH | Otolith height (width) in mm |
| OT | Otolith thickness in mm |

## REFERENCES

1. Svetovidov, A.N. Fauna SSSR: Ryby, (n.s. 34), Vol. 9, Fasc. 4; Treskoobraznye [Gadiformes]: Zoologicheskii Institut Akademi Nauk SSSR, Moscow-Leningrad, Russia, 1948; p. 294.
2. Svetovidov, A.N. Fauna of the U.S.S.R., Fishes, Gadiformes; Zoologicheskii Institut, Akademii Nauk SSSR, Published by Israel Program for Scientific Translations: Jerusalem, Israel, 1962; Volume 9, vip. 304p.
3. Gaemers, P.A.M. New concepts in the evolution of the Gadidae (Vertebrata, Pisces), based on their otoliths. Meded. Werkgr. Tert. Kwart. Geol. 1976, 13, 3–32.
4. Nolf, D.; Steurbaut, E. Importance and restrictions of the otolith-based fossil record of gadiform and ophidiiform fishes. In Papers on the Systematics of Gadiform Fishes; Cohen, D.M., Ed.; Science Series; Natural History Museum of the Los Angeles County: Los Angeles, CA, USA, 1989; Volume 32, pp. 47–58.

5. Endo, H. Phylogeny of the order Gadiformes (Teleostei, Paracanthopterygii). Mem. Grad. School Fish. Sci. Hokkaido Univ. 2002, 49, 75–149.
6. Nelson, J.S. Fishes of the World, 4th ed.; John Wiley and Sons: Hoboken, NJ, USA, 2006; p. 624.
7. Teletchea, F.; Laudet, V.; Hänni, C. Phylogeny of the Gadidae (sensu Svetovidov, 1948) based on their morphology and two mitochondrial genes. Mol. Phylogen. Evol. 2006, 38, 189–199.
8. Roa-Varón, A.; Ortí, G. Phylogenetic relationship among families of Gadiformes (Teleostei, Paracanthopterygii) based on nuclear and mitochondrial data. Mol. Phylogen. Evol. 2009, 52, 688–704.
9. Howes, G.J. Anatomy of the Melanonidae (Teleostei: Gadiformes), with comments on its phylogenetic relationships. Bull. Nat. Hist. Mus. (Zool.) 1993, 59, 11–31.
10. Fahay, M.P. Early Stages of Fishes in the Western North Atlantic Ocean (Davis Strait, Southern Greenland and Flemish Cap to Cape Hatteras). Volume 1. Acipenseriformes through Syngnathiformes; NW Atl. Fish. Org.: Dartmouth, NS, Canada, 2007; p. 931.
11. Eschmeyer, W.N. (Ed.) Catalog of Fishes. Updated Internet Version of 31 October 2014. Catalog Databases of CAS Cited in FishBase. Available online: http://researcharchive.calacademy.org/research/ichthyology/catalog/fishcatmain.asp (accessed on 11 December 2014).
12. Gill, T.P. On the Family Ranicipitidae. Proceedings of the United States National Museum; U.S. Government Printing Office: Washington, DC, USA, 1890; Volume 13, pp. 1–4.
13. Jordan, A.D.; Møller, P.R.; Nielsen, J.G. Revision of the Arctic cod genus Arctogadus. J. Fish Biol. 2003, 62, 1339–1352.
14. Byrkjedal, I.; Rees, D.J.; Christiansen, J.S.; Fevolden, S.-E. The taxonomic status of Theragra finnmarchica Koefoed, 1956 (Teleostei: Gadidae): Perspectives from morphological and molecular data. J. Fish Biol. 2008, 73, 1183–1200.
15. Svetovidov, A.N. Gadidae. In Check-List of the Fishes of the North-Eastern Atlantic and of the Mediterranean; Hureau, J.-C., Monod, T., Eds.; UNESCO: Paris, France, 1973; Volume 1, pp. 303–320.
16. Svetovidov, A.N. Gadidae. In Fishes of the North-Eastern Atlantic and the Mediterranean; Whitehead, P.J.P., Bauchot, M.-L., Hureau, J.-C., Nielsen, J., Tortonese, E., Eds.; UNESCO: Paris, France, 1986; Volume 2, pp. 680–710.
17. Schmidt, E.J. Gadiculus argenteus and Gadiculus thori. In Mindeskrift i Anledning af Hundredaaret for Japetus Steenstrups Fødsel; En kreds af Naturforskere: Copenhagen, Denmark, 1913; Volume 14, p. 9.
18. Mattiangeli, V.; Bourke, E.A.; Ryan, A.W.; Mork, J.; Cross, T.F. Allozyme analyses of the genus Trisopterus: Taxonomic status and population structure of the poor cod. J. Fish Biol. 2000, 56, 474–494.
19. Mattiangeli, V.; Galvin, P.; Ryan, A.W.; Mork, J.; Cross, T.F. VNTR variability in Atlantic poor cod (Trisopterus minutus minutus) throughout its range: Single locus minisatellite data suggest reproductive isolation for the Faroe Bank population. Fish Res. 2002, 58, 185–191.
20. Mattiangeli, V.; Ryan, A.W.; Galvin, P.; Mork, J.; Cross, T.F. Eastern and western poor cod (Trisopterus minutus capelanus) populations in the Mediterranean Sea: Evidence from allozyme and minisatellites loci. Mar. Ecol. 2003, 24, 247–258.
21. Delling, B.; Noren, M.; Kullander, S.O.; González, J.A. Taxonomic review of the genus Trisopterus (Teleostei: Gadidae) with recognition of the capelan Trisopterus capelanus as a valid species. J. Fish Biol. 2011, 79, 1236–1260.
22. Gonzalez, E.G.; Cunha, R.L.; Sevilla, R.G.; Ghanavi, H.R.; Krey, G.; Bautista, J.M. Evolutionary history of the genus Trisopterus. Mol. Phylogenet. Evol. 2012, 62, 1013–1018.
23. Gaemers, P.A.M.; Crapon de Crapona, M.-D. Sexual dimorphism in otoliths of haplochromines (Pisces, Cichlidae). Ann. Mus. Roy. Afr. Cent. Sci. Zool. 1986, 251, 151–155.

24. Gaemers, P.A.M. The biostratigraphy of the Nieder Ochtenhausen borehole based on Gadidae otoliths and other fish remains, and the establishment of the Elbian, a new stage for the latest Late Miocene of NW Europe. Geol. Jahrb. 2001, 152, 301–339.
25. Härkönen, T. Guide to the Otoliths of the Bony Fishes of the Northeast Atlantic, 1st ed.; Danbiu ApS.: Hellerup, Denmark, 1986; p. 256.
26. Schwarzhans, W. Otolith-morphology and its usage for higher systematical units, with special reference to the Myctophiformes s.l. Meded. Werkgr. Tert. Kwart. Geol. 1978, 15, 167–185.
27. Nolf, D.; Steurbaut, E. Evidence from otoliths for establishing relationships between gadiforms and other groups. In Papers on the Systematics of Gadiform Fishes; Cohen, D.M., Ed.; Science Series; Natural History Museum of the Los Angeles County: Los Angeles, CA, USA, 1989; Volume 32, pp. 37–45.
28. Lambert, G.; Nielsen, J.R.; Larsen, L.I.; Sparholt, H. Maturity and growth population dynamics of Norway pout (Trisopterus esmarkii) in the North Sea, Skagerrak, and Kattegat. ICES J. Mar. Sci. 2009, 66, 1899–1914.
29. Lambert, G. Norway pout (Trisopterus esmarkii) life history strategy: Growth and maturity as evidence of possible spawning mortality. In Mémoire de fin d'études Pour l'obtention du Diplôme d'Agronomie Approfondie » (DAA) Spécialisation Halieutique; Agrocampus Rennes: Rennes, France, 2007; p. 75.
30. Raitt, D.F.S. Synopsis of the Biological Data on the Norway Pout Trisopterus esmarkii (Nilsson, 1855); FAO Fish Synopsis: Rome, Italy, 1968; Volume 33, pp. 1–36.
31. Saemundsson, B. On the age and growth of the coalfish (Gadus virens L.), the Norway pout (Gadus esmarkii Nilsson) and the poutassou (Gadus poutassou Risso) in Icelandic waters. Meddel. Komm. Danm. Fisk. Havunders. 1929, 8, 1–37.
32. Baranenkova, A.S.; Khokhlina, N.S. Distribution of eggs, larvae and adults of the Norway pout off northwestern Norway and in the Barents Sea. Rap. Proc.-Verb. Réun. Cons. Internatl. Explor. Mer 1968, 158, 90–100.
33. Magnussen, E.; Magnussen, M.D. Ecology of poor-cod (Trisopterus minutus) on the Faroe Bank. Mar. Biol. Res. 2009, 5, 133–142.
34. Cooper, A. The reproductive biology of poor cod, Trisopterus minutus L., whiting, Merlangius merlangus L., and Norway pout, Trisopterus esmarkii Nilsson, off the west coast of Scotland. J. Fish Biol. 1983, 22, 317–334.
35. Merayo, C.R. Reproduction and fecundity of the bib Trisopterus luscus (Linnaeus, 1758) (Pisces, Gadidae) in the central region of the Cantabrian Sea (northern Spain). Bol. Inst. Esp. Oceanogr. 1996, 12, 17–29.
36. Labarta, U.; Ferreiro, M.J. Age and Growth of the Galician Coast Pouting (Trisopterus luscus L.) Preliminary Data; CM1982/g:65; ICES: Copenhagen, Denmark, 1982.
37. Alonso-Fernández, A.; Domínguez-Petit, R.; Bao, M.; Rivas, C.; Saborido-Rey, F. Spawning pattern and reproductive strategy of female pouting Trisopterus luscus (Gadidae) on the Galician shelf of north-western Spain. Aquat. Living Resour. 2008, 21, 383–393.
38. Recordlijst zeevissen Nederland per 1 Januari 2013 (Nederland. Continentaal Plat). [Record List of Sea Fishes Netherlands at 1 January 2013, Continental Shelf of The Netherlands]. Available online: http://www.ncrz.nl/pdf/NCRZ_lijst_NL.pdf (accessed on 11 December 2014).
39. Ragonese, S.; Bianchini, M.L. Growth, Mortality and Yield-per-Recruit of the Poor Cod Trisopterus minutus Capelanus, from the Strait of Sicily; Naga, Iclarm Quarterly Fishbyte Section: Penang, Malaysia, 1998; pp. 61–70.
40. Šantić, M.; Pallaoro, A.; Mikulandra, I.; Rađa, B.; Jardas, I. Age, growth and mortality of poor cod (Trisopterus minutus L.) from the eastern Adriatic Sea. Arch. Biol. Sci. 2015, 11.
41. Metin, G.; Ilkyaz, A.T.; Kınacıgil, H.T. Growth, mortality, and reproduction of poor cod (Trisopterus minutus Linn., 1758) in the Central Aegean Sea. Turk. J. Zool. 2008, 32, 43–51.

42. Biagi, F.; de Ranieri, S.; Viva, C. Recruitment, length at first maturity and feeding of poor-cod, Trisopterus minutus capelanus, in the northern Tyrrhenian Sea. Ital. J. Zool. 1992, 59, 87–93.
43. Carpenter, K.E.; Stump, E. Trisopterus capelanus. In The IUCN Red List of Threatened Species; Version 2014.3; International Union for Conservation of Nature and Natural Resources: Gland, Switzerland, 2014.[Google Scholar]
44. Albert, O.T. Biology and ecology of Norway pout (Trisopterus esmarki Nilsson, 1855) in the Norwegian Deep. ICES J. Mar. Sci. 1994, 51, 45–61.
45. Wheeler, A. Key to the Fishes of Northern Europe; Frederick Warne Ltd.: London, UK, 1978; xixp. 380p.
46. Potter, I.C.; Claridge, P.N.; Warwick, R.M. Consistency of seasonal changes in an estuarine fish assemblage. Mar. Ecol. Progr. Ser. 1986, 32, 217–228.
47. Albert, O.T. Biology and Ecology of Norway Pout (Trisopterus esmarkii) in the Norwegian Deep; C.M. 1991/G:44; ICES Demersal Fish Comm.: Copenhagen, Denmark, 1991; p. 31.
48. Gibson, R.N.; Robb, L.; Burrows, M.T.; Ansell, A.D. Tidal, diel and longer term changes in the distribution of fishes on a Scottish sandy beach. Mar. Ecol. Progr. Ser. 1996, 130, 1–17.
49. Cohen, D.M.; Inada, T.; Iwamoto, T.; Scialabba, N. FAO species catalogue. Gadiform fishes of the world (Order Gadiformes). In An Annotated and Illustrated Catalogue of Cods, Hakes, Grenadiers and other Gadiform Fishes Known to Date; FAO Fisheries Synopsis: Rome, Italy, 1990; Volume 125, xp. 442p.
50. Froese, R.; Pauly, D. (Eds.) FishBase. World Wide Web Electronic Publication, Version 2015. Trisopterus minutus (Linnaeus, 1758), Poor Cod.. Available online: http://www.fishbase.org/summary/481 (accessed on 19 August 2015).
51. Jardas, I. Adriatic Ichthyofauna; Školska knjiga d.d.: Zagreb, Kroatia, 1996; p. 523. (In Croatian)
52. Di Natale, A.; Molinari, A.; Öztúrk, B.; Srour, A. The IUCN Red List of Threatened Species. Trisopterus capelanus. Available online: http://www. iucnredlist. org/details/49911704/0 (accessed on 19 August 2015).
53. Labropoulou, M.; Papaconstantinou, C. Community structure of deep-sea demersal fish in the North Aegean Sea (northeastern Mediterranean). Hydrobiology 2000, 440, 281–296.
54. Politou, C.-Y.; Papaconstantinou, C. Population biology of Trisopterus minutus capelanus (Gadidae) from the eastern coast of Greece. Cybium 1991, 15, 69–81.
55. Gaemers, P.A.M. New gadiform otoliths from the Tertiary of the North Sea Basin and a revision of some fossil and recent species. Leidse Geol. Meded. 1976, 49, 507–537.
56. Raitt, D.F.S. The Food and Feeding of Gadus Esmarkii (Nilsson) Compared with 0-Group Haddock and Whiting; ICES CM 1961/Gadoid Committee Mar. Res., No. 101; ICES: Copenhagen, Denmark, 1961.
57. Raitt, D.F.S. Observations on the population dynamics of the Norway pout in the North Sea. Rap. Proc. Verb. Cons. Perm. Internatl. Explor. Mer 1968, 158, 85–90.
58. Linnaeus, C. Systema Naturae per Regna Tria Naturae, Secundum Classes, Ordinus, Genera, Species, Cum Characteribus, Differentiis, Synonymis, Locis, 10th ed.; Lars Salvi: Stockholm, Sweden, 1758; Volume 1, p. 824.
59. Artedi, P. Ichthyologia, sive opera omnia de piscibus scilicet: Bibliotheca ichthyologica. Philosophia ichthyologica. Genera Piscium, Synonymia specierum, Descriptiones specierum. Omnia in hoc genere perfectiora, quam antea ulla. Posthuma. In Vindicavit, Recognovit, Coaptavit & Editit Carolus Linnæus, Med. Doct. & Ac. Imper. N.C.; Conradum Wishoff: Leiden, The Netherlands, 1738; Volume 3, ivp. 88p.
60. Müller, O.F. Zoologiae Danicae Prodromus, seu Animalium Daniae et Norvegiae Indigenarum Characteres, Nomina, et Synonyma Imprimis Popularium; Hallager: Copenhagen, Denmark, 1776; xxxiip. 282p.
61. Lacepède, B. Histoire Naturelle des Poisons; Plassan: Paris, France, 1800; Volume 2, lxivp. 632p.

62. Rafinesque-Schmaltz, C.S. Précis des Découvertes et Travaux Somiologiques de Mr. C.S. Rafinesque-Schmaltz Entre 1800 et 1814; Ou choix raisonné de ses principales Découvertes en Zoologie et en Botanique pour servir d'introduction à ses ouvrages futurs; Royale Typographie Militaire: Palermo, Italy, 1814; p. 55.
63. Risso, A. Histoire Naturelle des Principales Productions de l'Europe Méridionale et Particulièrement de Celles des Environs de Nice et des Alpes Maritimes; Levrault: Paris et Strasbourg, Croatia, 1826; Volume 3, xvip. 486p.
64. Schwarzhans, W. The Otoliths from the Miocene of the North Sea Basin; Backhuys Publishers: Leiden, The Netherlands; Margraf Publishers: Weikersheim, Germany, 2010; vp. 352p.
65. Tuset, V.M.; Lombarte, A.; Assis, C.A. Otolith atlas for the western Mediterranean, north and central eastern Atlantic. Sci. Mar. 2008, 72S1, 7–198.
66. Fage, L. Le Capelan de la Méditerranée: Gadus capelanus (Risso) et ses rapports avec les espèces voisines: G. luscus Linné et G. minutus O. Fr. Müller. Arch. Zool. Exp. Génér. Hist. Nat. Morphol. Histol. Évol. Anim. 1911, 6, 257–282.
67. Chaine, J.; Duvergier, J. Distinction des Gadus capelanus, minutus et luscus par leur sagittae. Compt. Rend. Hebdom. Séanc. Acad. Sci. Paris 1927, 184, 977–978.
68. Merayo, C.R.; Villegas, M.L. Age and growth of Trisopterus luscus (Linnaeus, 1758) (Pisces, Gadidae) off the coast of Asturias. Hydrobiologia 1994, 281, 115–122.
69. Zhang, D.-X.; Smith, J.-A.; Rico, C.; Hewitt, G.M. Isolation and characterization of 10 microsatellite loci in poor cod Trisopterus minutus (L). Mol. Ecol. Notes 2001, 1, 50–52.
70. Carr, S.M.; Kivlichan, D.S.; Pepin, P.; Crutcher, D.C. Molecular systematics of gadid fishes: Implications for the biogeographic origins of Pacific species. Can. J. Zool. 1999, 7, 19–26.
71. Cadée, G.C. Molluscan biocoenoses and thanatocoenoses in the Ría de Arosa, Galicia, Spain. Zool. Verhand. Rijksmus. Nat. Hist. Leiden 1968, 95, 1–121.
72. Froese, R.; Pauly, D. (Eds.) FishBase. World Wide Web Electronic Publication, Version 2015. Trisopterus luscus (Linnaeus, 1758), Pouting. Available online: http://www.fishbase.org/summary/1367 (accessed on 19 August 2015).
73. FAO Aquatic Species Distribution Map of Trisopterus luscus. Available online: http://www.fao.org/fishery/species/2235/en (accessed on 19 August 2015).
74. Assis, C. Guia para a Identificação de Algumas Famílias de Peixes Ósseos de Portugal Continental, Através da Morfologia dos Seus Otólitos Sagitta; Prémio do Mar Rei D. Carlos, Câmara Municipal de Cascais: Cascais, Portugal, 2004; p. 190.
75. Andrade da Silva, D.M. Discrimination of Trisopterus luscus (Linnaeus 1758) stocks in the northern of Portugal using otolith elemental fingerprints. Master's Thesis, University of Porto, Porto, Portugal, 2010; p. 34.
76. Magnussen, M.D. Ecology of poor cod (Trisopterus minutus) on the Faroe Bank. Bachelor's Thesis, University of the Faroe Islands, Tórshavn, Faroe Islands, 2004; p. 37.
77. Froese, R.; Pauly, D. (Eds.) FishBase. Computer Generated Map for Trisopterus minutus (Poor Cod). Available online: www.aquamaps.org (accessed on 11 December 2014).
78. Hognestad, P.T.; Vader, W. Saltvannsfiskene i Nord-Norge (The species of marine fishes in North Norway). Tromura 1979, 6, 1–74.
79. Gomes, M.C.; Serrão, E.; de Fátima Borges, M. Spatial patterns of groundfish assemblages on the continental shelf of Portugal. ICES J. Mar. Sci. 2001, 58, 633–647.
80. Froese, R.; Pauly, D. (Eds.) FishBase, Computer Generated Map for Trisopterus esmarkii (Norway Pout). Available online: www.aquamaps.org (accessed on 11 December 2014).
81. Mikkelsen, G. Changes in Norway Pout (Trisopterus esmarkii) Abundance and Distribution under Warming Conditions in the Barents Sea. Master's Thesis, University of Bergen, Bergen, Norway, 2014.
82. ICES. Report of the ICES Advisory Committee, 2008. Book 7. Bay of Biscay and Western Iberian Seas; ICES: Copenhagen, Denmark, 2008; p. 122.

83. Nolf, D. The Diversity of Fish Otoliths, Past and Present; Steurbaut, E., Brzobohaty, R., Hoedemakers, K., Eds.; Roy. Belg. Inst. Nat. Sci.: Brussels, Belgium, 2013; p. 222, + 359 plates.
84. Gaemers, P.A.M. A biozonation based on Gadidae otoliths for the northwest European younger Cenozoic, with the description of some new species and genera. Meded. Werkgr. Tert. Kwart. Geol. 1978, 1, 141–161.
85. Gaemers, P.A.M. Otolith biostratigraphy and paleoecology of well Eidelstedt-1, Nordgetränke (Hamburg Area, F.R.G.), and comparison with some other Langenfeldian and Gramian otolith faunas in the North Sea Basin. Meded. Werkgr. Tert. Kwart. Geol. 1978, 2, 95–118.
86. Gaemers, P.A.M. Otoliths (Gadidae). The description of the interregional zonation (otolith zones) and its stratigraphical consequences. The regional distribution of otolith assemblages; correlation of the interregional zonation with the regional lithostratigraphic formations. In The Northwest European Tertiary Basin. Results of the International Geological Correlation Programme Project No 124; Vinken, R., Ed.; Geol. Jahrb., German Geological Survey BGR: Hannover, Germany, 1988; Volume 100, pp. 369–390.
87. Gaemers, P.A.M. Refined correlations by means of lithostratigraphy and gadid otolith zonation of the Rupelian of the North Sea Basin: A progress report. Bull. Soc. Belge Géol. 1994, 10, 147–157.
88. Hennig, W. Grundzüge einer Theorie der Phylogenetischen Systematik; Deutscher Zentralverlag: Berlin, Germany, 1950; p. 370.
89. Grantham, T. The role of fossils in phylogeny reconstruction: Why is it so difficult to integrate paleobiological and neontological evolutionary biology? Biol. Philos. 2004, 19, 687–720.
90. Gaemers, P.A.M. A revision of the Gadidae otoliths (Pisces) from the Redonian of the Ligerian Gulf (NW France) and the stratigraphic position of the Redonian. Tert. Res. 1987, 8, 105–125.
91. Lanckneus, J.; Nolf, D. Les otolithes des téléostéens redoniens de Bretagne (Néogène de l'Ouest de la France). Bull. Inst. Géol. Bassin Aquit. 1979, 25, 83–109.
92. Hoedemakers, K. Teleost fish otoliths from the Neogene of Mill-Langenboom (province of Noord-Brabant, The Netherlands. Cainozoic Res. 2013, 10, 35–52.
93. Tirard, C.; Berrebi, P.; Raibaut, A.; Frenaud, F. Parasites as biological markers—Evolutionary relationships in the heterospecific combination of helminths (monogeans) and teleosts (Gadidae). Biol. J. Linn. Soc. 1992, 47, 173–182.
94. Gaemers, P.A.M.; Poulsen, J.Y. Recognition and distribution of two Gadiculus species, G. argenteus and G. thori, based on otolith morphology, larval pigmentation and molecular evidence. J. Fish Biol. 2016, in press.

# CHAPTER 8

# Evaluation of Toxicity in Silver Catfish

**Cláudia Turra Pimpão[1], Ênio Moura[2], Ana Carolina Fredianelli[3], Luciana G. Galeb[3], Rita Maria V. Mangrich Rocha[4] and Francisco P. Montanha[5]**

*[1] Laboratory of Pharmacology and Toxicology/PUCPR, College of Agricultural, Environmental Sciences and Veterinary Medicine,Pontifícia Universidade Católica do Paraná, Curitiba, Brazil*

*[2] Service of Medical Genetics/PUCPR, College of Agricultural, Environmental Sciences and Veterinary Medicine,Pontifícia Universidade Católica do Paraná, Curitiba, Brazil*

*[3] College of Agricultural, Environmental Sciences and Veterinary Medicine,Pontifícia Universidade Católica do Paraná, Curitiba, Brazil*

*[4] Service of Veterinary Clinical Pathology/PUCPR, College of Agricultural, Environmental Sciences and Veterinary Medicine,Pontifícia Universidade Católica do Paraná, Curitiba, Brazil*

*[5] Laboratory of Pharmacology and Toxicology/FAMED, College of Agricultural, Environmental Sciences and Veterinary Medicine,Pontifícia Universidade Católica do Paraná, Curitiba, Brazil*

## 1. INTRODUCTION

The global population growth observed in the recent decades coupled with continual technological advancement and increase in the generation of new industrial products, including the manufacture of chemicals such as fertilizers and pesticides, has led to an expansion in the levels of xenobiotic compounds in aquatic ecosystems (Jesus; Carvalho, 2008).

The pollution of rivers and lakes with chemicals of anthropogenic origin may have adverse consequences: the waters become unsuitable for drinking and other household purposes, irrigation, and fish cultivation, and the animal communities living in them may suffer seriously. Massive fish kills are recorded rather frequently, and changes in the population of the fauna as a consequence of sublethal effects on ecologically important species have also been described (Koprocu; Aydin, 2004).

Insecticides are used extensively in agriculture and industry because it is easy to apply, cost effective, and in some situations, it is only a practical method of control. However, benefits of pesticides are not derived without consequences. They are one of the most potentially harmful chemicals and are released into the environment by direct applications, spraying, atmospheric deposition, and surface runoff. Given the fact that, insecticides are not selective and affect non target species as readily as target organisms, it isn't surprising that a chemical that acts on the insect's different systems will elicit similar effects in higher forms of life (Dogan; Can, 2011).

Levels of insecticides in superficial waters generally range far below lethal concentrations for aquatic organisms. However, sublethal adverse effects may result from exposure of aquatic organisms to insecticides at environmentally relevant concentrations (Das; Mukherjee, 2003).

Pesticides in the environment may be used as a model for the study of ecotoxicology, because they contaminate air, land and water, causing adverse effects that affect from bacteria to humans. It is well proven that these chemicals are toxic to aquatic arthropods, bees and fish (Santos et al., 2007). The effects of the use of pesticides are recognized worldwide and aggravated by misuse since part of this material is accumulated in plants and soil and much of it is transported to the rivers by rain (Tsuda et al., 1995; Wilson; Tisdell, 2001).

As a result of a great variety of human activities, the aquatic environment is becoming increasingly threatened by an alarming number of foreign chemicals or xenobiotics. Fish populations living in highly polluted areas often have high incidences of gross pathological lesions (Malins et al., 1988), associated with elevated levels of toxic contaminants in the sediments. However, pesticides applied to the land may be washed into surface waters and may kill or at least adversely influence the life of aquatic organisms.

Contamination of water with large amounts of pesticides leads to fish mortality or starvation by destruction of food organism, many toxicants have been shown to affect growth rate, reproduction and behavior, with evidence of tissue damage (Van Der Oost et al., 2003; Srivastav et al., 2002).

The poisoning of fish by pesticides can be acute or chronic and in general acute poisoning causes mass mortality. However, pollution is an often chronic process, apparently without any visible damage but sometimes producing several sublethal effects (Rodrigues, 2003).

The aim of this chapter was present a review based on some aspects of silver catfish's toxicology.

## 2. THE FISH

Silfvergrip (1996) conducted extensive taxonomic revision of the genus based on characters of internal morphology, and concluded that the genre *Rhamdia* consists of only 11 species among 100 previously described. According to the same author, quelen taxonomic division belongs to the following: Class: Osteichthues, Series: Teleostei, Order: Siluriformes, Family: Pimelodidae. Genre: Rhamdia, Species: quelen.

*Rhamdia quelen*, popularly known, as silver catfish is a species of fish found from southern Mexico to Argentina that displays the absence of teeth and scales with variable-length cylindrical wattles. This is an omnivorous species with an eating preference for fish, crustaceans, insects, plants and organic debris (Silfvergrip, 1996).

The silver catfish (*Rhamdia quelen,* Quoy & Gaimard) is an endemic South American fish species that with stands cold winters and presents fast growth rate in summer. These characteristics make catfish a suitable species for fish production in southern South America or any region with a temperate or subtropical climate. In aquaculture systems, at a density of two to four fish/m$^2$ catfish reach a 600-800 g body weight in eight months. Our unpublished observations in experimental field trials and at fish farms have shown that this weight is easily reached, but high mortality rates (40-50%) might occur if small fish (1-3 g) are used to initiate the culture. However, when beginning with heavier juveniles (30-60 g), the final weight will still range from 600 to 800 g, but with mortality rates not exceeding 5-10%. Thus, the more reasonable order in silver catfish culture is: hatchery, larviculture (1-6 g), nursery (from 5-6 g to 30-60 g), and termination (from 30-60 g to 600-800 g) (Barcellos et al., 2001). This species can be considered eurytermal because the fry acclimated to 31°C withstand temperatures from 15 to 34°C (Barcellos et al., 2003).

## 3. CHEMICAL CONTAMINANTS IN AQUATIC SYSTEMS

Toxic compounds or natural anthropogenic are called xenobiotics. With the onset of the epidemiological investigations was to confirm the hypothesis of many xenobiotics to be dangerous to living things, as well as their respective offspring, exerting toxic effects in the short, medium or long term (Reys, 2001). These substances are persistent in the environment eventually absorbed and accumulated by living organisms, toxic effects on various organs and systems. Thus, it was noted that the use of xenobiotics without evaluation of risks to the ecosystem, constituted a potential threat to the health of people, animals and plants (Sanches, 2006).

The introduction of toxic substances in the aquatic environment causes local and immediate effects, but can also lead to contamination of watersheds and commitment of underground reservoirs by infiltration through the soil. Numerous compounds have been detected in surface water, groundwater and water supply relating to agricultural activities and human cases of environmental contamination (Sanches, 2006).

## 3.1. Insecticides

The growing use of synthetic insecticides is intensifying global pollution risks. Insecticides are toxic and were designed to repel or kill unwanted organisms and when used for their different purposes they may be brought to water bodies killing or influencing the lives of aquatic organisms (El Sayed et al., 2007). The effects of the use of insecticides are recognized worldwide and compounded by their improper use (Tsuda et al., 1995; Wilson; Tisdell, 2001).

Organophosphates comprise a group of chemical compounds extensively used in farming as insecticides, which cause accidental poisoning in animals and men. The toxicity of these compounds is due especially to the respiratory and cardiac impairment in consequence of autonomic nervous system disorders.

The primary effect of organophosphates (Ops) on vertebrate and invertebrate organisms is the inhibition of the enzyme Acetylcholinesterase (AChE), which is responsible for terminating the transmission of the nerve impulse. OPs block the hydrolysis of the neurotransmitter acetylcholine (ACh) at the central and peripheral neuronal synapses, leading to excessive accumulation of ACh and activation of ACh receptors (Peña-Llopis et al., 2003). The overstimulation of cholinergic neurones initiates a process of hyperexcitation and convulsive activity that progresses rapidly to *status epilecticus*, leading to profound structural brain damage, respiratory distress, coma, and ultimately the death of the organism if the muscarinic ACh receptor antagonist atropine is not rapidly administered (Shih; Mcdonough, 1997).

Melo (2004) showed that *Rhamdia quelen* (silver catfish) juveniles exposed for 96 hours to a sublethal dose (0.01 mL/L) of Folidol® 600, was target for the toxicant action and some alterations became evident after 4 hours of exposure. The alterations observed in the liver were reduction in the density of melanomacrophages, focuses of necrosis, enhancement in the density of hepatocytes, loss of the cellular contour of the hepatocytes, cytoplasmic granulation, reduction of the cytoplasmic vacuolization, mithocondrial disruption, disorganization of the rough endoplasmic reticulum, nuclear heterochromatization and decharacterization of the endothelium. These alterations could diminish the liver metabolism, and as a consequence, they could cause damages to the health of the *R. quelen* juveniles.

Deltamethrin (DM) and other pyrethroids have proven to be toxic to aquatic organisms, mainly to fish. Due to its lipophilic characteristics it can be highly absorbed by the fish gills, which partially explains the high sensitivity of these animals to DM exposure in concentrations up to a thousand times lower than in mammals (Rodrigues, 2003).

Galeb et al. (2010) and Montanha (2010) studied the behavior of silver catfish exposed to sublethal concentrations of DM and CM, respectively, and they have presented loss of balance, swimming alteration, dyspnea (they kept their mouths and opercula open). *Post-mortem* signs observed in the animals exposed to DM and CM were mainly darkening of the surface of the body, tail and wattles erosion and hemorrhagic spots on the body surface.

The main behavioral changes observed are represented by respiratory and neurological manifestations. Such results corroborate with Polat et al. (2002) and Ylmaz et al. (2004), who have tested Cypermethrin in guppies (*Poecilia reticulata*); Borges (2007), who have tested CM in silver catfish (*Rhamdia quelen*). These changes can be attributed to the neurotoxic effect of DM/CM by blocking sodium channels and inhibiting the GABA receptors in the nervous filaments which results in an excessive stimulation of the central nervous system that sometimes can lead to brain hypoxia (El- Sayed et al., 2007).

Galeb et al. (2010) studied the haematological response of silver catfish exposed by DM, it was a significant increase in the total leukocyte counts. Similar results were reported by El-Sayed et al. (2007) in Nile tilapia (*Oreochromis niloticus*) and Pimpão et al. (2007) in catfish (*Ancistrus multispinis*). Montanha (2010) showed similar results in silver catfish exposed to CM. The leukocytosis showed that these pesticides can generate inflammatory or stress responses.

It was observed decrease of serum levels of ALT (alanine aminotransferase), AST (aspartate aminotransferase) and FA (alkaline phosphatase) were observed in silver catfish exposed to DM (Galeb, 2010) and CM (Montanha, 2010).

Fipronil is a non-systemic, chiral phenylpyrazole insecticide registered for use to control ants, beetles, cockroaches, fleas, mole crickets, ticks, termites, thrips, and other insects in a variety of agricultural and residential uses. Its mechanism of action involves non-competitive binding to the GABA receptor, effectively blocking the chloride channel and resulting in paralysis (USEPA, 2011).

Their toxicity to fish varies with species and it is highly toxic to *Lepomis macrochirus* ($LC_{50}$ = 85 g/L), the *Oncorhynchus mykiss* ($LC_{50}$ = 248 g/L), the *O. niloticus* ($LC_{50}$ = 42 mg/L), *Poecilia reticulata* ($LC_{50}$ less than 100 g/L) and *Cyprinus carpio* ($LC_{50}$ = 430 g/L). Low concentrations of fipronil are lethal to most species of fish that were tested, and is especially toxic to young fish. Studies indicate that the dose 15 mg/L reduced the growth of trout. In addition, this compound is also bioaccumulated fish.

### 3.2. Drugs

The occurrence of pharmaceuticals for human and veterinary use has been detected in surface waters, sediments and drain in worldwide. Various substances such as anti-inflammatories, analgesics, antibiotics, hormones and antidepressants represent them. Although they have been subjected to pharmacokinetic studies, little information exists about the environmental fate and toxic effects in several organisms of aquatic fauna and flora, certainly affected (Stumpf et al., 1999; Fent et al. 2006).

Chemical toxicity distribution approaches have been employed for identifying Thresholds of Toxicological Concern for many industrial chemicals (Kroes et al., 2005), comparing the sensitivities of in vitro and fish models for estrogenicity (Dobbins et al., 2008), and predicting aquatic concentrations of ecotoxicological concern for chemical with common MOAs in plant models and invertebrates and fish (Dobbins et al., 2008).

## 4. ASSESSMENT OF TOXICITY IN FISH BY BIOMARKERS

The biological response of an organism to xenobiotics following absorption and distribution starts with toxicant induced changes at the cellular and biochemical levels, leading to changes in the structure and function of the cells, tissues, physiology and behavior of the organism. These changes can perhaps ultimately affect the integrity of the population and ecosystem. For the biomonitoring and management of the aquatic ecosystems, these biological responses (biomarkers) have been proposed to complement and enhance the reliability of the chemical analysis data (Parvéz; Raisuddin, 2005).

There are very few pollutants that have been confirmed to cause adverse effects. In most cases, casual relationships have not been established to a large group of persistent pollutants, due the complex chemical contamination of environmental compartments, which difficulty to attribute harmful effects to any particular pollutant or category of pollutants.

Several definitions have been given for the term 'biomarker', which is generally used in a broad sense to include almost any measurement reflecting an interaction between a biological system and a potential hazard, which may be chemical, physical or biological (WHO, 1993). A biomarker is defined as a change in a biological response (ranging from molecular through cellular and physiological responses to behavioral changes) that can be related to exposure to or toxic effects of environmental chemicals (Van Der Oost, R et al., 2003).

A bioindicator is defined as organism giving information on the environmental conditions of its habitat by its presence or absence or by its behavior, and an ecological indicator is an ecosystem parameter, describing the structure and functioning of ecosystems.

According to the WHO (1993), biomarkers can be subdivided into three classes:

- **Biomarkers of exposure:** covering the detection and measurement of an exogenous substance or its metabolite or the product of an interaction between a xenobiotic agent and some target molecule or cell that is measured in a compartment within an organism;
- **Biomarkers of effect:** including measurable biochemical, physiological or other alterations within tissues or body fluids of an organism that can be recognized as associated with an established or possible health impairment or disease;
- **Biomarkers of susceptibility:** indicating the inherent or acquired ability of an organism to respond to the challenge of exposure to a specific xenobiotic substance, including genetic factors and changes in receptors which alter the susceptibility of an organism to that exposure.

## 4.1. Histopathology

Histopathological characteristics of specific organs express condition and represent time-integrated endogenous and exogenous impacts on the organism stemming from alterations at lower levels of biological organization. Therefore, histological changes occur earlier than reproductive changes and are more sensitive than growth or reproductive parameters and, as an integrative parameter, provide a better evaluation of organism health than a single biochemical parameter. Histopathological biomarkers signal the effects of exposures both acute and chronic to toxic agents in high or low levels. The pathological changes may be adaptive or degenerative and will determine the survival or death of the organism. The organs most commonly damaged by toxic agents are gills, liver and kidneys, which may be also affected by bacteria, viruses and parasites. That is why a health assessment of the animal is important to differentiate damage induced by toxic agents and diseases. Pharmaceuticals of many categories may be detected on the environment contaminating water flows. In a study for the evaluation of the effect of the analgesic dipyron in surface waters, Pamplona (2009) observed important histological disorders in the kidneys of silver catfish. The fish were exposed to three concentrations of dipyrone (0.5, 5 and 50 μg/L) in the water for 15 days and then evaluated. According to the author, the histological parameters of the intoxicated animals showed important tissue damage in the silver catfish kidneys, such as necrosis, fibrous deposition, vacuolization and constriction of blood vessels of the renal parenchyma, hyperplasia, and even necrosis of great renal vessels and glomeruli in the higher concentration of dipyron. Such alterations were not observed in the control group.

Pesticides are the substances most extensively researched in the aquatic ecotoxicology due to the large amounts used in agriculture and livestock in the whole world. Melo (2004) exposed the silver catfish to the organophosphate methyl parathion at the sublethal concentration of 6 mg/L for 96 hours with the aim of histological and ultrastructural analysis of liver tissue. Melo (2004) suggested that the histological changes are related to the intoxication by methyl parathion that is likely to bring the individuals metabolic, cellular and subcellular problems. It was observed in *Rhamdia quelen* loss of the contour of the hepatocytes and endothelial cells, as well as changes in the nuclear chromatin. According to Melo (2004), changes in the organization of rough endoplasmic reticulum and mitochondrial disruption within four hours of intoxication and strongly eosin stained granulation in the cytosol within 24 hours of exposure could be noticed. Ultrastructurally, a high incidence of lipid droplets was observed in the cytosol of hepatocytes when compared with control within 48 hours of exposure to fipronil. Increased foci of necrosis in the liver of silver catfish after 48 and 72 hours of exposure in relation to the control group, which presents only occasional foci of necrosis, were noted, among other structural changes. After 96 hours of exposure, Melo (2004) describes cells indistinguishable contour, presence necrosis focus, in addition of damaged blood vessels, vacuolization of the cytosol and the presence of an unknown material strongly eosin stained in the cytoplasm of the hepatocytes. Blood leukocyte infiltration and congestion were observed in all animals analyzed. Melanomacrophage were found in various locations between the hepatocytes as well as pyknotic nuclei cells.

Ghisi (2010) studied the effects of the phenyl pyrazolefipronil in the gills of the silver catfish after 60 days of intoxication in the sublethal concentrations 0.05; 0.10 and 0.23 μg/L. The author reports hyperplasia, lamellar fusion and aneurysms in all groups treated, including the control group that can impair the gill function. However, Ghisi (2010) considers the injuries of low severity and possible regression if the source of stress is eliminated, since the concentration of fipronil used was very low.

Cattaneo (2009) studied the effects of 2,4 - dichlorophenoxiacetic acid (2,4-D) herbicide in *Rhamdia quelen* fingerlings in an acute toxicity assay. After acclimation, silver catfish were intoxicated with concentrations of 0, 400, 600 and 700 mg/mL of 2,4 – D for 96 hours. The author states that the histological analysis showed alterations in the liver of silver catfish after exposure to 2,4-D herbicide, such as abnormal arrangement of hepatocytes cords, cell membrane rupture and hepatocytes vacuolation.

The effects of the herbicide clomazone in teleost fish *Rhamdia quelen* were studied by Crestani et al. (2007). The silver catfish were exposed to the concentrations 0.5 and 1.0 mg/L of clomazone for 12, 24, 48, 96 and 192 h. Histological analysis showed vacuolation in the liver after herbicide exposure, some of that were completely restored after a recovery period. Ferreira (2010) studied the sublethal effects of glyphosate (1.21 mg/L), methyl parathion (0.8 mg/L) and tebuconazole (0.88 mg/L) herbicides to silver catfish. The fingerlings were intoxicated during 96 hours and then sampled after anesthesia. The mapathologicalcal changes in liver histology were: diffuse hepatocyte degeneration, bile stagnation, granules on the cytoplasm, hyperemia and vacuoles in the cytoplasm and nucleus of fingerlings exposed to methyl parathion and tebuconazole. The liver of fish exposed to glyphosate did not show any visible histological changes.

## 4.2. Hematological And Biochemical Analyses

Hematological analysis in fish is not routinely used for fish diseases diagnosis. Hematology of fish lags behind that of other classes of vertebrates, but analysis of blood still can be informative about disease processes in teleosts and elasmobranchs (Clauss et al., 2008). The most important factor limiting accurate hematological analysis in different fish species is the variation in types, numbers, and appearance of leukocytes. In addition, the considerable variation in reported leukocyte values from healthy fish, even within a species, is also partly caused by differences in the methodology used as well as by subjective interpretation of cell types by the investigator. Nevertheless, some effort should be made to try to surpass these difficulties since these parameters can be a useful tool for evaluation of the effects of pesticides in the cellular components of blood and even in the immune system. Therefore, the analysis of hematological and biochemical parameters in fish can contribute to the assessment of the animal's health and also the habitat conditions (Pimpão et al., 2007).

Table 1 provides an overview of some hematological and blood chemistry results in catfish (*Rhamdia quelen*) exposed to toxic chemicals.

## 4.3. Enzymatic Analyses

One of the more sensitive effect biomarkers is alteration in levels and activities of biotransformation enzymes. The activity of these enzymes may be induced or inhibited upon in fish exposure to xenobiotic. Enzyme induction is an increase in the amount or activity of these enzymes, or both and inhibition is the opposite of induction. In this case, enzymatic activity is blocked, possibly due to a strong binding or complex formation between the enzyme and the inhibitors.

The according by Van Der Oost, R et al. (2003) enzymatic analyzes to fish are used more:

- Phase I enzymes: Total cytochrome P450 (cyt P450); Cytochrome P450 1A ; Cytochrome b5 (cyt b5); Ethoxyresorufin O-deethylase (EROD) and arylhydrocarbon hydroxylase (AHH); NADPH cytochrome P450 reductase (P450 RED);
- Phase II enzymes and cofactors: Reduced and oxidized glutathione (GSH and GSSG); Glutathione S-transferases (GSTs); UDP-glucuronyltransferases (UDPGTs);
- Oxidative stress parameters: Superoxide dismutase (SOD); Catalase (CAT); Glutathione peroxidase (GPOX); Glutathione reductase (GRED); Non-enzymatic antioxidants
- Biochemical indices of oxidative damage: Lipid peroxidation; DNA oxidation;

Table 2 provides an overview of some enzymatic results in catfish exposed to toxic chemicals.

**TABLE 1.** Some hematological and blood chemistry results in catfish (*Rhamdia quelen*) exposed to toxic chemicals.

| | Diets containing aflatoxins * | | | | Treatment with diflubenzuron in 24-hour immersion baths ** | | | | *** |
|---|---|---|---|---|---|---|---|---|---|
| Chemical substance | Control | 41ppb/ aflatoxin/ kg | 90ppb/ aflatoxin/ kg | 204ppb/ aflatoxin/ kg | Control | 0,01 mg/ l⁻¹diflubenzuron | 0,1 mg/ l⁻¹diflubenzuron | 1,0 mg/ l⁻¹diflubenzuron | None |
| Erythrocytes (x 10⁶/µL) | | | | | 2,71 ± 0,3 | 2,67 ± 0,4 | 2,56 ± 0,3 | 2,6 ± 0,4 | 1,95 ± 0,40 |
| Hemoglobin (g/dL) | 8,2 ± 0,5 | 5,5 ± 0,3 | 4,0 ± 0,3 | 3,5 ± 0,2 | | | | | 6,73 ± 1,15 |
| Hematocrit (%) | 24,51 ± 1,51 | 13,16 ± 1,47 | 11,75 ± 0,41 | 11,16 ± 0,88 | 32,7 ± 3,62 | 32,9 ± 2,61 | 31,1 ± 2,8 | 30,1 ± 1,7 | 26,5 ± 5,3 |
| Leucocytes (x 10³/µL) | | | | | 13,4 ± 2,7 | 13,2 ± 2,9 | 12,6 ± 2,1 | 12,2 ± 2,4 | |
| Lymphocytes (%) | | | | | 62,1 ± 5,7 | 61,9 ± 6,8 | 62,4 ± 6,3 | 62,6 ± 6,4 | 11,61 ± 6,57 |
| Neutrophils (%) | | | | | 25,3 ± 5,5 | 26,9 ± 5,8 | 25,9 ± 6,7 | 25,9 ± 6,7 | 6,02 ± 3,33 |
| Granulocytic especial cell (%) | | | | | 2,7 ± 1,2 | 2,0 ± 0 | 2,2 ± 0 | 4,0 ± 2,0 | 1,24 ± 1,96 |
| Monocytes (%) | | | | | 11,6 ± 2,9 | 10,9 ± 4,5 | 11,1 ± 3,5 | 10,6 ± 4,3 | 1,13 ± 1,09 |
| Eosinophils (%) | | | | | 2,0 ± 0 | 2,0 ± 0 | 2,0 ± 0 | 2,0 ± 0 | 0,02 ± 0,14 |
| Glucose (mg/dL) | 45,00 ± 6,28 | 36,40 ± 2,42 | 24,40 ± 6,69 | 23,20 ± 5,93 | | | | | |

* Vieira, et al. (2006).

**Mabilia; Souza (2006)

***Tavares-Dias et al. (2002)

**TABLE 2.** Laboratory studies on responses of organic trace pollutants on fish enzymes

| Species | Pollutants | cyt P450 | CYP1A | AHH | EROD | cyt b5 | P450 RED | References |
|---|---|---|---|---|---|---|---|---|
| Channel catfish | PCB Aroclor1254 | = | | ++ | ++ | = | = | Ankley et al., 1986 |
| *Ictaluruspunctatus* | BKME | = | | | ++ | | | Mather-Mihaich;DiGiulio, 1991 |
| | PAH (BNF) | | | | ++ | | | Hasspieler et al., 1994 |
| | 2,4D+ Picloram | | | | | | | Gallagher and Di Giulio, 1991 |
| | 2 Aminocanthracene (AA) | | | | - | | | Watson et al.,1995 |
| | PAH (BaP) | | + | + | | | | Ploch et al., 1998 |
| | PCB126 | | ++ | | ++ | | | Rice and Roszell, 1998 |
| **Species** | **Pollutants** | **SOD** | **GPOX** | **GSSG** | **CAT** | **LPOX** | **GSH** | **References** |
| Catfish*Ictalurusnebulosus* | OPP (dichlorvos) | = | = | | = | + | = | Hai et al., 1995 |
| Channel catfish *Ictaluruspunctatus* | BKME | = | = | | + | = | - | Mather-Mihaich and Di Giulio, 1991 |
| | PAH | | | + | + | | + | Di Giulio et al., 1989 |

Van Der Oost, R et al. (2003)

Symbols: --, strong inhibition (<20% of control); -, inhibition; =, no (significant) response; +, induction; ++, strong induction (>500% of control);

## 4.4. Genetic Analyses

Genotoxic substances produce chemical or physical changes inDNA, commonly measured as breaks or DNA adducts, respectively (Nacci et al., 1996). The exposure of an organism to genotoxic substances can induce a cascade of events such as changes in the structure of DNA and the expression of damaged DNA using modified products. Thus, biological consequences may be triggered in cells, organs, body and finally in community and population (Lee; Steinert, 2003; Van Der Oost et al., 2003).

Many biomarkers have been used as tools for the detection of exposure and to evaluate the genotoxic effects of pollution. Among the major genetic biomarkers, we can mention the evaluation of the frequency of chromosomal aberrations, analysis of the frequency of sister chromatid exchanges, formation of DNA adducts, DNA breakage assessed by measuring the comet assay and micronucleus frequency and nuclear morphological abnormalities (Bombail et al., 2001).

According to the studies of Pamplona (2009) by the Comet Assay, widely used to measure DNA damage in aquatic organisms, after the intoxication of silver catfish by dipyrone, the genetic damages occurred in the lowest concentration of dipyron (0.5 μg/L). DNA fragmentation may be triggered by the release of free radicals in response to a stressor such as a xenobionte that may damage the nucleus chromatin, or by lipoperoxidation that causes the loss of cellular membrane integrity, subjecting the attack of the nucleic acid by oxygen radicals.

In a study of genotoxicity by copper sulphate in four different doses (5, 30, 50 and 500 mg/Kg) using the catfish, Costa (2011) induced DNA breakage triggered by comet assay. The fish were intoxicated by trophic copper sulphate during 60 days and suffered genetic damage in brain tissue, kidney, liver and blood tissues. Blood was less sensitive tissue to levels of copper used in this study. The piscine micronucleus test developed by Costa (2011), however, showed no significant changes in relation to the negative control, as well as the polychromatic erythrocytes frequency in relation to the total number of erythrocytes. There was an increase in the frequency of viable cells in blood tissue, according to the author.

Piancini (2011) performed the piscine micronucleus test and the comet assay in erythrocytes and gill cells of the silver catfish to evaluate the genotoxic effects of atrazine in concentrations 2, 10 and 100 μg/L for 96 hours. The author observed an increased frequency of nuclear morphological changes at all concentrations, a dose-dependent effect of atrazine on DNA trough the comet assay in erythrocytes: damage was significantly higher on 10 and 100 μg/L concentrations compared to control. In gills, there was no difference between treatments. Piancini (2011) observed a dose-dependent effect of atrazine in relation to genotoxicity and its capacity of damaging the DNA of the silver catfish in concentration considered safe by the Brazilian legislation. The same author evaluated the genotoxic effects of copper on *Rhamdia quelen* through the piscine micronucleus test and the comet assay in erythrocytes and gills. The silver catfish were exposed to concentrations of 2.0; 9.0 and 18 μg/L of cuprous chloride for 96 hours. Dose-dependent breaks to the DNA were detected by the comet assays in all treatments, as well as in gills.

Ghisi (2010) did not observe micronuclei in the catfish cells, but did confirm nuclear morphologic changes, after 60 days of intoxication in the sublethal concentrations 0.05; 0.10 and 0.23 μg/L of fipronil. The control group was not significantly different from the group intoxicated by the lower concentration of the insecticide. The higher concentrations were showed significantly higher among of micronuclei compared to the control group, not intoxicated by fipronil, suggesting that the pesticide in higher concentrations (0.10 e 0.23 μg/L) may induce damages to the DNA of *Rhamdia quelen*. The comet assay, also performed by Ghisi (2010) in the same study, showed no significant difference between the intoxicated groups and the control, corroborating the results of histopathological analysis.

The aim of the study performed by Salvagni et al. (2011) was to determine, by micronucleus testing on silver catfish, the risk of genotoxic impact by the use of pesticides such as glyphosate, cyhalothrin, atrazine, simazine and azoxystrobinon in farms located in the Lambedor River watershed in Guatambu, State of Santa Catarina. Blood samples were collected from several species of fish, including *Rhamdia quelen*, captured in 10 different dams existing in agricultural rural properties. Micronucleate erythrocytes were found in blood samples of silver catfish close to the overall average of the other species. Salvagniet at. (2011) believe through *in vivo* piscine micronucleus testing, that water from the Lambedor watershed can be considered genotoxic, with emphasis on the degree of genotoxicity from pollution in the area, since spontaneous formation of micronuclei in fish is normally very rare. Ferraro (2009) also exposed the silver catfish to different molecules of pesticides and the combinations of them in the aim to determine the bioindicator potential of this teleost species. After the exposure of the fish to the pesticides glyphosate (1.58 e 3.16 mg/L), tebuconazole (0.4 e 0.8 mg/L) and a mixture of them (3.16 e 0.8 mg/L) for 5, 10 and 15 days, were proceeded the micronucleus test and the comet assay in blood samples collected from the fish. DNA damage was confirmed by both techniques in all treatments, pointing the suitability of the species for biomonioring. In a similar way, Ramsdorf (2011) studied the genotoxic effects of fipronil (0.05; 0.10 and 0.23 μg/L), lead nitrate (0.01; 0.03 and 0.10 mg/L) and naphthalene (0.005; 0.06 and 3 mg/L) in the water for *Rhamdia quelen* during 60, 30 and 28 days, respectively. In silver catfish the concentrations 0.10 and 0.23 μg/L of fipronil increased the frequency of micronucleus, nuclear morphological changes, and damages to DNA observed after comet assay. After the intoxication by lead nitrate, it was observed that the concentrations 0.03 e 0.1 mg/L increased the levels of DNA breaks, as well as in all concentration of naphthalene tested for the species.

## 5. IMPACT OF TOXIC AGENTS IN THE REPRODUCTIVE SYSTEM

The use of fish embryos is gaining popularity for research in the area of toxicology and teratology. Particularly embryos of the zebrafish, catfish and rainbow trout offer an array of different applications ranging from regulatory testing to mechanistic research.

Recently, much attention has been given to possible adverse effects resulting from exposure of aquatic animals to chemicals during the prenatal and perinatal (Tramujas et al., 2006; Montanha; Pimpão, 2012).

The impairment of reproductive function in humans and animal species has been of particular concern in recent years. Many factors can interact with the components and the reproductive function and cause infertility and other functional and structural changes. Illness, stress, hormonal changes and exposure to chemicals, are some factors that contribute to the emergence of reproductive system disorders (Sanchez, 2006).

Fish are one of the most thoroughly studied organisms in terms of effect of substance with estrogenic activity in the development of abnormalities in the reproductive system. According to Sumpter (1998), research on how estrogenic substances affect the sexual system of fish began in the 1980s (Bila; Dezotti, 2003).

The fecundity, the spawning period and type-specific characteristics are essential for the maintenance of any species of fish (Gomiero et al., 2007; Montanha; Pimpão, 2012). Reproduction in fish, as in other vertebrates, is affected by environmental factors, social and nutritional (Parra et al., 2008; Montanha, 2010). Reproductive parameters are more complex indicators of exposure and accumulation of chemicals, hampered for several reasons, the two main effects of pollutants on reproduction caused directly and indirectly and physiological process (Bernardi et al., 2008).

Few data exist regarding the potential sublethal effects of pesticides on reproduction and long-term viability of fish populations (Moore; Waring, 2001).

In 60 years, Rachel Carson published her famous book considered a landmark in the history of environmental pollution, Silent Spring, calling attention to the reproductive failure in birds and fish caused by bioaccumulation of persistent organochlorinepesticides such as diclorofodifeniltricloroetano (DDT). This work brought out the fears of modern society in relation to the introduction of synthetic substances in the environment and renewed public interest in science and government toxicology (SANCHEZ, 2006).

In the mid-70s, researchers found that other chemicals, such as the Kepone and PCBs (polychlorinated biphenyls), also had hormonal effects. Thereafter the toxicological effects of the mixture of individual congeners have been studied mainly in fish, mammalian cells and even humans (Sanchez, 2006).

The structural integrity of the gonads can be altered by xenobiotics (Mayon et al., 2006). Chemical agents that can affect the endocrine system are called endocrine-active chemicals (ECAs), expression adapted to Portuguese as "endocrine disruptors" (Sanchez, 2006).

The Environmental Protection Agency of the United States (USEPA) alternatively defines endocrine disrupters as chemicals that lead to toxic results as various types of cancer and a wide range of adverse effects on the reproductive system (Sanchez, 2006).

The pollution in the aquatic environment can affect the reproductive potential, thus reducing the spread of fish species. This can occur by the possible interaction between the gametes and water pollutants, such as the blockage of the micropyle, which prevents the entry of sperm in the fertilization process (Hilbig et al., 2008). In addition, solutions containing certain levels of pollutants can directly interfere with sperm motility and morphology, and subsequently at fertilization (Hilbig et al., 2008; Witeck et al., 2011).

In teleost fish, such as quelen (catfish), with external fertilization, which occurs when spawning, the gametes are released into the environment for fertilization to occur. At that moment, the gametes are exposed to various contaminants in the water, including heavy metals mercury, zinc, lead, copper and cadmus. These trace elements, at certain levels, affect the motility of spermatozoa and fertilization of oocytes (Witeck et al., 2011).

The gonadotropins stimulate gonadal maturation and release of steroid hormones from the gonads. The steroid hormones and the pituitary determine the development of sexual characteristics and various influencing courtship and parental care (SANCHEZ, 2006).

In teleosts, there are two gonadotropin: R gonadotropin (GTH I) and II gonadotropin (GtH II). In females, the GTH I stimulates gonadal growth, gametogenesis and the entry of vitellogenin in the oocyte. GTH II is important for the final maturation of oocytes and spawning. In males, GTH II acts in the testis, acting on testosterone production (Sanchez, 2006).

Endocrine disrupters are chemicals, natural or synthetic compounds also contained in fungicides, pesticides and insecticides, which have estrogenic and antiandrogenic action. Some of these compounds can maintain their chemical nature for many years contaminating the waters, being accumulated in fish. Many are potentially harmful to health since they are difficult excretion, may accumulate in the body and cause changes in all systems, especially in the endocrine system (Lara et al., 2011).

Recently, the monitoring of drug residues in the environment has been gaining great interest due to the fact that many of these substances are frequently found in effluents from sewage treatment plants (STPs) and natural waters. Some groups of residual drugs deserve special attention, among them are estrogen, because of its potential to adversely affect the reproductive system of aquatic organisms such as the feminization of male fish found in rivers contaminated with effluent disposal TEE. This was observed in fish species such as *Cyprinus carpio* and *Rutilus rutilus*. Similar effects (induction of hermaphroditism or complete feminization) were also observed when fish species *Oryzias latipes* were exposed to estrogen 17 β-estradiol (Bila; Dezotti, 2003).

Currently, there is growing concern about the presence of pharmaceuticals in aquatic environments and their possible environmental impacts. Ternes et al. (1999) identified the presence of various estrogens in sewage effluent and wastwater treatment plant (WWTP) in Germany, Brazil and Canada (Bila; Dezotti, 2003).

These compounds are characterized by mimicking the structure and function of natural sex steroids leading to endocrine disorders, which can result in abnormal sexual and reproductive function in animals and humans (LARA et al., 2011).

Deltamethrin, a pyrethroid pesticide widely used, is listed by the USEPA as endocrine disruptor, interfering with the reproductive system (Tramujas et al., 2006). Concentrations of 0.01 and 0.1 mg/L of deltamethrin presented toxic to fertilization of eggs of silver catfish (*Rhamdia quelen*). Since the concentrations 0.1, 0.5 and 1.0 mg/L proved to be toxic in the hatching and survival rates of 12 and 24 hours for eggs and larvae of silver catfish (Montanha, 2010). Tramujas et al. (2006) reported histological changes in levels in the gonads of aquatic organisms subjected to different concentrations of deltamethrin.

In recent years, monitoring the impacts of pesticides and other pollutants on sperm production of various kinds of research has been aimed (Mabilia et al., 2008). According to Moore and Waring (2001), even at low levels, cypermethrin (synthetic pyrethroid) in water caused a negative effect on reproductive functions in Atlantic salmon (*Salmo salar*) (Begum, 2005; Mabilia et al., 2008; Montanha, 2010; Montanha; Pimpão, 2012), causing a reduction in fertilization rate (Moore; Waring, 2001). Tramujas et al. (2006) reported a reduction in egg hatching rate of zebrafish (*Danio rerio*) exposed to deltamethrin. Rodrigues (2007) reported hatching in early zebrafish (*Danio rerio*) exposed to dichlorodiphenyltrichloroethane (DDT), with consequent reduction in survival rate. Aydin et al. (2005) and Rodrigues (2007) reported mortality of larvae of carp (*Cyprinus carpio*) exposed to cypermethrin and zebrafish (*Danio rerio*) exposed to DDT, respectively after 12 hours of exposure.

DDT at concentrations of 50 μg/L, and cause premature death hatching of larvae of *Danio rerio*, as well as parathion, from which 20 μg/L cause a decrease in the growth rates of fish of the same species (Rodrigues, 2007).

Cypermethrin proved toxic to the catfish during embryonic development in concentrations of 0.001, 0.01 and 0.1 mg/L relative to the fertilization rate and hatching. As regards the survival rate at 12 hours of exposure to cypermethrin after hatching, the concentration 10 mg/L performed toxic to larvae of catfish and 24 hours, concentrations of 0.001 to 10 mg/L (Montanha, 2010).

Some studies have reported the influence of heavy metals in fish breeding, referring to the influences of contaminants on the quality of the gametes on the fertilization of the oocytes and the quality of the larvae (Hilbig et al., 2008).

The water contamination by lead, the moment of fertilization to the first 8 hours of incubation eggs artificially catfish gray (quelen) at a concentration greater than 0.25 mg/L cause a deleterious effect on the percentage of fertilized eggs (Hilbig et al., 2008).

Water contamination fertilization with cadmus at levels up to 28.65 mg/L does not reduce rates of fertilization and hatching. However, concentrations above 20 mg/L cause reduction in sperm motility catfish (quelen) (Witeck et al., 2011).

In the study by Routledge et al. (1998), two fish species, *Oncorhynchus mykiss* and *Rutilus rutilus*, were exposed for 21 days at concentrations of 17 β-estradiol and estrone environmentally relevant. According to these researchers, the results confirmed that estrogens identified in effluents are present in sufficient amounts to cause the synthesis of vitellogenin, a protein that plays an important role in the reproductive system of female oviparous vertebrates, in fish species. Kang et al. (2002) clearly show that exposure to estrogen concentration of 17 β-estradiol environmentally relevant (in the range of 30-500 ng/L), for three weeks, induces high levels of vitellogenin hermaphrodite and the incidence of male fish species *Oryzias latipes* (Bila; Dezotti, 2003).

According to studies by Panter et al. (1998), low concentrations of 17 β-estradiol and estrone caused effects in male fish species *Pimephales promelas*, as vitellogenin synthesis and testicular inhibition (Bila; Dezotti, 2003).

Exposure to pesticides and other toxic substances during the prenatal and perinatal, can alter the reproductive system of fish without compromising growth and viability of offspring, but cause functional changes that become apparent later in adulthood (Tramujas et al., 2006; Montanha; Pimpão2012). Many xenobiotics are known to affect reproduction in many organisms (Mayonet al., 2006; Montanha, 2010).

# 6. THE FISH AND THE FISH EMBRYO TOXICITY TEST AS AN ANIMAL ALTERNATIVE METHOD IN HAZARD AND RISK ASSESSMENT AND SCIENTIFIC RESEARCH

Biological and ecological responses to contaminant stressors may range from changes at the molecular level, where genetic integrity and subcellular processes are evaluated, to population and community levels where dynamics and structure of entire foodwebs can be affected. Indicators of stress at several levels of biological organization have been used to evaluate effects of contaminants at the organism level.

A variety of *in vitro* and *in vivo* assays with fish are being used as model systems for toxicological, biochemical and developmental studies (Powers, 1989). Fish species such as rainbow trout *(Oncorhynchus mykiss),* carpas, catfish and Japanese medaka *(Oryzias latipes)* have been used extensively as test organisms for studies of carcinogenesis. A variety of teleost species have been used to study the development of early life stages of fish and the teratogenic effects of environmental contaminants (Wisk; Cooper, 1990). Fish respond in a manner similar to mammalian test species to chemicals that induce peroxisome proliferation in hepatocytes (Yang et al., 1990), and oxidative damage in hepatocytes (Washburn; Di Giulio, 1989). Hepatic cytochrome P-450 monooxygenases in fish also metabolize many carcinogens in a manner analogous to mammalian organisms (Stegeman; Lech, 1991). The advantages of using fish as model organisms include the ease with which teleosts, especially small aquarium species, can be held in the laboratory and exposed to toxic chemicals. Since fish often respond to toxicants in a manner similar to higher vertebrates, they can be used to screen for chemicals that have the potential to cause teratogenic and carcinogenic effects in humans. However, the main application for model systems with fish is to determine the distribution and toxic effects of chemical contaminants in the aquatic environment. Industrial and urban discharges are responsible for high concentrations of toxic substances in the aquatic environment, although in many countries, regulatory activity is beginning to limit point-source discharges of toxic chemicals to our water resources.

Animal alternatives research has historically focused on human safety assessments and has only recently been extended to environmental testing. This is particularly for those assays that involve the use of fish. A number of alternatives are being pursued by the scientific community including the fish embryo toxicity (FET) test, a proposed replacement alternative to the acute fish test. Discussion of the FET methodology and its application in environmental assessments on a global level was needed. With this emerging issue in mind, the ILSI Health and Environmental Sciences Institute (HESI) and the European Centre for Ecotoxicology and Toxicology of Chemicals (ECETOC) held an International Workshop on the Application of the Fish Embryo Test as an Animal Alternative Method in Hazard and Risk Assessment and Scientific Research in March, 2008.

# 7. CONCLUSION

Environmental pollution is of current environmental concern. It is therefore necessary to find solutions to determine environmental contamination, because the organisms are continuously exposed to a variety of anthropogenic toxicants daily released to the environmental.

The silver catfish have been used for bioindicator of environmental contamination for many researches and can be used to aquatic biological systems. Moreover, a considerable expansion of knowledge in using various model species which are easily cultured and have lesion induction times of relatively short duration, such silver catfish.

# REFERENCES


1. Ankley, G. T., Blazer, V. S., Reinert, R. E., Agosin, M. (1986). Effects of Aroclor 1254 on cytochrome P450-dependent monooxygenase, glutathione S-transferase, and UDP-glucuronosyltransferase activities in channel catfish liver. Aquat. Toxicol. 9, 91-103.
2. Aydin, R. ; Koprucu, K. ; Dorucu, M. ; Koprucu, S. S. ; Pala, M. (2005) Acute toxicity of synthetic pyrethroidcypermethrin on the common carp (*Cyprinus carpio* L. ) embryos and larvae. Aquaculture International, v. 13, p. 451-458.
3. Barcellos, L. J. G., Kreutz, L. C., Rodrigues, L. B., Fioreze, I., Quevedo, R. M., Cericato, L., Conrad, J., Soso, A. B., Fagundes, M., Lacerda, L. A. & Terra, S. (2003) Haematological and biochemical characteristics of male jundiá (*Rhamdia quelen* Quoy & *Gaimardpimelodidae*): changes after acute stress. Aquaculture Research, *34:* 1465-1469.
4. Barcellos, L. J. G., Woehl, V. M., Wassermann, G. F., Krieger, M. H., Quevedo, R. M. & Lulhier, F. (2001) Plasma levels of cortisol and glucose in response to capture and tank transference in *Rhamdia quelen* (Quoy & Gaimard), a South American catfish. Aquaculture Research,32(3):123-125.
5. Begum, G. In vivo biochemical changes in liver and gill of *Clarias batrachus* during cypermethrin exposure and following cessation of exposure. (2005) Pesticide Biochemical and Physiology,v. 82, p. 185-196.
6. Bernardi, M. M. ; Moraes, R. C. ; Varoli, F. M. F. ; Osti, S. C. Ecotoxicologia. In: Spinosa, H. de S. ; Gómiak, S. L. ; Palermo-Neto, J. Toxicologia aplicada à medicina veterinária. São Paulo: Manole, 2008. 942 p.
7. Bila, D. M. ; Dezotti, M. (2003) Fármacos no meio ambiente. Quim. Nova. v. 26, n. 4, p. 523-530.

8. Bombail, V. ; AW, D. ; Gordon, E. & Bartty, J. (2001). Aplication of the comet and micronucleus assays to butterfish (*Pholis gunnellus*) erythrocyte from the Firth of Forth, cotland, Chemosphere, 44: 383–392.
9. Borges, A. (2007) Changes in hematological and serum biochemical values in Jundiá *Rhamdia quelen* due to sublethal toxicity of cipermethrin. Chemosphere, v. 69, p. 920-926.
10. Cattaneo, R. (2009) Parâmetros metabólicos e histológicos de jundiás (*Rhamdia quelen*) expostos à formulação comercial do herbicida 2,4 – diclorofenoxiacético (2,4 – D). Santa Maria. *Thesis*, Programa de Pós Graduação em Ciências Biológicas, Universidade Federal de Santa Maria, Santa Maria, 2009.
11. Clauss, T. M., Dove, A. D. M., Arnold, J. E. (2008) Hematologic Disorders of Fish. Vet Clin Exotic Anim. 11, 445-462.
12. Costa, P. M. (2011) Avaliação do efeito tóxico de sulfato de alumínio e sulfato de cobre em bioensaio de contaminação subcrônica via trófica no bioindicador *Rhamdia quelen* (siluriforme). Curitiba. Thesis. Área de Concentração em Genética, Universidade Federal do Paraná, Curitiba, 2011.
13. Crestani, M. ; Menezes, C. ; Glusczak, L. ; Miron, D. D. ; Spanevello, R. ; Silveira, A. ; Gonçalves, F. F. ; Zanella, R. ; LORO, V. L. (2007) Effect of clomazone herbicide on biochemical and histological aspects of silver catfish (*Rhamdia quelen*) and recovery pattern. Chemosphere, 67: 2305-2311.
14. Das, B. K. ; Mukherjee, S. C. (2003) Toxicity of cypermethrin in *Labeo rohita* fingerlings: biochemical, enzymatic and hematological consequences. Comparative Biochemistry and Physiology, n. 134, p. 109-121.
15. Di Giulio, R. T., Washburn, P. C., Wenning, R. J., Winston, G. W., Jewell, C. S. (1989). Biochemical responses in aquatic animals: a review of determinants of oxidative stress. Environ. Toxicological Chemistry. 8, 1103-1123.
16. Dobbins, L. L., Brain, R. A., Brooks, B. W. (2008) Comparison of the sensitivities of common *in vitro* and *in vivo* assays of estrogenic activity: application of chemical toxicity distributions. Environmental Toxicological Chemistry, 27(12):2608-16.
17. Dogan D. ; Can, C. (2011) Hematological, biochemical, and behavioral responses of *Oncorhynchus mykiss* to dimethoate. Fish Physiology Biochemistry, 37:951–958.
18. El-Sayed, Y. S. ; Saad, T. T. ; El-Bahr, S. M. (2007) Acute intoxication of deltamethrin in monosex Nile Tilapia, *Oreochromis niloticus* with special reference to the clinical, biochemical and hematological effects. Environmental Toxicology and Pharmacology, n. 24, p. 212-217.
19. Fent, K. ; Weston, A. A. ; Caminada, D. (2006) Ecotoxicology of human pharmaceuticals. Aquatic Toxicology, v. 76, p. 122 - 159.
20. Ferraro, M. V. M. (2009) Avaliação de três espécies de peixes – *Rhamdia quelen*, *Cyprinus carpio* e *Astyanax bimaculatus*, como potenciais bioindicadores em sistemas hídricos através dos ensaios: cometa e dos micronúcleos. Curitiba, 2009. Thesis, Curso de Pós-Graduação em Genética, Setor de Ciências Biológicas, Universidade Federal do Paraná, Curitiba.
21. Ferreira, D. (2010) Parâmetros de estresse oxidativo e estudo de lesões histopatológicas em jundiás (*Rhamdia quelen*) expostos a agroquímicos. Thesis, Programa de Pós-Graduação em Farmacologia, Universidade Federal de Santa Maria, Santa Maria, RS.
22. Gallagher, E. P., Di Giulio, R. T. (1991) Effects of 2,4-dichlorophenoxyacetic acid and picloram on biotransformation, peroxisomal and serum enzyme activities in channel catfish (*Ictalurus punctatus*). Toxicological Letters. 57, 65-72.
23. Galeb, L. A. G. (2010) Avaliação dos efeitos toxicológicos da deltametrina em uma espécie de peixe fluvial nativo jundiá (*Rhamdia quelen*). Curitiba, 2010. 70 f. Thesis, Programa de Pós-Graduação em Ciência Animal, Pontifícia Universidade Católica do Paraná, 2010.
24. Galeb, L. A. G., Fredianelli, A. C. ; Montanha, F. P.,Mikos, J. D., Fam, A. L. D.,Rocha, R. M. V. M., Pimpão, C. T. (2011) Determination of lethal and sublethal concentrations of

deltamethrin in Jundia (*Rhamdia quelen*) during 96 hours of exposure. Rev. Acad. (in press).

25. Ghisi, N. C. (2010) Avaliação genotóxica em *Rhamdia quelen* após contaminação sub-crônica com fipronil. Thesis, Pós-Graduação em Genética, Setor de Ciências Biológicas, Universidade Federal do Paraná, Curitiba, 2010.
26. Gomiero, L. M. ; Souza, U. P. ; Braga, F. M. S. (2007) Reprodução e alimentação de *Rhamdia quelen* (Quoy & Gaimard, 1824) em rios do Núcleo Santa Virgínia, Parque Estadual da Serra do Mar, São Paulo, SP. Biota Neotropical, v. 7, n. 3, p. 127-133.
27. Hai, D. Q., Varga, I. S., Matkovics, B., 1995. Effects of an organophosphate on the antioxidant systems of fish tissues. Acta Biol. Hungar. 46, 39-50.
28. Hasspieler, B. M., Behar, J. V., Carlson, D. B., Watson, D. E., DI Giulio, R. T. (1994) Susceptibility of channel catfish (*Ictalurus punctatus*) and brown bullhead (*Ameriurus nebulosus*) to oxidative stress: a comparative study. Aquatic Toxicological. 28, 53-64.
29. Hilbig, C. C. ; Bombardelli, R. A. ; Sanches, E. A. ; Oliveira, J. D. S. ; Baggio, D. M. ; Souza, B. E. (2008) Efeito do chumbo sobre a fertilização artificial e incubação de ovos de Jundiá cinza (*Rhamdia quelen*). Acta Sciencie Animal Science, v. 30, n. 2, p. 217-224.
30. Jesús, T. B., Carvalho, C. E. V. (2008) Utilização de biomarcadores em peixes como ferramenta para avaliação de contaminação ambiental por mercúrio (Hg). Oecologia Brasiliensis, 12:7.
31. Kang, I. J. ; Yokota, H. ; Oshima, Y. ; Tzuruda, Y. ; Yamaguchi, T. ; Maeda, M. ; Imada, N. ; Tadokoro, H. ; Honjo, T. (2002) Effect of 17β-estradiol on the reproduction of Japanese medaka (*Oryzias latipes*). Chemosphere., 47, 71-80.
32. Koprucu, K. ; Aydin, R. (2004) The toxic effects of pyrethroiddeltamethrin on the common carp (*Cyprinus carpio L.* ) embryos and larvae. Pesticide Biochemistry and Physiology, 80, 47–53.
33. Kroes, R. ; Kleiner, J. ; Renwick, A. (2005) The Threshold of Toxicological Concern Concept in Risk Assessment. Toxicological Science, 86 (2): 226-230.
34. Lee, R. F. ; Steinert, S. (2003) Use of the single cell gel electrophoresis/comet assay for detecting DNA damage in aquatic (marine and freshwater) animals. Mutation Research, 544: 43-64.
35. Lara, L. A. S. ; Duarte, A. A. F. ; Reis, R. M. (2011) Impacto dos disruptores endócrinos na função reprodutiva e sexual de homens e mulheres. Editorial.
36. Mabilia, R. G. ; Souza, S. M. G. (2006) Efeito do tratamento com diflubenzuron na hematologia de jundiás, *Rhamdia quelen* (Pimelodidae) infestados por *Lernaea cyprinacea* (Copepoda) em banhos de imersão de 24 horas. Acta Sci. Biol. Sci. v. 28, n. 2, p. 159-163.
37. Mabilia, R. G. ; Souza, S. M. G. ; Oberst, E. R. (2008) Efeito do banho de imersão de diflubenzuron no sêmen do jundiá *Rhandia quelen* parasitado por *Lernaea cyprinacea*. Semina: Ciências Agrárias. v. 29, n. 3, p. 661-666.
38. Malins, D. C. ; Mccain, B. B. ; Landahl, J. T. ; Myers, M. S. ; Krahn, M. M. ; Brown, D. W. ; Chan; S. L. ; Roubal, W. T. (1988) Neoplastic and other diseases in fish in relation to toxic chemicals: an overview. Aquatic Toxicology,11, 43-67,43.
39. Mather-Mihaich, E., Di Giulio, R. T. (1991) Oxidant, mixed-function oxidase and peroxisomal responses in channel catfish exposed to a bleached kraft mill effluent. Archives Environmental Contamination Toxicological, 20, 391-397.
40. Mayon, M. ; Bertrand, A. ; Leroy, D. ; Malbrouck, C. ; Mandiki, S. N. M. ; Silvestre, F. ; Goffart, A. ; Thomé, J-P. ; Kestemont, P. (2006) Multiscale approach of fish responses to different types of environmental contaminations: a case study. Science of the Total Environmental, v. 367, p. 715-731.
41. Melo, G. C. (2004) Efeitos subletais da ação do organofosforado folidol600 no fígado do peixe de água doce jundiá *Rhamdia quelen* (QUOY & GAIMARD, 1824): uma análise histopatológica. Thesis. UFPR, Curitiba, Pr, 2004. Dissertação (Mestrado em

Biologia Celular e Molecular) - Programa de Pós-Graduação em Biologia Celular e Molecular, Setor de Ciências Biológicas, Universidade Federal do Paraná, Curitiba.
42. Montanha, F. P. (2010) Efeitos toxicológicos de piretróides (Cipermetrina e Deltametrina) em Jundiá (*Rhamdia quelen*): fase adulta e desenvolvimento embrionário. Curitiba, 2010. Thesis. Pontifícia Universidade Católica do Paraná (Mestrado em Ciência Animal).
43. Montanha, F. P. ; Pimpão, C. T. (2012) Efeitos toxicológicos de piretróides (Cipermetrina e Deltametrina) em peixes – revisão. Revista Científica Eletrônica de Medicina Veterinária. ISSN: 1679-7353, Ano IX, n. 18.
44. Moore, A. ; Waring, C. P. (2001) The effects of a synthetic pyrethroid pesticide on some aspects of reproduction in Atlantic Salmon (*Salmo salar* L. ). Aquatic Toxicology, v. 52, p. 1-12.
45. Nacci, D. E. ; Cayula, S. ; Jackim, E. (1996) Detection of DNA damage in individual cells from marine organisms using the single cell gel assay. Aquatic Toxicology, 35: 197-210.
46. Pamplona, J. H. (2009) Avaliação dos efeitos tóxicos da dipirona sódica em peixe Rhamdia quelen: estudo bioquímico, hematológico e histopatológico. Curitiba, 2009. Thesis. Programa de Pós-graduação em Farmacologia, Setor de Ciências Biológicas, Universidade Federal do Paraná, Curitiba, 2009.
47. Panter, G. H. ; Thompson, R. S. ; Sumpter, J. P. (1998) Aquatic Toxicology. 1998, 42, 243.
48. Parra, J. E. G. ; Neto, J. R. ; Veiverbeg, C. A. ; Lazzari, R. ; Bergamin, G. T. ; Pedron, F. A. ; Rossato, S. Sutili, F. J. (2008) Alimentação de fêmeas de Jundiá com fontes lipídicas e sua relação com o desenvolvimento embrionário e larval. Ciência Rural, Santa Maria, v. 38, n. 7, p. 2011-2017.
49. Parvez, S. ; Raisuddin, S. (2005) Protein carbonyls: novel biomarkers of exposure to oxidative stress-inducing pesticides in freshwater fish *Channa punctata* (Bloch). Environmental Toxicological and Pharmacological., 20:112–117.
50. Peña-Llopis, S., Ferrando, M. D., Peña, J. (2003) Fish tolerance to organophosphate-induced oxidative stress is dependent on the glutathione metabolism and enhanced by *N*-acetylcysteine. Aquatic Toxicological . 65, 337–360.
51. Piancini, L. D. S. (2011) Utilização de biomarcadores genéticos na avaliação aguda do efeito mutagênico dos contaminantes atrazina e cloreto de cobre em *Rhamdia* quelen (Siluriformes, Heptapteridae). Curitiba, 2011. Thesis, Programa de Pós-Graduação em Genética, Setor de Ciências Biológicas, Universidade Federal do Paraná, Curitiba, 2011.
52. Pimpão, C. T. ; Zampronio, A. R. ; Silva DE Assis, H. C. (2007) Effects os deltamethrin on hematological parameters and enzymatic activity in *Ancistrus multispinis* (Pisces, Teleostei). Pesticide Biochemistry and Physiology, v. 88, p. 122-127.
53. Ploch, S. A., King, L. C., Kohan, M. J., Di Giulio, R. T. (1998) Comparative in vitro and in vivo benzo[a]pyrene_/DNA adduct formation and its relationship to CYP1A activity in two species of ictalurid catfish. Toxicol. Appl. Pharmacol. 149, 90-98.
54. Polat, H. ; Erkoç, F. U. ; Viran, R. ; Koçak, O. (2002) Investigation of acute toxicity of beta-cypermethrin on guppies *Poecilia reticulata*. Chemosphere, v. 49, p. 39-44.
55. Powers, D. A., (1989). Fish as model systems. Science 246, 352/358. Power, M., McCarty, L. S., 1997. Fallacies in ecological risk assessment
56. Ramsdorf, W. (2011) Avaliação da toxicidade dos compostos fipronil, nitrato de chumbo e naftaleno em peixes. Thesis, Pós-Graduação em Ciências Biológicas da Universidade Federal do Paraná, Curitiba, PR, 2011.
57. Reys, L. L. (2011) Tóxicos ambientais desreguladores do sistema endócrino. RFML, Grupo de Medicina Preventiva e Ciências Sociais, Faculdade de Medicina de Lisboa, Série III, p. 213-225.
58. Rice, C. D., Roszell, L. E. (1998) Tributyltin modulates 3,3?,4,4?,5-pentachlorobiphenyl (PCB 126)-induced hepatic CYP1A activity in channel catfish, *Ictalurus punctatus*. Journal Toxicological Environmental Health, 55, 197-212.

59. Rodrigues, L. C. (2003) Estudo das Glutation S-Transferases hepáticas solúveis do peixe Piaractus mesopotamicus Holmberg, 1887 (Pacu). Thesis. UERJ, RJ, Brazil.
60. Rodrigues, B. K. (2007) Avaliação dos impactos de agrotóxicos na região do Alto Mogi-Guaçu (MG) por meio de ensaios laboratoriais com *Danio rerio* (Cypriniformes, Cyprinidae). São Carlos, 2007. 138 f. Thesis, Escola de Engenharia de São Carlos, Universidade de São Paulo, 2007.
61. Routledge, E. J. ; Sheahan, D. ; Desbrow, C. ; Brighty, G. C. ; Waldock, M. ; Sumpter, J. P. (1998) Identification of estrogenic chemicals in STW effluent. 1. Chemical fractionation and *in vitro* biological screening. Environmental Science Technololy. 32 (11): 1549-58.
62. Salvagni, J. ; Ternus, R. Z. ; Fuentefria, A. M. (2011) Assessment of the genotoxic impact of pesticides on farming communities in the countryside of Santa Catarina State, Brazil. Genetics and Mollecular Biology, 34(1): 122–126.
63. Sanches, D. C. O. (2006) Desreguladores endócrinos na indução da vitelogenina em peixes nativos. Thesis. UFPR, Curitiba, PR, 2006.
64. Santos, M., Reyes, F., Areas, M. (2007) Piretróides–uma visão geral. Alimentos e Nutrição Araraquara, v18, n3.
65. Shih, T. M., Mcdonough Jr., J. H. (1997) Neurochemical mechanisms in soman-induced seizures. J. Appl. Toxicol. 17, 255–264.
66. Silfvergrip, A. M. C. (1996) A systematic revision of the neotropical catfish genus Rhamdia (Teleostei, Pimelodidae). . 156 f. Thesis (PhD in Zoology)–Stockholm University, Stockholm.
67. Srivastav, A. K. ; Srivastava, S. K. ; Mishra, D. ; Srivastav, S. ; Srivastav, S. K. (2002) Ultimobranchial Gland of Freshwater Catfish, *Heteropneustes fossilis* in Response to Deltamethrin Treatment. Bulletin Environmental Contamination Toxicology. 68:584–591.
68. Stegeman, J. J., Lech, J. J., (1991) Cytochrome P450 momooxygenase systems in aquatic species: carcinogen metabolism and biomarkers for carcinogen and pollutant exposure. Environ. Health Perspect. 90, 101/109.
69. Stumpf, M. ; Ternes, T. A. ; Wilken, R-D; Rodrigues, S. V. ; Baumann,W. (1999) Polar drug residues in sewage and natural waters in the state of Rio de Janeiro, Brazil. The Science Total Environmental, v. 225, p. 135-141.
70. Sumpter, J. P. (1998) Xenoendorine disrupters--environmental impacts. Toxicology Letters. 102-103, 337.
71. Tavares-Dias, M., Melo, J. F. B., Moraes, G., Moraes, F. R. (2002). Haematological characteristics of brazilian teleosts. vi. Parameters of jundiá *Rhamdia quelen* (Pimelodidae). Ciência Rural, v. 32, n. 4, 2002.
72. Ternes, T. A. ; Stumpf, M. ; Mueller, J. ; Haberer, K. ; Wilken, R. D. ; Servos, M. (1999) Behavior and occurrence of estrogens in municipal sewage treatment plants I. Investigation in Germany, Canada and Brazil. Science Total Environmental. 225, 81.
73. Tramujas, F. F. ; Fávaro, L. F. ; Pauka, L. M. ; Silva De Assis, H. C. (2006) Aspectos reprodutivos do peixe-zebra, *Danio rerio,* exposto a doses subletais de deltametrina. Archives of Veterinary Science, v. 11, n. 1, p. 48-53.
74. Tsuda, T., Aoki, S., Inoue, T., Kojima, M. (1995) Accumulation and excretion of diazinon, fenthion and fenitrothion by killifish: Comparison of individual and mixed pesticides. Water Res, 29:455-458.
75. United States Environmental Protection Agency (US EPA). (2011) Fipronil Summary Document Registration Review: Initial Docket June 2011 EPA-HQ-OPP-2011-0448, Washington, D. C. http://www. regulations. gov/> documentDetail;D=EPA-HQ-OPP-2011-0448-0003. 2011. Accessed 30 July 2012.
76. Van Der Oost, R. et al. (2003) Fish bioaccumulation and biomarkers in environmental risk assessment: a review. Journal Environmental Toxicology and Pharmacology, v. 13, p. 57- 149.
77. Vieira, V. L. P., Randunz Neto, J., Lopes, P. R. S., Lazari, R., Fonseca, M. B., Menezes, C. (2006). Alterações Metabólicas e Hematológicas em Jundiás (*Rhamdia quelen*)

Alimentados com Rações Contendo Aflatoxinas. Ciência Animal Brasileira. Goiânia, v. 7, n. 1, p. 49-55.
78. Washburn, P. C. ; Di Giulio, R. T. (1989) Biochemical responses in aquatic animals: A review of determinants of oxidative stress. ToxicolChem, 8(12) 1103-1123.
79. Watson, D. E., Menard, L., Stegeman, J. J., Di Giulio, R. T. (1995) Aminoanthracene is a mechanism-based inactivator of CYP1A in channel catfish hepatic tissue. Toxicological Apply Pharmacological. 135, 208-215.
80. WHO International Programme on Chemical Safety (IPCS), 1993. Biomarkers and risk assessment: concepts and principles. Environmental Health Criteria 155, World Health Organization, Geneva.
81. Wilson, C., Tisdell, C. (2001) Why farmers continue to use pesticides despite environmental, health and sustainability costs. Ecolog Econom.,39:449-462.
82. Wisk, J. D. ; Cooper, K. R. (1990) The stage specific toxicity of 2,3,7,8-tetrachlorodibenzo-*p*-dioxin in embryos of the japanese medaka (*Oryzias latipes*). Environ ToxicolChem, 9 (9), 1159-1169.
83. Witeck, L. ; Bombardelli, R. A. ; Sanches, E. A. ; Oliveira, J. D. S. ; Baggio, D. M. ; Souza, B. E. (2011) Motilidade espermática, fertilização dos ovócitos e eclosão de ovos de jundiá em água contaminada por cádmo. R. Bras. Zootec. v. 40, n. 3, p. 477-481.
84. Yang, C. S., Yoo, J. -S. H., Ishizaki, H., Hong, J. (1990). Cytochrome P450IIE1: Roles in nitrosamine metabolism and mechanisms of regulation. Drug Metabol. Rev. 22, 147–159.
85. Yilmaz, M. ; Gul, A. ; Erbasli, K. (2004) Acute toxicity of alpha-cypermethrin to guppy (*Poecilia reticulate,* Pallas, 1859). Chemosphere, v. 56, p. 381-385.

# CHAPTER 9

# Capture and Collection of Lampreys: The State of The Science

**Mary L. Moser.JoAnne M. Butzerin. Douglas B. Dey**


## ABSTRACT

The need for native lamprey conservation, improved lamprey fishery management, and better assessment of non-indigenous lamprey control measures has resulted in increased effort to survey lamprey populations and assess their status. Depending on the study objectives and target species/life stage, collection methods vary dramatically. We therefore provide a comprehensive review of sampling considerations and techniques used to capture, collect, handle, and enumerate both juvenile and adult lamprey life stages. Surveys for lamprey are often constrained by the lack of basic biological information, such as reliable characters for field identification of larvae (ammocoetes), migratory timing of anadromous forms, and spawning/nest building behavior of adults. However, there are a number of studies that have documented habitat preferences of the relatively sedentary ammocoetes. Consequently, existing sampling protocols have focused on the development of stratified sampling that targets optimal ammocoete habitat. In addition to this approach, we discuss methods and gear that can be used to survey migratory life stages, lamprey nests, and difficult-to-sample, deepwater ammocoete habitats.


**Keywords:** Lampetra Petromyzontiformes Gear Sampling Handling Anesthetic

# INTRODUCTION

Development of accurate methods for assessing lamprey status is critically needed to manage lamprey fisheries, conserve endangered/threatened lamprey stocks, and assess the efficacy of control measures. Lampreys have historically supported economically and culturally important fisheries worldwide. In many English and Finnish rivers, lampreys are still taken in commercial fisheries, despite dramatic declines in abundance (Tuunainen et al. 1980; Valtonen 1980; Masters et al. 2006). In northwestern North America, lamprey continue to be important to indigenous peoples, even though declines in abundance have limited the use of lamprey for food, medicinal, and ceremonial purposes (Close et al. 2002, 2004).

Over half of all lamprey species are considered to be endangered, vulnerable, or extinct in at least a portion of their range (Renaud 1997). Declines in native lamprey abundance have resulted primarily from habitat degradation or reduction, and poisoning to control non-native lamprey (Vladykov 1973; Renaud 1997; Filcek et al. 2005). The need for lamprey conservation has resulted in calls for standardized sampling methods to assess lamprey status (Kirchhofer 1995; Moser and Close 2003; Harvey and Cowx 2003). Much of the methodology currently in use for capture and collection of lamprey was originally developed in the 1950s during efforts to enumerate and eradicate non-indigenous sea lamprey *Petromyzon marinus* in the North American Great Lakes (Braem and Ebel 1961). Sea lamprey became established in the Great Lakes in the early 1900s, and their proliferation resulted in the development of methods to assess abundance and evaluate the effects of various control measures (reviewed in Christie and Goddard 2003; Slade et al. 2003).

The unique life history of lampreys presents both challenges and opportunities for capture and collection. Adult lamprey typically spawn in gravel/cobble substrate, building a discrete nest in which the eggs are laid (Fig. 1). After hatching, the larvae (ammocoetes) settle in silt/sand substrate, where they assume a largely sedentary lifestyle. The degree to which ammocoetes move during this life stage, which may last 3–7 years, is not well documented. However, some ammocoetes clearly make both downstream and upstream excursions, occupying new habitat as a result of either active or passive displacement (Potter 1980a; Maitland 2003; White and Harvey 2003; Quintella et al. 2004).

**Figure. 1** Pacific lamprey *Lampetra tridentata* construct a discrete nest using 20–40 mm substrate. It covers an average area of 0.1–0.2 $m^2$ and is excavated to a mean depth of 5–10 cm. *Photo courtesy of the US Fish and Wildlife Service*

Of the 38 recognized lamprey species, 18 feed parasitically as adults and may participate in extensive anadromous or potamodromous migrations (Potter and Gill 2003). The parasitic forms metamorphose (becoming macrophthalmia), emigrate from freshwater nursery areas, are parasitic for 1–2 years, and then participate in free-swimming, spawning migrations. In contrast, the non-parasitic species do not range far from their natal habitat. Because all lampreys exhibit protracted freshwater residence as larvae and are typically more concentrated in freshwater streams as adults, lamprey capture and collection has targeted sedentary ammocoetes, emigrating macrophthalmia, and migrating adults mainly in freshwater rivers and streams.

A number of techniques are currently used to capture, collect, and enumerate various lamprey species and life stages. Moreover, several independent protocols for sampling lamprey have been developed (Harvey and Cowx 2003; Slade et al. 2003). Collection methods vary dramatically depending on the objective of sampling, the target species, and/or the life stage of interest. Our objective is to provide a comprehensive review of techniques that have worked in a variety of settings that provides guidance for future survey development and refinement of existing protocols.

## CONSIDERATIONS FOR SAMPLING DESIGN

Much of the available data on lamprey distribution and abundance has been collected during surveys for other fish species. Consequently, the timing of collections, gear efficiency, and site selection have not been ideally suited to assessing lamprey status (Heard 1966; Bond et al. 1983; Todd and Kelso 1993). However, modification of existing surveys to provide useful lamprey information is possible, and this may often be the only recourse for obtaining lamprey data (Harvey and Cowx 2003).

As is the case for any fish survey, the appropriate sampling design for lamprey is dependent upon the objective of the survey: documentation of presence/absence, enumeration (density, absolute or relative abundance), or population dynamics (size distributions, trends in abundance over time). However, development of appropriate sampling designs for lamprey can be complicated by two factors: species identification of ammocoetes, and lack of basic biological information.

Distinguishing ammocoetes of closely related species can be extremely difficult (Heard 1966; Brown and Moyle 1993; Gardiner 2003; Maitland 2003; Filcek et al. 2005; Meeuwig and Bayer 2005; Fig. 2). Non-parasitic and parasitic species pairs (Potter 1980b) or satellite species (Vladykov and Kott 1980) co-occur in many river systems. Often, members of paired and satellite species are not even genetically distinguishable (Docker et al. 1999). Due to unreliable species identification techniques, researchers have resorted to holding specimens until metamorphosis (Heard 1966; Beamish 1980), conducting post-hoc genetic or morphological identification (Filcek et al. 2005; Meeuwig and Bayer 2005), or sampling only above impassable obstacles, where anadromous parasitic forms do not occur (Maitland 2003).

The lack of basic life history information can also be a factor in sampling design. For example, the timing of macrophthalmia emigration or the degree of movement by ammocoetes must be established before accurate reach-specific estimates of population growth or mortality can be made. Similarly, identification of preferred habitat is needed to develop ammocoete sampling programs, while information on the incidence of multiple nest building or occupation of a single nest by multiple spawners is needed to interpret the results of nest surveys (Jang and Lucas 2005; Mundahl and Sagan 2005; Stone and Brandt 2005; Stone In press).

**Figure. 2** There are no reliable keys to distinguish many larval lamprey species. For example, adult Pacific lamprey *L. tridentata* and western brook lamprey *L. richardsoni* ammocoetes co-occur and are difficult to distinguish in the field. *Photo courtesy of the Confederated Tribes of the Umatilla Indian Reservation*

# SELECTION OF SAMPLING LOCATIONS

## Juveniles

The first step in assessing ammocoete abundance is to classify and quantify habitat within the study area. Ammocoetes are patchily distributed in freshwater streams and rivers. A number of studies have focused on defining ammocoete habitat preferences by assessing the relative effects of environmental variables on ammocoete abundance: water depth and velocity; light levels; substrate grain size, depth, and organic content; and chlorophyll levels (reviewed in Hardisty and Potter 1971a; Ojutkangas et al. 1995). In these studies, habitats are randomly sampled, and all environmental variables are measured. The relative importance of each variable to lamprey abundance is then quantified statistically (e.g., Malmqvist 1980; Potter et al. 1986; Young et al. 1990; Beamish and Jebbink 1994; Beamish and Lowartz 1996;

Jellyman and Glova 2002; Torgeson and Close 2004; Stone and Brandt 2005).

In most studies that have defined optimal ammocoete habitat on small spatial scales, substrate grain size and water velocity were the most important indicators of larval lamprey abundance (Malmqvist 1980; Beamish and Jebbink 1994; Beamish and Lowartz 1996; Almeida and Quintella 2002; Sugiyama and Goto 2002). However, other variables, such as water depth, proximity to adult spawning areas, and riparian canopy can be important on larger spatial scales (Almeida and Quintella 2002; Torgeson and Close 2004). In addition, the relative importance of habitat variables can change with ammocoete size (Young et al. 1990; Almeida and Quintella 2002; Sugiyama and Goto 2002).

After defining the characteristics of optimal ammocoete habitat, sampling areas are classified into broad categories. For example, in tributaries of the Great Lakes, habitat suitability in sea lamprey surveys is qualitatively assessed as preferred (Type I = a loosely compacted mixture of sand and fine organic matter in depositional areas), acceptable (Type II = shifting sand or gravel with little fine organic matter), or unacceptable (Type III = bedrock, rubble or gravel) (Mullett and Bergstedt 2003; Slade et al. 2003). Harvey and Cowx (2003) recommend delineation of habitat as either optimal (stable, fine sediment or sand >15 cm deep, low water velocity with organic detritus present) or sub-optimal (transient shallow sediment interspersed with coarse substrate such as sediment trapped by tree roots, detritus overlying bedrock, or submerged vegetation rooted in silt or sand).

Following classification of habitats, ammocoete abundance in the optimal habitat is estimated by multiplying the observed density in samples from optimal habitat by the amount of optimal habitat. The degree to which sub-optimal habitats are sampled varies depending on the objectives, scope, and funding available for the survey. In some cases, only optimal habitat is sampled to reduce costs (Slade et al. 2003). Total ammocoete abundance is obtained by assuming that larval density and variance of sub-optimal habitat are in constant proportion to that of optimal habitat (Slade et al. 2003). However, Hansen et al. (2003) found that larval density varies substantially in sub-optimal habitats, and that a fixed-ratio model produced inaccurate estimates of ammocoete abundance in streams with large amounts of sub-optimal habitat. Consequently, sampling of both optimal and sub-optimal habitats is recommended (Harvey and Cowx 2003; Hansen et al. 2003).

## Adults

Lamprey nests or spawning adults can be enumerated; however such surveys are often fraught with difficulties. Lamprey construct nests in gravel and cobble substrates similar to those of salmonids (Hardisty and Potter 1971b; Takayama 2002; Fig. 1). As with salmonid redd surveys, mapping lamprey nests requires adequate water depth and clarity, as well as prior knowledge of preferred habitat types and spawning times (Takayama 2002; Jang and Lucas 2005; Stone In press). Enumeration of nests to produce estimates of adult abundance is problematic in that multiple adults may participate in nest building, or a single adult may build more than one nest (Farlinger and Beamish 1984; Jang and Lucas 2005; Mundahl and Sagan 2005). Moreover, mixed species aggregations of spawning lamprey have also been documented (Cochran and Pettinelli 1988; Cochran and Gripentrog 1992). Nevertheless, nest surveys can be used to identify likely ammocoete rearing areas (Torgeson and Close 2004), evaluate changes in quality and quantity of spawning habitat (Cochran and Gripentrog 1992; Takayama 2002; Jang and Lucas 2005; Mundahl and Sagan 2005), and assess populations with very low abundance or recruitment (Farlinger and Beamish 1984).

Trapping and visual observations of migrating adult lamprey have been used as an index of abundance in a variety of systems (e.g., Beamish 1980; Schuldt and Heinrich 1982; Stier and Kynard 1986a; Cochran and Marks 1995; Moser and Close 2003; Mullet et al. 2003). For this type of sampling, site selection was limited to areas where adults are concentrated during spawning migrations, typically at man-made barriers such as dams or weirs. Development of appropriate survey methodology and accurate interpretation of the results require some prior knowledge of migration times, environmental effects on movement, and specific lamprey behaviors (Moser and Close 2003).

# SAMPLING FREQUENCY AND TIMING

## Juveniles

Ammocoete populations are sampled annually in most surveys (e.g., Weise and Pajos 1998; Quintella et al. 2003; Slade et al. 2003). This infrequent sampling arises from the assumption that ammocoetes do not move much between rearing areas in a given year. This is thought to be the case particularly for non-anadromous forms or populations that settle in very low gradient streams (Potter 1980a). However, a recent mark-recapture study of ammocoetes in their native range indicated that lamprey ammocoetes readily move both upstream and downstream, and that they traveled up to 27 m during a 7-week period (Quintella et al. 2004). Another consideration is the degree of temporal variation in the amount of optimal habitat. In tributaries of the Great Lakes, the proportion of optimal habitat (Type I) showed no significant interannual variation, but the amount of sub-optimal habitat (Type II) varied significantly among years (Slade et al. 2003). Therefore, more frequent sampling may be required to accurately document reach-specific population dynamics (i.e., recruitment, growth, metamorphosis, or mortality rates).

The seasonal timing of ammocoete surveys should allow for collection of all size classes of interest. Although sampling for small ammocoetes (<40 mm) is less efficient than for larger size classes (Pajos and Weise 1994), the presence of age-0 lamprey is important in surveys where evidence of recent recruitment is desired (Moser and Close 2003; Harvey and Cowx 2003). Such surveys should occur only in the summer or fall after larval settlement. For anadromous species in most systems, sampling in late summer and fall also allows for collection of metamorphosing lamprey prior to their emigration from freshwater habitats (Harvey and Cowx 2003; Slade et al. 2003).

Larval lamprey surveys typically use electrofishers in either single pass (e.g., Malmqvist 1980; Potter et al. 1986; Slade et al. 2003; Almeida and Quintella 2002) or depletion sampling (consecutive samples collected at the same location; e.g., Pajos and Weise 1994; Beamish and Lowartz 1996; Harvey and Cowx 2003; Jellyman and Glova 2002; Torgeson and Close 2004; Stone and Brandt 2005). For depletion sampling, Pajos and Weise (1994) electrofished 1–2 m wide transects every 15 min until no additional lamprey were collected, while Harvey and Cowx (2003) recommended a 5-min resting period when electrofishing a 1-$m^2$ quadrat. Extended intervals between repeated electrofishing sessions may be necessary to offset the effects of narcosis on ammocoetes buried in the sediment (Pajos and Weise 1994).

## Adults

While relatively infrequent sampling for ammocoetes can produce a reliable snapshot of lamprey abundance in a given location, repeated sampling is required for migrating adults to account for seasonal, daily, or even hourly variation in run timing. Timing of the adult lamprey spawning migration is dictated by a variety of environmental cues: temperature, discharge, photoperiod, and presence of olfactory cues (Hardisty and Potter 1971b; Bjerselius et al. 2000; Moser et al. 2005). The temporal variability in these environmental cues will dictate adult lamprey sampling intervals. In some cases, nearly continuous sampling throughout a protracted migration period is needed to produce reliable estimates of adult lamprey abundance (Stier and Kynard 1986a; Moser and Close 2003; Mullett et al. 2003).

Spawning ground surveys generally occur over a shorter time period; however, the timing of such surveys is critical. The peak of spawning activity may last less than one week and can end abruptly (Cochran and Pettinelli 1988). In addition, nests may only be visible for little over one month following their initial construction (Stone In press). Ideally, surveys of lamprey nests should be conducted soon after spawning, when the nest material is cleanest and most easily identified. In most systems, spawning occurs in spring or early summer. Due to funding constraints, lamprey nest surveys are often added to salmonid redd surveys. Unfortunately, the timing of these surveys may be either before or after the peak of lamprey spawning activity.

## ACTIVE CAPTURE TECHNIQUES

### Juveniles

Most juvenile lamprey surveys rely on the use of a backpack or shore-based electrofisher in waters less than 0.8 m deep (Fig. 3). Sampling often involves a two-stage method, such as that detailed by Weisser and Klar (1990). First, 90–125 V direct current with a 10–25% duty cycle is applied at a slow rate of 3 pulses/s to induce ammocoetes to emerge from the sediment. A pattern of three slow pulses followed by a skipped pulse (3:1 pulse pattern) also helps to encourage emergence. Second, immediately after the ammocoetes emerge, a fast pulse setting of 30 pulses/s is used to immobilize them (Slade et al. 2003). Extended exposure to electrofishing can result in electronarcosis of buried ammocoetes and failure to emerge (Pajos and Weise 1994). Moreover, lamprey size and density, as well as water depth, temperature, and conductivity, all affect capture efficiency (Bowen et al. 2003; Steeves et al. 2003). Consequently, capture efficiencies of various electrofishers and personnel should be tested under a range of field conditions prior to development of standardized sampling protocols (Steeves et al. 2003).

**Figure. 3** Backpack electroshocking is one of the most common methods for assessing larval lamprey abundance. A typical first step in developing the sampling design is to identify and quantify optimal larval rearing habitat. *Photo courtesy of the Confederated Tribes of the Umatilla Indian Reservation*

While electrofishing in shallow water relies on capture of ammocoetes in dip nets or seines, electrofishing in deep water is often coupled with either a suction pump or trawl to bring immobilized larvae to the surface (McLain and Dahl 1968; Bergstedt and Genovese 1994; Fodale et al. 2003). Bergstedt and Genovese (1994) tested the capture efficiency of a deepwater electrofisher in water depths of 1–2 m. Power for this device was supplied by a standard backpack electrofisher. Immobilized larvae were suctioned from the sediment surface via a 7.6-cm hose and passed into a collection basket without going through the suction pump. Larvae were alive and unharmed following collection. Overall capture efficiency was 75%, but it decreased with increasing larval lamprey length. The efficiency of this device has not been determined in areas with significant current velocity (R. Reed, Karuk Tribal Fisheries Program, personal communication). The Wisconsin Department of Natural Resources personnel electrofish with multiple cabled electrodes from several boats. A large crew of people dipnet the stunned lamprey, thereby allowing greater coverage of large streams (P. Cochran, Saint Mary's University, personal communication). Validated methods for collecting ammocoetes in deep water are desperately needed, as recent research indicates that some of the highest ammocoete densities can occur in 5–7 th order reaches, where water depths exceed 1 m (S. van de Wetering, Siletz Indians, personal communication.).

A variety of other active gear types have been employed to collect ammocoetes. These include use of a shovel or suction dredge to remove lamprey from the sediment (Kainua and Valtonen 1980; Beamish and Youson 1987; Ojutkangas et al. 1995) or towed nets to collect both migrating ammocoetes and macrophthalmia (Heard 1966; Beamish and Youson 1987; Beamish and Levings 1991; Gadomski and Barfoot 1998). Lee and Weise (1989) used a manned submersible to visually quantify lamprey in deep lentic habitats. They also determined ammocoete abundance via surface collection of dead and dying lamprey that had been exposed to a larvicide. Finally, Quintella et al. (2004), described the detection of ammocoetes tagged with passive integrated transponder (PIT) tags in waters less than 1 m deep using a portable PIT-detection system. While labor intensive, this method allows for documentation of fine-scale movement patterns and microhabitat use without handling the lamprey (Quintella et al. 2004).

## Adults

Adult lamprey, particularly resident forms, are regularly captured in electrofishing surveys designed to sample other fish species. Consequently, this is an important technique for documentation of lamprey occurrence, and much of the historical data on lamprey distribution is based on electrofishing survey data. Adult lampreys are vagile and exhibit cryptic behaviors, hiding under boulders or other structures (Cochran and Gripentrog 1992; Kelso and Glova 1993). Due to low capture efficiency, electrofishing for adult lampreys does not provide accurate abundance estimates.

As is the case for eel (*Anguilla* sp.; Tesch 2003), almost every known fishing technique has been used to collect either adult lamprey or hosts with attached lamprey. These include jigging (Beamish and Levings 1991), use of towed nets and seines (Pletcher 1963; Beamish 1980; Bond et al. 1983), dip netting, and collection by hand (Heard 1966; Cochran 1987, 1989; Close et al. 2004). Parasitic-phase adults can be obtained via opportunistic capture in commercial and recreational fishing operations that target host species (Johnson and Anderson 1980; Cochran et al. 2003a, b; Cochran and Lyons 2004). Finally, visual counts of lamprey nests and spawning aggregations can yield valuable data on spawning habitat, timing of spawning events, and relative abundance (Cochran and Gripentrog 1992; Takayama 2002; Jang and Lucas 2005; Stone In press).

# PASSIVE GEAR

## Juveniles

A variety of passive gear types have been employed to capture migrating ammocoetes or macrophthalmia. Low capture efficiency renders these gear types of limited use for abundance estimation, but they can be important tools for determining the presence of upstream spawning or rearing areas, timing of juvenile migration, or relative abundance among sites or years. Gear types include rotary screw traps (Fig. 4), floating inclined plane traps (Beamish and Levings 1991), anchored nets (Heard 1966; Long 1968; Johnston 1997; Gadomski and Barfoot 1998; White and Harvey 2003), and rotating cooling water intake screens at power stations (Claridge et al. 1986).

**Figure. 4** Rotary screw traps used to collect emigrating salmonid smolts can be used to collect juvenile lamprey and provide information on the distribution of spawning and rearing areas. *Photo by James P. Reed*

## Adults

For migrating adults, passive gear types are often more successful than active collection methods, both for research and in fisheries. Trap designs vary, but they are typically set at obstacles to upstream passage or in dam

**Figure. 5** A lamprey trap installed in a fishway is used to capture adult lamprey as they pass over a weir at Bonneville Dam, Columbia River, USA. *Photo courtesy of the National Marine Fisheries Service*

fishways, where adult lamprey are concentrated (e.g., Beamish 1980; Schuldt and Heinrich 1982; Stier and Kynard 1986a; Cochran and Marks 1995; Moser and Close 2003; Mullett et al. 2003; Cochran and Lyons 2004; Fig. 5). As is the case for eels (*Anguilla* sp.; Tesch 2003), a broad range of passive gear has also been used in unobstructed streams and rivers to capture adult lamprey (Smith and Elliott 1953; Morris and Maitland 1987; Beamish and Levings 1991; Kelso and Glova 1993; Masters et al. 2006; Jang and Lucas 2005, Fig. 6). In most cases, passive gear is used in short-term, mark-recapture studies and to obtain specimens for research (Smith and Elliott 1953; Beamish 1980; Beamish and Levings 1991; Bergstedt and Seelye 1995; Moser and Close 2003; Masters et al. 2006; Jang and Lucas 2005) or in fisheries (Tuunainen et al. 1980; Valtonen 1980; Sjoberg 1980). However, long-term adult trapping efforts in the Great Lakes have been used to develop estimates of lamprey abundance and survival (Bergstedt et al. 2003; Mullett et al. 2003).

**Figure. 6** A simple lamprey pot allows collection of adult lamprey during the spawning migration in Cedar Creek, a tributary of the Lewis River in Washington State. *Photo courtesy of the US Fish and Wildlife Service*

Passive gear has also been used successfully to collect parasitic- and spawning phase lamprey. In most cases, parasitic-phase lamprey are captured incidentally during trapping or pound netting operations for host species (Johnson and Anderson 1980; Cochran and Marks 1995; Harvey 2001). However, baiting traps with potential hosts has also been used to collect lamprey when they are abundant and attachment rates are high (Hall 1963). Use of a sex pheromone to draw spawning-phase lamprey into traps has also been used with some success (Johnson et al. 2006).

Visual enumeration of adult lamprey passage has traditionally been conducted at dam count stations (Stier and Kynard 1986a; Haro and Kynard 1997; Moser and Close 2003). While this method can provide a useful index of abundance, it is labor-intensive and prone to error. Migrating adult lamprey are primarily nocturnal. Consequently, counts taken during the day at many facilities will underestimate adult lamprey passage. In addition, lamprey can pass via uncounted routes, mill back and forth in front of count stations, and fall back over dams (Haro and Kynard 1997; Moser and Close 2003). These behaviors can cause both over- and underestimation of lamprey passage. Therefore, dam counts should be used with caution when assessing lamprey status (Moser and Close 2003).

## HANDLING CONSIDERATIONS

Adult and juvenile lamprey are notoriously active and difficult to handle without anesthesia. Nonetheless, to save time, many lamprey surveys do not anesthetize for enumeration and measurement. A V-shaped measuring trough or curved pipe fitted with a ruler is useful for controlling lamprey during measurement (Harvey and Cowx 2003). In instances where species identification is difficult, use of magnifiers and a white background for examination of ammocoetes can be very helpful (Harvey and Cowx 2003; Gardiner 2003). While lamprey are hardy and can be subjected to extended holding without significant mortality, they are also quite susceptible to physical injury. This is particularly true of juveniles, where loss of mucous can result in subsequent fungal infection, particularly in warm water temperature (Mueller et al. 2006; C. Schreck, U. S. Geological Survey, Oregon Cooperative Fish and Wildlife Research Unit, Oregon State University, personal communication). Therefore, every effort should be made to reduce mucous loss and physical trauma during handling.

Anesthesia is recommended for the more extensive handling associated with tagging, very accurate determinations of length, or meristic measurements for species identification. A wide variety of anesthetics have been used with success. Tricaine methanesulfonate (MS222) has been widely used at concentrations of 30–70 ppm (e.g., Stier and Kynard 1986b; Weise and Pajos 1998; Moser and Close 2003). Quintella et al. (2004) used 1 ml $l^{-1}$ 2-phenoxyethanol with no mortality of ammocoetes and transforming juveniles. Ethyl 3-aminobenzoate (Benzocaine) at 50 mg $l^{-1}$ is also recommended, although care must be taken to avoid deep anesthesia that can result in an overly slow recovery period (Harvey and Cowx 2003). Eugenol (clove oil, 60 ppm) is recommended for adult lamprey due to the rapid recovery period and apparent lack of effects on orientation.

Regardless of the anesthetic used, lamprey should be allowed to recover fully prior to release. Before release, juveniles are often held in a bucket with sediment until all cohorts have recovered enough to burrow into the substrate (C. Claire, Idaho Department of Fish and Game, personal communication).

## CONCLUSIONS

Several protocols have been developed for long-term assessment of lamprey for both conservation (Harvey and Cowx 2003) and eradication (Slade et al. 2003). These protocols provide useful guidelines for development of ammocoete sampling programs using an electrofisher in waters <0.8 m in depth. In addition, Harvey and Cowx (2003) provide sample data sheets and estimates of person-days required to conduct the work. Continuing steps toward refinement of survey protocols are (1) testing for sampling bias (e.g., Steeves et al. 2003), (2) modeling to illuminate survey changes that can reduce sampling effort without loss of statistical power (e.g., Hansen et al. 2003; Slade et al. 2003), and (3) developing sampling methods for new species and habitats (e.g., Fodale et al. 2003). Many native lampreys have recently been identified as species of concern for conservation. Moreover, continued assessment for control of non-indigenous lampreys and any future invasions is needed (Balon et al. 1986; Farlinger and Beamish 1984; Christie and Goddard 2003). Therefore, the demand for information to help develop lamprey sampling protocols in all parts of the world is likely to increase in the coming years.

## ACKNOWLEDGEMENTS

Discussions with Christopher Claire, David Close, Ron Reed, and Jennifer Stone provided useful information about lamprey collection methods. These colleagues and Jim Reed were also kind enough to provide photos to help illustrate the text. Craig Wilson helped to identify references. Mike Hayes, Craig Robinson, Christian Torgeson, and Stan van de Wetering provided critical reviews of a draft manuscript. We also thank Phil Cochran for his review comments and for providing valuable information on parasitic-phase lamprey collection methods.

# REFERENCES


1. Almeida PR, Quintella BR (2002) Larval habitat of the sea lamprey (*Petromyzon marinus*) in the River Mondego (Portugal). In: Collares-Pereira MJ, Coelho MM, Cowx IG (eds) Freshwater fish conservation: options for the future. Fishing News Books, Blackwell Science, Oxford, pp 212–130
2. Balon EK, Crawford SS, Lelek A (1986) Fish communities of the upper Danube River (Germany, Austria) prior to the new Rhein-Main-Donau connection. Environ Biol Fishes 15:243–271
3. Beamish FWH, Jebbink JA (1994) Abundance of lamprey larvae and physical habitat. Environ Biol Fishes 39:209–214
4. Beamish FWH, Lowartz S (1996) Larval habitat of American brook lamprey. Can J Fish Aquat Sci 53:693–700
5. Beamish RJ (1980) Adult biology of the river lamprey (*Lampetra ayresi*) and the Pacific lamprey (*Lampetra tridentata*) from the Pacific Coast of Canada. Can J Fish Aquat Sci 37:1906–1923
6. Beamish RJ, Levings CD (1991) Abundance and freshwater migrations of the anadromous parasitic lamprey (*Lampetra tridentata*), in a tributary of the Fraser River, British Columbia. Can J Fish Aquat Sci 48:1250–1263
7. Beamish RJ, Youson JH (1987) Life history and abundance of young adult *Lampetra ayresi* in the Fraser River and their possible impact on salmon and herring stocks in the Strait of Georgia. Can J Fish Aquat Sci 44:525–537
8. Bergstedt RA, Genovese JH (1994) New technique for sampling sea lamprey larvae in deepwater habitats. N Am J Fish Manage 14:449–452
9. Bergstedt RA, McDonald RB, Mullett KM, Wright GM, Swink WD, Burnham KP (2003) Mark-recapture population estimates of parasitic sea lampreys (*Petromyzon marinus*) in Lake Huron. J Great Lakes Res 29:226–239
10. Bergstedt RA, Seelye JG (1995) Evidence for lack of homing by sea lampreys. Trans Am Fish Soc 124:235–239
11. Bjerselius R, Li W, Teeter JH, Seelye JG, Johnsen PB, Maniak PJ, Grant GC, Polkinghorne CN, Sorensen PW (2000) Direct behavioral evidence that unique bile acids released by larval sea lamprey (*Petromyzon marinus*) function as a migratory pheromone. Can J Fish Aquat Sci 57:557–569
12. Bond CE, Kan TT, Myers KW (1983) Notes on the marine life of the river lamprey, *Lampetra ayresi*, in Yaquina Bay, Oregon and the Columbia River estuary. Fish Bull 81:165–167
13. Bowen AK, Weisser JW, Bergstedt RA, Famoye F (2003) Response of larval sea lampreys (*Petromyzon marinus*) to pulsed DC electrical stimuli in laboratory experiments. J Great Lakes Res 29:174–182
14. Braem RA, Ebel WJ (1961) A back-pack shocker for collecting lamprey ammocoetes. Prog Fish Cult 23:87–91
15. Brown LR, Moyle PB (1993) Distribution, ecology, and status of the fishes of the San Joaquin River drainage, California. Calif Fish Game 79:96–114
16. Christie GC, Goddard CJ (2003) Sea Lamprey International Symposium (SLIS II): advances in the integrated management of sea lamprey in the Great Lakes. J Great Lakes Res 29:1–14
17. Claridge PN, Potter IC, Hardisty MW (1986) Seasonal changes in movements, abundance, size composition, and diversity of the fish fauna of the Severn Estuary. J Mar Biol Assoc UK 66:229–258
18. Close DA, Fitzpatrick MS, Li HW (2002) The ecological and cultural importance of a species at risk of extinction, Pacific lamprey. Fisheries 27:19–25
19. Close DA, Jackson AD, Conner BP, Li HW (2004) Traditional ecological knowledge of Pacific lamprey (*Entosphenus tridentatus*) in northeastern Oregon and

southeastern Washington from indigenous peoples of the Confederated Tribes of the Umatilla Indian Reservation. J Northwest Anthropol 38:141–161
20. Cochran PA (1987) The southern brook lamprey (*Icthyomyzon gagei*) in the St. Croix River drainage of Wisconsin and Minnesota. Copeia 1987:443–446
21. Cochran PA (1989) Maintaining parasitic lampreys in closed laboratory systems. Am Biol Teacher 51:115–119
22. Cochran PA, Gripentrog AP (1992) Aggregation and spawning by lampreys (genus *Ichthyomyzon*) beneath cover. Environ Biol Fishes 33:381-387
23. Cochran PA, Hodgson JY, Kinziger AP (2003a) Change in energy density of the sea lamprey (*Petromyzon marinus*) during its parasitic phase: implications for modeling food consumption and growth. J Great Lakes Res 29:297–306
24. Cochran PA, Lyons J (2004) Field and laboratory observations on the ecology and behavior of the silver lamprey (*Ichthyomyzon unicuspis*) in Wisconsin. J Freshwater Ecol 19:245–253
25. Cochran PA, Lyons J, Gehl MR (2003b) Parasitic attachments by overwintering silver lampreys, *Ichthyomyzon unicuspis*, and chestnut lampreys *Ichthyomyzon castaneus*. Env Biol Fish 68:65–71
26. Cochran PA, Marks JE (1995) Biology of the silver lamprey, *Ichthyomyzon unicuspis*, in Green Bay and the lower Fox River, with a comparison to the sea lamprey, *Petromyzon marinus*. Copeia 1995:409–421
27. Cochran PA, Pettinelli TC (1988) Northern and southern brook lampreys (*Ichthyomyzon fossor* and *I. gagei*) in Minnesota. J Minn Acad Sci 54:23–27
28. Docker MF, Youson JH, Beamish RJ, Devlin R (1999) Phylogeny of the lamprey genus *Lampetra* inferred from mitochondrial cytochrome *b* and ND3 gene sequences. Can J Fish Aquat Sci 56:2340–2349
29. Farlinger SP, Beamish RJ (1984) Recent colonization of a major salmon producing lake in British Columbia by Pacific lamprey (*Lampetra tridentata*). Can J Fish Aquat Sci 41:278–285
30. Filcek KB, Gilmore SA, Scribner KT, Jones ML (2005) Discriminating lamprey species using multilocus microsatellite genotypes. N Am J Fish Manage 25:502–509
31. Fodale MF, Bronte CR, Bergstedt RA, Cuddy DW, Adams JV (2003) Classification of lentic habitat for sea lamprey (*Petromyzon marinus*) larvae using a remote seabed classification device. J Great Lakes Res 29:190–203
32. Gadomski DM, Barfoot CA (1998) Diel and distributional abundance patterns of fish embryos and larvae in the lower Columbia and Deschutes rivers. Environ Biol Fishes 51:353–368
33. Gardiner R (2003) Identifying lamprey: a field key for sea, river, and brook lamprey. In: Conserving Natura 2000 Rivers. Conservation Techniques Series No. 4 English Nature, Peterborough. Available via the Life In UK Rivers project, http://www.english-nature.org.uk/lifeinukrivers/publications/lamprey_key.pdf, Cited 22 March 2006
34. Hall JD (1963) An ecological study of the chestnut lamprey, *Ichthyomyzon castaneus* (Girard), in the Manistee River, Michigan. PhD Thesis, University of Michigan, Ann Arbor, MI 101 pp
35. Hansen MJ, Adams JV, Cuddy DW, Richards JM, Fodale MF, Larson GL, Ollila DJ, Slade JW, Steeves TB, Young RJ, Zerrenner A (2003) Optimizing larval assessment to support sea lamprey control in the Great Lakes. J Great Lakes Res 29:766–782
36. Hardisty MW, Potter IC (1971a) The behaviour, ecology, and growth of larval lampreys. In: Hardisty MW, Potter IC (eds) The biology of lampreys, vol 1. Academic Press, London, pp 85–125
37. Hardisty MW, Potter IC (1971b) The general biology of adult lampreys. In: Hardisty MW, Potter IC (eds) The biology of lampreys, vol 1. Academic Press, London, pp 249–277
38. Haro A, Kynard B (1997) Video evaluation of passage efficiency of American shad and sea lamprey in a modified Ice Harbor fishway. N Am J Fish Manage 17:981–987

39. Harvey CJ (2001) A stable isotope analysis of food web structure in Lake Superior. PhD Thesis, University of Wisconsin, Madison, 137 pp
40. Harvey J, Cowx I (2003) Monitoring the river, brook, and sea lamprey, *Lampetra fluviatilis*, *L. planeri*, and *Petromyzon marinus*. In: Conserving Natura 2000 Rivers. Conservation Techniques Series No. 5 English Nature, Peterborough. Available via the Life In UK Rivers project, http://www.english-nature.org.uk/lifeinukrivers/publications/lamprey_monitoring.pdf. Cited 23 March 2006
41. Heard WR (1966) Observations on lampreys in the Naknek River system of southwest Alaska. Copeia 1966:332–339
42. Jang M-H, Lucas MC (2005) Reproductive ecology of the river lamprey. J Fish Biol 66:499–512
43. Jellyman DJ, Glova GJ (2002) Habitat use by juvenile lampreys (*Geotria australis*) in a large New Zealand river. NZ J Mar Freshwater Res 36:503–510
44. Johnson BGH, Anderson WC (1980) Predatory-phase sea lamprey (*Petromyzon marinus*) in the Great Lakes. Can J Fish Aquat Sci 37:2007–2020
45. Johnson NS, Luehring MA, Siefkes MJ, Li W (2006) Mating pheromone reception and induced behavior in ovulating female sea lamprey. N Am J Fish Manage 26:88–96
46. Johnston TA (1997) Downstream movements of young-of-the-year fishes in Catamaran Brook and the Little Southwest Miramichi River, New Brunswick. J Fish Biol 51:1047–1062
47. Kainua K, Valtonen T (1980) Distribution and abundance of European river lamprey (*Lampetra fluviatilis*) larvae in three rivers running into Bothnian Bay, Finland. Can J Fish Aquat Sci 37:1960–1966
48. Kelso JRM, Glova GJ (1993) Distribution, upstream migration and habitat selection of maturing lampreys, *Geotria australis*, in Pigeon Bay Stream, New Zealand. Australian J Mar Freshwater Res 44:749–759
49. Kirchhofer A (1995) Concept of conservation for the European brook lamprey (*Lampetra planeri*) in Switzerland. Fischokologie 8:93–108
50. Lee DS, Weise JG (1989) Habitat selection of lentic larval lampreys: preliminary analysis based on research with a manned submersible. J Great Lakes Res 15:156–163
51. Long C (1968) Diel movement and vertical distribution of juvenile anadromous fish in turbine intakes. Fish Bull 66:599–609
52. Maitland PS (2003) Ecology of the river, brook, and sea lamprey. In: Conserving Natura 2000 Rivers. Conservation Techniques Series No. 5 English Nature, Peterborough. Available via the Life in UK rivers project, http://www.english-nature.org.uk/lifeinukrivers/publications/lamprey.pdf. Cited 23 March 2006
53. Malmqvist B (1980) Habitat selection of larval brook lampreys (*Lampetra planeri*, Bloch) in a south Swedish stream. Oecologia 45:35–38
54. Masters JEG, Jang M-H, Ha K, Bird PD, Frear PA, Lucas MC (2006) The commerical exploitation of a protected anadromous species, the river lamprey (*Lampetra fluviatilis* (L.)), in the tidal River Ouse, north-east England. Aquat Conserv Mar Freshw Ecosyst 16:77–92
55. McLain AL, Dahl FM (1968) An electric beam trawl for the capture of larval lampreys. Trans Am Fish Soc 97:289–293
56. Meeuwig HM, Bayer JM (2005) Morphology and aging precision of statoliths from larvae of Columbia River Basin lampreys. N Am J Fish Manage 25:38–48
57. Morris KH, Maitland PS (1987) A trap for catching adult lampreys (Petromyzonidae) in running water. J Fish Biol 31:513–516
58. Moser M.L., Close D.A. (2003) Assessing Pacific lamprey status in the Columbia River Basin. Northwest Sci 77:116–125
59. Moser ML, Zabel RW, Burke BJ, Stuehrenberg LC, Bjornn TC (2005) Factors affecting adult Pacific lamprey passage rates at hydropower dams: using "time-to-event" analysis of radiotelemetry data. In: Spedicato MT, Marmulla G, Lembo G (eds) Aquatic telemetry: advances and applications. FAO-COISPA, Rome, pp 61–70

60. Mueller RP, Moursund RA, Bleich MD (2006) Tagging juvenile Pacific lamprey with passive integrated transponders: methodology, short-term mortality, and influence on swimming performance. N Am J Fish Manage 26:361–366
61. Mullett KM, Bergstedt RA (2003) Agreement among observers classifying larval sea lamprey (*Petromyzon marinus*) habitat. J Great Lakes Res 29:183–189
62. Mullett KM, Heinrich JW, Adams JV, Young RJ, Henson MP, McDonald RB, Fodale MF (2003) Estimating lake-wide abundance of spawning-phase sea lampreys (*Petromyzon marinus*) in the Great Lakes: extrapolating from sampled streams using regression models. J Great Lakes Res 29:240–252
63. Mundahl ND, Sagan RA (2005) Spawning ecology of the American brook lamprey, *Lampetra appendix*. Env Biol Fish 73:283–292
64. Ojutkangas E, Aronen K, Laukkanen E (1995) Distribution and abundance of river lamprey (*Lampetra fluviatilis*) ammocoetes in the regulated River Perhonjoki. Regul Rivers Res Manage 10:239–245
65. Pajos TA, Weise JG (1994) Estimating populations of larval sea lamprey with electroshocking sampling methods. N Am J Fish Manage 14:580–587
66. Pletcher FT (1963) The life history and distribution of lampreys in the Salmon and certain other rivers in British Columbia, Canada. MS thesis, University of British Columbia
67. Potter IC (1980a) Ecology of larval and metamorphosing lampreys. Can J Fish Aquat Sci 37:641–1657
68. Potter IC (1980b) The Petromyzoniformes with particular reference to paired species. Can J Fish Aquat Sci 37:595–1615
69. Potter IC, Gill HS (2003) Adaptive radiation of lampreys. J Great Lakes Res 29:95–112
70. Potter IC, Hilliard RW, Bradley JS, McKay RJ (1986) The influence of environmental variables on the density of larval lampreys in different seasons. Oecologia 70:433–440
71. Quintella BR, Andrade NO, Almeida PR (2003) Distribution, larval stage duration and growth of the sea lamprey ammocoetes, *Petromyzon marinus* L., in a highly modified river basin. Ecol Freshw Fish 12:286–293
72. Quintella BR, Andrade NO, Espanhol R, Almeida PR (2004) The use of PIT telemetry to study movements of ammocoetes and metamorphosing sea lampreys in river beds. J Fish Biol 65:1–10
73. Renaud CB (1997) Conservation status of Northern Hemisphere lampreys (Petromyzontidae). J Appl Ichthyol 13:143–148
74. Schuldt RJ, Heinrich JW (1982) Portable trap for collecting adult sea lampreys. Prog Fish-Cult 44:220–221
75. Sjoberg K (1980) Ecology of the European river lamprey (*Lampetra fluviatilis*) in northern Sweden. Can J Fish Aquat Sci 37:1974–1980
76. Slade JW, Adams JV, Cuddy DW, Neave FB, Sullivan WP, Young RJ, Fodale MF, Jones ML (2003) Techniques and methods for estimating abundance of larval and metamorphosed sea lampreys in Great Lakes tributaries, 1995–2001. J Great Lakes Res 29:130–136
77. Smith BR, Elliott OR (1953) Movement of parasitic-phase sea lampreys in lakes Huron and Michigan. Trans Am Fish Soc 82:123–128
78. Steeves TB, Slade JW, Fodale MF, Cuddy DW, Jones ML (2003) Effectiveness of using backpack electroshocking gear for collecting sea lamprey (*Petromyzon marinus*) larvae in Great Lakes tributaries. J Great Lakes Res 29:161–173
79. Stier K, Kynard B(1986a) Abundance, size, and sex ratio of adult sea-run sea lampreys, *Petromyzon marinus*, in the Connecticut River. Fish Bull 84:476–480
80. Stier K, Kynard B (1986b) Movement of sea-run sea lampreys, *Petromyzon marinus*, during the spawning migration in the Connecticut River. Fish Bull 84:749–753
81. Stone J (In press) Observations of nest characteristics, spawning habitat, and spawning behavior of Pacific and western book lamprey in a Washington stream. NW Sci

82. Stone J, Brandt S (2005) Spatial distribution and habitat use of Pacific lamprey (*Lampetra tridentata*) ammocoetes in a Western Washington stream. J Freshwater Res 20:171–185
83. Sugiyama H, Goto A (2002) Habitat selection by larvae of a fluvial lamprey, *Lethenteron reissneri*, in a small stream and an experimental aquarium. Ichthyol Res 49:62–68
84. Takayama M (2002) Spawning activities and physical characteristics of the spawning ground of *Lethenteron reissneri* at the headstream of the Himekawa River, central Japan. Ichthyol Res 49:165–170
85. Tesch F-W (2003) The eel. Blackwell Science, Ltd., Oxford, 408 pp
86. Todd PR, Kelso JRM (1993) Distribution, growth and transformation timing of larval Geotria australis in New Zealand. Ecol Freshwater Fish 2:99–107
87. Torgeson CE, Close DA (2004) Influence of habitat heterogeneity on the distribution of larval Pacific lamprey (*Lampetra tridentata*) at two spatial scales. Freshw Biol 49:614–630
88. Tuunainen P, Ikonen E, Auvinen H (1980) Lampreys and lamprey fisheries in Finland. Can J Fish Aquat Sci 37:1953–1959
89. Valtonen T (1980) European river lamprey (*Lampetra fluviatilis*) fishing and lamprey populations in some rivers running into Bothnian Bay, Finland. Can J Fish Aquat Sci 37:1967–1973
90. Vladykov VD (1973) North American nonparasitic lampreys of the family Petromyzonidae must be protected. Can Field-Nat 87:235–239
91. Vladykov VD, Kott E (1980) Description and key to metamorphosed specimens and ammocoetes of Petromyzonidae found in the Great Lakes Region. Can J Fish Aquat Sci 37:1616–1625
92. Weise JG, Pajos TA (1998) Intraspecific competition between larval sea lamprey year-classes as Salem Creek was recolonized, 1990–1994, after a lampricide application. N Am J Fish Manage 18:561–568
93. Weisser JW, Klar GT (1990) Electric fishing for sea lamprey (*Petromyzon marinus*) in the Great Lakes Region of North America. In: Cowx IG (ed) Developments in electric fishing. Fishing News Books, Oxford, pp 59–64
94. White JL, Harvey BC (2003) Basin-scale patterns in the drift of embryonic and larval fishes and lamprey ammocoetes in two coastal rivers. Env Biol Fishes 67:369–378
95. Young RJ, Kelso JRM, Weise JG (1990) Occurrence, relative abundance, and size of landlocked sea lamprey (*Petromyzon marinus*) ammocoetes in relation to stream characteristics in the Great Lakes. Can J Fish Aquat Sci 47:773–778

# CHAPTER 10

# Partial Characterization of Digestive Proteases in the Green Cichlid, Cichlasoma Beani

**Leonardo Martínez-Cárdenas [1], Carlos A. Álvarez-González [2], Oscar U. Hernández-Almeida [1], Carlos A. Frías-Quintana [3,*], Jesús T. Ponce-Palafox [4] and Sergio Castillo-Vargasmachuca [4]**

[1] *Secretaria de Investigación y Posgrado, Universidad Autónoma de Nayarit, Tepic 63155, Nayarit, México*

[2] *Laboratorio de Acuicultura Tropical, División Académica de Ciencias Biológicas, Universidad Juárez Autónoma de Tabasco, Villahermosa 86150, Tabasco, México*

[3] *División Académica Multidisciplinaria de Jalpa de Méndez, Universidad Juárez Autónoma de Tabasco, Jalpa de Méndez 86205, Tabasco, México*

[4] *Escuela Nacional de Ingeniería Pesquera, Tepic 63000, Nayarit, México*

## ABSTRACT

This study undertakes the characterization of digestive proteases in the juvenile green cichlid, *Cichlasoma beani.* The results obtained showed a higher activity of alkaline proteases ($0.14 \pm 0.01$ U mg protein$^{-1}$) compared to acid proteases ($0.07 \pm 0.01$ U mg protein$^{-1}$) in this species. The optimum temperature of the alkaline proteases was 65 °C and these enzymes were more thermostable to temperature changes than the acid proteases, characterized by an optimal temperature of 55 °C. The pH optimum was 2 for acid proteases, and 11 for alkaline proteases, which were also more stable to changes in pH between 8 and 10. The use of specific inhibitors showed an acid protease inhibition of 88% with pepstatin A as inhibitor. In the zymogram SDS-PAGE analysis of alkaline proteases, five active fractions were revealed, indicating the presence of serine proteases. These results confirm that both alkaline and acid proteases are involved in the digestion of *C. beani*, and suggest that this species is omnivorous with carnivorous tendencies. The present study contributes to our knowledge about the digestive physiology of *C. beani*, and can be applied towards improved understanding of the kinds of protein sources that could be used in the development of inerts diets.

**Keywords:** *Cichlasoma beani*; digestive enzymes; inhibitors; proteases; SDS-PAGE

# 1. INTRODUCTION

The development of fish aquaculture in Mexico is mainly based on the culture of introduced species such as the Nile tilapia *Oreochromis niloticus*, despite the existence of a variety of native species with high potential for aquaculture [1]. It has recently become of interest to develop new cultures of species with regional commercial potential, in response to the depletion of wild populations, mainly caused by overfishing, pollution and habitat alteration [2].

The green cichlid or mojarra Sinaloa *Cichlasoma beani* is a euryhaline species with potential for farming in Northwest Mexico. In this context, a study by [3] focused on the in situ reproductive population structure and its relation to the ecology of this species. Prior studies examined the effects of stocking density on growth and survival of this species under culture conditions [4,5]. There are currently no studies on biochemistry and digestive physiology of this species, although studies have been conducted on digestive biochemistry in species of the genus *Cichlasoma* [6,7,8], which contribute to the knowledge of the digestive physiology of these cichlids for studies related to other species and future references. Given the above consideration, understanding of species-specific digestive physiology in fish, i.e., the composition and factors that affect the activity of digestive enzymes, provides information useful for the formulation of diets, i.e., selection of the nutrients and the type of protein ingredient required for growth maximization. This depends on the enzymatic activity in the digestive system and knowledge of the species capacity to hydrolyze protein, the most costly component in fish diets [9,10,11]. Description of the operational parameters of digestive proteases is essential to understand their types and form of action, as well as their activity level. Determination of pH and temperature optima and stability to these factors, allows long-term maximization of enzymatic activity [12,13,14].

Studies of enzymatic characterization have been supplemented by the use of specific inhibitors in the application of biochemical and electrophoretic techniques, which affect the enzyme active centers and thus decrease enzymatic activity, allowing determination of the sensitivity and the mechanisms of action of digestive enzymes [11,13,14,15]. Due to the importance of the effect of inhibitors found in many ingredients (mainly vegetable), the characterization of digestive proteases have been studied in marine and freshwater fish species such as spotted sand bass *Paralabrax maculatofasciatus* [12], common snook *Centropomus undecimalis* [16], Atlantic bluefin tuna *Thunnus thynnus* [17], Asian bonytongue *Scleropages formosus* [18], Senegal sole *Solea senegalensis* [19], Nile tilapia *Oreochromis niloticus* [20], halibut *Scophthalmus maximus* [21], bay snook *Petenia splendida* [22], Albacore *Thunnus alalunga* [23], tropical gar *Atractosteus tropicus* [24] and three-spot cichlid *Cichlasoma trimaculatum* [25]. These studies have demonstrated that acidic proteases commonly have an optimal pH level of 2 and optimum temperatures between 35 and 65 °C. In contrast, most of the alkaline proteases have an optimal pH between 9 and 10, and optimum temperatures in the range of 45–65 °C. In some carnivorous species inhibition of the enzymatic activity of alkaline proteases are highly affected by inhibitors and can be reduced by 100%. Alkaline protease inhibition varies among species and can be influenced by the type of food item in the habitat [12]. Therefore, the main objective of this study was to characterize the digestive proteases in juvenile *C. beani*, and thus to generate basic information on this species' digestive physiology, namely the determination of the optimal conditions of pH and temperature of digestive proteases, their susceptibility to changes in these parameters, and the effects of specific inhibitors on enzymatic activity. This information will allow the selection of appropriate ingredients, and formulation of diets that meet the nutritional requirements of the species under culture conditions.

## 2. RESULTS

The optimum pH for the activity of acid proteases was 2 (Figure 1A). With respect to changes in pH stability, the remaining enzyme activity was 56% for pH 3, while pH 4–7 had remaining activities around 15%–20%, starting at 90 min pre-incubation (Figure 1B). For alkaline proteases activity the pH optimum was 11 (Figure 1C). The changes in pH stabilities in alkaline proteases occurred at pH 8–11, reaching a stability around 60%–100% in 90 min of pre-incubation, while pH 12 showed a decrease in its residual activity until reaching to about 5% at 90 min of pre-incubation (Figure 1D).

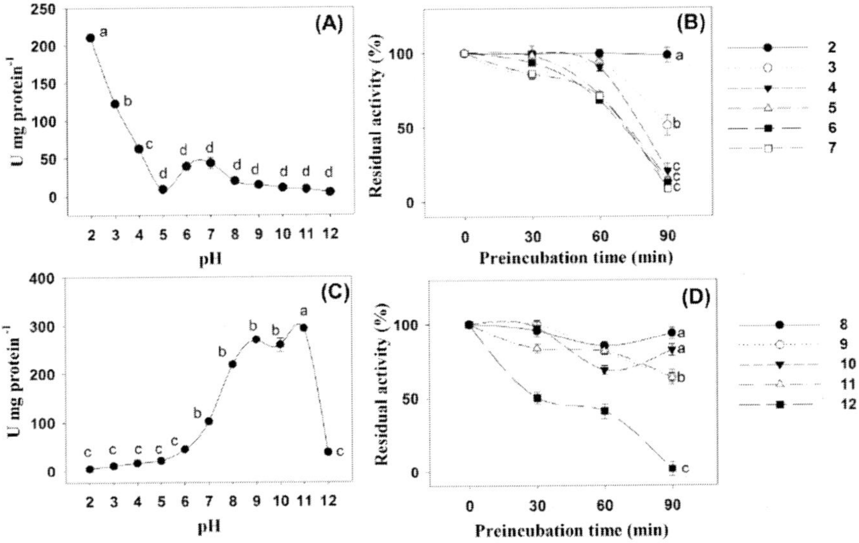


**Figure 1.** Effect of pH on the enzymatic activity of digestive proteases of juvenile Cichlasoma beani for the determination of: (**A**) optimal pH of acidic proteases; (**B**) pH stability of acidic proteases; (**C**) optimal pH of alkaline proteases; and (**D**) pH stability of alkaline proteases (mean ± SD, n = 3). Mean values denoted with unlike letters are statistically different, $p < 0.05$.

The optimum temperature for acidic proteases was 55 °C (Figure 2A). The stability of the acidic proteases presented thermal variations around 60% at 55 °C, 50% at 45 °C, with the temperature of 65 °C showing a decrease of <10% at 90 min of pre-incubation (Figure 2B).The optimum temperature of alkaline protease activity increased rapidly above 35 °C, presenting two peaks of maximum activity at 45 and 65 °C (Figure 2C). Finally, residual alkaline protease activity decreased by 70% at 45 °C, while temperatures of 55 and 65 °C decreased approximately 35%–40% at 90 min of pre-incubation (Figure 2D).

Analysis of the effects of inhibitors indicated that acid protease was inhibited by 85% by pepstatin A, a specific inhibitor of pepsin type A enzymes. Inhibitors of the activity of alkaline proteases showed an inhibition of 53% for TPCK, a specific inhibitor of chymotrypsin; inhibition of 74% for TLCK, a specific inhibitor of trypsin, 88% and 57% for Ovalbumin and SBT1, respectively, which inhibit serine proteases, a reduction of 87% and 55% for the metalo-protease inhibitors Phenanthroline and EDTA, respectively, and inhibition of 60% for PMSF, a general inhibitor of serine proteases (Figure 3).

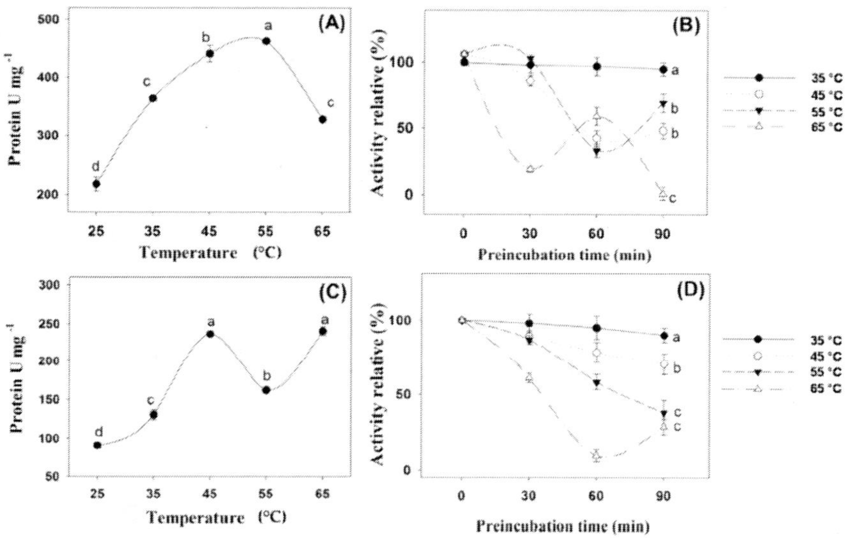


**Figure 2.** Effect of temperature on digestive proteases of juvenile of Cichlasoma beani: (**A**) optimum temperature of acidic proteases; (**B**) thermal stability of acidic proteases; (**C**) optimum temperature of alkaline proteases; and (**D**) thermal stability of alkaline proteases (mean ± SD, n = 3). Mean values denoted with unlike letters are statistically different, $p < 0.05$.

The analysis of SDS-PAGE zymograms for alkaline proteases revealed five active fractions in control without inhibitor (92.3, 60.1, 56.7, 38.2 and 35.4 kDa). Ovalbumin and TPCK inhibited three active fractions (92.3, 56.7 and 35.4 kDa), while PMSF and SBT1 inhibited only one active fraction (60.1 kDa). The TLCK inhibited three active fractions (92.3, 60.1 and 56.7 kDa). Phenanthroline and EDTA inhibited four active fractions (92.3, 60.1, 56.7 and 35.4 kDa, respectively) (Figure 4).

# 3. DISCUSSION

Results obtained in the present study with the cichlid *C. beani*, show that this species exhibits a relatively high activity of digestive proteases, and that the alkaline protease activity was higher compared to that of acid proteases, which indicates that this species has omnivorous habits unlike other fish such as the Asian bonytongue *Scleropages formosus* [18] and turbot *Scophthalmus maximus* [21]. The latter two species have carnivorous habits characterized by high acid protease activities (> 1500 UEA (Unit Enzymatic Activity) mg of protein$^{-1}$) in the stomach, the most important organ for the digestion of proteins, and a lower activity of alkaline proteases in the intestine (<250 UEA mg protein$^{-1}$), where fine hydrolysis of proteins to peptides and amino acids, absorbed by enterocytes, is performed [26].

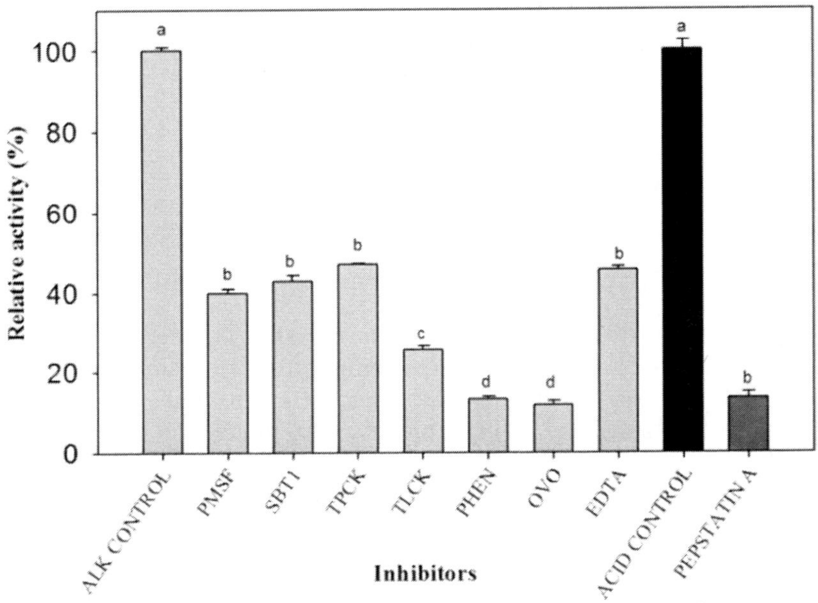


**Figure 3.** Residual activity (%) of digestive proteases using several inhibitors on multienzyme extracts of juvenile Cichlasoma beani. Alkaline proteases with no inhibitor (Alk control), tosylphenylanyl-chloromethyl ketone (TPCK), phenanthroline (PHEN), ethyl-diamine tetra-acetic acid (EDTA), tosyl-lysyl-chloromethyl ketone (TLCK), ovalbumin (OVO), soybean trypsin inhibitor (SBT1), phenyl methyl sulphonyl fluoride (PMSF), acidic proteases with no inhibitor (Acid control), pepstatin A (mean ± SD, n = 3). Columns with different letters show a significant difference ($p < 0.05$).

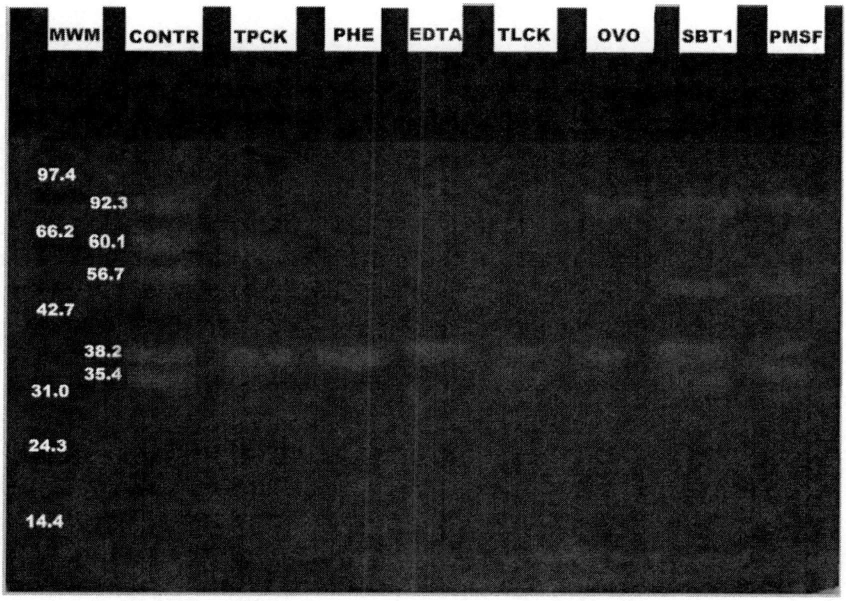


**Figure 4.** SDS-PAGE zymogram of alkaline digestive proteases of Cichlasoma beani: Molecular Weight Marker (MWM), rabbit phosphorylase B (97.4 kDa), bovine serum albumin (66.2 kDa), ovalbumin (42.7 kDa), carbon anhydrase (31.0 kDa) and lysozyme (14.4 kDa); control (without inhibitor); inhibitors were the same as in Figure 3.

In the present study, acid proteases reached their maximum activity at a pH level of 2, which is consistent with findings of the majority of prior studies for species such as Senegal sole *Solea senegalensis* [27], Blue disc *Symphysodon aequifasciata* [28], *Diplodus puntazzo* [29], coelacanth *Latimeria chalumnae* [30] *Nemipterus* spp. [31] and *Acipenser naccarii* [32]. The bay snook *Petenia splendida* and rainbow trout *Oncorhynchus mykiss* have optimum pH values of 5 and 4, respectively, uncommon for acid digestion [22,32]. For proteases the pH at which activity is maximized, is also that at which the enzymes' active centers have adequate ion formation to perform highly efficient catalysis [33].

In this study we found only a single active fraction of acid digestion using the inhibitor Pepstatin A. This protease acts first in protein hydrolysis under acidic conditions in the stomach, followed by digestion and nutrient absorption under alkaline conditions in the intestine. In *C. beani*, alkaline proteases showed a pH optimum of 11, which differed from the optimal pH (9 and 10) recorded for other marine and freshwater species, such as *Brycon orbignyanus* [26], *Symphysodon aequifasciata* [28]; *Catla*

*catla*, *Labeo rohita* and *Hypophthalmichthys molitrix* [34]; *Thunnus orientalis* [35], *Ctenopharyngodon idella* [36], and *O. niloticus* × *O. aureus* [37], and *D. puntazzo* [29]. Some species, however, have been reported to exhibit different pH optimum; even in the same species two pH values of maximum activity may be recorded, as in *P. maculatofasciatus* [33] and *C. undecimalis* [16]. This is possible because alkaline proteases are composed of various types of enzymes, which have different pH optima [38]. Finding of a well-defined optimum pH in *C. beani* suggests a single type of enzyme or a major contribution of various types of enzymes that share the same pH optimum [35]. This could be the case of carboxypeptidase A, elastase or collagenase activities, which show maximum value in the pH range of 9–11 [35]. In *C. beani*, the greatest inhibition of enzymatic activity (55%) was found for phenanthroline and EDTA, which are specific inhibitor and chelating agent of metalloproteases.

Acid proteases are stable at acid pH and their activity decreases with increasing pH. In contrast alkaline proteases are stable at alkaline pH and are resistant to slightly acidic pH, conditions similar to those reported in the cichlid *P. splendida* [22]. In other marine fish species prone to carnivorous feeding behavior, however, acid proteases are stable for some time to alkaline pH, as reported for *T. thynnus* [17] and *S. maximus* [21]. This has also been found in omnivore marine species such as *S. senegalensis* [19] and carnivorous freshwater species such as *A. tropicus* [24]. From a physiological standpoint, the fact that acid proteases are resistant to alkaline pH, allows them to continue acting in the degradation of food when it passes into the intestine, until alimentary bolus alkalize, leads to hydrolysis by alkaline proteases. This could hardly occur in *C. beani* because acid proteases lose activity at a neutral or alkaline pH. This loss of activity is compensated, however, by the hydrolysis of alkaline proteases at an acidic pH, which is also compensated by a long intestine and strong sphincters that allow longer gut residence of the food, in addition to the joint action of digestive hormones such as cholecystokinin that causes peristalsis, increasing surface contact with the enzymatic substrate [33]. Thus, alkaline digestive proteases may start acting from the moment the food enters the intestine even at an acidic pH.

It is noteworthy that alkaline proteases are more stable at different pH values, because they are comprised by a greater number of isoforms (trypsin, chymotrypsin, L-aminopeptidase and carboxypeptidase A). This is shown in the zymogram by the detection of five active fractions with activity relative to the control without inhibitor, which increases the possibility of hydrolyzing food proteins despite the pH variation [38]. Instead, acid proteases are often more susceptible to pH changes, since they only have an isoform corresponding to pepsin, which when subjected to different pH values, undergoes changes in its structural configuration or denaturation that leads to partial or total loss of activity [39]. These variations in digestive enzyme activity differ among species, depending on genetic makeup and habitat conditions. In this respect, the optimum temperature of alkaline proteases is typically 10 °C higher than the optimum temperature of acid proteases, and it is more thermostable. Usually, the difference between the optimum temperatures of alkaline and acid proteases have been described in carnivorous fish species such as *T. thynnus* [17], *B. orbignyanus* [26] and *P. maculatofasciatus* [33]. The optimum temperature of acid proteases in *C. beani* was 55 °C, comparable to that described in the hybrid *O. niloticus* × *O. aureus* [37], but higher than that found for *S. senegalensis* [19], *T. orientalis* [35] and *C. trimaculatum* [8], with temperature optima from 35 to 55 °C. The activity of acid proteases is stable up to 45 °C, resulting in values >100% for some of the incubation times, but over 60% of the activity is lost above 55 °C. This could be due to denaturation of pepsin at temperatures >55 °C, in agreement with results described by [32]. In contrast, alkaline proteases have an optimal temperature of 65 °C, which is greater than that reported for the crevalle jack *Caranx hippos*, spotted goatfish *Pseudupeneus maculatus*, parrotfish *Sparisoma* sp., trahira *Hoplias malabaricus* [40] and *O. niloticus* [20], species that show an optimum temperature ranging from 50 to 55 °C. The high thermostability of the alkaline proteases in *C. beani* is similar to that reported for *C. undecimalis* [16], *A. tropicus* [24] and *C. trimaculatum* [25]. It is noteworthy that the temperature optima and the thermal stability of the digestive proteases are only operational parameters of enzymes, rather than physiologically relevant results, since activities may vary with enzyme configuration, habitat, environment and genetic aspects of the species [30,32,38]. This deficiency in activity is compensated, however, by a longer retention of food in the digestive system to perform the hydrolysis of proteins at lower temperatures than optimal [10,32,39,41]. Determination of the number of isoforms that constitute acid proteases in *C. beani* was performed using the specific inhibitor pepstatin A, which inhibited the activity of acidic proteases by 86%. This reduction is lower than that found in carnivorous species such as *S. maximus* [21], the Mandarin fish *Siniperca chuatsi* [42] and Skipjack tuna *Katsuwonus pelamis* [39], in which the purified pepsin enzyme was completely inhibited by pepstatin A, thus indicating that acid hydrolysis is carried out mainly by a pepsin-type enzyme.

The seven types of specific inhibitors reduced the activity of alkaline proteases by at least 53%. Of these, the group of serine proteases was inhibited by 60% by Phenyl methyl sulphonyl fluoride (PMSF), and by 88% and 57% with ovalbumin and soybean trypsin inhibitor (SBT1), respectively. This pattern in which serine proteases were more markedly affected than the rest of the enzymes was similar to findings in *S. aequifasciata* [28], *T. thynnus* [17], *C. idella* [36] and *C. trimaculatum* [8]. The other enzyme group subject to high metal inhibition was that of the proteases, with 87% inhibition by Phenanthroline and 55% by ethyl-diamine tetra-acetic acid (EDTA). This indicates that via alkaline digestion *C. beani* has the ability to hydrolyze proteins from the carboxyl and amino ends, thus releasing amino acids. This must be considered in diet formulation to determine the composition of ingredients that do not contain inhibitors of metalloproteases or serine proteases. The tosylphenylanyl-chloromethyl ketone (TPCK) was inhibited 53% of chymotrypsin activity and tosyl-lysyl-chloromethyl ketone (TLCK) inhibited 74% of trypsin activity. This result was similar to that found in *B. orbignyanus* [26] and *S. formosus* [18], in which greater inhibition of trypsin than of chymotrypsin by TLCK was documented. Thus, it is confirmed the determination of the types of enzymes that comprise the digestive proteases of a species using specific inhibitors allows identification of dietary ingredients that may reduce the activity of specific enzymes and thus limit feed digestibility [17].

Specific inhibitors of alkaline proteases was performed on SDS-PAGE, five active fractions were detected in the control well, differentiating two groups of isoforms: the first group of three active fractions with molecular weights between 92.3, 60.1 and 56.7 kDa, and the second group of two active fractions with molecular weights of 38.2 and 35.4 kDa. The number of active fractions found in *C. beani* was similar to that reported in marine species such as *P. maculatofasciatus* [33], and freshwater species such as *P. splendida* [22], *C. undecimalis* [16], and *A. tropicus* [24]. In other omnivorous species such as the South American pilchard *Sardinops sagax caerulea* [43] and round sardinella *Sardinella aurita* [44], trypsin showed active fractions with different molecular weights ranging between 35 and 38.8 kDa, while purified chymotrypsin from *S. sagax caerulea* had a molecular weight of 35.5 kDa [45]. The inhibitors SBT1 and PMSF, specific for serine protease, presented the highest percentages of inhibition as they eliminated three active fractions in *C. beani*, which could be trypsin-like or chymotrypsin-like enzymes [42,44,45,46]. On the other hand the inhibitors specific for metalloproteases phenanthroline and EDTA, inhibited two active fractions in the first group of enzymes. Therefore, future studies will be conducted with purified enzymes to determine the specific type of enzymes present in *C. beani*.

# 4. MATERIALS AND METHODS

## 4.1. Sample Preparation

Samples were obtained from juveniles donated by the Universidad Autónoma de Nayarit, with an average wet weight of 28.71 ± 5.35 g (n = 35), which were euthanized with an overdose of methanesulfonate tricaine (MS-222, Sigma-Aldrich, Saint Louis, MI, USA). Weight and total body length were recorded for each juvenile and then the stomachs and intestines were extracted with ice; the stomachs and intestines were homogenized separately in a 200 mg $mL^{-1}$ ratio, with 100 mmol $L^{-1}$glycine-HCl buffer at pH 2 for the stomachs and buffer Tris-HCl 12.5 mmol $L^{-1}$ + $CaCl_2$ $L^{-1}$ 30 mmol $L^{-1}$ at pH 7.5, for the intestines respectively. The homogenate was centrifuged at 16,000× g for 30 min at 4 °C; the multienzyme extract thus obtained was stored at −80 °C for further analysis. The concentration of soluble protein in the multienzyme extracts of stomach and intestine was determined using the [47] technique and using bovine serum albumin as standard.

## 4.2. Biochemical Analyses

For measurement and determination of the pH and temperature optima, and enzymatic stability as well as inhibition assays, the enzymatic extracts of ten stomachs were used, and for the alkaline proteases enzymatic extracts were used from ten juvenile intestines respectively.

The total activity of the acid proteases was determined using 1% hemoglobin in a glycine-HCl 100 mM buffer at pH 2 following the method of [48]. The activity of the alkaline proteases was evaluated following the method of [25], as modified by [49], using 1% casein in a Tris-HCl 100 mmol $L^{-1}$ + $CaCl_2$ 10 mmol $L^{-1}$ buffer at pH 9. One unit of activity was defined as 1 μg of tyrosine liberated per minute with a molar extinction coefficient (MEC) of 0.005 $mL^{-1}$ $\mu g^{-1}$ $cm^{-1}$.

## 4.3. Effects of pH and Temperature on Digestive Enzyme Activity

The optimum pH of the acid and alkaline proteases was evaluated using a universal buffer [50] with a pH range of 2–12 following the previously described procedure to determine protease activity. The optimum temperature of the acid and alkaline proteases was established by testing different temperatures (35, 45, 55, 65 °C). The effect of pH and temperature on the stability of the acid and alkaline proteases was determined by pre-incubating the multienzyme extracts for 0, 30, 60 and 90 min with a buffer at the assay temperature [50]. For determination of the stability to pH and temperature the pre-incubation time 0 min value was used as a control to determine the residual activity expressed as a percentage of residual activity comparing with the extracts that were incubated at different pH and temperatures values.

## 4.4. Zymograms and Effects of Inhibitors

The enzymes in the multienzyme extracts were characterized following the method described by [51]. Pepstatin A 1 mmol $L^{-1}$ was used as inhibitor in the case of acid proteases, and the following inhibitors were used for the alkaline proteases: phenyl-methanesulphonyl fluoride 100 mmol $L^{-1}$ (PMSF), *N*-*q*-tosyl-llysine-chloromethyl ketone hydrochloride 10 mmol $L^{-1}$ (TLCK), soybean trypsin inhibitor 250 mmol $L^{-1}$ (SBT1), *N*-*q*-tosyl-L-phenylalanine chloromethyl ketone 10 mmol $L^{-1}$ (TPCK), 1,10-phenanthroline 10 mmol $L^{-1}$ (Phen), trypsin inhibitor, type II-T: turkey egg ovoalbumine 250 mmol $L^{-1}$ (Ovo) and ethylene diamine tetra acetic acid 10 mmol $L^{-1}$ (EDTA). Extracts were incubated in the presence of each inhibitor for 1 h (1:1 *v*/*v*) at pH 7.5, the value of intestine extracts, and at pH 2 for stomach extracts during sample preparation. In both tests a control without inhibitor was used as 100% of total activity and the residual activity was expressed as a percentage of reduction of the activity with the different inhibitors. All assays were conducted in triplicate.

The effect of inhibitors on the alkaline proteases was performed by SDS-PAGE electrophoretic analysis using a discontinuous polyacrylamide gel, which included a 4% stocking gel and 10% stocking gel, according to the methods of [51], as modified by [26]. The seven inhibitors: TPCK, PHEN, EDTA, TLCK, Ovo, SBT1, PMSF, described by [15] were used. They were pre-incubated in a 1:1 ratio (enzyme: inhibitor) for 1 h. The molecular weight marker Bio Basic Inc., BM523 and the program Quantity One 1-D Analysis Software de Bio-Rad® were used to calculate the molecular weight of each active fraction.

## 4.5. Statistical Analysis

All the enzymatic activities of pH and optimum temperature and stabilities, as well as the inhibitors were carried out in triplicate and were evaluated by means of non-parametric statistical analysis of Kruskal-Wallis test and a posteriori Nemenyi test as the data were not normally distributed. Statistical tests were conducted with the Statistica software v. 7.0 (Statsoft, Tulsa, OK, USA) using a significance value of 0.05.

## 5. CONCLUSIONS

In conclusion, results of the present study show that digestion in C. beani, as inferred from enzymatic activity, is carried out either by acid proteases or alkaline proteases; a pepsin-like enzyme is responsible for most of the hydrolysis, but it shows little resistance to pH and temperature. In contrast, digestion of alkaline proteases is more stable to these parameters. These biochemical conditions allow C. beani to hydrolyze proteins in formulated foods. Results of the present study also suggest that C. beani is an omnivorous species, with some tendency to carnivore, results which can be used to guide the development of specific diets for culture.

## ACKNOWLEDGMENTS

The authors thank the Autonomous University of Nayarit for the donation of specimens, the Universidad Juárez Autónoma de Tabasco for the development of this collaborative research project, Susan Mitchell and Jackob Parker for their kind assistance in the improvement of earlier manuscript drafts.

## REFERENCES

1. Baltazar, P.M. La tilapia en el Perú: Acuicultura, mercado y perspectivas. *Rev. Peru. Biol.* **2007**, *13*, 267–273.
2. Márquez, C.G. Producción por acuicultura sustentable de pegelagarto en Comalcalco, Tabasco. Available online: https://sgp.undp.org/ index.php? option= com_ sgpprojects&view=projectdetail&id=15073&Itemid=272 (accessed on 28 November 2016).
3. García-Lizárraga, M.A.; Soto-Franco, F.E.; Velazco-Arce, J.M.; Velázquez-Abunader, J.I.; Ramírez-Pérez, J.; Peña-Messina, S.E. Population structure and reproductive behavior of Sinaloa cichlid *Cichlasoma beani* (Jordan, 1889) in a tropical reservoir. *Neotrop. Ichthyol.* **2011**, *9*, 593–599.
4. Aragón-Flores, E.A.; Valdez-Hernández, E.F.; Martínez-Cárdenas, L.; Castañeda-Chávez, M.R.; Gonzales-Díaz, A.A.; Soria-Barreto, M.; Peña-Messina, E. Effect of Stocking Density on Growth, Survival, and Condition of the Mexican Cichlid *Cichlasoma beani*. *J. World Aquac. Soc.* **2014**, *45*, 447–453.
5. Martínez-Cárdenas, L.; Valdez-Hernández, E.F.; González-Díaz, A.A.; Soria-Barreto, M.; Castañeda-Chávez, M.R.; Ruiz-Velazco, J.M.; Robles-Bermúdez, A.

Preliminary observations on *Cichlasoma beani* in culture conditions. *Lat. Am. J. Aquat. Res.* **2014**, *42*, 639–643.
6. López-Ramírez, G.; Cuenca-Soria, C.A.; Álvarez-González, C.A.; Tovar-Ramírez, D.; Ortiz-Galindo, J.L.; Perales-García, N.; Márquez-Couturier, G.; Arias-Rodríguez, L.; Indy, J.R.; Contreras-Sánchez, W.M.; et al. Development of digestive enzymes in larvae of Mayan cichlid *Cichlasoma urophthalmus*. *Fish Physiol. Biochem.* **2011**, *37*, 197–208.
7. Cuenca-Soria, C.A.; Álvarez-González, C.A.; Ortiz-Galindo, J.L.; Nolasco-Soria, H.; Tovar-Ramírez, D.; Guerrero-Zárate, R.; Gisbert, E. Partial characterization of digestive proteases of the Mayan cichlid *Cichlasoma urophthalmus*. *Fish Physiol. Biochem.* **2014**, *40*, 689–699.
8. Toledo Solís, F.J.; Márquez, G.; Uscanga, A.; Guerrero-Zárate, R.; Perales- García, N.; Martín García, R.; Contreras, W.; Camarillo-Coop, S.; Álvarez-González, C.A. Partial characterization of digestive proteases of the three spot cichlid *Cichlasoma trimaculatum* (Günter 1867). *Aquac. Nutr.* **2015**, *22*, 1230–1238.
9. Kolkovski, S. Digestive enzymes in fish larvae and juveniles-implications and applications to formulated diets. *Aquaculture* **2001**, *200*, 181–201.
10. Lazo, J.P.; Mendoza, R.; Holt, G.J.; Aguilera, C.; Arnold, C.R. Characterization of digestive enzymes during larval development of red drum (*Sciaenops ocellatus*). *Aquaculture* **2007**, *265*, 194–205.
11. Jonas, E.; Ragyanski, M.; Olah, J.; Borros, L. Proteolytic digestive enzymes of carnivorous (*Silurus glanis*, L.), herbivorous (*Hypophthalmichthys molitrix*, VAL) and omnivorous (*Cyprinus carpio*, L.) fishes. *Aquaculture* **1983**, *30*, 145–154.
12. Moyano, F.J. Bioquímica Digestiva en Especies Acuicultivadas: Aplicaciones en Nutrición. In Proceedings of the VIII Simposium Internacional de Nutrición Acuícola, Universidad Autónoma de Nuevo León, Monterrey, Nuevo León, México, 15–17 November 2006; Cruz, S.E., Ricque, M.R., Tapia, S.M., Nieto, L.M., Villareal, C.D., Puello, C.A., García, O.A., Goytortua-Bores, E., Eds.; UANL: Monterrey, Nuevo León, México, 2006.
13. Álvarez-González, C.A.; Márquez-Couturier, G.; Arias-Rodríguez, L.; Contreras-Sánchez, W.M.; Uscanga-Martínez, A.; Perales-García, N.; Moyano-López, F.J.; Hernández-Jiménez, R.; Civera-Cerecedo, R.; et al. Avances en la Fisiología Digestiva y Nutrición de la Mojarra Tenguayaca Petenia splendida. In Proceedings of the Memoria del IX Avances en Nutrición Acuícola y IX Simposio Internacional de Nutrición Acuícola, Universidad Autónoma de Nuevo León, Monterrey, Nuevo León, México, 24–27 November 2008; Cruz, S.E., Ricque, M.D., Tapia, S.M., Nieto, L.M., Villareal, C.D., Lazo, J.P., Viana, T.V., Eds.; UANL: Monterrey, Nuevo León, México, 2008; pp. 135–235.
14. Salvensen, G.; Nagase, H. Inhibition of proteolytic enzymes. In *Proteolytic Enzymes*; Beynon, R., Bond, J.S., Eds.; IRL Press: Oxford, UK, 1989; pp. 83–104.
15. García-Carreño, F.L.; Dimes, L.E.; Haard, N.F. Substrate-gel electrophoresis for composition and molecular weight of proteinases or proteinaceous proteinase inhibitors. *Anal. Biochem.* **1993**, *214*, 65–69.
16. Álvarez-González, C.A.; Gabriela-Cortés, G.; Jiménez-Martínez, L.D.; Sánchez-Zamora, A.; Arena-Ortiz, G.; Martínez-Bruguete, T.; Tovar-Ramírez, D.; Concha-Frías, B.; Márquez-Couturier, G.; Perales-García, N.; et al. Avances en la fisiología digestiva del robalo blanco (*Centropomus undecimalis*) en Tabasco, México. In Proceedings of the Memorias del X Avances en Nutrición Acuícola y X Simposio Internacional de Nutrición Acuícola, Universidad Autónoma de Nuevo León, Monterrey, México, 8–10 November 2010; Cruz-Suarez, L.E., Ricque-Marie, D., Tapia-Salazar, M., Nieto-López, M.G., Villarreal-Cavazos, D.A., Eds.; Gamboa-Delgado: Monterrey, Nuevo León, México, 2010; pp. 99–232.
17. Essed, Z.; Fernández, I.; Alarcón, F.J.; Moyano, F.J. Caracterización de la actividad proteasa digestiva de atún rojo *Thunnus thynnus* (Linnaeus, 1758). *Bol. Inst. Español. Oceangr.* **2002**, *18*, 99–107.

18. Natalia, Y.; Hashim, R.; Ali, A.; Chong, A. Characterization of digestive enzymes in a carnivorous ornamental fish, the Asian bony tongue *Scleropages formosus* (Osteoglossidae). *Aquaculture* **2004**, *233*, 305–320.
19. Sáenz de Rodrigáñez, M.; Alarcón, F.J.; Martínez, M.I.; Ruiz, F.; Díaz, M.; Moyano, F.J. Caracterización de las proteasas digestivas del lenguado senegalés *Solea senegalensis* Kaup, 1858. *Bol. Inst. Español. Oceangr.* **2005**, *21*, 95–104.
20. Bezerra, S.R.; Lins, J.F.; Alencar, B.R.; Paiva, M.G.; Chaves, E.C.; Coelho, C.B.; Carvalho, B.L. Alkaline proteinase from intestine of Nile tilapia (*Oreochromis niloticus*). *Process Biochem.* **2005**, *40*, 1829–1834.
21. Wang, H.; Wang, Y.; Wang, Q.; Xue, C.; Sun, M. Purification and characterization of stomach protease from the turbot (*Scophthalmus maximus* L.). *Fish Physiol. Biochem.* **2006**, *32*, 179–188.
22. Perales-García, N. Ontogenia enzimática de la mojarra tenguayaca *Petenia splendida*. Master's Thesis, Universidad Juárez Autónoma de Tabasco, Villahermosa, Tabasco, México, 24 November 2010; p. 94.
23. Nalinanon, S.; Benjakul, S.; Kishimura, H. Biochemical properties of pepsinogen and pepsin from the stomach of albacore tuna (*Thunnus alalunga*). *Food Chem.* **2010**, *121*, 49–55.
24. Guerrero-Zarate, R.; Álvarez-González, C.A.; Olvera-Novoa, M.A.; Perales-García, N.; Frías-Quintana, C.A.; Martínez-García, R.; Contreras-Sánchez, W.M. Partial characterization of digestive proteases in tropical gar *Atractosteus tropicus* juveniles. *Fish Physiol. Biochem.* **2014**, *40*, 1021–1029.
25. Anson, M.L. The estimation of pepsin, tripsin, papain and cathepsin with hemoglobin. *J. Gen. Physiol.* **1938**, *22*, 79–89.
26. García-Carreño, L.F.; Albuquerque-Cavalcanti, C.; Navarrete del Toro, A.M.; Zaniboni-Filho, E. Digestive proteinases of *Brycon orbignyanus* (Characidae, Teleostei): Characteristics and effects or protein quality. *Comp. Biochem. Physiol.* **2002**, *132*, 343–352.
27. Yúfera, M.; Darias, M.J. Changes in the gastrointestinal pH from larvae to adult in Senegal sole (*Solea senegalensis*). *Aquaculture* **2007**, *267*, 94–99.
28. Chong, A.S.C.; Hashim, R.; Chow-Yang, L.; Ali, A.B. Partial characterization and activities of proteases from the digestive tract of discus fish (*Symphysodon aequifasciata*). *Aquaculture* **2002**, *203*, 321–333.
29. Tramati, C.; Savona, B.; Mazzola, A. A study of the pattern of digestive enzymes in *Diplodus puntazzo* (Cetti, 1777) (Osteichthyes, Sparidae): Evidence for the definition of nutritional protocols. *Aquac. Int.* **2005**, *13*, 89–95.
30. Tanji, M.; Yakabe, E.; Kageyama, T.; Shin-ichi, Y.; Ichinose, M.; Miki, K.; Ito, H.; Takahashi, K. Purification and characterization of pepsinogens from the gastric mucosa of African coelacanth, *Latimeria chalumnae*, and properties of the major pepsins. *Comp. Biochem. Physiol.* **2007**, *146*, 412–420.
31. Nalinanon, S.; Benjakul, S.; Visessanguan, W.; Kishimura, H. Tuna Pepsin: Characteristics and its use for collagen extraction from the skin of threadfin bream (*Nemipterus* spp.). *J. Food Sci.* **2008**, *73*, C413–C419.
32. Furné, M.; Hidalgo, M.C.; López, A.; García-Gallego, M.; Morales, A.E.; Domezain, A.; Domezainé, J.; Sanz, A. Digestive enzyme activities in Adriatic sturgeon *Acipenser naccarii* and rainbow trout *Oncorhynchus mykiss*. A comparative study. *Aquaculture* **2005**, *250*, 391–398.
33. Álvarez-González, C.A. Actividad enzimática digestiva y evaluación de dietas para el destete de larvas de la cabrilla arenera *Paralabrax maculatofasciatus* (Percoidei: serranidae). Ph.D. Thesis, Centro Interdisciplinario de Ciencias Marinas, La Paz, Baja California Sur, México, 28 November 2003.
34. Kumar, S.; García-Carreño, F.L.; Chakrabarti, R.; Toro, M.A.; Córdova-Murueta, J.H. Digestive proteases of three carps *Catla catla*, *Labeo rohita* and *Hypophthalmichthys molitrix*: Partial characterization and protein hydrolysis efficiency. *Aquac. Nutr.* **2007**, *13*, 381–388.

35. Matus, P.A.; Rosas, A.; Lazo, J.P.; Viana, M.T. Partial characterization of the digestive enzymes of Pacific bluefin tuna *Thunnus orientalis* under culture conditions. *Fish Physiol. Biochem.* **2007**, *33*, 223–231.
36. Liu, Z.Y.; Wang, Z.; Xu, S.Y.; Xu, L.N. Partial characterization and activity distribution of proteases along the intestine of grass carp, *Ctenopharyngodon idella* (Val.). *Aquac. Nutr.* **2008**, *14*, 31–39.
37. Li, J.S.; Li, J.L.; Wu, T. Ontogeny of protease, amylase and lipase in the alimentary tract of hybrid juvenile tilapia (*Oreochromis niloticus* × *Oreochromis aureus*). *Fish Physiol. Biochem.* **2006**, *32*, 295–303.
38. Alarcón, F.J.; Díaz, M.; Moyano, F.J.; Abellan, E. Characterization and functional properties of digestive proteases in two sparids; gilthead seabream (*Sparus aurata*) and common dentex (*Dentex dentex*). *Fish Physiol. Biochem.* **1998**, *19*, 257–267.
39. Nalinanon, S.; Benjakul, S.; Kishimura, H. Purifcation and biochemical properties of pepsins from the stomach of skipjack tuna (*Katsuwonus pelamis*). *Eur. Food Res. Technol.* **2010**, *231*, 259–269.
40. Alencar, B.R.; Biondi, M.M.; Paiva, M.V.; Carvalho-Junior, B.L.; Bezerra, S.R. Alkaline proteases from the digestive tract of four tropical fishes. *Braz. J. Food Technol.* **2003**, *6*, 279–284.
41. Uscanga, A.; Moyano, F.J.; Álvarez, C.A. Assessment of enzymatic efficiency on protein digestion in the tilapia *Oreochromis niloticus*. *Fish Physiol. Biochem.* **2010**, *36*, 1079–1085.
42. Zhou, Q.; Liu, G.; Huang, Y.; Weng, L.; Hara, K.; Su, W.; Cao, M. Pepsinogens and pepsins from mandarin fish (*Siniperca chuatsi*). *J. Agric. Food Chem.* **2008**, *56*, 5401–5406.
43. Castillo-Yáñez, F.J.; Pacheco-Aguilar, R.; García-Carreño, F.L.; Navarrete-Del Toro, M.A. Isolation and characterization of trypsin from pyloric caeca of Monterey sardine *Sagax caerulea caerulea*. *Comp. Biochem. Physiol. B***2005**, *140*, 91–98.
44. Ben, K.H.; Jellouli, K.; Souissi, N.; Ghorbel, S.; Barkia, A.; Nasri, M. Purification and characterization of three trypsin isoforms from viscera of sardinelle (*Sardinella aurita*). *Fish Physiol. Biochem.* **2011**, *37*, 123–133.
45. Castillo-Yáñez, F.J.; Pacheco-Aguilar, R.; Lugo-Sánchez, M.E.; García-Sánchez, G.; Quintero-Reyes, I.E. Biochemical characterization of an isoform of chymotrypsin from the viscera of Monterey sardine (*Sardinops sagax caerulea*), and comparison with bovine chymotrypsin. *Food Chem.* **2009**, *112*, 634–639.
46. Simpson, B.K. Digestive Proteases from Marine Animals. In *Seafood Enzymes*; Haard, N.F., Simpson, B.K., Eds.; Marcel Dekker: New York, NY, USA, 2000; pp. 191–213.
47. Dunn, B.M. Determination of protease mechanism. In *Proteolytic Enzymes: A Practical Approach*; Beynon, R.J., Bond, J.S., Eds.; IRL Press: Oxford, UK, 1989; pp. 57–58.
48. Walter, H.E. Proteinases: Methods with hemoglobin, casein and azocoll as substrates. In *Methods of Enzymatic Analysis*; Bergmeyer, H.J., Ed.; Verlag Chemie: Weinheim, Germany, 1984; pp. 270–277.
49. Kunitz, M. Crystalline soybean trypsin inhibitor II. General properties. *J. Gen. Physiol.* **1947**, *30*, 291–310.
50. Stauffer, C. *Enzyme Assays for Food Scientists*; Van Nostand Reinhold/AVI: New York, NY, USA, 1989; p. 317.
51. Laemmli, U.K. Cleavage of structural proteins during the assembly of the head of bacteriophage T4. *Nature* **1970**, *227*, 680–685.

# CHAPTER 11

# Functional and Structural Differences in Atria Versus Ventricles in Teleost Hearts

**Christine Genge[1], Leif Hove-Madsen[2] and Glen F. Tibbits[1, 3]**

[1] *Molecular Cardiac Physiology Group, Simon Fraser University, Burnaby, Canada*

[2] *Cardiovascular Research Centre CSIC-ICCC, Hospital de Sant Pau, Barcelona, Spain*

[3] *Cardiovascular Sciences, Child and Family Research Institute, Vancouver, Canada*

## 1. INTRODUCTION

Across vertebrates, the fish heart is structurally relatively simple. The heart of teleosts is unique in structure, composed of four chambers in series: venous sinus, atrium, ventricle and bulbus arteriosus. The two chambers acting as pumps are the atrium and ventricle, a simplified version of that seen in tetrapods. These chambers develop from a simple linear tube [1] and differ not only morphologically but also physiologically with different characteristic rates of contractility [2]. Fish heart design is reflective of the needs for oxygen delivery to working skeletal muscle in an often oxygen poor environment. The heart is the first definitive organ to develop and become functional, as embryological survival depends on its proper function. The vertebrate-specific development of multiple chambers with enough muscle to generate higher systemic pressures allowed for perfusion of larger and more complex tissues.

Fish show great variability in the development of the heart and modification of heart performance capacities to meet tissue perfusion demands imposed by differences in life history. Research has tended to focus on inotropic or chronotropic responses of the heart to stress such as temperature in larger fish such as trout, or on development through the use of zebrafish mutants. Between these sources much has been revealed about the contractile properties of the two chambers. This research will be reviewed in the context of roles of the atrium and ventricle in fish in achieving variability in myocardial contractility.

## 2. GENE EXPRESSION

In mammals, 23% of cardiac-specific genes show differences in expression between the atria and the ventricle [3]. Key morphogenetic events during heart ontogenesis are conserved across vertebrates [4]. Gene expression programs specific to each chamber are likely regulating key aspects of chamber formation. The focus on fish heart developmental patterns has primarily been due to the use of zebrafish as a model organism of vertebrate development. The morphological differences created during chamber development are guided by changes in gene expression patterns [1], a process well conserved in vertebrates, regardless of whether there are one or two atria or ventricles. Genetic screens in zebrafish have revealed genes and pathways underlying development in the heart. Since many stages of development are conserved across vertebrates, this information has generated considerable insight into congenital heart defects in humans. In the context of this chapter, timing and localization of gene expression can also contribute to our understanding of the differentiation of chambers both in morphology as well as through variation in contractility.

The T-box genes (Tbx) are responsible, in large part, for patterning that distinguishes chamber myocardium from non-chamber myocardium [5]. Subregionalization is important for atrial or ventricular identity, either via selective transcription or repression of genes. The physical interaction of Nkx2.5, Gata4 and Tbx5 in particular activates the expression of chamber-specific genes to stimulate the differentiation of primary myocardial cells into chamber myocardium [1]. By the time cells are restricted to cardiac lineage, they are already destined for either the atrium or the ventricle. During early gastrulation, myocardial progenitors already have axial coordinates that lead to chamber assignment [6]. There are two sources of these myocardial progenitors, or two heart fields. In the first heart field, cardiomyocyte differentiation initiates in the ventricle and allows continuous addition of new cardiomyocytes to the venous pole of the heart tube progressing towards the atrium [7]. Differentiation at the arterial pole in the second heart field occurs only in the later stages of cardiac development. By 72 hours post-fertilization (hpf) individual heart chambers are morphologically differentiated with the ventricular wall being thicker and ventricular cardiomyocytes being larger than those of the atrium [8].

The chambers express different subsets (or paralogs) of sarcomeric proteins such as myosin heavy chains (MHC) and myosin light chains (MLC) [9]. Variation in paralog expression (frequently referred to in the literature as isoforms) may confer differing contractile properties to the atrium and the ventricle, such as with altered MHC expression being correlated with altered myofibrillar ATPase activity (crucian carp - [10]). MHC expression is chamber-specific by 22 hpf, preceding any obvious chamber formation but corresponding to the initiation of the heart beat [6]. These isoforms are not only divided spatially but also by differences in timing of expression. Atrial-specific myosin gene expression is induced after ventricle-specific myosin gene expression in zebrafish [11]. This time-dependent expression may be due to repression of the promoters of these genes [1].

The timing of these developmental pathways is critical, as the development of one chamber can influence the other. Ventricular morphology can change with developmental atrial dysfunction seen with zebrafish mutations in atrial-specific MHC [11]. For example, in the wea mutant, a lack of a functional atrial MHC results in defects in atrial myofilament organization and contractility. Despite the mutation being seen only in a chamber-specific protein, the ventricle changes as well in response to these atrial defects. Effective embryonic circulation requires appropriate relative sizes of the atrium and ventricle. Bone morphogenetic protein (BMP) signaling has been shown to be involved with the regulation of chamber size in zebrafish [12]. However, mutations affecting BMP signaling only reduce atrium size but not the ventricle. During this stage of development, the size of chambers may be relatively independent of one another.

## 3. MORPHOLOGY

In fish the heart is arranged in series with the ventricle primarily filled by contraction of the atrium [13], rather than the mammalian system in which the thin-walled atrium contributes only a small amount to ventricular filling under resting conditions. While the atrium and ventricle compose the main muscular components of the fish heart, there are other chambers with important roles in guiding appropriate blood flow through the heart. Generally in fish blood flows through the sinus venosus to the atrium to the ventricle and then out to the bulbus anteriosus. More recently, the atrioventricular region and conus arteriosus have been identified as morphologically discrete segments present throughout modern teleosts as well [14, 15].

## 3.1 General Morphology

As with most vertebrates, the cardiac tube is S-shaped with a dorsally positioned atrium and ventrally positioned ventricle to allow for appropriate momentum of the bloodstream towards the arterial pole. The morphology of the heart sets up an alternating arrangement of slow conducting segments (e.g. inflow tract, atrioventricular region and outflow tract) with fast conducting contractile segments (atrium and ventricle). Blood first enters the thin-walled sinus venosus. This is an independent, but non-muscular chamber in fish that acts as a drainage pool for the venous system. The composition of the sinus venosus varies greatly between species, from primarily connective tissue in *Danio rerio* (zebrafish) to mainly myocardium in *Anguilla anguilla* (European eel) to even smooth muscle cells in *Cyprinus carpio* (carp) [16, 17, 18]. The presence of the sinus venosus has thus been suspected to have allowed the development of the atrium as the principal driving force for ventricular filling [19]. Blood then moves into the atrium, an active contractile chamber in fish that again shows considerable variability in size and shape between species [18]. A general trend throughout teleost species is the similarity in volume of the two main muscular chambers is indicative of the importance of the role of the atrium in filling the ventricle. While atrial volume is closer in size to that of the ventricle in fish, the atrium is still less developed muscularly and possibly representative of a more primitive heart [20-22].

Separating the atrium and the ventricle is the atrioventricular region, a ring of cardiac tissue supporting the AV valves. The degree of isolation of this region from the surrounding musculature varies between species [23] but in all species it plays a critical role in regulating the pattern of electrical conduction. The ventricle has a mass and wall thickness up to five times that of the atrium [18]. The thickly muscular ventricle varies in terms of structural organization, histology, coronary distribution and relative mass between species as well. This variation reflects the fact that different modes of heart performance are attributed to different ventricle types since this chamber is responsible for the active cardiac output [18]. This classification will be discussed further in this section. From the ventricle, blood moves into the outflow tract leading into the beginning of the dorsal aorta. In more basal fish there are two segments: the proximal muscular conus arteriosus and the distal arterial-like bulbus arteriosus [24]. While later teleosts were thought to have lost the conus arteriosus, more recent evidence shows it is a distinct segment in many families [14, 15]. The bulbus arteriosus is more prominent and serves as the main component of the outflow tract. This elastic chamber expands to store large portions of the stroke volume, and gradually releases a steady non-damaging flow to gill vascular [18]. The role of these segments is variable between species and has been reviewed in more detail [25].

The AV-region, bulbus arteriosus and conus arteriosus are unique to fish though homologous regions may be seen in early embryological stages of higher vertebrates. While, the atrium and ventricle are present across vertebrates, their diversity in morphology and function and how they contribute to contractility will be the focus of the remainder of this Chapter.

### 3.2. Myocardial Arrangement

The teleost myocardium is composed of two distinct layers: compact and spongy. The atrium tends to be primarily spongy, with more open space between fibers and greater variability in orientation [26]. The external rim is myocardium surrounding a complex network of pectinate muscles. Collagen provides support, being predominant both in surrounding the myocardium and the internal trabecular network [15]. Meanwhile, while the ventricle composition varies greatly between species, the periphery of the ventricle tends to be primarily compact tissue with the luminal being primarily spongy [15, 27]. The compact layer is composed of typically well-organized fibers composing well-arranged bundles covering the inner ventricle. The spongy layer is composed of long finger-like branching cells in close proximity to luminal fluids. Not all fish species have distinct layers in the ventricle to the same degree as the salmonids and scombroids in which most of the experiments characterizing the morphology of the chambers has been done. For example, the relatively sedentary flatfish has a heart with virtually no lumen and is composed almost entirely of spongy tissue [28]. These phenotypes have been subdivided into four subclasses (Types I-IV). A Type I ventricle consists of an entirely spongy ventricle, while Type II has both compact and spongy myocardium but with coronary vessels restricted to the compact layer. Type III and Type IV have vascularized trabeculae, with coronary vessels penetrating spongy and compact myocardium. Types III and IV are distinguished from one another by Type III having less than 30% of the heart mass composed of compact myocardium (exemplified by many elasmobranchs) and Type IV representing ventricles with greater amounts of compact myocardium [29].

### 3.3. Cardiomyocytes

On the cellular level, cardiomyocytes are more cuboidal in the ventricle compared to the more squamous shape seen in the atrium [30]. Ventricular myocytes from ecothermic vertebrates more closely resemble myocytes from neonatal mammals rather than adults in terms of structure and function. They are long (>100 μm), thin (<10 μm) and generally lacking in t-tubules (trout - [18]; trout - [31]; neonatal rabbit - [32]). These myocytes appear in electron microscopy images to have less extensive sarcoplasmic reticulum (SR) than mammals [33] but some peripheral couplings with sarcolemmal membrane are observed [34]. Atrial muscle cells contain well-developed sarcomeric reticulum, peripheral couplings but no t-tubules [35]. The narrow extended shape of teleost cardiomyocytes from both chambers decreases the diffusion distance from the sarcolemma to the center of the cell [36-38]. This high surface-to-volume ratio of the cells may be related to increased efficacy of E-C coupling. These morphological distinctions contribute to the chamber-specific contractile properties which will be discussed further in this Chapter.

## 3.4. Lifestyle Demands

These morphological differences between species are reflective of a diversity of lifestyle demands. Active fish have a larger relative heart mass than that of less active species; the relative heart mass in the skipjack tuna is 10 times larger than in the sluggish flounder [39]. However, a similar atrium to ventricle mass ratio exists in species with very different lifestyles (0.2 in active Atlantic salmon, Atlantic cod and inactive winter flounder - [28]). More active fish also show a difference in whole heart shape with active fish (e.g. salmonid and scombrid families) exhibiting pyramid-shaped hearts whereas inactive fish hearts are more rounded and sac-like in shape [40]. The pyramidal shape normally correlates with a thicker myocardial wall and high cardiac output. This allows for high levels of stroke work and the maintenance of relatively high blood pressures [18].

The amount of compact myocardium is generally increased in fish with greater cardiac output demands. With pyramidal ventricles, the myocardial wall is normally formed primarily by the compact layer, which is complex and thicker in these higher performance fish [18] although this phenotype is also found in the relatively sedentary Antarctic teleosts [26]. This suggests that shape and degree of compact myocardium do not always correlate with lifestyle and in fact the shape of the chamber does not always correlate with the inner architecture [41]. This active phenotype may also display plasticity, as trout that are rendered inactive will display an abnormal rounded shape [42]. This may also be the case in aquaculture conditions, where fish become less active and may be overfed [43]. The proportion of compact myocardium has also been shown to vary with seasons [18] and growth [44].

Active fish rely on coronary circulation to allow for increased cardiac output. More compact tissue in the ventricle generally requires greater $O_2$ perfusion. In general, sluggish species lack this coronary circulation and can rely completely on $O_2$ contained in the venous circulation [45]. Hence thicker ventricles seen in more active fish have a more developed coronary distribution to the compact myocardium, and a greater percentage of compact myocardium. Very active fish like scombroids and elasmobranchs also have a coronary supply to the spongy myocardium and atrium [46]. However, this increased demand for $O_2$ supply for cardiac function means that these species are at a disadvantage in hypoxic water. The more primitive hagfish heart composed almost entirely of spongy myocardium can meet cardiac ATP requirements for resting conditions even in anoxic water [47].

# 4. FILLING PATTERNS

Atrial filling occupies a substantial proportion (up to 75%) of the cardiac cycle and as discussed above, the atrial chamber volume in fish may be comparable to that of the ventricle. Substantial debate exists in the literature as to the mechanism of both atrial and ventricular filling. Though noninvasive echocardiography has greatly contributed to research on flow patterns, it does require anaesthetization and restraint of the fish. Because the heart is acutely sensitive to venous pressure, and venous pressure is acutely sensitive to gravitational forces and extramural pressure, considerable variation may be seen from study to study. Certain studies have shown a more monophasic filling of both atrium and ventricle via a vis-a-tergo mechanism (e.g. trout - [48]). In this model, ventricular pressure is derived from arterial flow and venous capacitance is the primary determinant of cardiac output [13]. Other studies have demonstrated a more biphasic filling pattern via a vis-a-front mechanism in which cardiac suction plays a prominent role in ventricular pressure (e.g. elasmobranchs - [49]). Even triphasic patterns have been observed, with a third wave of filling possibly involving sinus contraction (e.g. leopard shark - [49]).

These filling patterns of the chambers may be both a reflection of phylogeny and external conditions. Less active fish such as the sea raven require above-ambient filling pressures that may require vis-a-tergo filling [50]. The venous pressure control mechanisms are not seen to be as developed in elasmobranchs as in teleosts resulting in a greater necessity of vis-a-front filling [51]. Meanwhile with increased stroke volumes during exercise resulting in more positive filling pressures, teleosts may further shift to a vis-a-tergo cardiac filling mode for increased activity [52]. However, the relative proportion of early to late phase filling (or passive vs. atrial active) suggests that the late phase is greater than the early phase in biphasic [49] and the third phase in triphasic does not change this pattern. With late phase (active atrial contraction) playing such an important role regardless of filling mode, the trend still holds that fish hearts are still more dependent on atrial contraction for ventricular filling than mammals.

# 5. PHYSIOLOGICAL DIFFERENCES BETWEEN CHAMBERS

It has been proposed that a key regulator of cardiac performance is the end diastolic volume (EDV) in both mammals and fish. Many of the morphological differences between chambers may have evolved in order to meet the high physiological demands of variation in metabolic rate and increased tolerance to a wide range of environmental conditions. Several of these will be highlighted in this section.

## 5.1. Force Generation

Cardiac output may be modulated through variations in stroke volume and heart rate, but the relative contributions vary substantially among fish species. In general fish respond to greater hemodynamic loads by increasing cardiac output mainly through increased stroke volume rather than heart rate. Stroke volume can vary up to 2-3 fold with exercise (salmonids - [53]) but not all species exhibit the same ability [54-55]. End systolic volume (ESV) of the fish heart relative to stroke volume is very small [56-57] unlike that of the mammalian heart. However, end diastolic volume in fish is greatly determined by atrial contraction rather than central venous pressure seen in mammals [13]. The modulation of atrial contractility and Frank-Starling mechanisms are thus particularly important in determining ventricle filling.

The whole heart displays a greater sensitivity to filling pressure than in mammals, in part attributed to greater myocardial extensibility coupled with an increase in myofilament calcium sensitivity over a large range of sarcomere lengths [58]. The Starling principle dictates that the strength of contraction is, in part, a function of the length of the muscle fiber (and hence sarcomere length) prior to contraction (i.e. preload). Stroke volume increases as a function of end diastolic volume since muscle fibers and sarcomeres within the fibers are stretched towards their optimum. Between various species of fish there is considerable variation in length-dependent sensitivity. For example, flounder heart is very sensitive to filling pressures, needing smaller increases in preload to achieve higher values of stroke volume. This more pronounced Starling curve is further impacted by the high distensibility of its chambers based on pressure-volume curves [28]. This length-dependent strength of contraction is also directly affected by temperature in certain teleost species (e.g. trout- [58]). With fish cardiac myocytes showing enhanced length-dependent activation, and calcium sensitivity increasing with sarcomere length [59], calcium handling is an important consideration in the basis of the length dependence of the strength of cardiac contraction in fish. The unique properties of E-C coupling in the chambers of the fish heart will be discussed in greater detail further in this Chapter. The reduced relative duration of diastole seen with active atrial contraction in fish may have evolved in response to the relatively slow rate of the excitation-contraction coupling processes.

The importance of strong atrial contraction is reflected in the fact that atrial muscle generates greater maximal force ($F_{max}$) than ventricular muscle in a variety of teleost species (e.g. rainbow trout- [60]; yellowfin tuna - [61]; brown trout - [62]). Although systolic blood pressure measurements show the atrium to be a weaker pump, the higher force generated in excised strips suggests the relative weakness of the atrium may simply be reflective of the anatomical difference in wall thickness between the chambers. The greater $F_{max}$ of the atrial cardiomyocytes individually does not account for the difference in overall force generation by the greater quantity of cardiomyocytes in the thicker ventricle wall.

Cardiac performance is directly related to the contractile properties of cardiomyocytes. As mentioned previously, ventricular and atrial muscles express different paralogs (or isoforms) of myosin heavy chain (MHC) [63]. The maximum rate of force produced by muscle fibers is determined to a large degree by MHC paralog content. Alterations in paralog composition therefore are associated with changes in myofibrillar function as well as myosin ATPase activity [64]. Because of this, the myosin ATPase reaction is higher in atrial than in ventricular cardiomyocytes [65]. Across vertebrates, from the primitive dogfish right up to mammals, the atrium has a higher concentration of MHC protein content than the ventricle [4]. The increased levels of MHC may be contributing to the increased $F_{max}$of individual atrial cardiomyocytes relative to ventricular. However, MHC expression pattern differences between chambers may also be species-specific. In goldfish, both chambers have electrophoretically similar MHC while in other fish species such as pike, ventricular myocardium exhibited electrophoretically faster MHC [65]. Linking chamber-specific isoform usage with functional differences across a broader phylogenetic range as well as temperatures may help to clarify the importance of this heterogeneity.

## 5.2. Action Potential

As in other vertebrates, the teleost action potential can be divided into a rapid upstroke of the action potential (phase 0), a rapid initial repolarization of the AP (phase 1) followed by a plateau phase (phase 2). The plateau phase is followed by repolarization of the action potential (phase 3) returning it to the resting membrane potential (phase 4). Teleost atrial and ventricular myocytes also have distinct AP morphologies with the atrial action potential having a shorter plateau phase and a slower repolarization of the AP. Among the ionic currents contributing to the differences in fish atrial and ventricular AP morphologies, ventricular myocytes have prominent inwardly rectifying potassium current $I_{K1}$ and a small $I_{Kr}$ while the opposite is true for atrial myocytes [66-67]. Other currents expected to affect the AP plateau such as the L-type calcium current ($I_{Ca}$) or the $Na^+/Ca^{2+}$ exchanger (NCX) current ($I_{NCX}$) have not been compared systematically in the teleost heart.

Action potentials have been recorded in a number of teleost species (carp - [68]; trout - [34]; zebrafish - [69]) with a considerable variability in AP morphology with the atrial AP duration ranging from 197 ms in carp atrium at 27 °C to 281 ms in trout atrium at 15 °C, 601 ms in carp at 11 °C and 1284 ms at 4 °C and ventricular AP duration ranging from 407 ms at 10 degrees to 143 ms at 20 °C in trout ventricle [71-72] and from 1042 ms at 11 °C to 737 ms at 18 °C in Burbot ventricle [70,72], likely reflecting species-dependent differences as well as variations in physiological-experimental conditions such as temperature or salinity. Although these variations in AP-duration are larger than 10-fold, they allow the teleost heart to adapt to a wide temperature range [18] while the mammalian heart becomes refractory when temperature is lowered. The greater adaptability of the teleost heart to temperature is also reflected in the properties of ionic channels and ion transporters the determine excitability and the action potential shape. For example the change in trout NCX activity is only weakly modified by temperature changes ($Q_{10}$ = 1.2 - 1.5) as compared to mammals ($Q_{10}$ = 2.5) [73-74] and calcium uptake and leak from the sarcoplasmic reticulum is slowed in parallel, resulting in preservation of calcium release from the SR in trout [75-76]

In recent years, action potentials have also been recorded in myocytes isolated from teleost atrial and ventricular myocytes with several reports showing that the resting membrane potential in atrial myocytes is significantly more depolarized than in ventricular myocytes [66, 77]. It was proposed that this could be due to the low $I_{K1}$ density in trout atrial myocytes [66]. However, membrane potentials recorded in atrial tissue preparations [71] are comparable to values recoded in trout ventricular preparations [70, 78] under comparable experimental conditions. Studies have showed that the resting membrane potential recorded with current-clamp technique in trout atrial myocytes was highly sensitive to the holding current applied [70]. Moreover, recordings of potassium conductance in these myocytes suggested that the loss of the resting membrane potential in isolated trout atrial myocytes is likely due to the lack of activation of the acetyl choline-activated potassium current ($I_{KACh}$).

## 5.3. Excitation-Contraction Coupling

Calcium handling is an important consideration as the primary basis for regulation of cardiac contraction in all species, including fish, since it is determined by delivery and removal of calcium to and from cardiac troponin C (cTnC), the thin filament protein which senses changes in cytosolic $Ca^{2+}$concentrations and initiates the contractile process. Therefore, the amplitude and kinetics of the cytosolic calcium transient will modulate the kinetics and force of contraction, and this in turn will set the limit for the contraction frequency. Mammalian cardiac muscle has intracellular calcium stores in the SR in close proximity to T-tubules that provide the majority of the calcium that binds to cTnC. In ectotherms, the activation of calcium release from the SR and its contribution to contractile activation is controversial with some reports showing that the majority of calcium is derived from extracellular space [79, 36, 34, 58] while others report a considerable contribution of the SR both to the activation of contraction [38, 80] and to calcium removal from the cytosol during relaxation [58, 80-81]. Interestingly, calcium handling by the SR has also been reported to differ between fish atrial and ventricular myocytes and these differences as well as species dependent differences are likely to account for part of this controversy on the role of the SR in the teleost heart. Additionally, a significant number of reports examining the role of the SR in the teleost heart relied heavily on pharmacological inhibition of the SR calcium release channel (also termed ryanodine receptor or RyR2) in multicellular tissue preparations. As discussed below, this approach may not be optimal to evaluate the contribution of the SR to the activation of contraction if other calcium delivering mechanisms are capable of compensating for inhibition of the RyR2. Instead, quantification of the contribution of each mechanism involved in the delivery and removal of calcium to and from cTnC is necessary to estimate its capacity and contribution.

## 5.4. Activation of Contraction

In mammalian cardiomyocytes the key contributors to the activation of contraction are the L-type calcium channel that mediates calcium entry across the sarcolemma and the SR, which releases the majority of the calcium from the SR. During the upstroke of the action potential, reverse mode NCX may also contribute [82]. Since the initial reports on $I_{Ca}$ measurements in fish cardiomyocytes [73, 83-84] there is now substantial information on $I_{Ca}$ density in atrial and ventricular myocytes from a number of fish species under different experimental conditions [77, 85]. Again, differences in experimental conditions and species are abundant and there are substantial differences in the measurements of calcium entry though $I_{Ca}$ among species at different temperatures; Burbot : $I_{Ca}$, 0.81±0.13 pA $pF^{-1}$ at 4°C, and 1.35±0.18 pA $pF^{-1}$ at 11°C [55]; trout: 1, 2 and 4 pA $pF^{-1}$ at 7, 14 and 21 °C, respectively [77]. Among the experimental conditions that have may have the largest impact on $I_{Ca}$ measurements, ruptured-patch technique with a high (5 mM) EGTA-buffered patch pipette solution has been shown to result in overestimation of $I_{Ca}$ when compared to measurements performed with low EGTA (25 μM) or perforated patch-clamp technique [75]. Therefore, estimations of the contribution of the calcium carried by $I_{Ca}$ to the activation of contraction also varies from as little as 25-30% in trout atrial myocytes using perforated patch technique [75,86], to more than 50% in zebrafish ventricular myocytes subjected to perforated patch-clamp technique [87] and close to 100% using ruptured patch-clamp technique with EGTA buffered patch pipettes in carp ventricular myocytes [37] and in trout atrial myocytes subjected to β-adrenergic stimulation [87]. The latter does not, however, necessarily imply that $I_{Ca}$ is the sole source of calcium delivery in these preparations but merely indicates that it is sufficiently large to activate contraction if no other calcium sources are available. As mentioned above, chamber specific differences in the $I_{Ca}$ density have not been systematically investigated in the teleost heart.

Theoretically, reverse mode NCX could also contribute to the activation of contraction at depolarized membrane potentials and/or when sodium channels produce a large local increase in the subsarcolemmal Na concentration. Several studies have shown that fish cardiomyocytes have a high NCX activity than mammalian cardiomyocytes, and the direct contribution of the NCX to the activation of contraction has been estimated to range from 25-45% under physiological conditions [86-87, 90]. However, the relative contribution of $I_{NCX}$ to total sarcolemmal calcium entry depends critically on the membrane potential and the intracellular $Na^+$ concentration [74, 85-86]. Moreover, there is evidence that reverse mode NCX can trigger calcium release from the SR in trout atrial myocytes [90].

Finally, calcium release from the SR has also been reported to contribute to the activation of contraction in fish cardiac myocytes with a contribution that varies between 0 and more than 50% depending on species, tissue and experimental temperature. Some of the first publications to examine the effects of ryanodine, a compound that modulates the opening of the SR calcium release channel, on contraction in multicellular tissue preparations were [91-92] but the reported effects on contraction were only marginal at physiological temperatures and stimulation frequencies. Ryanodine was, however, reported to inhibit post rest potentiation, a phenomenon caused by calcium release from the SR [78], and in 1992 it was reported to have a strong effect on contraction in trout [93] and skipjack tuna [94] at 25 °C, suggesting a temperature- and/or species-dependent component of the effect of ryanodine. In 1998, it was shown that steady-state and maximal SR calcium content in trout ventricular myocytes was about 10 times higher than values reported in human cardiomyocytes and that uptake rates were sufficiently fast for the SR to participate in the regulation of calcium handling on a beat-to-beat basis [75] and in 1999 it was shown that brief membrane depolarizations were capable of eliciting calcium release from the SR in trout atrial myocytes corresponding to a contribution of up to 50% of the total calcium required for the activation of contraction [80]. Subsequently it was shown that this contribution of the SR to the activation of contraction was equally important at 21 and 7°C in trout atrial myocytes [95]. These findings were confirmed in trout using simultaneous recordings of membrane currents and calcium transients [89] and with confocal calcium imaging, which also demonstrated that local spontaneous calcium release from the SR (known as calcium sparks in mammalian cardiomyocytes) are also present in trout atrial and ventricular myocytes as well as in zebrafish ventricular myocytes [96]. This study also showed that the spark frequency in trout atrial myocytes was comparable to that recorded in human atrial myocytes under similar experimental conditions [97], but twice as fast as in trout ventricular myocytes. In line with this, the frequency of spontaneous membrane depolarizations, induced by SR calcium release, were also about 6-fold more frequent in trout atrial than ventricular myocytes, suggesting that SR calcium is more labile in atrial than ventricular myocytes. A higher contribution of the SR $Ca^{2+}$ release has also been reported for the tuna heart [38, 67].

Experiments attempting to identify the triggers of SR calcium release showed that both ICa and reverse mode NCX are capable of triggering calcium release from the SR [90] and, unexpectedly, this study showed that the gain of $I_{Ca}$ and $I_{NCX}$-induced calcium release was similar in trout atrial myocytes. Moreover, it was shown that the relative contribution of the two mechanisms depended strongly on the membrane potential and the intracellular $Na^+$-concentration. Considering that an intracellular $[Na^+]$ between 13 and 14 mM has been reported in trout cardiomyocytes [98] reverse mode NCX could may be expected to contribute significantly to the triggering of calcium release from the SR in trout. While in mammals $[Na^+]_i$ is higher in atrial than ventricular cells [99], no difference has been seen between chambers in rainbow trout [98]. However, with loss of function of NCX in mutant zebrafish (tre mutant) more cardiac fibrillation is seen in the atrium than the ventricle [100]. This suggests that the atrium is more susceptible to spontaneous $Ca^{2+}$ release from the SR as a result of disruptions of normal calcium transients.

## 5.5. Relaxation of Contraction

Under steady-state conditions, the calcium delivered to the myofilaments during the activation of contraction has to be removed from the cytosol during relaxation. Moreover, if equilibrium is to be maintained, the calcium entering across the sarcolemma must be extruded from the cell and calcium released from intracellular organelles must be re-sequestered by the same organelle. Based on the estimates of sarcolemmal calcium entry in the fish heart, sarcolemmal calcium extrusion would therefore be expected to vary between 50 and 100% in fish cardiomyocytes depending on the species and tissue. In the mammalian heart, forward mode NCX generally accounts for the vast majority of the sarcolemmal calcium extrusion [101] and experiments using rapid caffeine application in 0Na/0Ca-solution showed that this is also true for trout ventricular myocytes [84]. Assuming that the tail current elicited upon repolarization of the membrane potential is primarily carried by forward mode NCX, the time integral of this current can be used to estimate calcium extrusion by the NCX [75]. With this approach calcium extrusion by the NCX was estimated to account for about 55% of the total increase in calcium during activation of contraction suggesting that intracellular recycling of calcium during contraction must account for about 45% of the total calcium transient [86]. On the other hand, measurements of the NCX activity showed a forward NCX-rate under physiological conditions of 36 amol $pF^{-1}\,s^{-1}$, which is sufficient to remove the total calcium required for the activation of contraction in a few hundred ms [76].

In agreement with the notion that calcium cycling by the SR takes place on a beat-to-beat basis and could contribute to the activation of contraction, measurements of SR calcium load with rapid caffeine application and specific protocols used to measure steady-state and maximal SR calcium loading, showed that the SR calcium content in trout ventricular and atrial myocytes can be up to 10 times higher than in mammalian myocytes [58,61,77,84], and that the rate of SR calcium uptake under physiological conditions is also sufficient to remove the total calcium required for the activation of contraction between beats at physiological beating rates and temperatures in trout atrial and ventricular myocytes [75,95]. In the presence of β-adrenergic stimulation, the relationship between SR calcium uptake and the cytosolic calcium level was shifted towards lower calcium concentrations, but when normalized to the total calcium transport SR calcium uptake was only slightly increased from 35 to 41% by β-adrenergic stimulation [102].

It is interesting to note that the SR calcium load in fish cardiomyocytes is more than 10-fold higher than the amount of calcium released on each beat [84, 95, 103]. The reason for such an excessive SR calcium load compared to the calcium requirements on a beat-to-beat basis remains elusive, but it definitely provides fish species with an intracellular calcium reservoir that allows the heart to function under adverse conditions where SR calcium refilling is impaired for extended periods of time. This also explains why the inhibition of SR calcium uptake (with cyclopazonic acid) or increasing SR calcium leak (with ryanodine) may have a much slower effect on contraction in fish than in mammalian hearts.

Atrial cardiomyocytes have both greater SR $Ca^{2+}$ loads [105-106] and make greater use of SR $Ca^{2+}$than ventricular cardiomyocytes [60, 105]. The degree of atrio-ventricular difference in SR $Ca^{2+}$ load and usage is also species specific, being larger in rainbow trout than crucian carp or burbot [105].

The reliance on intracellular $Ca^{2+}$ leads to more rapid contraction kinetics of the fish atrium due to faster calcium release and reuptake rates via the SR relative to ventricular tissue in a variety of teleost species (e.g. rainbow trout - [60]; burbot - [105]). Lowering of cytosolic $Ca^{2+}$ concentration in this manner is via sarco-endoplasmic reticulum $Ca^{2+}$-ATPase (SERCA), and thus the higher SERCA protein expression found in atrial contributes to this shorter duration of isometric contraction relative to ventricular myocardium [107]. In addition, the strength of contraction in teleost atrial muscle has greater ryanodine sensitivity than that in ventricles [60, 62] and is similar in magnitude to what is typically seen in many mammals [82].

Interestingly, the maximal capacity and steady-state SR $Ca^{2+}$ load does not necessarily correlate with SERCA activity or ryanodine sensitivity in all ectotherms. Inverse relationships between ryanodine sensitivity and SR $Ca^{2+}$ load have been found in rainbow trout, burbot and crucian carp [105]. While atrial contraction is more strongly inhibited by ryanodine than ventricular contraction [60], the increased activity of SR $Ca^{2+}$ ATPase may indicate that atria are able to load the SR with $Ca^{2+}$ in the time between contractions. These factors may contribute to a faster recovery of myocardial contractility or increased mechanical restitution rates in atrial muscle. This may mean increased importance of SR $Ca^{2+}$ for contraction during periods of environmental stress. More studies are necessary to clarify the role of SR $Ca^{2+}$ handling in fish cardiomyocytes, especially with the species-specific and chamber-specific vagaries.

Differences in chamber-specific expression of the contractile element responsible for calcium sensitivity may also guide differences between contractility of the atrium and ventricle. Myofilament $Ca^{2+}$ sensitivity is a critical factor in regulating cardiac contractility. The cardiac TnT subunit of the zebrafish troponin complex has been shown to be required for the contraction of both the atrium and the ventricle [106] as it is in mammals. However, zebrafish possess two copies of the gene for cardiac troponin C, with one demonstrated to be predominantly expressed in the heart of the zebrafish embryo. Loss of function of this gene results in abnormal atrial and ventricular chamber morphology, but the second cTnC gene enables function to be regained in the atrium [107]. This suggests a possible difference in expression of cTnC between chambers that could potentially lead to variation in calcium sensitivity. Gene duplication has led to multiple copies of sarcomeric proteins, and variation between these fish-specific paralogs may be a method of coping with varying environmental conditions.

## 6. ENVIRONMENTAL INFLUENCES

The interspecific variation of fish cardiac anatomy and physiology is guided by evolutionary adaptation to different habitats, modes of life and activity levels. Because ectothermic fish are exposed to different temperatures either seasonally or via the vertical thermocline, temperature-dependent regulation of cardiac contractility is crucial. Many fish face unpredictable temperature changes, meaning that fish hearts must have intrinsic mechanisms or rely on post-translational modifications to protect against acute temperature changes when there is not enough time for transcriptional regulation to take effect [108, 34].

Cardiac performance may change through independent changes in stroke volume and heart rate, with relative contributions of each varying substantially among fish. During exercise, more athletic fish modulate cardiac output preferentially by increasing SV, while less active species preferentially modulate HR in different species [54-55]. With environmental hypoxia, reflex bradycardia occurs and increased stroke volume is necessary to maintain cardiac output [109-110]. Higher temperatures have been shown to increase heart rate while colder temperatures decrease stroke volume in crucian carp [10]. For this review, the impact of temperature on cardiac contractility will be focused on. Mechanisms of control of both stroke volume and heart rate must be considered when looking at temperature sensitivity of the fish heart.

There appears to be seasonal variation with the activity of pacemaker cells (marine plaice - [111]). Heart rate is maintained despite the lower temperatures without the need for extrinsic modification. Changes in heart rate seen with temperature are not solely dependent on pacemaker cells. Therefore, changes that are seen may be due to ion channel temperature sensitivity. Chronic temperature stress can also modify $K^+$ channel conductances and reduce action potential duration to enable higher heart rates (trout - [112]). Thermal acclimation modifies functional properties and subunit composition of $K_{IR2}$channels [113]. Cold temperatures induce decreased $I_{K1}$, possibly increasing the excitability of cardiomyocytes but increased $I_{Kr}$ leading to limited AP duration and decreased refractoriness of the heart [114]. These changes in $I_K$ appear when it is the predominant current, namely in ventricular cardiomyocytes [66] while the changes in $I_{Kr}$ appear in both atrial and ventricular cardiomyocytes [114]. The density in particular of ventricular $I_{K1}$ increases in warm-acclimated vs. cold acclimated fish (trout - [113]).

The plasticity of E-C coupling in fish has been suggested to be based, at least in part, on these temperature-induced changes in $I_K$ [114]. These modifications lead to changes in $Ca^{2+}$ influx and hence variation in E-C coupling. $Ca^{2+}$ pumps and ion channels themselves are also known to be sensitive to acute temperature change [103]. The large SR calcium stores and low calcium sensitivity of ryanodine receptors have been proposed to play a critical role in maintaining contraction at lower temperatures than mammals [81]. Higher SR $Ca^{2+}$ content may be necessary to initiate and maintain SR $Ca^{2+}$ release events at lower temperatures [80,115].

Different species may have variations in response to temperature depending on their evolutionary history. Eurythermal fish such as trout must be able to cope with seasonal temperature fluctuations as well as more acute temperatures through the thermal gradient. Other stenothermal species are only able to cope with a more limited temperature range, though this range may be considered on the extreme end of the range for a eurythermal fish (e.g. Antarctic fish). Interspecies variation in excitation-contraction coupling seen with temperature acclimation may be indicative of the capacity for thermal niche expansion. This substantial variation in SR $Ca^{2+}$ release relative to SL $Ca^{2+}$ influx ratio exists between species as well as being dependent on acute and chronic temperature changes. Cold-tolerant active fish have a larger capacity to store $Ca^{2+}$ in the SR than mammals (e.g. active teleosts - [115]; scombroids - [116]). The atrio-ventricular differences in $Ca^{2+}$ loading are also more pronounced in cold-acclimated than warm acclimated fish (trout - [112].

However, the contribution of the SR to cytosolic $Ca^{2+}$ management varies with temperature in fish hearts. With cold acclimation, SR function is also enhanced more strongly in atrial than ventricular myocytes (trout - [60]). More cold tolerant species may also possess greater SR $Ca^{2+}$ content to begin with (bluefin tuna and mackerel - [112]). The degree of modification of $I_{Ca}$ to offset the negative effects of colder temperatures appears to be species specific, with many teleosts such as trout demonstrating relatively temperature-insensitive $Ca^{2+}$ flux [103] whereas certain scombroids show significant reductions in $I_{Ca}$ with decreasing temperatures [112]. This temperature sensitivity may differ between atrial and ventricular cardiomyocytes, with ventricular maximum SR $Ca^{2+}$ load varying only in the ventricle of most scombrids [112].

As previously mentioned, isoform compositional changes can lead can lead to variation in contractility. Modulation of MHC isoform expression patterns with temperature also leads to changes in myofibrillar ATPase activity (crucian carp - [10]). This may also be a species-related difference, since not all species display the same changes. The degree to which each chamber varies is still unclear. Differing paralog usage between the chambers may be linked to temperature-induced $Ca^{2+}$ sensitivity differences. Examples of this may exist in important of the contractile element. Troponin I, the inhibitory subunit of troponin, has multiple paralogs expressed in cardiac tissue. The relative expression of these paralogs in the ventricle shifts with temperature, possibly leading to changes in the calcium affinity of the entire troponin complex [117]. However, while paralog profiles vary between tissues (heart, slow skeletal) no information is known about the atrium. TnC isoform differences between chambers [107] may also lend insight into $Ca^{2+}$ sensitivity changes with temperature as the affinity of cTnC in trout is temperature dependent [118]. As previously mentioned, the very presence of multiple fish-specific paralogs of important components of the contractile element may be indicative of how fish have evolved mechanisms to thrive in variable environments. Incorporating the chamber-specific variation with temperature response reveals the complexity of these mechanisms in achieving survival.

# 7. SUMMARY

The unique structure of the fish heart accommodates the variability in function that have allowed species to exploit many different environments. The increased importance of the atrial contraction in guiding changes in cardiac output indicates that some of this physiological versatility may be due to atrio-ventricular differences.These two main chambers of the fish heart differ in terms of contractile properties from the molecular level of E-C coupling and AP morphology, and the whole heart morphology and function reflect this. The fish heart is representative of early embryological stages of higher vertebrates, these studies on both the development and functioning of chamber-specific properties lends insight into proper functioning of the heart not only in fish. However, the degree of atrio-ventricular differences vary between species making it very difficult to generalize species-specific results across phylogeny. The species-specific differences in these two chambers lend insight into how evolutionary history guides responses to environmental factors such as ambient temperature. Fish-specific genome duplications may have allowed for multiple chamber-specific genes with variable functions that allow for flexibility in function. Further work is required to determine how genome duplication has shaped the structure and function of the fish heart and to clarify how this wide-range of species-specific phenotypes has been maintained.

# REFERENCES


1. Moorman, A.F. and Christoffels, V.M. (2003) Cardiac chamber formation: development, genes and evolution. Physiological Reviews 83: 1223-1267
2. Satin, J., Fujii, S. and DeHaan, R. (1988) Development of cardiac beat rate in early chick embryos is regulated by regional cues. Developmental Biology 129: 103-113.
3. Tabibiazar, R., Wagner, R., Liao, A. and Quertermous, T. (2003) Transcriptional profiling of the heart reveals chamber-specific gene expression patterns. Circulation Research 93: 1193-1201.
4. Franco, D., Gallego, A., Habets, P., Sans-Coma, V. and Moorman, A. (2002) Species-specific differences of myosin content in the developing cardiac chambers of fish, birds and mammals. Anatomical Record 268: 27-37.
5. Evans, S., Yelon, D., Conlon, F. and Kirby, M. (2010) Myocardial lineage development. Circulation Research 107: 1428-1444.
6. Stanier,D. and Fishman, M. (1992) Patterning the zebrafish heart tube: acquisition of anteroposterior polarity. Developmental Biology 153: 91-101.
7. de Pater, E., Clijsters, L., Marques, S., Lin, Y., Garavito-Aquilar, Z., Yelon, D. and Bakkers, J. (2009) Distinct phases of cardiomyocyte differentiation regulate growth of the zebrafish heart. Development 136, 1633-1641.
8. Warren, K., Wu, J., Pinet, F. and Fishman, M. (2000) The genetic basis of cardiac function: dissection by zebrafish (Danio rerio) screens. Phil. Trans. Royal Society of London B 355: 939-944.
9. Yelon, D., Horne, S. and Stainier, D. (1999) Restricted expression of cardiac myosin genes reveals regulated aspects of heart tube assembly in zebrafish. Developmental Biology 214: 23-37.
10. Vornanen, M. (1994) Seasonal adaptation of crucian carp (Carassius carassius L.) heart: glycogen stores and lactate dehydrogenase activity. Canadian Journal of Zoology, 1994, 72(3): 433-442.
11. Berdougo, E., Coleman, H., Lee, D.H., Stanier, D.Y. and Yelon, D. (2003) Mutation of weak atrium/atrial myosin heavy chain disrupts atrial function and influences ventricular morphogenesis in zebrafish. Development 130: 6121-6129.
12. Marques, S. and Yelon, D. (2009) Differential requirement for BMP signaling in atrial and ventricular lineages establishes cardiac chamber proportionality. Developmental Biology 328: 432-482.
13. Cotter, P.A. Han, A.J., Everson, J.J. and Rodnick, K.J. (2008) Cardiac hemodynamics of the rainbow trout (Oncorhynchus mykiss) using simultaneous Doppler echocardiography and electrocardiography. Journal of Experimental Zoology Part A: Ecological Genetics and Physiology 309: 243-254.
14. Schib, J.L., Icardo, J.M., Durán, A.C., Guerrero, A., López, D., Colvee, E., de Andrés, A.V. and Sans-Coma, V. (2002) The conus arterious of the adult gilthead seabream (Sparus auratus). Journal of Anatomy 201:395–404.
15. Icardo, J.M. (2006) Conus arteriosus of the teleost heart: dismissed, but not missed. Anatomical Record A 288: 900-908.
16. Santer, R. and Cobb, J. (1972) The fine structure of the heart of the teleost, Pleuronectes platessa L. Zeitschrift für Zellforschung und mikroskopische Anatomie 131: 1-14.
17. Yamauchi, A. (1980) Fine structure of the fish heart. In: Bourne G (ed) Heart and Heart-like Organs, vol 1. Academic, New York, pp 119–148.

18. Farrell A.P. and Jones, D.R. (1992) The heart. In: Hoar WS, Randall DJ, Farrell AP (eds) Fish physiology, Vol XII, The cardiovascular system, Part A. Academic, San Diego, pp 1–87.
19. Johansen, K. (1965) Cardiovascular dynamics in fishes, amphibians, and reptiles. Annals of the New York Academy of Sciences 127 Comparative Cardiology: 414-442.
20. Forster, M. and Farrell, A. (1994) The volumes of the chambers of the trout heart. Comparative Biochemisty and Physiology A.109: 127-132.
21. Sudak, F.N. (1965) Intrapericardial and intercardiac pressures and the events of the cardiac cycle in Mustelus canis (Mitchill). Comparative Biochemical Physiology 14, 689-705.
22. Hu, N., Yost, H.J. and Clark, E. (2001) Cardiac morphology and blood pressure in the adult zebrafish. Anatomical Record 264: 1-12.
23. Icardo, J.M. and Colvee, E. (2011) The atrioventricular region of the teleost heart. A distinct heart segment. Anatomical Record 294: 236-242.
24. Icardo, J.M., Colvee, E., Cerra, M. and Tota, B. (2002) The structure of the conus arteriosus of the sturgeon (Acipenser naccarii) heart II. The myocardium, the subepicardium, and the conus-aorta transition. Anatomical Record 268: 388-398.
25. Icardo, J.M. (2012) The teleost heart: A morphological approach. In Ontogeny and Phylogeny of the Vertebrate Heart ed. Wang, T. and Sedmera D. Springer-Verlag New York Inc.
26. Santer, R.M., Walker, M.G., Emerson, L. and Witthames, P.R. (1983) On the morphology of the heart ventricle in marine teleost fish (Teleostei). Comparative Physiology and Biochemistry 76A:453–457.
27. Icardo, J., Imbrogno, S., Gattuso, A., Colvee, E. and Tota, B. (2005) The heart of Sparus auratus: a reappraisal of cardiac functional morphology in teleosts. Journal of Experimental Zoology 303A: 665-675.
28. Mendonca, P., Genge, A., Deitch, E. and Gamperl, A. (2007) Mechanisms responsible for the enhance pumping capacity of the in situ winter flounder heart (Pseudopleuronectes americanus). American Journal of Physiology Regulatory and Integrative Comparative Physiology 293:2112-2119.
29. Tota, B. (1989) Myoarchitecture and vascularization of the elasmobranch heart ventricle. Journal of Experimental Zoology:122–135.
30. Rohr, S., Otten, C. and Abdelilah-Seyfried, S. (2008) Asymmetric involution of the myocardial field drives heart tube formation in zebrafish. Circulation Research 102: 12-19.
31. Vornanen, M. (1998) L-type Ca2+ current in fish cardiac myocytes: effects of thermal acclimation and β-adrenergic stimulation. Journal of Experimental Biology 201, 533–547.
32. Huang, J., Hove-Madsen, L. and Tibbits, G. (2004) Na+/Ca2+ exchange activity in neonatal rabbit ventricular myocytes. American Journal of Physiology: Cell Physiology 288: C195-C203.
33. Santer, R.M. (1974) The organization of the sarcoplasmic reticulum in teleost ventricular myocardial cells. Cell and Tissue Research 151: 395-402.
34. Vornanen, M., Shiels, H. and Farrell, A. (2002) Plasticity of excitation-contraction coupling in fish cardiac myocytes. Comparative Biochemical Physiology 132A, 827-846.
35. Saetersdal, T.S., Justesen, N.P. and Krohnstad, A.W. (2003) Ultrastructure and innervation of the teleostean atrium. Journal of Molecular and Cellular Cardiology 6: 415-437.
36. Tibbits G.F., Moyes C.D. and Hove-Madsen L. (1992) Excitation–contraction coupling in the teleost heart. In: Hoar W.S., Randall D.J., Farrell A.P. (eds) Fish physiology. Academic Press, New York, pp 267–296.
37. Vornanen, M. (1997). Sarcolemmal Ca2+ influx through L-type Ca channels in ventricular myocytes of a teleost fish. American Journal of Physiology 272, R1432–R1440.

38. Di Maio, A. and Block, B.A. (2008) Ultrastructure of the sarcoplasmic reticulum in cardiac myocytes from Pacific bluefin tuna. Cell and Tissue Research 334:121-34.
39. Brill, R., and Bushnell, P. 1991 Effects of open and closed system temperature changes on blood oxygen dissociation curves of yellowfin tuna (Thunnus albacares) and skipjack tuna (Katsuwonus pelamis). Canadian Journal of Zoology 69:1814-1821.
40. Santer, R.M. (1985) Morphology and innervation of the fish heart. Advances in Anatomy, Embryology and Cell Biology 89, 1–102.
41. Simoes, K., Vicentini, C.A., Orsi, A.M. and Cruz, C. (2002) Myoarchitecture and vasculature of the heart ventricle in some freshwater teleosts. Journal of Anatomy 200:467–475.
42. Claireaux, G., McKenzie, D., Genge, A., Chatelier, A., Aubin, J. and Farrell, A. (2005) Linking swimming performance, cardiac pumping ability and cardiac anatomy in rainbow trout. Journal of Experimental Biology 208: 1775-1784.
43. Gamperl, A. and Farrell, A. (2004) Cardiac plasticity in fishes: environmental influences and intraspecific differences. Journal of Experimental Biology 207: 2539-2550.
44. Cerra, M.C., Imbrogno, S., Amelio, D., Garofalo, F., Colvee, E., Tota, B., and Icardo, J.M. (2004) Cardiac morphodynamic remodelling in the growing eel (Anguilla Anguilla L.). Journal of Experimental Biology 207:2867–2875.
45. Graham, M.S. and Farrell, A.P. (1990) Myocardial oxygen consumption in trout acclimated to 5°C and 15°C. Physiological Zoology 63:536-554.
46. Davie, P. and Farrell, A. (1991) The coronary and luminal circulations of the myocardium of fishes. Canadian Journal of Zoology 69: 1993-2001.
47. Forster, M., Axelsson, M., Farrell, A. and Nilsson, S. (1991) Cardiac function and circulation in hagfishes. Canadian Journal of Zoology 69: 1985-1992.
48. Minerick, A., Chang, H., Hoagland, T. and Olsen, K. (2003) Dynamic synchronization analysis of venous pressure-driven cardiac output in rainbow trout. American Journal of Physiology Regulatory and Integrative Comparative Physiology 285: 889-896.
49. Lai, N., Shebatai, R., Graham, J., Hoit, B., Sunndrhagen, K. and Bhargava, V. (1990) Cardiac function of the leopard shark, Triakis, semifasciatus. Journal of Comparative Physiology: Biochemical and System Environmental Physiology 160: 259-268.
50. Farrell, A.P. (1984). A review of cardiac performance in the teleost heart: intrinsic and humoral regulation. Canadian Journal of Zoology 62, 523-536.
51. Sandblom, E., Cox, G., Perry, S. and Farrell, A. (2009) The role of venous capacitance, circulating catecholamines and heart rate in the hemodynamic response to increased temperature and hypoxia in the dogfish. American Journal of Physiology: Regulatory and Integral Comparative Physiology 296: 1547-1556.
52. Sandblom, E. and Axelsson, M. (2006) Adrenergic control of venous capacitance during moderate hypoxia in the rainbow trout (Oncorhynchus mykiss): role of neural and circulating catecholamines. American Journal of Physiology: Regulatory and Integral Comparative Physiology 291: 711-718.
53. Kiceniuk, J.W. and Jones, D.R. (1977) The oxygen transport system in trout (Salmo gairdneri) during exercise. Journal of Experimental Biology 69, 247-260.
54. Cooke, S., Grant, E., Schreer, J. Philipp, D. and DeVries, A. (2003) Low temperature cardiac response to exhaustive exercise in fish with different levels of winter quiescence. Comp. Biochem. Physiol. Mol. Integr. Physiol. 134: 157-165.
55. Clark, T. and Seymour, R. (2006) Cardiorespiratory physiology and swimming energetics of a high-energy demand teleost, the yellowtail kingfish (Seriola lalandi). Journal of Experimental Biology 209: 3940-2951.
56. Franklin, C. E. and Davie, P. S. (1992). Dimensional analysis of the ventricle of an in situ perfused trout heart using echocardiography. Journal of Experimental Biology 166, 47-60.

57. Coucelo, J., Joaquim, N. and Couvelo J. (2000) Calculation of volumes and systolic indices of heart ventricle from Halobatrachus didactylus: echocardiographic noninvasive method. Journal of Exp. Zool. 286: 585-595.
58. Shiels, H.A. and White, E. (2005) Temporal and spatial properties of cellular Ca2+ flux in trout ventricular myocytes. American Journal of Physiology: Regulatory and Integral Comparative Physiology 288: R1756-R1766.
59. Patrick, S., White, E. and Shiels, H. (2010) Rainbow trout myocardium does not exhibit a slow inotropic response to stretch. Journal of Experimental Biology 214: 1118-1122.
60. Aho, E. and Vornanen, M. (1999) Contractile properties of atrial and ventricular myocardium of the heat of the rainbow trout Oncorhynchus mykiss: effects of thermal acclimation. Journal of Experimental Biology 202: 2663-267.
61. Shiels H.A., Freund E.V., Farrell A.P. and Block B.A. (1999) The sarcoplasmic reticulum plays a major role in isometric contraction in atrial muscle of yellowfin tuna. Journal of Experimental Biology 202:881–890.
62. Mercier, C., Axelsson, M., Imbert, N., Claireaux, G., Lefrancois, C., Altimiras, J. and Farrell A.P. (2002) In vitro cardiac performance in triploid brown trout at two acclimation temperatures. Journal of Fish Biology 60: 117-133.
63. Karasinski, J. (1993) Diversity of native myosin and myosin heavy chain in fish skeletal muscles. Comparative Biochemisty and Physiology Part B: Comparative Biochemistry 106: 1041-1047.
64. Bottinelli, R., Canepari, M., Pellegrino, M. and Reggiani, C. (1995) Force-velocity properties of human skeletal muscle fibres: myosin heavy chain isoform and temperature dependence. Journal of Physiology 495: 573-586.
65. Karasinski, J., Sokalski, A. and Kilarski, W. (2001) Correlation of myofibrillar ATPas activity and myosin heavy chain content in ventricular and atrial myocardium of fish heart. Folia Histochem Cytobiol 39: 23-28.
66. Vornanen, M., Ryokkynen, A. and Nurmi, A. (2002) Temperature-dependent expression of sarcolemmal K+ currents in rainbow trout atrial and ventricular myocytes. American Journal of Physiology Regulatory and Integrative Comparative Physiology 282: R1191-R1199.
67. Galli, G., Lipnick, M. and Block, B. (2009) Effect of thermal acclimation on action potentials and sarcolemmal K+ channels from Pacific bluefin tuna cardiomyocytes. American Journal of Physiology: Regulatory and Integrative Comparative Physiology 297:502-509.
68. Vornanen, M. and Tuomennoro, J. (1999) Effects of acute anoxia on heart function in crucian carp: effects of cholinergic and purinergic control. American Journal of Physiology 277: R465-R475.
69. Nemtsas, P., Wettwer, E., Christ, T., Weidinger, G. and Ravens, U. (2009) Adult zebrafish heart as a model for human heart? An electrophysiological study. Journal of Molecular and Cellular Cardiology 48:161-71.
70. Moller-Nielsen, T. and Gesser, H. (1992) Sarcoplasmic reticulum and excitation contraction coupling at 20 and 10 degrees in rainbow trout myocardium. Journal of Comparative Physiology B 162: 526-534.
71. Molina, C., Gesser, H., Llach, A. and Hove-Madsen, L. (2007) Modulation of membrane potential by an acetylcholine-activated potassium current in trout atrial myocytes. American Journal of Physiology: Regulatory and Integrative Comparative Physiology 292: R388-R395.
72. Haverinen, J. and Vornanen, M. (2009) Responses of action potential and K+ currents to temperature acclimation in fish hearts: phylogeny or thermal preferences? Physiological Biochemical Zoology 82: 468-482.
73. Tibbits G. F., Philipson K. D. and Kashihara H. (1992) Characterization of myocardial Na+-Ca2+ exchange in rainbow trout. American Journal of Physiology: Cell Physiology 262: C411–C417.
74. Xue, X., Hryshko, L., Nicoll, D., Philipson, K.D. and Tibbits, G.F. (1999) Cloning, expression and characterization of the trout cardiac Na+/Ca2+ exchanger

75. Hove-Madsen, L. and Tort, L. (1998). L-type Ca2+ current and excitation-contraction coupling in single atrial myocytes from rainbow trout. American Journal of Physiology: Regulatory and Integrative Comparative Physiology 275, R2061-R2069.
76. Hove-Madsen, L. and Tort, L. (2001). Characterization of the relationship between Na+-Ca2+ exchange rate and cytosolic calcium in trout cardiac myocytes. Pflugers Arch. 441, 701-708.
77. Shiels, H., Vornanen, M. and Farrell, T. (2002) Effects of temperature on intracellular [Ca2+] in trout atrial myocytes. Journal of Experimental Biology 205: 3641-3650.
78. Hove-Madsen, L. and Gesser, H. (1989) Force frequency relation in the myocardium of rainbow trout. Effects of K+ and adrenaline. Journal of Comparative Physiology B 159: 61-69.
79. Fabiato, A. (1983) Calcium-induced release of calcium from the cardiac sarcoplasmic reticulum. American Journal of Physiology: Cell Physiology 245, C1-C14.
80. Hove-Madsen, L., Llach, A. and Tort, L. (1999) Quantification of calcium release from the sarcoplasmic reticulum in rainbow trout atrial myocytes. Pflügers Arch 438: 545–552.
81. Shiels, H. A., Thompson, S. and Block, B. A. (2011) Warm fish with cold hearts: thermal plasticity and excitation-contraction coupling in bluefin tuna. Proceedings of the Royal Society of London B: Biological Sciences 278, 18-27.
82. Bers, D. (2002) Cardiac excitation-contraction coupling. Nature 415, 198-205.
83. Vornanen, M. (1997). Sarcolemmal Ca2+ influx through L-type Ca channels in ventricular myocytes of a teleost fish. American Journal of Physiology: Regulatory and Integrative Comparative Physiology 272, R1432–R1440.
84. Hove-Madsen, L., Llach, A. and Tort, L. (1998) Quantification of Ca2+ uptake in the sarcoplasmic reticulum of trout ventricular myocytes. American Journal of Physiology: Regulatory and Integrative Comparative Physiology 275: R2070-R2080.
85. Haverinen, J. and Vornanen, M. (2005) Significance of Na+ current in the excitability of atrial and ventricular myocardium of the fish heart. Journal of Experimental Biology 209: 549-557.
86. Hove-Madsen, L., Llach, A. and Tort, L. (2000) Na+/Ca2+ exchange activity regulates contraction and SR Ca2+ content in rainbow trout atrial myocytes. American Journal of Physiology: Regulatory and Integrative Comparative Physiology 279:R1856-R1864.
87. Zhang, P.C., Llach, A., Sheng, X.Y., Hove-Madsen, L. and Tibbits, G.F. (2011) Calcium handling in zebrafish ventricular myocytes. American Journal of Physiology: Regulatory and Integrative Comparative Physiology 300: R56-R66.
88. Shiels, H.A., Vornanen, M. and Farrell, A.P. (2003) Acute temperature change modulates the response of ICa to adrenergic stimulation in fish cardiomyocytes. Physiological and Biochemical Zoology 76(6):816–824.
89. Vornanen M (1999) Na+/Ca2+ exchange current in ventricular myocytes of fish heart: contribution to sarcolemmal Ca2+ influx. Journal of Experimental Biology 202:1763–1775.
90. Hove-Madsen, L., Llach, A., Tibbits, G.F. and Tort, L. (2003) Triggering of sarcoplasmic reticulum Ca2+ release and contraction by reverse mode Na+/Ca2+ exchange in trout atrial myocytes. American Journal of Physiology: Regulatory and Integrative Comparative Physiology 284: R1330–R1339.
91. Driedzic, W. and Gesser, H. (1988) Differences in force-frequency relationships and calcium dependency between elasmobranch and teleost hearts. Journal of Experimental Biology 140, 227-241.
92. el-Sayed, M.F and Gesser, H. (1989) Sarcoplasmic reticulum, potassium, and cardiac force in rainbow trout and plaice. American Journal of Physiology: Regulatory and Integrative Comparative Physiology 257, R599–R604, 1989.
93. Hove-Madsen, L. (1992) The influence of temperature on ryanodine sensitivity and the force-frequency relationship in the myocardium of rainbow trout. Journal of Experimental Biology 167: 47-60.

94. Keen, J.E., Farrell, A.P. Tibbits, G.F. and Brill, R.W. (1992) Cardiac physiology in tunas II: Effect of ryanodine, calcium, and adrenaline on force–frequency relationships in atrial strips from skipjack tuna, Katsuwonus pelamis. Canadian Journal of Zoology 70(6): 1211-1217.
95. Hove-Madsen, L., Llach, A. and Tort, L. (2001) The function of the sarcoplasmic reticulum is not inhibited by low temperatures in trout atrial myocytes. American Journal of Physiology: Regulatory and Integrative Comparative Physiology 281: R1902-R1906.
96. Llach, A., Molina, C., Alvarez-Lacalle, E., Tort, L., Benitez, R. and Hove-Madsen, L. (2011) Detection, properties, and frequency of local calcium release from the sarcoplasmic reticulum in teleost cardiomyocytes. PLoS ONE 6(8): e23708. doi:10.1371/journal.pone.0023708.
97. Christ T., Boknik P., Wohrl S. (2004) L-type Ca2+ current down-regulation in chronic human atrial fibrillation is associated with increased activity of protein phosphatases. Circulation. 110:2651–2657.
98. Birkedal, R. and Shiels, H. (2007) High [Na+]i in cardiomyocytes from rainbow trout. American Journal of Physiology Regulatory and Integrative Comparative Physiology 293: 861-866.
99. Wang, G., Schmied, R., Ebner, F. and Korth, M. (1993) Intracellular sodium activity and its regulation in guinea-pig atrial myocardium. Journal of Physiology 465: 73-84.
100. Langenbacher, A., Dong, Y., Shu, X., Choi, J., Nicoll, D., Goldhaber, J., Philipson, K. and Chen J. (2005) Mutation in sodium-calcium exchanger 1 (NCX1) causes cardiac fibrillation in zebrafish. Proceedings of the National Academy of Sciences of the United States of America 102: 17699-17704.
101. Bassani J., Bassani R. and Bers D.M. (1994) Relaxation in rabbit and rat cardiac cells: species-dependent differences in cellular mechanisms. Journal of Physiology 476: 279-293.
102. Llach, A., Huang, J., Sederat, F., Tort, L., Tibbits, G.F. and Hove-Madsen, L. (2004) Effect of beta-adrenergic stimulation on the relationship between membrane potential intracellular [Ca2+] and sarcoplasmic reticulum Ca2+ uptake in rainbow trout atrial myocytes. Journal of Experimental Biology 207: 1369-1377.
103. Shiels H.A., Vornanen M. and Farrell A.P. (2000) Temperature-dependence of L-type Ca2+ channel current in atrial myocytes from rainbow trout. Journal of Experimental Biology 203:2771–2780.
104. Rantin, F., Guerra, C., Kalinin, A. and Glass, M. (1998) The influence of aquatic surface respiration (ASR) on the cardio-respiratory function of the serrasalmid fish Piaractus mesopotamicus. Comparative Biochemistry and Physiology 119A: 991-997.
105. Haverinen, J. and Vornanen, M. (2009) Comparison of sarcoplasmic reticulum calcium content in atrial and ventricular myocytes of three fish species. American Journal of Physiology: Regulatory and Integral Comparative Physiology 297, R1180-R1187.
106. Sehnert, A.J., Huq, A., Weinstein, B.M., Walker, C., Fishman, M. and Stanier, D.Y. (2002) Cardiac troponin T is essential in sarcomere assembly and cardiac contractility. Nature Genetics 31: 106-110.
107. Sogah, VM, Serluca, FC, Fisherman, MC, Yelon, DL, Macrae, CA and Mably, JD (2010) Distinct troponin C isoform requirements in cardiac and skeletal muscle. Developmental Dynamics 239 (11), 3115-3123.
108. Tiitu, V. and Vornanen, M. (2003) Ryanodine and dihydropyridine receptor binding in ventricular cardiac muscle of fish with different temperature preferences. Journal of Comparative Physiology B 173, 285-291.
109. Gamperl, A.K. and Driedzic, W.R. (2009) The cardiovascular systems. In Fish physiology vol 27 edited by J.G. Richards, A.P. Farrell and C.J. Brauner. Academic Press, San Diego, Calif. pp.487-503.

110. Perry, S. and Desforges, P. (2006) Does bradycardia or hypertension enhance gas transfer in rainbow trout (Oncorhynchus mykiss)? Comparative Biochemistry and Physiology 144A: 163-172.
111. Harper, A., Newton I. and Watt, P. (1995) The effect of temperature on spontaneous action potential discharge of the isolate sinus venosus from winter and summer plaice (Pleuronectes platessa). Journal of Experimental Biology 198: 137-140.
112. Haverinen, J. and Vornanen, M. (2009) Responses of action potential and K+ currents to temperature acclimation in fish hearts: phylogeny or thermal preferences? Physiological Biochemical Zoology 82: 468-482.
113. Hassinen, M., Paajanen and Vornanen, M. (2008) A novel inwardly rectifying K+ channel Kir2.5, is upregulated under chronic cold stress in fish cardiac myocytes. Journal of Experimental Biology 211: 2162-2171.
114. Aho, E. and Vornanen, M. (2002) Cold acclimation increases basal heart rate but decreases its thermal tolerance in rainbow trout (Oncorhynchus mykiss). Journal of Comparative Physiology B 171, 173–179.
115. Hassinen, M., Laulaja, S., Paajanen, V., Haverinen, J. and Vornanen, M. (2011) Thermal adaptation of the crucian carp (Carassius carassius) cardiac delayed rectifier current, IKs, by homomeric assembly of K v7.1 subunits without MinK. American Journal of Physiology Regulatory and Integrative Comparative Physiology 301: R255–R265.
116. Galli, G., Lipnick, M., Shiels, H. and Block, B. (2011) Temperature effects on Ca2+ cycling in scombrid cardiomyocytes: a phylogenetic comparison. Journal of Experimental Biology 214, 1068-1076.
117. Alderman, S., Klaiman, J., Deck, C. and Gillis, T. (2012) Effect of cold acclimation on troponin I isoform expression in striated muscle of rainbow trout. American Journal of Physiology Regulatory and Integrative Comparative Physiology 303: R168-R176.
118. Gillis, T.E., Marshall, C.R., Xue, X.H., Borgford, T.J. and Tibbits, G.F. (2000) Ca2+ binding to cardiac troponin C: effects of temperature and pH on mammalian and salmonid isoforms. American Journal of Physiology Regulatory and Integrative Comparative Physiology 279:R1707–R1715.

# CHAPTER 12

# Spiny Lobster Development: Where Does Successful Metamorphosis To The Puerulus Occur?: A Review

**Bruce Frank Phillips Æ Paulette S. McWilliam**

*Department of Environmental and Aquatic Sciences, Curtin University of Technology, G.P.O. Box U1987, Perth, WA 6085, Australia*

## ABSTRACT

This review re-addresses the question: Where does metamorphosis to the puerulus mainly take place among the shallow-water palinurids? A decade ago we reviewed this ecological question in a paper that focused on phyllosomal development of the western rock lobster, *Panulirus cygnus*. The main region of occurrence of its metamorphosis was found to be in the slope region beyond the shelf break. Because the puerulus of *P. cygnus* is a non-feeding stage, it was hypothesised that metamorphosis will not occur until the final phyllosoma has reached some critical, and specific, level of stored energy reserves. For late larval development and successful metamorphosis of *P. cygnus*, the richest food resources seem to be located in the slope waters adjoining the shelf break off Western Australia. This, like most shelf break areas, is a region of higher zooplankton and micronekton biomass than is usually found further offshore, and is dominated (in winter-spring months) by the warm south-flowing Leeuwin Current. In this new review, distribution and abundance data of final phyllosomas and pueruli are examined from, *Panulirus argus*, *Panulirus cygnus*, *Panulirus japonicus*, *Panulirus ornatus* and *Jasus edwardsii*, and where possible, related to features of the satellite imagery of the areas in which they occur. We hypothesise that metamorphosis will occur where the final stages have partaken of sufficient, appropriate nutrition to provide them with a reserve of bioenergetic resources, and this can occur where oceanographic fronts effect greater planktonic productivity and concentrations of food organisms. This may be near the shelf-break, or out to large distances offshore, because of large-scale oceanographic events such as the prevailing current system, its off-shoots, mesoscale eddy fronts, counter-currents, etc. However, we contend that, in terms of population recruitment, metamorphosis in most shallow-water palinurid species occurs mainly in the slope waters adjoining the shelf break of the region to which the species is endemic. Although some final phyllosomas may metamorphose much further offshore, it is unlikely that these pueruli will reach the shore, let alone settle and successfully moult to the juvenile stage. All of the data indicate that successful metamorphosis from the final-stage phyllosoma to the puerulus stage in all species occurs offshore but close to the continental shelf.

**Keywords:** Location of metamorphosis Nutrition Ocean fronts Phyllosoma Puerulus Spiny lobsters

## INTRODUCTION

The final-stage phyllosomal larva of spiny (or rock) lobsters (Decapoda, Palinuridae) metamorphoses into the puerulus (the postlarva). An outline of phyllosomal development and metamorphosis in these shallow-water palinurids is given in the Introduction in McWilliam and Phillips (2007), which is included in this two-part review.

As reported by Booth and Phillips (1994) there is evidence that most metamorphosis takes place mainly near the continental shelf break because that is where final stages are usually most abundant. The puerulus is a transitional, nektonic stage (well-adapted for strong, horizontal, directional swimming) (see also Jeffs and Holland 2000), which bridges the planktonic and benthic phases of the life cycle. It is a short-lived (ca. 3–4 weeks) non-feeding, stage (ca. 30–35 mm in total length) which swims across the continental shelf towards the shore. Because the puerulus is a non-feeding, postlarva, it does not aggregate at the surface at night. It is sparsely distributed in the plankton and has only been caught in small numbers by plankton sampling (Phillips et al. 1978). How it navigates shoreward has not yet been determined (see Jeffs et al. 2005). When the puerulus reaches shallow water near the shore, it settles. Its exoskeleton becomes pigmented and calcified just prior to its moult to the first benthic juvenile stage which resumes feeding.

It is a decade since we first reviewed the question of what induces metamorphosis to the puerulus, and where it mainly occurs (McWilliam and Phillips 1997). Our 1997 review focused on phyllosomal development of the western rock (spiny) lobster, *P. cygnus*, endemic to Western Australia, and extrapolation from results of laboratory rearing of other larval decapods as well as other species of spiny lobsters.

The stimulus for a re-examination of the question was an invitation by the editors of *The Lobster Newsletter* ( http://www.odu.edu/ ~mbutler/ newsletter /index.html) for readers to make comments on the following topics:

Where does metamorphosis to the puerulus stage mainly take place among the shallow-water palinurids, and what factors – internal and/or external – are implicated in its progression?

Five articles responding to this invitation were published in the December 2005 issue of *The Lobster Newsletter* (Jeffs 2005; Yeung 2005; Dennis 2005; Booth and Chiswell 2005; Yoshimura 2005).

The second (physiological) question of the topic in *The Lobster Newsletter* (2005) has been examined as a review of the molecular mechanisms inducing metamorphosis to the puerulus by McWilliam and Phillips (2007).

In this paper we address the first (ecological) question. We concentrate on data on the shallow-water, palinurid species discussed in the Newsletter articles, viz. *J. edwardsii*, *P. argus*, *P. cygnus*, *P. japonicus*, and *P. ornatus.* The purpose of this review is to assemble and assess all relevant, available information on the location where metamorphosis may mainly occur, thereby increasing the background knowledge for future fisheries research into these commercially valuable species: also to provide a background literature search, including cases; such as *P. argus*, where there is a paucity of sampling data for final-stage phyllosomas. It is important to know where metamorphosis takes place because if we are to develop models of lobster fisheries we need to model larval behaviour, including what factor(s) induces metamorphosis, when it is established. Clearly this factor determines where metamorphosis occurs, and we need such observations to help 'tune' the model of larval behaviour.

## TERMINOLOGY

Before proceeding, it is necessary to define several terms, because of the ambiguities of meaning in their use in the general literature.

## Metamorphosis

This occurs in spiny lobsters at the last larval moult when the final phyllosomal stage emerges as a puerulus which has a body plan resembling the adult and is vastly different from the phyllosomal morph; it also undergoes a change in behaviour, locomotion and metabolism from the larval stages. The term is used in this complete sense here. The puerulus is morphologically adapted as a strong swimmer. Whereas the long, thoracic appendages are used for (mainly vertical) swimming in the zoeal form of larva (i.e. the phyllosoma), these become shorter, uniramous, walking-type legs in the puerulus, which has relatively large pleopods, both rami similar in size and fringed with plumose setae, plus appendices internae on their endopods. These appendices, hooked to the opposite pleopod pair by cincinnuli, enhance forward swimming by synchronous sweeping movements. These morphological features are unique to the puerulus of spiny lobsters and non-functional in succeeding developmental stages, being gradually diminished through the juvenile instars as the pleopodal endopod is reduced and resorbed (see Abrunhosa and Kittaka 1997, Figs. 1, 3–8). Also, in all palinurids, metamorphosis occurs *before* settlement of the postlarva (see Gore 1985; Felder et al. 1985). This is true also of the postlarva of most known decapods that do not show abbreviated development (Felder et al. 1985). It is also true of the Scyllaridae (slipper lobsters) another family in the Achelata (formerly Palinura) whose postlarva is called a nisto.

## Shelf break

This is where the continental shelf joins the continental slope. In most continents, including Australia, this in general occurs at the 200 m isobath, so the latter is generally regarded as the demarcation line for the edge of the continental or insular shelf, i.e. the 'shelf break', and this marker is used in the present paper. The width of the continental shelf varies geographically within and among continents. In the latitudinal distribution of *P. cygnus* off Western Australia it ranges from ca. 45 to 80 km but is mainly narrower in the regions of New Zealand and Japan inhabited by *J. edwardsii* and *P. japonicus*, respectively. In these regions, there is a more sudden 'drop-off', rather than a slope, just beyond the shelf break, as there is off the Great Barrier Reef (GBR) of north eastern Australia where the phyllosomas and pueruli of *P. ornatus* occur (Dennis et al. 2001). In the Florida Keys, where early developmental stages of *P. argus* have been sampled, the shelf ranges from ca. 3 to 30 km in width (Yeung et al. 2001).

## Shore, onshore, offshore

In strict parlance, the 'shore' or 'shoreline' (also 'coastline') is where the land meets the sea. 'Onshore' and 'offshore' generally mean, respectively, 'towards the shore' and 'away from the shore' and are usually applied to wind direction, However, in this paper we use such terms regionally—'onshore' meaning in the region from shoreline to shelf break; that is, 'over the shelf' or in shelf waters (i.e. strictly coastal waters), and 'offshore' meaning 'in waters beyond the shelf break', (i.e. oceanic waters). These definitions use the shelf break as the line of demarcation between shelf and oceanic sampling stations. Yeung (2005) argued for a line at 100 m in the central West Atlantic, but whatever is used, it should be clearly defined by each researcher to avoid confusion.

In this paper we have not attempted to guess what authors may have meant by 'coast', etc. Nor have we tried to recalculate distances offshore from particular locations. However, where we are uncertain, or would suggest such measures could be modified to produce more accurate information, we have marked such instances by [sic].

## Sources of data

All conclusions about distributions of phyllosomas and pueruli in the wild are based on the results of sampling and this sampling depends on the objectives of the particular study. However, there are many gaps in information because, usually, the sampling has not been designed to answer the above specific question—which requires sampling from the shelf break to well offshore in order to assess the locations, relative to the shelf break, where final-stage phyllosomas of each species are generally most abundant. A severe biasing problem is introduced if samples are taken only on the shelf, and not beyond the shelf break. In addition, many cruises have not included oceanic sampling stations that were a long way from the shelf.

Table 1 summarises the relevant data that are available on the phyllosomas and pueruli of the species discussed in this paper that have been caught in the plankton.

**Table 1** Data relevant to the general location of metamorphosis from final phyllosoma to the puerulus of spiny lobsters in the wild

| Species | Location of final stage phyllosomas caught | Number of final stage phyllosomas caught | Location where pueruli were caught | Number of pueruli caught | Width of shelf (to 200 m isobath) | Maximum distance pueruli caught offshore of the shelf break | Author(s) |
|---|---|---|---|---|---|---|---|
| *Panulirus argus* | Offshore-beyond the shelf break | 100 | N/A | 393 | N/A | N/A | Richards and Potthoff (1980) |
| | Offshore-beyond the shelf break | N/A | N/A | 0 | N/A | Not applicable | Alfonso et al. (1991) |
| | Not applicable | 0 | In the shallow (shelf) channels between cays 1.3 and 3 m | >347 | N/A | Not applicable | Yeung et al. (2001) |
| | Offshore-30 and 54 km from the shelf break | 2 | N/A | 0 | 1–9 km | N/A | Briones-Fourzán et al. (1988) |
| *Panulirus cygnus* | Offshore beyond the shelf break | 100 | N/A | 11 | 46–80 km | Between 280 and 460 km | Chittleborough and Thomas (1969) |
| | Offshore beyond the shelf break | N/A | On the shelf or within 75 km of the shelf break, cast exuviae of 19 stage IX caught, at 165–320 km from shelf break | 31 | 46–80 km | About 80 km | Ritz (1972) |
| | Offshore of the shelf break except for 1 caught just on the outer edge of the shelf | 123 | Both on and off the shelf, but most sampling was within 60 km of shelf break | 301 | 46–80 km | <60 km | Phillips et al. (1978) |
| | At 7 stations from 180 to 1,360 km offshore along 29°30′S) | 63 spread along the whole transect | Not applicable | 0 | 46–80 km | Not applicable | Phillips et al. (1979) |
| | Offshore-beyond the shelf break within the Leeuwin Current | Not specified | Not applicable | 0 | 46–80 km | Not applicable | Pearce et al. (1992) |
| | Offshore-beyond the shelf break in the area adjacent to or in the Leeuwin Current | 898 | 2 on the shelf, rest within the Leeuwin Current or slightly beyond its seaward rim | 10 | 46–80 km | N/A | Phillips and Pearce (1997) |
| *Panulirus japonicus* | Offshore (54–63 km) south of the Kuroshio Current | 84 of which 3 were metamorphosing phyllosoma | Most near shore or in the Kuroshio Current | 43 | N/A | 63 km offshore | Yoshimura et al. (1999) |
| | Offshore in waters north of 24°N | 30 | Not applicable | 0 | N/A | Not applicable | Inoue and Sekiguchi (2001) |
| | Offshore-beyond the shelf break within the Kuroshio Current | 1 phyllosoma metamorphosing | Most within the Kuroshio Current | 8 | N/A | N/A | Yoshimura (2005) |
| *Panulirus ornatus* | Offshore beyond the shelf break but most numerous adjacent to the GBR shelf | ca. 323 | Offshore beyond the shelf break but mostly numerous adjacent to the GBR shelf | 161 | N/A | Most were 35–40 km from shelf break. 6 caught 600 km beyond the GBR shelf | Dennis et al. (2001) |
| Species | Location of final stage phyllosomas caught | Number of final stage phyllosomas caught | Location where pueruli were caught | Number of pueruli caught | Width of shelf (to 200 m isobath) | Maximum distance pueruli caught offshore of the shelf break | Author(s) |
| *Jasus edwardsii* | Offshore-beyond the shelf break but at the inshore edge of the Wairarapa Eddy | Not specified | Offshore beyond the shelf break | 3 | Not specified | Not specified | Booth and Stewart (1991) |
| | Offshore-beyond the shelf break but at the inshore edge of the Wairarapa Eddy | Not specified | Most found onshore within the shelf, but some offshore and <50 km of the shelf break | 133 | Not specified | 50 km | Booth (1994) includes data from Booth and Stewart (1991) |
| | Offshore-beyond the shelf break | 12 | Not specified | Not specified | Not specified | 53 km | Bruce et al. (2000) |
| | Most abundant offshore beyond the shelf break but inshore of the edge of the Wairarapa Eddy | Not specified | 33 pueruli with soft carpaces found between 24 and 216 km offshore | 383 | N/A | 284 km | Chiswell and Booth (1999)and Jeffs et al. (2001) |
| | Offshore-beyond the shelf break at the inshore edge of the Wairarapa Eddy | N/A | Close to but a little beyond the shelf break | 431 (includes the 383 from Chiswell and Booth (1999)) | N/A | 330 km from the shore | Booth and Chiswell (2005) |

N/A = not advised; Shelf break = the 200 m isobath

# Distribution and abundance of phyllosomas and pueruli

## Phyllosomas

A number of studies to examine the larval recruitment of *J. edwardsii* in New Zealand were undertaken between the 1970s and 1990s. The results of these investigations were summarised by Booth (1994) who reported that 'advanced' phyllosomas (stages V to XI, the final stage in this species) were found almost exclusively seaward of the continental slope and widely distributed from about 20 km seaward of the shelf edge out to 900 km from the mainland. The highest abundance of advanced phyllosomas was off the east coast of the North Island, south of East Cape, and far fewer larvae of any stage were taken off the South Island. Booth (1994) found 2,184 advanced stage phyllosomas near the Louisville Ridge, 1,300 km off the North Island of New Zealand.

Additional sampling on five cruises between June 1988 and April 2001 (Chiswell and Booth 2005) showed that, on four of the five cruises, late-stage (IX, X, and XI) phyllosomas were found further inshore (49 km) than mid-stage phyllosomas. Mid-stage (V to VIII) larvae appeared to be contained within a large semi-permanent eddy found offshore, known as the Wairarapa Eddy, whereas the late stages have been found along the inshore rim of the eddy (Chiswell and Booth 1999, Fig. 1).

**Pueruli**

Booth (1994) reported that most pueruli (>400) of *J. edwardsii* have been caught either over the shelf or just seaward of the shelf break.

Chiswell and Booth (1999) conducted a sampling program off the east coast of New Zealand in February 1998. They caught 383 pueruli of which 33 had soft-carapaces, so it was assumed that they had recently moulted from the last phyllosomal instar (see Fig. 1). This 1998 cruise, which occurred during part of the 1997–1998 El Nino year, was also regarded as anomalous in that both newly metamorphosed and nektonic pueruli were caught well offshore. Jeffs et al. (2001) discussed the 33 new pueruli, collected by Chiswell and Booth in 1998. These new pueruli were captured between 24 and 216 km offshore (mean of 94 km), indicating that most were caught somewhere in between these distances, but still well beyond the shelf break (Jeffs et al. 2001, Fig. 1d). The eastern shelf of the North Island is very narrow—typically ca. 30 km wide, so most of the pueruli caught would have been off the shelf.

Chiswell and Booth (2005) carried out an extensive review of all of the data from five cruises between June 1988 and April 2001, including the cruise in February 1998. In the one cruise in which they could test late-stage phyllosomas against newly metamorphosed pueruli, they found no statistical difference in the distributions. The pueruli sampled had a mean distance offshore of 71 km, and the 33 new pueruli were caught over the same distributional range (mean distance offshore, 93 km) as the late stage phyllosomas (mean distance offshore 104 km). They concluded that metamorphosis to the puerulus takes place throughout the spatial range of the late-stage phyllosomas, which is clearly offshore of the shelf break.

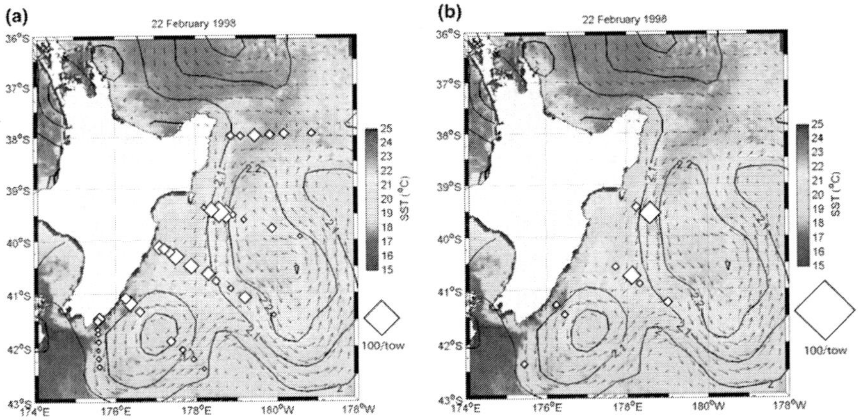


**Figure. 1** Distribution of *Jasus edwardsii* **A** pueruli, **B** newly metamorphosed pueruli caught off the east coast of New Zealand in February 1998 superimposed on dynamic height calculated from AVISO satellite data for the same time period. The data are from Chiswell and Booth (1999, 2005) and S.M. Chiswell (pers. com.) NIWA, New Zealand

## *Panulirus argus*

### Phyllosomas

Through the 1960s and 1970s there have been many studies of the numbers of phyllosomas and pueruli of this species caught in plankton sampling (both deliberate and incidental) in the Florida Keys and adjoining regions (see review by Lyons (1981), and citations in Yeung and McGowan (1991)). Although many were concerned with establishing the mainly 'downstream' Caribbean larval recruitment to Florida stocks, most of these studies up to the 1990s have suffered from two factors that would preclude the assessment of where metamorphosis of this species generally occurs:

- Very few sampling stations were located beyond the shelf off the Florida Keys.
- Until recently it was difficult to separate the late stage larvae, as well as early and middle stages, of the three congeneric species occurring in the region, viz *P. argus*, *Panulirus guttatus* and *Panulirus laevicauda*.
- This means that all earlier phyllosomal counts could have been biased as they were lumped as "*Panulirus* spp.", or attributed to *P. argus*, because the latter was regarded as likely to be the most abundant species (Sims and Ingle 1966). Similarly Richards and Potthoff (1980) reported on an examination of 13,770 specimens of phyllosomas and

pueruli held by the Florida Department of Natural Resources, but were unsure of their identification. This uncertainty also includes the data of Yeung and McGowan (1991) whose sampling transects were mostly on the shelf and extended "from the edge of the reef tract (2–4 km offshore) to the approximate edge of the shelf (30 km offshore), or less". Therefore their 'shore' meant 'shoreline', and their Table 1 indicates only nine stations were at 200 m or more depth (262 m at the furthest offshore station). They did not catch any final-stage (XI) phyllosomas, and only one subfinal stage (X).

Studies of distribution of phyllosomas of *P. argus* in other regions of the Caribbean, e.g. Cuba (Baisre 1976; Alfonso et al. 1991), have indicated that, as with other palinurid species discussed here, mid- and late-stage larvae were almost exclusively found off the continental or island shelves, *ergo* metamorphosis of this species primarily occurs beyond the shelf break, as had been surmised by earlier researchers (see review by Lyons 1981). However, very few, if any, final-stage larvae of this species have been caught in any of these regions.

Manzanilla-Dominguez and Gasca (2004, Table II) working in the Gulf of Mexico and Yucatan waters, reported the catch of one subfinal stage (X) specimen of *P. argus* in the southern Gulf of Mexico and stated that "the largest specimens (stages VIII–X) of *P. argus* [collected] were recorded on the shelf and shelf border areas of the Campeche Bank". In addition, 10 late -stage phyllosomas are shown at four mid-shelf sampling sites.

Since the distribution of the final stages is vital in answering the question, identification and separation of the late-stages of *P. laevicauda* (see Baisre and Ruiz de Quevedo 1982), from those of *P. guttatus* (see Baisre and Alfonso 1994), and the separation of the final stage of the two former species from that of *P. argus*, is now facilitated because the latter species has been cultured from egg to juvenile and its phyllosomal stages described (Goldstein et al. 2006, 2008).

## Pueruli

All of the data on the distribution of the pueruli of *P. argus* in the Florida Keys are based on collections made too far inshore, and without appropriate offshore sampling, to provide information on the location of metamorphosis from the final phyllosomal stage (e.g. Yeung 2005; Yeung et al. 2001).

However, the study of Briones-Fourzan et al. (2008) off the Yucatan Peninsula in Mexico clearly showed that the pueruli of *P. argus* are in the oceanic waters beyond the shelf and that the physical forces of waves and swell, including those caused by the action of hurricanes, are part of the forces transporting them into the coastal back-reef, coastal lagoons and shallow bays. From this it may be inferred that metamorphosis of this species occurs offshore.

## *Panulirus cygnus*

### Phyllosomas

*Panulirus cygnus* in Western Australia is the species in which the oceanic phase of the life cycle has been studied most thoroughly. Extensive oceanic sampling of *P. cygnus* in the 1960s through to the 1990s was aimed initially at an understanding of the distribution and ecology of the phyllosomas and then focussed on understanding the ecology of the puerulus stage in the oceanic recruitment of this species. Sampling was conducted over a vast area from 9°00′S to 34°36′S and from the west coast of Australia out to 100°39′E (see Fig. 2). The biological program was supported by oceanographic studies of the area off Western Australia which have continued right to the present day (see Waite et al. 2007a).

Spatial and temporal field studies by Chittelborough and Thomas (1969), Ritz (1972), Phillips et al. (1978, 1979) Phillips and Pearce (1997) and the review by Phillips (1981), showed that the larvae of *P. cygnus* were widespread in the eastern Indian Ocean. Using the nine stages described by Braine et al. (1979) the phyllosoma were assigned to stages I, II and III as early stages, IV and V as mid-stages and VI, VII, VIII and IX as late-stage phyllosomas, stage IX in this species being the final stage which metamorphoses to the puerulus.

In a study to examine the distribution of phyllosoma larvae on and off the shelf in August, September and October 1974, only 9 out of 1,169 late-stages (stages VI to IX) phyllosomas were caught on the shelf and these were caught towards the outer edge of the shelf (Phillips et al. 1978). Mid- and late-stage larvae in the Phillips et al. (1979) study were found in greatest numbers offshore but near the edge of the continental shelf and were normally absent in shelf waters or, when present, in very small numbers and confined to the outer edge of the shelf.

A wide spatial survey in 1976 and 1977 caught 19,744 phyllosomas and, during midwinter, there were no stations beyond the continental shelf between 24°30′S, and 35°00′S, out to 99°00′E at which at least some phyllosoma were caught (Phillips et al. 1979, Fig. 4). In May 1976 the greatest concentration of late-stage (VI to IX) phyllosomas along 29°30′S was between 103°00′E and 109°00′E (500–1,080 km) offshore (Phillips 1981); and along 29°30′S in September 1976, final-stage phyllosomas, were present at all stations from 99°70′E (approximately 1,500 km offshore and 1,360 km from the shelf), the maximum distance offshore sampled during this cruise, to 111°80′E (Phillips et al. 1979, Fig. 5). No pueruli were caught at any of these stations.

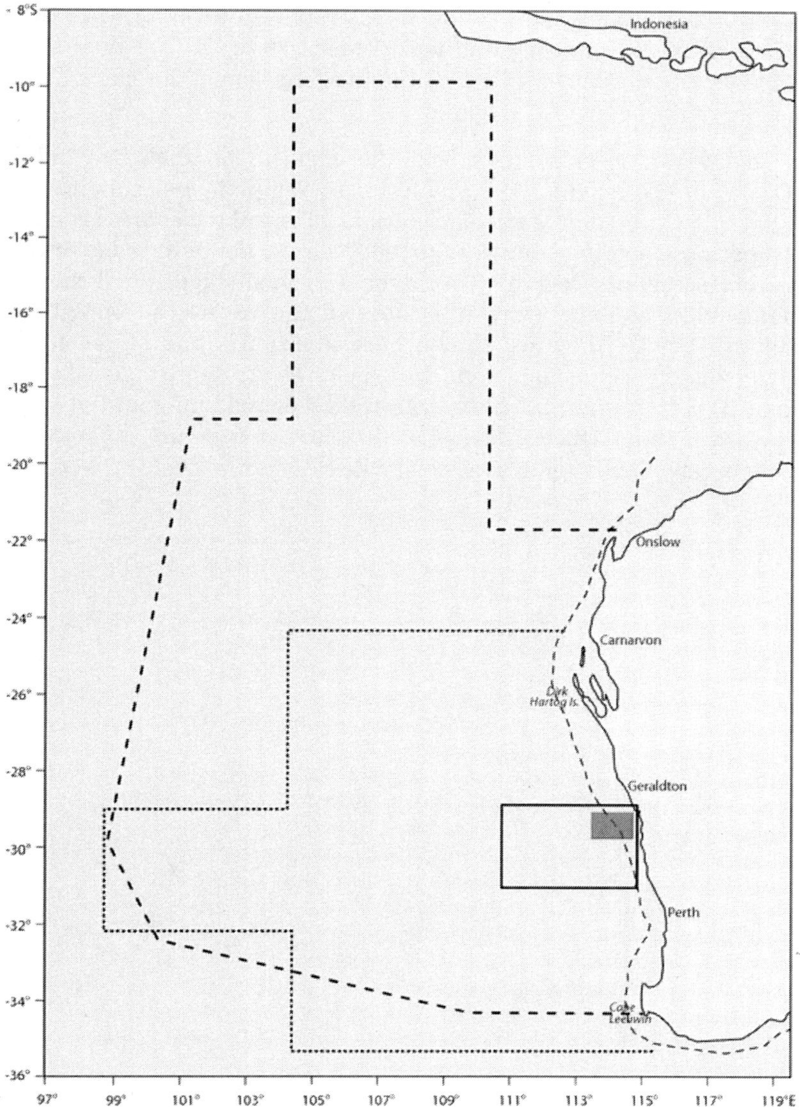


**Figure. 2** The approximate extent of larval sampling for *Panulirus cygnus* that has been undertaken off Western Australia by Chittelborough and Thomas (1969) (*dashed black line*); Ritz (1972) (*black rectangle*): Phillips et al. (1979) (*dotted black line*), and Phillips and Pearce (1997) *light purple rectangle*

The study by Pearce et al. (1992) in August/September 1987, consisting of three transects across the continental shelf at 28°S, 29°07′S and 29°45′S, found a strong dependence on the water masses associated with the Leeuwin Current. This current was first named by Cresswell and Golding (1980). Phyllosomas were found exclusively off the shelf, i.e. in the outer region of the Current, and all were stages VI to IX. The final-stage (IX) larvae were all concentrated at the stations along the shelf break just inshore of the Leeuwin Current.

In September, 1991 a study near the edge of the shelf by Phillips and Pearce (1997), caught 3,199 phyllosomas. Sampling was conducted at the surface and sub-surface for distributions of the phyllosomas (and pueruli) in a period when the Leeuwin Current overran the outer edge of the shelf in the sampling region. The data on final-stage phyllosomas and zooplankton biomass were presented in the review by McWilliam and Phillips (1997). They are shown here in Fig. 3. None of the final-stage phyllosomas was caught on the shelf. Most final-stage phyllosomas (Fig. 3b) were caught offshore, within the Leeuwin Current but nearer its seaward margin which, because of the current overrun, was close to the shelf break. This region was also where biomass values were highest (Fig. 3a).

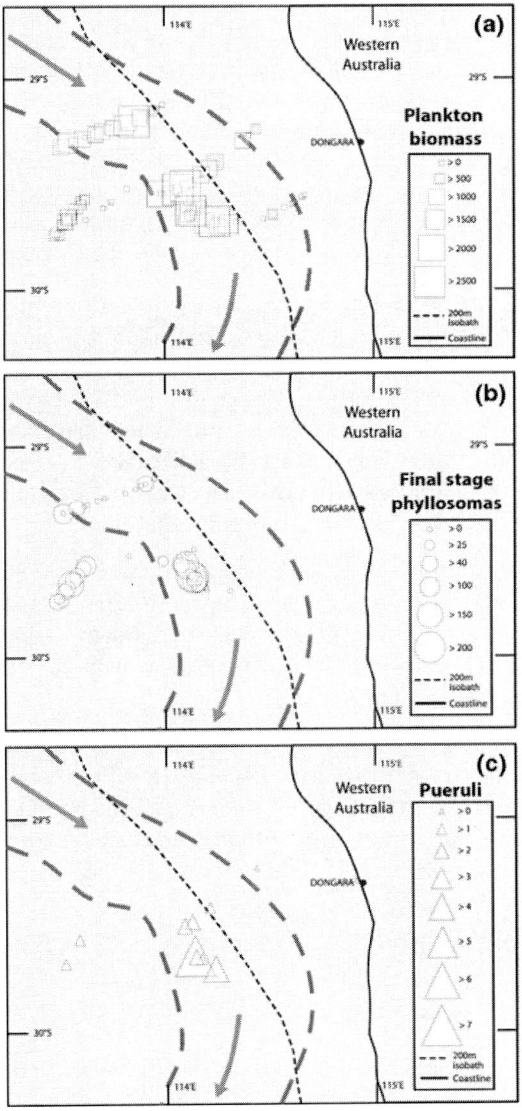


**Figure. 3** **A** Plankton biomass, measured as wet weight in cubic metre of water filtered, from all samples. **B** Numbers of final-stage (IX) phyllosomas of *Panulirus cygnus* per 10,000 m$^3$of water filtered, and **C** Numbers of pueruli of *Panulirus cygnus* 10,000 m$^3$ of water filtered, collected in these samples from the shelf-break region off Western Australia, September 1991. All data are from Phillips and Pearce (1997). *Dashed lines* indicate approximate boundaries of the Leeuwin Current as determined from satellite images; *arrows* indicate direction of current flow. The 200 m isobath is a *black dashed line*

### Pueruli

Most of the >500 pueruli of *P. cygnus* caught in the plankton have been taken in shelf waters or slope waters near the shelf-break (Phillips et al. 1978). However, small numbers have been caught further offshore. Chittelborough and Thomas (1969) recorded one specimen 130–259 km offshore, two at somewhere between 259 and 389 km and one at somewhere between 359 and 519 km. The sampling methods used at that time did not permit greater accuracy as to the distance offshore.

Ritz (1972) also caught 19 cast exuviae of final-stage phyllosomas at more than 161 km (ranging from 215 to 400 km) from the shore [sic] of Western Australia (see Ritz 1972, Fig. 8). It is possible that these exuviae may have been transported some of the distance offshore by the surface currents in this area which are known to transport the early stage I phyllosomas offshore (Phillips 1981) or perhaps by a westward propagating warn-core eddy (Waite et al. 2007a). However, Ritz also pointed out that 80% of the pueruli were caught within 60 km of the coast [sic].

In the August/September 1987 study by Pearce et al. (1992) no pueruli were caught in the waters over the shelf and it was suggested that metamorphosis to the puerulus occurs beyond the shelf (and perhaps in the Leeuwin Current) and that movement towards the shore by the pueruli is subsurface (Pearce et al. 1992).

In the September 1991 study by Phillips and Pearce (1997), most pueruli were found offshore near the edge of the shelf break, but they were found in greatest numbers close to and offshore of the shelf (Fig. 3c) in the same location where the final-stage phyllosomas were most abundant and biomass values highest.

## *Panulirus japonicus*

### Phyllosomas

A number of studies of the distribution of phyllosomas of *P. japonicus* have been conducted, including those of Yoshimura et al. (1999), Inoue and Sekiguchi (2001) and Yoshimura et al. (2002).

Inoue and Sekiguchi (2001) sampled beyond the shelf, but adjacent to it, in waters south of the Kuroshio Subgyre which is south of Kyushu, Japan. They clearly showed that the phyllosomas of *P. japonicus* were beyond the shelf and most of the 30 final-stage larvae caught were in the northern part of the waters east of the Ryuku Archipelago (Inoue and Sekiguchi 2001, Fig. 3).

The numbers of phyllosomas (especially the final stage) caught during these and earlier cruises were low and as reported by Yoshimura et al. (2002) "Very little is known about the scale and processes of transport of *P. japonicus* larvae. The problem lies in the ambiguity of species identification of larvae and the small number of samples collected from the plankton."

This identification problem should now be largely overcome because of recent molecular species identification and descriptions of mid- and late-stage phyllosomas of *Panulirus* species occurring in Japanese waters by Chow et al. (2006a, b) and Shirai et al. (2006).

Further biological/oceanographic studies might reveal that the interannual and decadal changes in the path of the Kuroshio Current along the shelf break of southern Japan (sometimes straight, sometimes with a small meander and sometimes with a large meander offshore (Tomczak and Godfrey 2005)) play an important role in the distribution and abundance of final-stage phyllosomas and puerulus settlement in this species.

**Pueruli**

The paper by Yoshimura et al. (1999) contains the most relevant data on the pueruli of this species. Between 1992 and 1995, on four research cruises, 43 pueruli of *P. japonicus* were caught. In 1992, 35 pueruli were captured in three hauls from an area off the edge of the shelf off south-western Japan. A further two pueruli were taken slightly further off shore in or south of the estimated position of the Kuroshio Current. In 1993 two pueruli were caught in six hauls in the same general location; and a final stage phyllosoma that was moulting to the puerulus stage was caught in the Kuroshio Current. None was caught north of the Current. These data are combined and are shown in Fig. 4.

Yoshimura et al. (1999) suggested that the final-stage phyllosomas of *P. japonicus* metamorphose to pueruli in the Kuroshio Current and the pueruli then swim across the Kuroshio Current to coastal areas to settle. But it is of interest that Yoshimura (2005) reported that "when the Kuroshio is more than 100 miles [about 160 km] from the shore virtually no settlement of pueruli occurs."

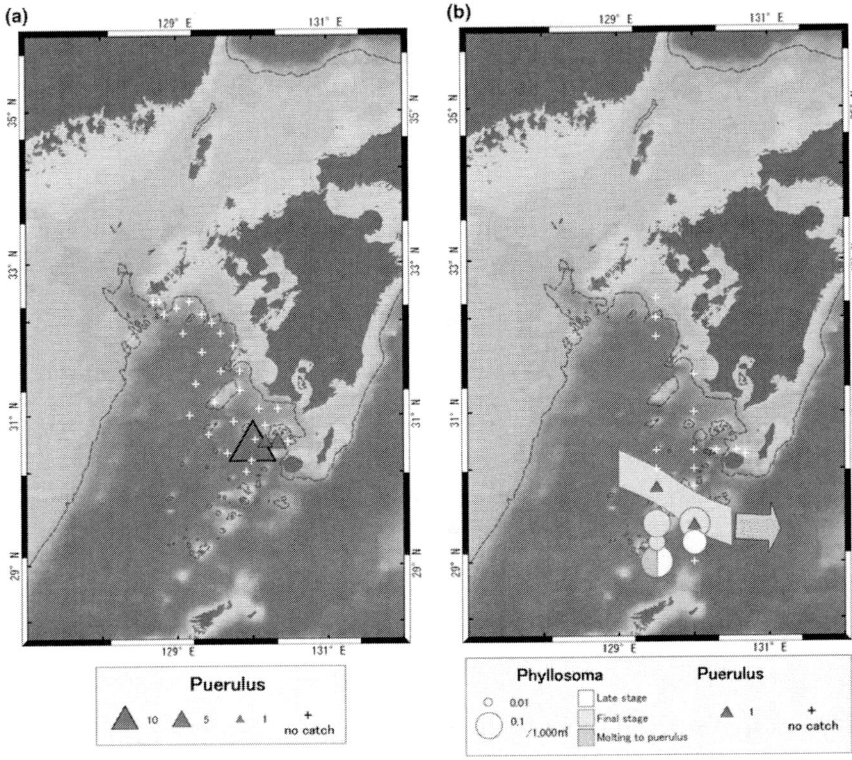


**Figure. 4** **A** Distribution of *Panulirus japonicus* pueruli collected off southern Japan in 1992. **B** Distribution of *Panulirus japonicus* pueruli and phyllosomas collected off southern Japan in 1993. The *pink shaded area* indicates the area of the Kuroshio Current axis estimated from the temperature distribution at a depth of 200 m. All data are from Yoshimura et al. (1999) who also kindly supplied the figure

## *Panulirus ornatus*

### Phyllosomas

The only study of the distribution of the phyllosomas of *P. ornatus* is by Dennis et al. (2001) for the Coral Sea in May 1997. They caught 4,270 phyllosomas and 161 pueruli. Sampling was not conducted on the shelf and in most cases was restricted to about 30 km beyond the edge of the GBR. The phyllosomas were generally restricted to the North West Coral Sea. Final-stage phyllosomas were most abundant at a station about 300 km directly east of the Queensland coast and also at locations adjacent to the GBR shelf break, mainly north of Princess Charlotte Bay (see Dennis et al. 2001, Fig. 5c).

In the *P. ornatus* larval modelling investigation by Pitcher et al. (2005), they stated: "[it was] assumed that once phyllosomas are at least 120 days old, they can metamorphose to the puerulus if they find themselves in water <100 m deep." In addition, they assumed that. "The continental shelf is extremely narrow throughout the eastern Papua New Guinea region, so a model larva of sufficient age that encounters the 100 m isobath (triggering metamorphosis) [sic]also encounters the 20 m isobath, triggering settlement [of the puerulus]." Both of these assumptions were introduced into the modelling process to allow the model to function, but they have no basis in fact, since the 'trigger' for metamorphosis in palinurid phyllosomas of any species has not yet been determined (see McWilliam and Phillips 2007); nor is the preferred substratum for settlement of pueruli of *P. ornatus* known. Therefore, their modelling study is invalid.

**Pueruli**

The only study of the distribution of the puerulus of *P. ornatus* is that reported by Dennis et al. (2001) for the Coral Sea in May 1997. Pueruli were most abundant at two locations (14 and 15) both near the GBR shelf break and south of Princess Charlotte Bay. About 45 of the total of 161 pueruli were caught at locations very close to the shelf break northeast of Cairns, but 6 pueruli were caught about 600 km east of the GBR. These data are shown here in Fig. 5.

## Integration of data for phyllosomas, pueruli and ocean circulation

In this review we examine five regions:

- The South Pacific Ocean off the North Island of New Zealand
- The waters off the Florida Keys, USA and the Caribbean Sea
- The southeastern Indian Ocean off Western Australia
- The northwest Pacific Ocean off Kyushu, Japan.
- The Coral Sea between Papua New Guinea and Queensland, Australia

The general oceanography of all of these regions is now relatively well understood compared with our knowledge in the 1960s and 1970s, the times when most of the larval sampling was undertaken. This is largely because of the significant development of technology which includes satellite images, recording current metres, etc. and the use of computers to model these data. Unfortunately satellite images are not available for most

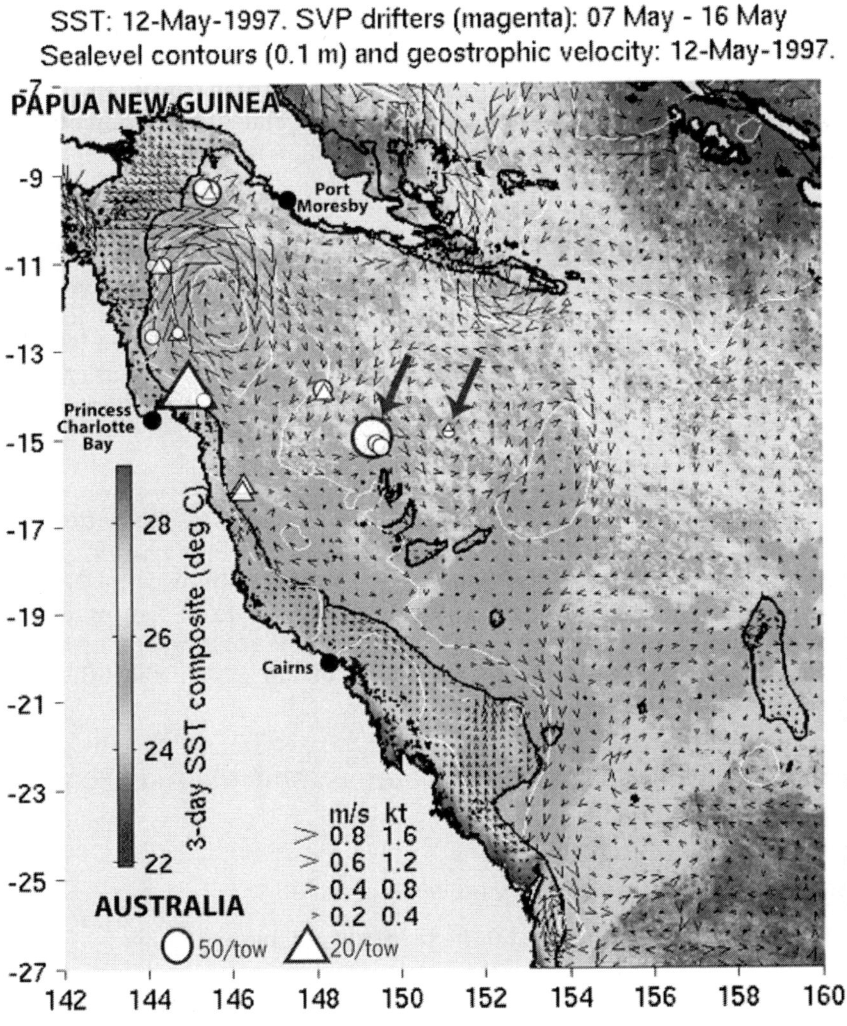


**Figure. 5** Distribution of the final phyllosomas (*circles*) and the pueruli (*triangles*) of *Panulirus ornatus* collected in the north-west Coral Sea off North East Australia in May 1997 by Dennis et al. (2001). The *heavy black line* skirting Queensland (GBR) and New Guinea coasts and Coral Sea reefs is the 200 m isobath. The *arrows* indicate location 4 where final stages were most numerous (i.e. 48% of all final stages were caught here—300 km from the GBR shelf break) and the other = Location 5, about 600 km from the GBR shelf break, where 6 pueruli were caught. The satellite image taken on 12 May 1997 used in Fig. 5 was kindly provided by the CSIRO Division of Marine and Atmospheric Research

of the periods of the sampling discussed above because the satellites had not then been launched. However, images showing relevant features of the circulation pertaining to each region are now available for all regions. The images used in this paper are from a variety of sources, each of which is acknowledged on the illustration.

The nature and extent of the circulation, including eddies and fronts shown in the images, and their relationships to the data on the location of final-stage phyllosomas and pueruli will be discussed in relation to the points raised in the *Lobster Newsletter* (2005).

## *Jasus edwardsii*

The most conspicuous water circulation off the east coast of New Zealand is within the East Cape Current system. The area is marked by recirculation and eddying off the North Island. A major feature of the area is the large semi-permanent Wairarapa Eddy, which is believed to play a part in keeping phyllosomas close enough to the coast to allow successful recruitment (Chiswell and Booth 2005).

Booth and Chiswell (2005) and Jeffs (2005), dealing with *J. edwardsii* phyllosomas and pueruli off the southeast coast of the North Island of New Zealand (Chiswell and Booth 1999; Jeffs et al. 2001), stated that most pueruli caught by plankton sampling have been caught close to shore around the 1,000 m depth contour, i.e. out to a little beyond the shelf break, which is within 50–100 km of the coast (=shoreline), and that metamorphosis takes place near the shelf break. However, some pueruli were taken up to 330 km offshore. Later in the article, they added the comment that most metamorphosis in this species occurs inshore of the Wairarapa Eddy (see Jeffs et al. 2001, Fig. 1).

Jeffs (2005) stated that the only 'hard' evidence for the location of metamorphosis is the presence of newly moulted (i.e. soft-shelled) pueruli in samples. Jeffs et al. (2001), had further discussed the series of data on the *J. edwardsii* pueruli collected by Chiswell and Booth (1999) over a wide stretch of offshore waters, (between 24 and 216 km beyond the shelf break, mean of 94 km). A statistical analysis comparing the characteristics of the water masses in which the newly moulted pueruli were caught and those in which they were not caught, was non significant (Jeffs et al. 2001). However, Jeffs (2005) reiterated that in *J. edwardsii*, metamorphosis of this species was associated with the inshore margins of the Wairarapa Eddy and its offshoots, as he had previously suggested (Jeffs et al. 2001).

The largest catch of newly moulted pueruli in the 1998 cruise, re-examined by Jeffs et al. (2001), was 12 out of 33 pueruli, taken at the locations shown here in Fig. 1b. This station was 72 km offshore. If we examine a representation of the circulation of the region at this time (Fig. 1), we see that this location is a jet of the East Cape Current located in the shoreward perimeter of the Wairarapa Eddy. New pueruli appeared to be generally distributed around the inshore margins of the Wairarapa Eddy, and absent from areas offshore and towards the centre of the eddy (Jeffs et al. 2001). For clarity, this is not the same representation of the circulation as presented by Jeffs et al. (2001, Fig. 1) which was based on the dynamic height from shipborne observations. This figure is based on dynamic height calculated from AVISO satellite data for the same time period.

## *Panulirus argus*

*Panulirus argus* has such a wide distribution in the central western Atlantic that it is impossible to summarise the oceanic environment in which it occurs. However, the ocean circulation off the Florida Keys is marked by the warm Florida Current which passes close to the reef tract, and coastal eddies and counters currents (Yeung et al. 2001). Larval transport of *P. argus* may be driven by ocean currents, eddies and wind-driven currents, but the exact mechanisms are uncertain (Yeung and Lee 2002).

The article by Yeung (2005) addressed the topic with respect to *P. argus* stocks in the Florida Keys. She reported that metamorphosis is believed to occur "offshore", but was thwarted in answering the question about the specific location because the earlier studies, as well as the study by Yeung and McGowan (1991) only sampled onshore, to approximately the edge of the shelf, which is approximately 30 km from the edge of the reef tract, and they did not catch any final-stage larvae (XI). Yeung (2005) therefore focussed on discussing larval transport pathways across the coastal eddies in a circulation model of the Florida Keys and offshoots of the Florida Current and the Loop Current variations in the Gulf of Mexico (Lee et al. 1994).

## *Panulirus cygnus*

A dominant feature of the ocean circulation off Western Australia is the Leeuwin Current, an anomalous poleward-flowing eastern boundary current that carries warm, low-salinity water southward along the shelf break of Western Australia (Waite et al. 2007a). This Current is stronger in the austral winter months, and its eddies are more frequent at this time of year, than in summer months, when its flow is weaker due to prevailing southerly winds.

McWilliam and Phillips (1997) found that the main region of occurrence of metamorphosis in *P. cygnus* in Western Australia could be inferred from the distribution and abundance of final phyllosomas, together with the location of the relative numbers of pueruli. Based on all the data available it was usually found to be in the slope region, offshore of the shelf break. It is important to realise that the seaward current border also represents a "front" and therefore also probably a more productive part of the offshore region.

In Western Australia satellite images and buoy tracks have now provided a completely different perspective to the knowledge of what may be happening with the Leeuwin Current (which was not even named until 1980) and the other parts of the circulation of the south-eastern Indian Ocean, to what was understood during the time period when most of the sampling was undertaken. Figure 6 shows the latest simplified schema of the surface currents off Western Australia (adapted from Waite et al. 2007a, Fig. 1).

It is clear from Fig. 6 that the Leeuwin Current can influence the ocean circulation over a wide area, and create a large "eddy field". This includes the production of mesoscale eddies, both anticyclonic warm-core (WC) and cyclonic cold-core (CC) eddies (Waite et al. 2007b). Both types of LC eddies propagate westward, the WC eddies have a poleward bias and CC eddies an equator-ward bias. A series of sea surface temperature fronts can be expected to occur over a wide area. Also each of these could potentially create localised conditions of higher plankton biomass such as the area identified by McWilliam and Phillips (1997) in the slope region offshore of the shelf break. Under such conditions "competent" final-stage phyllosomas might metamorphose into pueruli. The data from the widespread survey conducted by Phillips et al. (1979, Fig. 4) are shown in Fig. 7. All of the phyllosoma data on this survey could have been influenced by the Leeuwin Current, or its eddies if these conditions had prevailed.

Recent extensive studies of both types of LC eddies, including a counter-rotating pair of larger ones, have shown that primary productivity is much higher in WC eddies, as is the abundance and biomass of mesozooplankton, compared with those of CC eddies of the LC. This has been attributed partly to the origin of the water masses involved—e.g. entrainment of some nutrient-richer and more productive shelf/shelf break waters into the WC eddy during its formation as the Current runs south along the edge of the shelf (Strzelecki et al. 2007; Waite et al. 2007b). Whereas the CC eddy studied, in 2003 had a cap of warm Indian Ocean water, that seems to have limited upwelling between 1,000 and 500 m (Waite al 2007a)—and it is well-known that these offshore Indian Ocean waters are oligotrophic cf. slope and shelf waters (Tranter and Kerr 1969; Strzelecki et al. 2007). However, there was also a suggestion, from modelling results (Rennie et al. 2007) that, in general, the CC eddies may be sourced from the deeper LC Undercurrent.

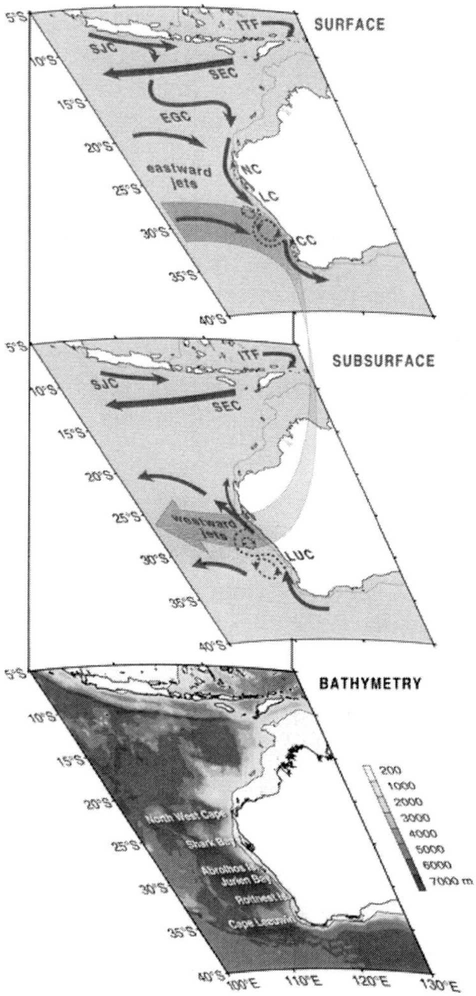


**Figure. 6** Simplified schema of the surface currents off Western Australia emphasising the connection between the Leeuwin Current (LC) and Leeuwin Undercurrent (LUC) as a curved vertical arrow. Indicated is (1) the bathymetry and key coastal landmarks (*bottom*), (2) the subsurface (200–400 m) northward-flowing LUC, showing westward jets offshore, (3) the LC at the surface, fed by the eastward jets transporting Subtropical Water/South Indian Central Water into the LC and then the LUC. *Abbreviations*: ITF, Indonesian Throughflow; SJC, South Java Current; CC, Capes Current; and NC, Ningaloo Current. Figure adapted from Waite et al. (2007a), Fig. 1, and based on the circulation patterns described in Domingues et al. (2007), kindly supplied by Dr. Catia Domingues, CSIRO Marine and Atmospheric Research

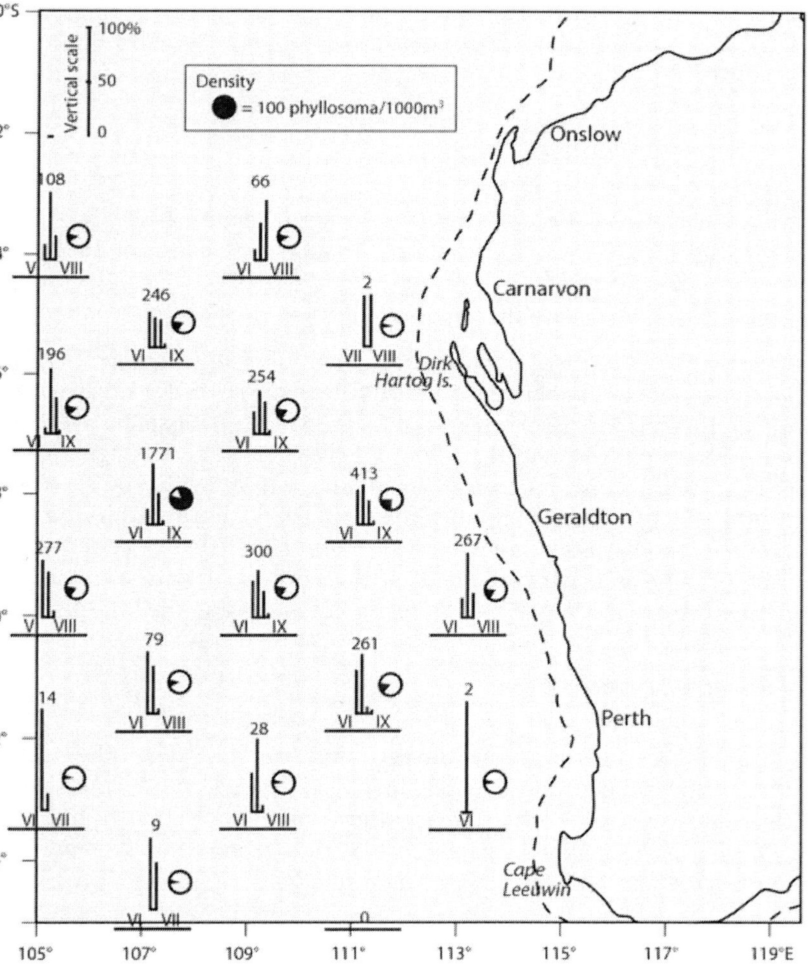


**Figure. 7** Sampling areas for percentages of *Panulirus cygnus* phyllosomal stages in samples and total percentages of phyllosomas caught each night during June and July 1976. Numbers of phyllosomas caught are shown above histogram; transect length by *horizontal lines* below. All phyllosomal data are from Phillips et al. (1979, Fig. 4)

This higher production found in WC eddies of this anomalous, eastern boundary current, is contrary to the findings of higher production in CC eddies (cf WC eddies) of western boundary currents such as the Gulf Stream off the USA. The mesoscale WC Leeuwin Current eddy was also found to have a healthier, although similar mesozooplankton community to that of the CC eddy, but with greater lipid storage (Waite et al. 2007b), which is an interesting fact in relation to the essential diet and known prey of late-stage phyllosomas (see McWilliam and Phillips 2007 and references therein).

## *Panulirus japonicus*

The larval cycle of *P. japonicus* appears to be intimately related to the activities of the Kuroshio Current (Inoue and Sekiguchi 2001) and the distribution of final phyllosomas, as well as puerulus settlement, is probably influenced by the above-mentioned changes in its pathway.

Yoshimura (2005) had data (4 specimens) of final-stage phyllosomas of *P. japonicus* metamorphosing to pueruli in relation to the Kuroshio Current off southern Japan. Because mid- and late-stage phyllosomas were found in the ocean south of the Kuroshio, whereas most final-stage larvae were caught within this current (Yoshimura et al. 1999, 2002), Yoshimura (2005) inferred that metamorphosis occurs offshore within the Kuroshio Current as discussed above. Yoshimura (2005) also added that their metamorphosis occurs, at night, at the surface or within 40 m of it, just after sunset or midnight, and between full moon and the last quarter.

## *Panulirus ornatus*

Oceanic transport in the northwest Coral Sea is largely influenced by the influx of warm equatorial water flowing westward in the South Equatorial Current that enters between the Solomon Islands and Vanuatu (Church 1987), and bifurcates east of the Queensland Plateau (between about 16°S and 18°S). The East Australian Current branch flows southward along the GBR to enter the Tasman Sea, while the northward-flowing branch forms a clockwise coastal current (the Hiri Current) along the shelf of Papua New Guinea and a partially closed gyre in the Gulf of Papua, with an exit into the Solomon Sea (Burrage 1993, Fig. 1).

Dennis (2005) stated that the majority of final-stage phyllosomas and pueruli of *Panulirus* spp. were caught at depths less than 25 m (presumably meaning '25 m from the surface in offshore waters') and most were caught near the surface (in night sampling presumably). Therefore he (Dennis) considered that the "larvae" [perhaps he really meant pueruli?] would be transported on-shore [sic] by surface currents produced by the prevailing South-East Trade Winds. This led to his conclusion that in *P. ornatus*, it is likely that most metamorphosis occurs near the shelf break and near the surface.

The question of the presence of numerous final-stage phyllosomas of *P. ornatus* at about 300 km offshore was raised by Dennis (2005). Many final (gilled) stages of *P. ornatus* occurred along the shelf break in the NW Coral Sea in May 1997, just prior to the time of peak settlement of this species. However, about 50 of the final-stage phyllosomas were collected about 300 km east of the GBR shelf break (see Fig. 5) in the South Equatorial Current which flows toward the Queensland coast. Five pueruli were also caught at this location, and 6 pueruli were caught slightly further offshore (where about 5 final-stage phyllosomas were also caught) about 600 km east of the GBR. The positions of these sampling sites are indicated by the arrows on Fig. 5. The image used is for 12 May 1997, kindly made available by the CSIRO Division of Marine and Atmospheric Research.

The occurrence of these final-stage phyllosomas and pueruli of *P. ornatus* this far offshore is of interest, but needs to be put into perspective. It is not unusual to find late-stage, including final-stage, phyllosomas well offshore. Chittelborough and Thomas (1969) reported a total of 51 final stage phyllosomas of *P. cygnus* at varying distances offshore including seven which were about 1,130 km from the coast [sic] in May 1966.

Sampling off Western Australia in 1976 by Phillips et al. (1979) (see Table 2) yielded final-stage phyllosomas of *P. cygnus* at every station sampled. The data in Table 2 show final stage phyllosoma at all stations out to 1,360 km offshore. However, they may have occurred much farther offshore as the furthest offshore sampling site was at 99°70′E. Similarly Booth (1994) found advanced phyllosomal stages of *J. edwardsii* 900 km offshore.

**Table 2** Final stage phyllosoma larvae (Stage IX) of *P. cygnus* caught along 29°30′S off Western Australia in September 1976

| **Sampling location** | **99°70′E** | **101°′85E** | **103°80′E** | **105°80′E** | **107°80′E** | **109°90′E** | **111°80′E** | **Width of shelf** |
|---|---|---|---|---|---|---|---|---|
| Distance from shelf break | 1,360 km | 1,140 km | 955 km | 760 km | 570 km | 360 km | 180 km | 76 km |
| Number of final stage phyllosoma caught | 16 | 21 | 34 | 91 | 16 | 169 | 121 | |

It is clear that reaching the final phyllosoma stage can occur in many distant offshore locations as shown by both *P. cygnus* and *P. ornatus*, but is no guarantee of metamorphosis to the puerulus stage. The situation with *P. ornatus* requires further comment because in addition to the final-stage phyllosomas of this species, some pueruli were caught at the same locations. This suggests that there may have been some sort of front occurring in this area, at this time.

It is interesting to note that the sampling data on *P. ornatus* reported by Dennis et al. (2001) came only from collections in May 1997, and Pitcher et al. (2005) remarked that this happened to be an "unusual year" because of the effects of Tropical Cyclone "Justin". It was also an El Nino year. The sampling data of *J. edwardsii* collected by Chiswell and Booth (1999), and used by Jeffs et al. (2001), are stated to have been collected in an "anomalous" kind of year (February 1998), when water temperatures were very warm. It is known that the above sampling data for both species was obtained in the strong El Nino year 1997–1998. However, the biological significance of either of these "unusual conditions" is unknown.

## DISCUSSION

The questions to be answered are:

*Where does the majority of metamorphosis from final phyllosoma to puerulus occur?*

A simple answer to this question is: beyond the shelf break (at the 200 m isobath) probably within ca. 100–150 km from it (depending on dimensions and distance offshore of the major local warm, ocean current(s) and in the more productive offshore areas where the final phyllosomal instars are most abundant. The shelf break is an area where fronts of large extent commonly occur and usually these result in high concentrations of plankton (see also Owen 1981). It was never suggested by McWilliam and Phillips (1997) that metamorphosis was associated with the shelf break per se. Although we have discussed only a few species in this review, the puerulus distribution of another shallow-water species, *Jasus lalandii*, in South Africa (Pollock 1986) is similar to that of *J. edwardsii* pueruli, being caught near to or on the continental shelf.

The above generalisation is drawn from quantitative, or semi-quantitative, sampling of final phyllosomas and/or pueruli of four shallow-water species: *P. cygnus* (Phillips and Pearce 1997), *P. japonicus* (Yoshimura et al. 1999), *P. ornatus* (Dennis et al. 2001) and *J. edwardsii* (Chiswell and Booth 1999; Jeffs et al. 2001). The larval transport pathways of all four species are associated with a warm current—several in the case of the tropical species, *P. ornatus* in the Coral Sea. The latter also has a much shorter pelagic larval phase (4–7 months) than the subtropical and temperate-water species (*P. cygnus* and *J. edwardsii*, respectively), as does also the tropical species, *P. argus*, ca. 5–7 months in culture (Goldstein et al. 2006, 2008). The case for the main location of metamorphosis in *P. argus* in the Florida Keys region is inconclusive because of the complexities in defining 'offshore', plus the small number of samples made beyond the reef complex (Yeung 2005). However, sampling in other Caribbean waters (e.g. Baisre 1976; Alfonso et al. 1991) has indicated that this species probably also metamorphoses offshore.

These currents, which bring warmer tropical or subtropical waters to cooler regions, generate more productive (and extensive) frontal zones, eddies, and often nutrient enrichment by increased mixing with shelf waters or other water masses such as counter-currents. Compared with the oligotrophic waters further out to sea, and the usual patchiness of zooplankton, bioenergetic stores necessary for phyllosomal development and metamorphosis (see also McWilliam and Phillips 2007) can be readily accumulated by late-stage larvae feeding in the relatively more eutrophic waters nearer to the shelf. Slightly raised temperatures in the currents there will also result in shorter intermoult periods, *ergo* faster development (McWilliam and Phillips 1997).

For *P. cygnus*, the highest densities of final-stage larvae, and by inference, their metamorphosis, seem mainly to occur over the slope, within or near the warm Leeuwin Current which flows close to the shelf break. Thus the main region of their metamorphosis seems to be associated with more productive fronts near the borders of the Leeuwin Current (Phillips and Pearce 1997; Griffin et al. 2001), its offshoots and the periphery of the mesoscale Warm Core eddies which it often forms. Eddies may also be influential not only in aggregation of zooplankton (including the late phyllosomal stages and their suitable prey) near their border (see Owen 1981), but perhaps also, larval retention (Caputi et al. 2001, Fig. 2; Sponaugle et al. 2002). It is likely that this pattern of metamorphosis, mainly occurring within richer feeding areas closer to (but not on) the shelf, will pertain to other species of shallow-water palinurids. This seems to be apparent even from the distribution of newly metamorphosed pueruli of *J. edwardsii* (Jeffs et al. 2001, Fig. 1c, d, and is also indicated by Booth and Chiswell (2005).

All planktotrophic decapod larvae require exogenous food sources to attain metamorphic competency (Anger 2001). Therefore we think that, for all shallow-water palinurid species, including *P. ornatus* and *P. argus*, whose larval transport pathways are effected by multi-current systems (Dennis et al. 2001; Yeung et al. 2001), metamorphosis will occur mainly in, or adjoining, regions where there are sustained and richer food resources for instars of the penultimate and final stage phyllosomas at critical periods of their moulting cycles (McWilliam and Phillips 1997). This nutritional effect has been found experimentally for other decapod larvae (see Anger 2001, Chap. 4), and has been discussed more fully in our recent review (McWilliam and Phillips 2007). But we also think metamorphosis of these phyllosomas will only occur in those richer areas where water depths are greater than the normal daytime depth-distribution of the late larval stages of each species, for the phyllosoma form is not adapted for encounters with the natural substratum—hence this is another role for the transitional puerulus stage.

In the case of *P. cygnus*, we would also suggest that where metamorphosis seems mainly to occur is not entirely random, but rather due to selection pressures operating on all phases of development in this species: the behaviour of spawning females of *P. cygnus*, the larval cycle, and larval behaviour effective in larval transport, are geared to the seasonal (bi-annual) variation in strength of the Leeuwin Current (George 1997, 2006). Nanoplankton analyses have shown that a strong Proto–Leeuwin Current flowed down the Western Australian coast several times during the Eocene and middle to late Oligocene (Shafik 1990).

Nevertheless, it is unrealistic to suppose that more than a small proportion of any individual brood of a shallow-water palinurid species will recruit to its natal population. Why else have they evolved with such large numbers of eggs per brood? Having a lengthy, pelagic, larval phase is like entering a lottery—the more units invested, the greater the chance of a win. The chance of larval attrition, by advection too far out to sea, (especially if developing in an eddy) loss downstream if within the boundary Current), death by starvation or predation, or other factors, must be much greater than that of successful metamorphosis—and not every resulting puerulus may reach settlement. (see Yoshimura 2005).

*Where else does metamorphosis from final phyllosomal stage to puerulus occur, and how often?*

It is obvious that some metamorphosis occurs on some occasions far out to sea and as pointed out by Jeffs (2005) metamorphosis is not "restricted to the shelf break [region]".

Booth and Chiswell (2005) provided a plausible explanation for the occasional 'hard evidence' of metamorphosis of *J. edwardsii* occurring several hundreds of kilometres beyond the shelf break. They suggested that "every now and then some (late-stage phyllosomas) are transported into more productive waters. Here, where prey abundance is higher, the phyllosomas can stock up on food to reach some critical energy reserve level, and hence metamorphose". However, they concluded that metamorphosis "will usually occur near the shore [sic], in the region of the shelf break".

Jeffs et al. (2001), talking of *J. edwardsii*, stated a similar opinion when they said that there was a suggestion that metamorphosis could be associated with the inshore margins of oceanic eddies which are a dominant oceanographic feature of this region. We think Jeffs et al., had the right idea. For *J. edwardsii* it may be the inshore margins of the oceanic eddies. In the case of *P. cygnus* it is the inshore edge of the Leeuwin Current running down the edge of the shelf that is the most productive in biomass, and it is in this area that most pueruli are produced. However, any oceanic current or eddy could produce similar areas of high zooplankton biomass at their borders (i.e. frontal regions, where there is confluence of two different water masses) and if final phyllosomas in a suitable state of development are present, a number may metamorphose to pueruli. Not only the seaward edge of the Current, but also the periphery of the mesoscale, anticyclonic eddies produced frequently by the Leeuwin Current, provides frontal regions further offshore, and this may account for some pueruli being found so far offshore.

*What is the likely fate of the pueruli found long distances offshore, hence what is their relevance to the recruitment process?*

Jeffs et al. (1999, 2001) and Phillips et al. (2006) have all dealt with energy stores in the puerulus stage and discussed the ability of pueruli to swim to settlement sites and successfully moult into the first juvenile stage of either *J. edwardsii* or *P. cygnus*. Jeffs et al. (2001) estimated a maximum distance of 200 km for *J. edwardsii*, but Booth and Chiswell (2005) suggested that it was probably more like 100 km.

It is hard to believe that pueruli such as the single specimen of a puerulus reported by Chittelborough and Thomas (1969) at somewhere between 359 and 519 km offshore, can successfully reach the shelf. The distances offshore from which the four species can successfully recruit are unknown, but presumably vary with species and the time of year, because of differences in the ocean circulation and water temperatures in a particular region. However, it is likely that few of the offshore pueruli survive and settle.

A study of pueruli of *P. cygnus* by Limbourn et al. (2008) showed that in addition to the energy reserves that are necessary to reach a settlement site near the shore, reserves are needed to sustain the animal until it has moulted to the juvenile stage and located adequate food to continue its growth and development. This further suggests that few of the animals that become pueruli well offshore will survive to adulthood.

# CONCLUSIONS

Despite our attempts to define a number of terms at the beginning of this review, we have still found that the definitions of some of them are open to question. For example, what is "coastal", a term used by some authors? There are "coastal" currents such as the Leeuwin Current which, through its meanders, westward- propagating WC eddies, and jets, can influence the processes such as lobster recruitment by advecting final-stage phyllosomas out to at least 760 km beyond the shelf break (see Fig. 7) and perhaps a lot further offshore.

McWilliam and Phillips (1997) concluded that the main region of occurrence of metamorphosis in *P. cygnus*, is over the slope near the continental shelf break. However, while this was the picture to emerge from the sampling programs which were undertaken on several occasions, the situation may well vary with the scale of the ocean climate. Booth and Chiswell (2005) also considered changed climatic variations in February 1998 "a particularly warm February", with higher than average sea temperatures, could have accounted for some pueruli being swept up to 330 km from shore [sic], whereas most were caught within and about 50 km seaward of the shelf break (Chiswell and Booth 1999; Jeffs et al. 2001).

Bradford et al. (2005) summarised the vertical distribution of the phyllosomas of a number of spiny lobster species including *P. cygnus*, *P. argus*, and *J. edwardsii* discussed in this paper. Each species had different vertical behaviour patterns and it seems likely the pueruli of these species would also exhibit different diurnal vertical migration behaviour adding to the complexity of the transport processes and contributing to the difficulty of understanding them.

There is a lot of discussion about the retention of phyllosomas in eddies, but we doubt that this situation is always as beneficial to recruitment to natal populations as supposed—unless the eddy is a semi-permanent feature of the region, like the Wairarapa Eddy of New Zealand (Chiswell and Booth 1999). Consider the dynamics of a bunch of say sub-final and final-stage phyllosomas located near the rim of an eddy, e.g. the Wairarapa, or a Leeuwin Current one, and slowly circulating with the eddy (i.e. if their daytime vertical distribution is such that they still remain within it). From all the studies on the warm-core, mesoscale eddies of the East Australian Current, it seems that zooplankton abundance is poorer in the middle than at the edges, and richest just outside the rim of eddies. However, in their study of the pair of LC eddies, Strzelecki et al. (2007) found that the mesozooplankton was more abundant and had higher biomass in the WC eddy than in the CC eddy. Moreover, within each of these eddies, biomass at the perimeter and at the centre were not significantly different.

So the pueruli metamorphosing from phyllosomas feeding at the WC eddy perimeter have a swimming advantage, for if they happen to metamorphose where and when their part of the eddy rim is nearer or nearest to the shelf, then the chances of successful settlement are better than if they had metamorphosed on the seaward rim some 150–200 km or greater in the opposite direction. The same applies if metamorphosis had occurred on the shoreward rather than seaward edge of the particular current. We think this has some bearing on why more pueruli are found near the shelf, than well offshore. If the final phyllosomal instars do not reach metamorphic competency when that "gate" is open, then they are likely to be lost out to sea, or downstream, if still in an eddy or current.

What is most needed now, to advance the topic of this review, is a new multidisciplinary approach with new collections of late-stage phyllosomas, other zooplankton, biomass data and concurrent oceanographic data. This would probably need to be conducted by two vessels and obtaining funding is likely to be a major obstacle to such a project.

## ACKNOWLEDGMENTS

The authors acknowledge the assistance of Sandy Clarke from the Department of Fisheries in Western Australia in preparing the illustrations. Also a special thanks to Dr Yoshimura Taku (Seikai National Fisheries Research Institute) and Dr Steve Chiswell (NIWA, New Zealand) for supplying the figures used to illustrate the areas off Japan and New Zealand, respectively. We also thank Louise Bell for preparing the tif file of Fig. 6 which Dr Catia Domingues (CSIRO Marine and Atmospheric Research, Tasmania) kindly made available. The comments by Dr Roy Melville-Smith (Department of Fisheries, Western Australia) and Dr Steve Chiswell on an earlier draft of this manuscript are also appreciated. The satellite image used in Fig. 5 was kindly provided by the CSIRO Division of Marine and Atmospheric Research.

## OPEN ACCESS

# REFERENCES


1. Abrunhosa FA, Kittaka J (1997) The morphological development of juvenile western rock lobster, *Panulirus cygnus* George, 1952 (Decapoda, Palinuridae) reared in the laboratory. Bull Mar Sci 61:81–96
2. Alfonso I, Frias P, Baisre J (1991) Distribución y abundancia de larvas de la langosta *Panulirus argus* en aguas alrededor de Cuba. Rev Inv Mar 2(1–3):3–19
3. Anger K (2001) The biology of decapod crustacean larvae. A.A. Balkema Publishers, Lisse Abingdon
4. Baisre JA (1976) Distribucion de las larvas de *Panulirus argus* y *Scyllarus americanus* (Crustacea, Decapoda) en aguas alrededor de Cuba. Rev Inv Nac Pesca Cuba 2:277–297
5. Baisre JA, Alfonso I (1994) Late stage larvae of *Panulirus guttatus* (Latreille, 1804) (Decapoda, Palinuridae) with notes on the identification of phyllosomata of *Panulirus* in the Caribbean Sea. Crustaceana 66:32–43. doi:10.1163/156854094X00134
6. Baisre JA, Ruiz de Quevedo ME (1982) Two phyllosome larvae of *Panulirus laevicauda* (Latreille, 1817) (Decapoda, Palinuridae) from the Caribbean Sea with a discussion about larval groups within the genus. Crustaceana 43:147–153. doi:10.1163/156854082X00470
7. Booth J (1994) *Jasus edwardsii* larval recruitment off the east coast of New Zealand. Crustaceana 66(3):295–317. doi:10.1163/156854094X00044
8. Booth J, Chiswell S (2005) Food is the factor. Lobster Newsl 18(2):15–18. http://www.odu.edu/~mbutler/newsletter/index.html
9. Booth JD, Phillips BF (1994) Early life history of spiny lobster. Crustaceana 66:271–294. doi:10.1163/156854094X00035
10. Booth JD, Stewart RA (1991) Distribution of phyllosoma larvae of the red rock lobster *Jasus edwarsii* off the east coast of New Zealand in relation to the oceanography. In: Hancock DA (ed) Larval biology, Bureau of Rural Resources Proc No. 15, Aust Soc Fish Biology, Workshop, Hobart, 20 August 1991
11. Bradford RW, Bruce BD, Chiswell SM (2005) Vertical distribution and diurnal migration patterns of *Jasus edwardsii* phyllosomas off the east coast of the North Island, New Zealand. N Z J Mar Freshw Res 39:593–604
12. Braine SJ, Rimmer DW, Phillips BF (1979) An illustrated key and notes on the phyllosoma stages of the western rock lobster *Panulirus cygnus* George. CSIRO Division of Fisheries and Oceanography, Report No. 102
13. Briones-Fourzan P, Candela J, Lozano-Alvarez E (2008) Postlarval settlement of the spiny lobster *Panulirus argus* along the Caribbean coast of Mexico: patterns, influence of physical factors, and possible sources of origin. Limnol Oceanogr 53:970–985
14. Briones-Fourzán P, Lozano-Álvarez E, Colinas-Sánchez F, Negrete-Soto F (1988) Distribución de larvas filosomas de la familia Palinuridae (langostas espinosas) en al Caribe mexicano y plataforma de Yucatán. In: Biología y dinámica poblacional de las langostas del Caribe mexicano. Informe final, Proyecto PCECBNA-021927, CONACYT/Instituto de Ciencias del Mar y Limnología, UNAM, pp 46–70
15. Bruce B, Bradford R, Griffin D (2000) A synthesis of existing data on larval rock lobster distribution in southern Australia. Final Report, Project No. 96/107, Fisheries Research and Development Corporation, Hobart, CSIRO Marine Research
16. Burrage DM (1993) Coral Sea currents. Corella 17:135–145
17. Caputi N, Chubb C, Pearce A (2001) Environmental effects on recruitment of the western rock lobster, *Panulirus cygnus*. Mar Freshw Res 52:1167–1174. doi:10.1071/MF01180

18. Chiswell SM, Booth JD (1999) Rock lobster *Jasus edwardsii* larval retention by the Wairarapa Eddy off New Zealand. Mar Ecol Prog Ser 183:227–240. doi:10.3354/meps183227
19. Chiswell SM, Booth JD (2005) Distribution of mid- and late-stage *Jasus edwardsii* phyllosomas: implications for larval recruitment processes. N Z J Mar Freshw Res 39:1157–1170
20. Chittelborough RG, Thomas LR (1969) Larval ecology of the Western Australian marine crayfish, with notes upon other panulirid larvae from the eastern Indian Ocean. Aust J Mar Freshw Res 20(3):199–223. doi:10.1071/MF9690199
21. Chow S, Suzuki N, Imai H et al (2006a) Molecular species identifications of spiny lobster phyllosoma larvae of the genus *Panulirus* from the Northwestern Pacific. Mar Biotechnol 8:260–267. doi:10.1007/s10126-005-5151-9
22. Chow S, Yamada H, Suzuki N (2006b) Identification of mid to final stage lobster phyllosoma larvae of the genus *Panulirus* White 1847, collected in Ryuku Archipelago. Crustaceana 79(6):745–764. doi:10.1163/156854006778026771
23. Church JA (1987) East Australian Current adjacent to the Great Barrier Reef. Aust J Mar Freshw Res 38:671–683. doi:10.1071/MF9870671
24. Cresswell GR, Golding TJ (1980) Observations of a south-flowing current in the southeastern Indian Ocean. Deep-Sea Res 27A:449–466. doi:10.1016/0198-0149(80)90055-2
25. Dennis DM (2005) Coral Sea *Panulirus ornatus*. Lobster Newsl 18(2):4–15. http://www.odu.edu/~mbutler/newsletter/index.html
26. Dennis DM, Pitcher CR, Skewes TD (2001) Distribution and transport pathways of *Panulirus ornatus* and *Panulirus* spp. larvae in the Coral Sea, Australia. Mar Freshw Res 52:1175–1185. doi:10.1071/MF01186
27. Domingues CM, Maltrud ME, Wijeffels SE et al (2007) Simulated Langrangian pathways between the Leeuwin current system and the upper-ocean circulation of the southeast Indian Ocean. Deep Sea Res Part II Top Stud Oceanogr 54:797–817. doi:10.1016/j.dsr2.2006.10.003
28. Felder DL, Martin JW, Goy JW (1985) Patterns in early postlarval development of decapods. In: Wenner AM (ed) Larval growth. A.A. Balkema, Rotterdam
29. George RW (1997) Tectonic plate movements and the evolution of *Jasus* and *Panulirus* spiny lobsters. Mar Freshw Res 48:1121–1130. doi:10.1071/MF97202
30. George RW (2006) Tethys Sea fragmentation and speciation of *Panulirus* spiny lobsters. Crustaceana 78:1281–1309. doi:10.1163/156854005776759780
31. Goldstein J, Matsuda H, Butler M (2006) Success! Caribbean spiny lobster, *Panulirus argus* is cultured from egg to juvenile for the first time. Lobster Newsl 19(1):3–4. http://www.odu.edu/~mbutler/newsletter/index.html
32. Goldstein JS, Matsuda H, Takenouchi T (2008) A description of the complete development of larval Caribbean spiny lobster, *Panulirus argus* (Latreille, 1804) in culture. J Crust Biol 28(2):306–327. doi:10.1651/0278-0372(2008)0 28[0306:TCDOLC]2.0.CO;2
33. Gore RH (1985) Molting and growth in decapod larvae. In: Wenner AM (ed) Larval growth. A.A. Balkema, Rotterdam
34. Griffin DA, Wilkin JL, Chubb CF et al (2001) Mesoscale oceanographic data analysis and data assimilative modelling with application to western Australian fisheries. FRDC Projecst No. 1997/1.39, CSIRO Marine Research, August 2001
35. Inoue N, Sekiguchi H (2001) Distribution of late-stage phyllosoma larvae of *Panulirus japonicus* in the Kuroshio Subgyre. Mar Freshw Res 52:1201–1209. doi:10.1071/MF01184
36. Jeffs AG (2005) Metamorphosis in spiny lobsters. Lobster Newsl 18(2):10–12. http://www.odu.edu/~mbutler/newsletter/index.html
37. Jeffs AG, Holland RC (2000) Swimming behaviour of the puerulus of the spiny lobster *Jasus edwardsii* (Hutton, 1875) (Decapoda, Palinuridae). Crustaceana 73:847–856. doi:10.1163/156854000504859

38. Jeffs AG, Willmott ME, Wells RMG (1999) The use of energy stores in the puerulus of the spiny lobster *Jasus edwardsii* across the continental shelf of New Zealand. Comp Biochem Physiol A 123:351–357. doi:10.1016/S1095-6433(99)00073-2
39. Jeffs AG, Chiswell SM, Booth JD (2001) Distribution and condition of pueruli of the spiny lobster *Jasus edwardsii* offshore from north-east New Zealand. Mar Freshw Res 52:1211–1216. doi:10.1071/MF01182
40. Jeffs AG, Montgomery J, Tindle CT (2005) How do spiny lobster post-larvae find the coast? N Z J Mar Freshw Res 39:605–617
41. Lee TN, Clarke ME, Williams E (1994) Evolution of the Totugas Gyre and it influence on recruitment in the Florida Keys. Bull Mar Sci 54:621–646
42. Limbourn AJ, Babcock RC, Johnston DJ (2008) Post-settlement energy reserves in *Panulirus cygnus*: experimental effects of starvation on survival and nutritional condition. Mar Biol (Berl) 153:445–456. doi:10.1007/s00227-007-0825-7
43. Lyons WG (1981) Possible sources of Florida's spiny lobster population. Proc Gulf Caribb Fish Inst 33:253–266
44. Manzanilla-Dominguez H, Gasca R (2004) Distribution and abundance of phyllosoma larvae (Decapoda, Palinuridae) in the southern Gulf of Mexico and the western Caribbean Sea. Crustaceana 77:75–93. doi:10.1163/156854004323037900
45. McWilliam PS, Phillips BF (1997) Metamorphosis of the final phyllosoma and secondary lecithotrophy in the puerulus of *Panulirus cygnus* George: a review. Mar Freshw Res 48:783–790. doi:10.1071/MF97159
46. McWilliam PS, Phillips BF (2007) Spiny lobster development: mechanism(s) inducing metamorphosis to the puerulus stage: a review. Rev Fish Biol Fish 17:615–632. doi:10.1007/s11160-007-9067-5
47. Owen RW (1981) Fronts and eddies in the sea: mechanisms, interactions and biological effects. In: Longhurst AR (ed) Analysis of marine ecosystems. Academic Press, London
48. Pearce AF, Phillips BF, Crossland CJ (1992) Larval distributions across the Leeuwin Current: report on RV *Franklin* Cruise Fr8/87 in August/September 1987. CSIRO Marine Laboratories Report No. 217
49. Phillips BF (1981) The circulation of the southeastern Indian Ocean and the planktonic life of the western rock lobster. Oceanogr Mar Biol Ann Rev 19:11–39
50. Phillips BF, Pearce AF (1997) Spiny lobster recruitment off Western Australia. Bull Mar Sci 61:21–41
51. Phillips BF, Rimmer DW, Reid DD (1978) Ecological investigations of the late-stage phyllosoma and puerulus larvae of the western rock lobster *Panulirus longipes cygnus*. Mar Biol (Berl) 45:347–357. doi:10.1007/BF00391821
52. Phillips BF, Brown PA, Rimmer DW et al (1979) Distribution and dispersal of the phyllosoma larvae of the western rock lobster, *Panulirus cygnus*, in the southeastern Indian Ocean. Aust J Mar Freshw Res 30:773–783. doi:10.1071/MF9790773
53. Phillips BF, Jeffs AG, Melville-Smith R et al (2006) Changes in lipid and fatty acid composition of late larval and puerulus stages of the spiny lobster (*Panulirus cygnus*) across the continental shelf of Western Australia. Comp Biochem Physiol Part B 143:219–228 (see my note 2). doi:10.1016/j.cbpb.2005.11.009
54. Pitcher CR, Turnbull C, Atfield J et al (2005) Biology, larval transport modelling and commercial logbook data analysis to support management of the NE Queensland rock lobster *Panulirus ornatus* fishery. CSIRO, Queensland Government Department of Primary Industries and Fisheries, Australian Government Fisheries Research and Development Corporation. Project No. 2002/008, September 2005
55. Pollock DE (1986) Review of the fishery for and biology of rock lobster *Jasus lalandii* with notes on larval recruitment. Can J Fish Aquat Sci 43:2107–2117
56. Rennie SJ, Pattiaratchi CB, Ivey GN (2007) Eddy formation through the interaction between the Leeuwin Current, Leeuwin Undercurrent and topography. Deep Sea Res Part II Top Stud Oceanogr 54:818–836. doi:10.1016/j.dsr2.2007.02.005

57. Richards WJ, Potthoff T (1980) Distribution and seasonal occurrence of larval pelagic stages of spiny lobsters (Palinuridae, *Panulirus*) in the western tropical Atlantic. Proc Gulf Caribb Fish Inst 33:244–252
58. Ritz DA (1972) Factors affecting the distribution of rock lobster larvae (*Panulirus longipes cygnus*) with reference to variability of plankton-net catches. Mar Biol (Berl) 13:309–317. doi:10.1007/BF00348078
59. Shafik S (1990) The Maastrichian and early tertiary record of the great Australian basin and its onshore equivalents on the Australian southern margin. J Aust Geol Geophys 11:473–497
60. Shirai SM, Yoshimura T, Konishi K et al (2006) Identification of phyllosoma larvae: a molecular approach for Japanese *Panulirus* lobsters (Crustacea: Decapoda: Palinuridae) using mitochondrial rDNA region. Species Divers 11:307–325
61. Sims HW, Ingle RM (1966) Caribbean recruitment of Florida's spiny lobster populations. Q J Fla Acad Sci 29:207–243
62. Sponaugle S, Cowen RK, Shanks A (2002) Predicting self-recruitment in marine populations: biophysical correlates. Bull Mar Sci 49:341–375
63. Strzelecki J, Koslow A, Waite A (2007) Comparison of mesozooplakton communities from a pair of warm- and cold-core eddies off the coast of Western Australia. Deep Sea Res Part II Top Stud Oceanogr 54:1103–1112. doi:10.1016/j.dsr2.2007.02.004
64. Tomczak M, Godfrey JS (2005) Regional oceanography: an introduction, 2nd edn. Daya Publishing House, Delhi
65. Tranter DJ, Kerr JD (1969) Seasonal variation in the Indian Ocean along 110°E. Aust J Mar Freshwat Res 20:77–814. doi:10.1071/MF9690077
66. Waite AM, Thomson PA, Pesant S et al (2007a) The Leeuwin Current and its eddies: an introductory overview. Deep Sea Res Part II Top Stud Oceanogr 54:89–796
67. Waite AM, Muhling BA, Holl CM et al (2007b) Food web structure in two counter-rotating eddies baed on $\delta^{15}$ N and $\delta^{13}$ C isotopic analyses. Deep Sea Res Part II Top Stud Oceanogr 54:1055–1107. doi:10.1016/j.dsr2.2006.12.010
68. Yeung C (2005) The Florida spiny lobster. Lobster Newsl 18(2):12–13. http://www.odu.edu/~mbutler/newsletter/index.html
69. Yeung C, Lee TN (2002) Larval transport and retention of the spiny lobster, *Panulirus argus*, in the coastal zone of the Florida Keys, USA. Fish Oceanogr 11(5):286–309. doi:10.1046/j.1365-2419.2002.00209.x
70. Yeung C, McGowan MF (1991) Differences in inshore–offshore and vertical distribution of phyllosoma larvae of *Panulirus*, *Scyllarus* and *Scyllarides* in the Florida Keys in May–June, 1989. Bull Mar Sci 49:699–714
71. Yeung C, Jones DL, Criales MM et al (2001) Influence of coastal eddies and counter-currents on the influx of spiny lobster, *Panulirus argus*, postlarvae in Florida Bay. Mar Freshw Res 52:217–232. doi:10.1071/MF01110
72. Yoshimura T (2005) The final-stage phyllosoma metamorphoses to the puerulus while caught up in the flow of the Kuroshio? Lobster Newsl 18(2):18–21 http://www.odu.edu/~mbutler/newsletter/index.html
73. Yoshimura T, Yamakawa H, Kozasa E (1999) Distribution of final stage phyllosoma larvae and free-swimming pueruli of *Panulirus japonicus* around the Kuroshio Current off southern Kyusyu, Japan. Mar Biol (Berl) 33:293–306. doi:10.1007/s002270050468
74. Yoshimura T, Morinaga K, Shirai S et al (2002) Palinurid phyllosoma larvae and their distribution in winter off the Pacific coast of Japan. Fish Sci 68(supp 1):194–197

# CHAPTER 13

# Lipids and Fatty Acids of the White Sea Herring *Clupea pallasi marisalbi* Berg (Clupeiformes, Clupeidae) from Different Habitats of the White Sea

**Svetlana A. Murzina [1,*], Zinaida A. Nefedova [1], Svetlana N. Pekkoeva [1], Tatiana R. Ruokolainen [1], Pauli O. Ripatti [1], Andrey V. Semushin [2] and Nina N. Nemova [1]**

[1]Institute of Biology, Karelian Research Centre of the Russian Academy of Sciences, Petrozavodsk 185910, Russia,,,,

[2]Knipovich Polar Research Institute of Marine Fisheries and Oceanography (PINRO), Arkhangelsk 163002, Russia

## ABSTRACT

Habitation conditions significantly affect the physiological and biochemical state of aquatic organisms, including the balance of lipids that performs important functions in cellular metabolism. The lipid and fatty acid profiles of White Sea herring in Dvina, Kandalaksha, and Onega bays, and in the waters of Tersky Shore (the mouth of the Varzuga River) in autumn (after the fattening period) were studied. Different methods—thin-layer, high-performance liquid and gas chromatography—were used for this purpose. Determined heterogeneity on studied parameters of lipid metabolism of fish from different habitats is associated with differences in trophic and hydrological conditions in these bays. Variations of lipid compositions of herring groups can be a reflection of qualitative and quantitative strategies of biochemical adaptations, aimed at compensating the lipid metabolism of fish under different trophic-ecological conditions of habitation in the White Sea.

**Keywords:** biochemical adaptations; biochemical indicators; ecological monitoring; fatty acids; lipids; the Arctic; White Sea herring

# 1. INTRODUCTION

The White Sea herring population, *Clupea pallasi marisalbi* Berg (Clupeiformes, Clupeidae), is one of the most similar forms to the Pacific herring, *Clupea pallasi*, and it is one of the most important commercial fish of the White Sea. Local herring herds are confined to certain hydrological conditions in different bays of the sea [1]. Previously, it was thought that White Sea herring from different bays of the White Sea belonged to a single population, a "single biological entity" [2,3]. However, there are other points of view, proponents of which consider that the forms of White Sea herring from different habitats in the sea have different origins [4,5,6]. It is also supposed that herring are a complex of genetically-isolated forms, related by origin [7]. In recent years, the growth of herring stock has been caused by gradual recovery of White Sea eelgrass, the decline of which happened in 1960 [8,9]. Habitation conditions significantly affect the physiological and biochemical states of aquatic organisms, including the balance of lipids that perform important functions in cellular metabolism. It is known that lipid parameters characterize—to a high degree of accuracy—the functional state of individual species and fish populations as a whole [10,11,12]. Changing of lipid composition is a part of adaptive reactions that provide the possibility of survival under different combinations of ecological factors (within the limits of optimal oscillations) [10,13].

The main aim of the present research was to determinate the lipid and fatty acid profiles of the White Sea herring in Dvina, Kandalaksha, and Onega bays, and in the waters of Tersky Shore (the mouth of the Varzuga River) in autumn (after the fattening period).

# 2. RESULTS

To study the combined effects of two factors—"fish length" (AB, cm) and "habitat" in the White Sea (complex of ecological environmental factors) on lipid indexes—total lipids, phospholipid, triacylglycerol, cholesterol ester and cholesterol (% of dry weight)—multifactor analysis of variance was applied. It was shown there was no effect of "fish length" factors on lipid indexes, despite the differences ($p \leq 0.05$) in size of herring from different capture locations in the White Sea (Table 1). The influence of the "habitat" factor ($p \leq 0.05$) was mainly determined for all lipid indexes (except for PL). The most notable influence of habitat factor was defined in terms of CHOL—30.1% ($p = 0.0000$), TL—19.5% ($p = 0.0001$), and TAG—12.6% ($p = 0.0076$) (Table 1).

**Table 1.** Total lipids and lipid classes (% dry weight) of the White Sea herring, *Clupea pallasi marisalbi*, collected from different bays of the White Sea.

| Parameter \ Sampling Place | Dvina Bay | Kandalaksha Bay | Tersky Shore | Onega Bay | MANOVA | | | | | |
|---|---|---|---|---|---|---|---|---|---|---|
| Number of fish | 20 | 21 | 15 | 25 | | | | | | |
| Fish length, cm | 11.9 ± 0.17 [2,3,4] (10.5–13.0) | 12.4 ± 0.13 [1,3,4] (11.0–13.5) | 11.2 ± 0.22 [1,2,4] (9.0–12.5) | 9.98 ± 0.39 [1,2,3] (8.0–15.0) | *p* | | % | | *F* | |
| % dry weight | | | | | | | | | | |
| TL | 34.5 ± 0.6 [2,4] | 41.7 ± 1.4 [1,3,4] | 34.7 ± 0.8 [2,4] | 31.8 ± 0.8 [1,2,3] | 0.7623 | 0.0001 | 12.1 | 19.5 | 0.75 | 8.50 |
| PLs | 19.1 ± 1.3 [4] | 18.4 ± 1.0 [4] | 17.2 ± 0.8 [4] | 14.3 ± 0.7 [1,2,3] | 0.5593 | 0.1608 | 21.4 | 5.9 | 0.93 | 1.78 |
| TAGs | 14.1 ± 1.2 [2] | 19.4 ± 0.9 [1,3,4] | 12.2 ± 1.0 [2,4] | 15.5 ± 0.9 [2,3] | 0.3771 | 0.0076 | 22.0 | 12.6 | 1.10 | 4.39 |
| EtCHOLs | 0.4 ± 0.1 [2,3] | 0.7 ± 0.1 [1] | 0.8 ± 0.1 [1] | 0.5 ± 0.1 | 0.3845 | 0.0301 | 26.1 | 10.9 | 1.09 | 3.20 |
| CHOL | 0.9 ± 0.1 [2,3,4] | 3.2 ± 0.3 [1,3,4] | 4.5 ± 0.9 [1,4] | 1.5 ± 0.1 [1,2,3] | 0.9976 | 0.0000 | 6.3 | 30.1 | 0.31 | 10.57 |
| CHOL/PLs | 0.05 ± 0.01 [2,3,4] | 0.19 ± 0.03 [1,4] | 0.29 ± 0.08 [1,4] | 0.11 ± 0.01 [1,2,3] | - | - | - | - | - | - |

Data means mean ± standard deviation for length of fish—m ± m (min–max). [1] Indicates significant differences ($p \leq 0.05$) in comparison to fish from Dvina Bay; [2] indicates significant differences ($p \leq 0.05$) in comparison to fish from Kandalaksha Bay; [3] indicates significant differences ($p \leq 0.05$) in comparison to fish from the Tersky Shore; and [4] indicates significant differences ($p \leq 0.05$) in comparison to fish from Onega Bay. TL: total lipids; PL: phospholipids; TAG: triacylglycerols; EtCHOL: cholesterol esters.

Biochemical analysis revealed that herring caught in the Kandalaksha Bay differed from those of the Dvina and Onega bays and the water area of the Tersky shore by having a higher level of total lipids (41.7%; 34.5%; 34.5%; 31.8% and 34.7% of dry weight, respectively) (Table 1).

Among the fish of the Kandalaksha and Onega bays, TAG dominated in the composition of TL (19.4% and 15.5% of dry weight, respectively), while, among the herring from Dvina Bay and the Tersky shore, structural PL dominated (19.1% and 17.2%, respectively). Among the herring from all the areas of the White Sea, the composition of total PL–PC was the largest in terms of quantity (within 10.1%–16.1% of dry weight). It was higher among the fish of Dvina Bay ($p \leq 0.05$), and it was lower in herring from Onega Bay (Table 2). The content of other individual PL: PI, PS, PEA, LysoPC and SFM, in herring from different habitats, was minor (from 0.15% to 0.95% of dry weight); although, in some cases, it was statistically significant ($p \leq 0.05$). Thus, the herring from the water area of the Tersky Shore, in comparison with other groups, differed by a reduced portion of LysoPC, and, from Dvina Bay, SFM (Table 2).

**Table 2.** Total phospholipids and individual phospholipids classes (% dry weight) of the White Sea herring *Clupea pallasi marisalbi*, collected from different bays of the White Sea.

| Sampling Place/Parameter | Dvina Bay | Kandalaksha Bay | Tersky Shore | Onega Bay |
|---|---|---|---|---|
| Number of Fish | 20 | 21 | 15 | 25 |
| PLs | $19.1 \pm 1.3^{4}$ | $18.4 \pm 1.0^{4}$ | $17.2 \pm 0.8^{4}$ | $14.3 \pm 0.7^{1,2,3}$ |
| PI | $0.52 \pm 0.08$ | $0.68 \pm 0.05^{3,4}$ | $0.38 \pm 0.04^{2}$ | $0.43 \pm 0.03^{2}$ |
| PS | $0.80 \pm 0.15$ | $0.79 \pm 0.10^{4}$ | $0.64 \pm 0.07$ | $0.54 \pm 0.04^{2}$ |
| PEA | $0.59 \pm 0.11$ | $0.82 \pm 0.07$ | $0.85 \pm 0.10$ | $0.62 \pm 0.12$ |
| PC | $16.06 \pm 1.07^{2,3,4}$ | $13.39 \pm 0.78^{1,4}$ | $12.93 \pm 0.84^{1,4}$ | $10.11 \pm 0.55^{1,2,3}$ |
| LysoPC | $0.55 \pm 0.12^{2,3,4}$ | $1.44 \pm 0.13^{1,3}$ | $0.25 \pm 0.02^{1,2,4}$ | $1.28 \pm 0.14^{1,3}$ |
| SFM | $0.15 \pm 0.05^{2,3,4}$ | $0.54 \pm 0.08^{1,4}$ | $0.95 \pm 0.23^{1}$ | $0.95 \pm 0.12^{1,2}$ |
| Unidentified | $0.39 \pm 0.10^{3}$ | $0.75 \pm 0.18^{4}$ | $1.20 \pm 0.25^{1,4}$ | $0.41 \pm 0.06^{2,3}$ |

PLs: phospholipids; PI: phosphatidylinositol; PS: phosphatidylserine; PC: phosphatidylcholine; PEA: phosphatidylethanolamine; LysoPC: lysophosphatidylcholine; and SFM: sphingomyelin. [1] Indicates significant differences ($p \leq 0.05$) in comparison to fish from Dvina Bay; [2] indicates significant differences ($p \leq 0.05$) in comparison to fish from Kandalaksha Bay; [3] indicates significant differences ($p \leq 0.05$) in comparison to fish from the Tersky Shore; and [4] indicates significant differences ($p \leq 0.05$) in comparison to fish from Onega Bay.

The content of other structural lipids—CHOL, and its reserve form EsCHOL—among the fish from Kandalaksha Bay and the water area of the Tersky shore—was higher ($p \leq 0.05$) than among fish from Dvina and Onega bays. At the same time, one of the key indexes of biomembrane microviscosity—CHOL/PL—was significantly higher (0.29) due to CHOL in the groups of herring from the water area of the Tersky shore than among the fish from other bays (0.05–0.19) (Table 1).

Thus, the group of herring from Kandalaksha Bay, which had the highest level of TL (due to reserve TAG), stands out among all the studied groups of White Sea herring, where habit is different in terms of trophoecological conditions. At the same time, in herring from Onega Bay, a reduced level of TL was shown; and among the fish of Dvina Bay, the portion of SFM decreased and the portion of PC increased. The highest index of CHOL/PL was revealed among herring from the water area of the Tersky shore; while in herring from Dvina bay, it was lowest.

The discriminant analysis of White Sea herring, inhabiting different areas of the White Sea, where the main classes of lipids—PL, TAG, EsCHOL and CHOL—were selected as variables, allowed to divide the studied groups of herring in accordance with three discriminant functions. The first function determined 63% of variability of the studied objects (Wilks' lambda $\lambda = 0.27$, $p = 0.0000$), the second function determined 29% ($\lambda = 0.58$, $p = 0.0000$), and the third determined 8% ($\lambda = 0.88$, $p = 0.0071$). When location of the research objects was in the space of the first and second discriminant functions (Figure 1), two groups—the herring of the Kandalaksha Bay and the Tersky shore—which slightly transgress with each other, clearly stood out.

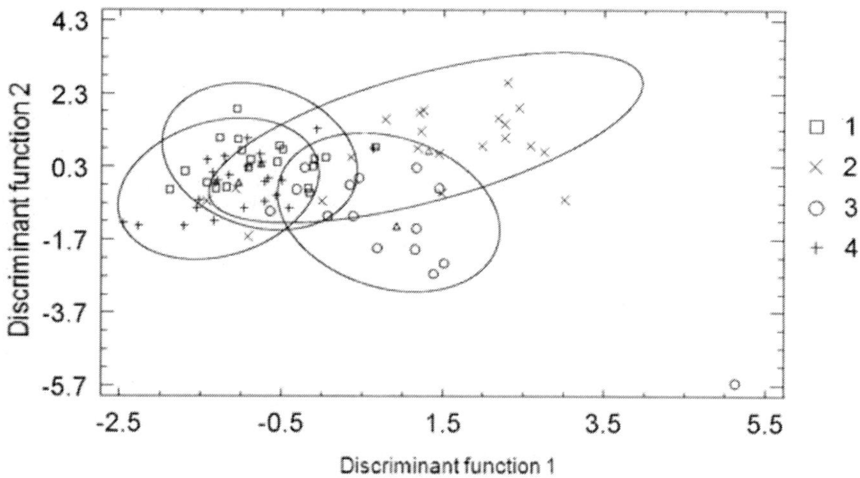


**Figure 1.** Discriminant analysis of the White Sea herring from different areas of the sea on contents of PLs (phospholipids), TAGs (triacylglycerols), EfCHOL (cholesterol esters), CHOL (cholesterol) in the cloud of two main discriminant functions. 1—herring from Dvina Bay, 2—herring from Kandalaksha Bay, 3—herring from Tersky Shore, 4—herring from Onega Bay. Ovals mean the density, location between two axes (strength of factor), and overlap of fish from different biotopes on studied lipid parameters.

The groups of herring from the Onega and Dvina bays showed a strong joint transgression. The first function was formed by CHOL (load: 0.85), PL (load: 0.79), TAG (load: 0.59), and EsCHOL (load: 0.46), which demonstrated and determined the direction of fish variability in terms of the qualitative and quantitative content of lipid classes—primarily structural. The second function was characterized by a significant contribution of TAG (factor load: 0.94) and PL (load: 0.62), which were opposed by ESCHOL (load: 0.10) and CHOL (load: 0.44). This result confirmed the conclusions of the biochemical analysis regarding different quantitative compositions of energetic and structural lipids in the groups of herring from different habitats of the sea. The third dispersion function indicated the direction and speed of energetic and structural metabolism, which was evidenced by the factor load of TAG (0.38), and made the greatest contribution to the formation of this axis, and opposed the factor load of PL (−0.71).

In the fatty acid profile of TL of White Sea herring in the fish groups from all studied areas, MUFA (monounsaturated) dominated (from 38.8% to 42.1% of total FA) in fish studied collected from all sites, among which

18:1(n-9) and 16:1(n-7) FA (within the limits of 18.9%–20.1% and 7.73%–11.5% of total FA, respectively) prevailed (Table 3). The fish from Kandalaksha Bay had increased ($p \leq 0.05$) content of 18:1(n-9) FA in relation to those of the Tersky shore and Onega Bay. At the same time, the herring from Onega Bay differed by a low ($p \leq 0.05$) level of 16:1(n-7) FA. It was determined that the amount of other MUFA—18:1(n-7), 20:1(n-9), 22:1(n-11), 22:1(n-7), 24:1(n-9)—did not exceed 5.0% of total FA, but it significantly ($p \leq 0.05$) differed in groups from different habitats. In the herring groups of the compared habitats, variations with minor specific FA, 20:1(n-9) and 22:1(n-11), 22:1(n-7), and 24:1(n-9), require additional studies (Table 3).

**Table 3.** Fatty acids profile (% sum of FAs) of the White Sea herring *Clupea pallasi marisalbi* collected from different bays of the White Sea.

| Fatty Acids/Sampling Place | Dvina Bay | Kandalaksha Bay | Tersky Shore | Onega Bay |
|---|---|---|---|---|
| Number of Fish | 20 | 21 | 15 | 25 |
| 14:0 | 10.53 ± 0.35 [2,4] | 12.40 ± 0.28 [1,3,4] | 11.01 ± 0.20 [2,4] | 8.80 ± 0.43 [1,2,3] |
| 16:0 | 21.66 ± 0.47 [3] | 21.40 ± 0.35 [3] | 20.43 ± 0.23 [1,2,4] | 22.26 ± 0.62 [3] |
| 17:0 | 0.47 ± 0.06 [3,4] | 0.37 ± 0.06 [4] | 0.31 ± 0.01 [1,4] | 0.90 ± 0.02 [1,2,3] |
| 18:0 | 2.25 ± 0.06 [2,4] | 1.66 ± 0.07 [1,3,4] | 2.10 ± 0.01 [2,4] | 3.21 ± 0.17 [1,2,3] |
| ΣSFA | 37.12 ± 0.80 | 37.23 ± 0.66 [3] | 35.23 ± 0.33 [2] | 37.26 ± 0.81 |
| 16:1(n-9) | 0.73 ± 0.06 | 0.40 ± 0.04 [3,4] | 0.68 ± 0.02 [2] | 0.74 ± 0.10 [2] |
| 16:1(n-7) | 11.31 ± 0.29 [4] | 11.45 ± 0.22 [4] | 11.11 ± 0.23 [4] | 7.73 ± 0.24 [1,2,3] |
| 17:1(n-7) | 0.35 ± 0.04 [4] | 0.34 ± 0.04 [4] | 0.40 ± 0.03 [4] | 0.59 ± 0.03 [1,2,3] |
| 18:1(n-9) | 19.80 ± 0.35 | 20.14 ± 0.33 [3,4] | 18.86 ± 0.25 [2] | 18.40 ± 0.73 [2] |
| 18:1(n-7) | 4.05 ± 0.28 [2,4] | 3.19 ± 0.26 [1] | 3.62 ± 0.06 | 3.39 ± 0.10 [1] |
| 18:1(n-5) | 0.82 ± 0.05 [2,3,4] | 1.09 ± 0.05 [1,4] | 1.07 ± 0.05 [1,4] | 0.69 ± 0.03 [1,2,3] |
| 20:1(n-11) | 0.24 ± 0.05 | 0.17 ± 0.02 | 0.26 ± 0.06 | 0.58 ± 0.24 |
| 20:1(n-9) | 1.46 ± 0.12 [2,3] | 2.10 ± 0.21 [1] | 2.50 ± 0.22 [1] | 1.93 ± 0.56 |
| 22:1(n-9) | 0.32 ± 0.02 | 0.32 ± 0.02 | 0.37 ± 0.02 | 0.79 ± 0.42 |
| 22:1(n-7) | 0.23 ± 0.01 [3,4] | 0.21 ± 0.01 [4] | 0.20 ± 0.01 [1,4] | 0.43 ± 0.03 [1,2,3] |
| 24:1(n-9) | 0.22 ± 0.08 [4] | 0.27 ± 0.08 [3,4] | 0.04 ± 0.00 [2,4] | 1.18 ± 0.04 [1,2,3] |
| ΣMUFA | 41.40 ± 0.75 | 42.06 ± 0.51 | 41.77 ± 0.49 | 38.81 ± 1.57 |
| 16:3(n-6) | 0.35 ± 0.03 [2,4] | 0.21 ± 0.03 [1,3,4] | 0.37 ± 0.01 [2,4] | 0.01 ± 0.00 [1,2,3] |
| 18:2(n-6) | 1.31 ± 0.05 [3,4] | 1.21 ± 0.02 [3,4] | 1.50 ± 0.06 [1,2,4] | 1.06 ± 0.05 [1,2,3] |
| 18:3(n-6) | 0.15 ± 0.01 [2,3] | 0.11 ± 0.01 [1,3,4] | 0.09 ± 0.01 [1,2,4] | 0.15 ± 0.01 [2,3] |
| 20:4(n-6) | 0.28 ± 0.02 [3] | 0.30 ± 0.01 [3] | 0.37 ± 0.01 [1,2,4] | 0.29 ± 0.02 [3] |
| Σ(n-6) PUFA | 3.52 ± 0.16 [2,4] | 2.70 ± 0.07 [1,3] | 3.50 ± 0.11 [2,4] | 2.58 ± 0.09 [1,3] |
| 18:3(n-3) | 0.76 ± 0.06 [2] | 0.92 ± 0.03 [1,3] | 0.74 ± 0.03 [2,4] | 0.93 ± 0.07 [3] |
| 20:5(n-3) | 6.51 ± 0.52 | 6.61 ± 0.31 | 6.78 ± 0.38 | 7.38 ± 0.47 |
| 22:6(n-3) | 5.25 ± 0.45 [3,4] | 5.11 ± 0.27 [3,4] | 6.53 ± 0.18 [1,2] | 7.01 ± 0.65 [1,2] |
| Σ(n-3) PUFA | 15.12 ± 1.15 [4] | 15.84 ± 0.68 [4] | 17.06 ± 0.49 | 19.19 ± 1.35 [1,2] |
| ΣPUFA | 21.48 ± 1.38 | 20.71 ± 0.74 [3] | 23.0 ± 0.50 [2] | 23.92 ± 1.48 |
| (n-3)/(n-6) | 4.29 ± 0.02 [2,4] | 5.86 ± 0.01 [1,3,4] | 4.87 ± 0.01 [2,4] | 7.44 ± 0.01 [1,2,3] |
| 18:3(n-3)/18:2(n-6) | 0.58 ± 0.04 [2,4] | 0.76 ± 0.02 [1,3,4] | 0.50 ± 0.02 [2,4] | 0.89 ± 0.05 [1,2,3] |
| 16:0/18:1(n-9) | 1.10 ± 0.31 [4] | 1.07 ± 0.27 [4] | 1.08 ± 0.32 [4] | 1.24 ± 0.05 [1,2,3] |

The amounts of 12:0, 15:0, 20:0, 24:0, 14:1(n-5), 16:1(n-5), 20:1(n-7), 22:1(n-11), 16:2(n-9), 18:2(n-9), 20:2(n-9), 22:2(n-9), 16:2(n-7), 16:2(n-6), 20:2(n-6), 18:4(n-3), and 20:4(n-3) FAs were minor and are not present in the table. [1] Indicates significant differences ($p \leq 0.05$) in comparison to fish from Dvina Bay; [2] indicates significant differences ($p \leq 0.05$) in comparison to fish from Kandalaksha Bay; [3] indicates significant differences ($p \leq 0.05$) in comparison to fish from the Tersky Shore; and [4] indicates significant differences ($p \leq 0.05$) in comparison to fish from Onega Bay.

It was noted that, in the composition of TL of White Sea herring of all groups, the proportion of SFA was high (35.2%–37.3% of total FA), mainly due to 16:0 and 14:0 FA—the levels of which were in the range of 20.4%–22.3% and 8.8%–12.4% of total FA, respectively. Low content of 18:0 FA (1.66%–3.21% of total FA) was noted in fish from all water areas (Table 3).

In White Sea herring, the content of total PUFA (polyunsaturated fatty acids) was in the range of 20.7%–23.9% of total FA, mainly due to FA of the (n-3) family (15.1%–19.2% of total FA) with a dominance of FAs 20:5(n-3) and 22:6(n-3). The herring from Onega Bay differed slightly, but significantly ($p \leq 0.05$), by a high level of these acids, and indexes of (n-3)/(n-6) PUFA, 18:3(n-3)/18:2(n-6), and 16:0/18:1(n-9) (Table 3).

Discriminant analysis of White Sea herring, inhabiting different places of the White Sea, was carried out in terms of the content of 11 fatty acids of total lipids—14:0, 16:0, 18:0, 16:1(n-7), 18:1(n-9), 18:1(n-7), 20:1(n-9), 18:2(n-6), 20:4(n-6), 18:3(n-3), and 22:6(n-3). It was defined that these fatty acids allowed determining a 95% accuracy, which, of the studied groups, this research object belongs to. The first discriminant function (73% of total variability, $\lambda = 0.02$, $p = 0.0000$) enables to select a group of herring from Onega Bay, which was characterized by maximal (for 18:0 and 22:6(n-3) FA) and minimal (for 16:1(n-7) FA) values (Figure 2, Table 3).

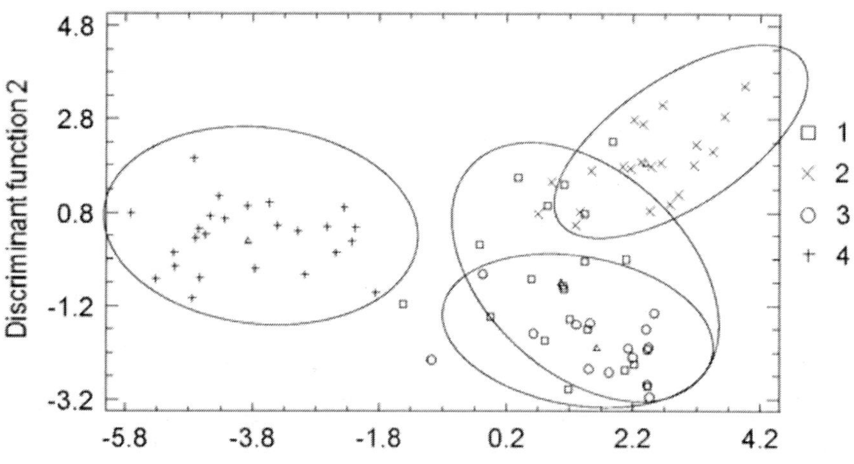


**Figure 2.** Discriminant analysis of White Sea herring from different areas of the sea, based on the contents of 11 fatty acids—14:0, 16:0, 18:0, 16:1(n-7), 18:1(n-9), 18:1(n-7), 20:1(n-9), 18:2(n-6), 20:4(n-6), 18:3(n-3), and 22:6(n-3), in the areas of the two main discriminant functions. 1—herring from Dvina Bay, 2—herring from Kandalaksha Bay, 3—

herring from the Tersky Shore, 4—herring from Onega Bay. Ovals mean the density, location between two axes (strength of factor) and overlap of fishes from different biotopes on studied fatty acids.

These FAs made the most significant contributions to the formation of the first discriminant function. The second discriminant function (19% of total variability, $\lambda$ = 0.19, $p$ = 0.0000) allows to select herring from Kandalaksha and Dvina bays and the Tersky shore. The herring from Dvina Bay showed uniform dissemination in a wide range, and mixing with herring from the Tersky Shore of the White Sea (Figure 2). The second discriminant axis is formed mainly by 18:2(n-6) and 18:3(n-3) FA, which have opposite factor loads: −1.15 and 1.75, respectively—the opposite correlation. Only 9% ($\lambda$ = 0.54, $p$ = 0.0000) refer to the third discriminant function. It was formed by the following FA: 18:1(n-9), 20:1(n-9), 20:4(n-6), and 22:6(n-3), which were opposed by 16:1(n-7), 18:1(n-7), 18:2(n-6), and 18:3(n-3). This was reasonably explained by the heterogeneity of ecological, particularly trophic, conditions in different habitats of the White Sea.

Thus, White Sea herring, from all the studied catch locations in the White Sea, were characterized by high levels of total MUFA ($p \geq 0.05$). The fish from Onega Bay were distinguished from the others in terms of variations of individual FA: With high contents of 17:0, 18:0, 17:1(n-7), 22:1(n-7), 24:1(n-9), 18:2(n-7), 22:6(n-3), and reduced contents of 14:0, 16:1(n-7), 18:1(n-5), 16:3(n-6), and 18:2(n-6). In addition to this, this group was characterized by high ratios of (n-3)/(n-6) PUFA, 18:3(n-3)/18: 3(n-6), and 16:0/18:1(n-9).

# 3. DISCUSSION

The results of multifactor analyses of variance show that, as a factor in these conditions, "fish length" has no significant effect on the lipid spectrum of White Sea herring. It is the "habitat", and more exactly, a complex of abiotic and biotic factors that have direct and indirect effects on the formation of differences in terms of lipid status groups inhabiting various places of the White Sea.

## 3.1. Total Lipids and Their Classes

A complex of factors, among which ecological factors—temperature regime and food base—play a leading role, influences the processes of accumulation and consumption of lipids, mainly TAG (storage lipids) in fish [13,14,15]. The high content of TAG in the herring (especially from Kandalaksha Bay) during feeding season (autumn) indicates that food-supply accumulation of energy reserves [13] is especially important for their overwintering. It is known that the hydrological conditions of Kandalaksha Bay contribute to intensive development of plankton. The biomass produced by planktonic bacteria in July–September is characterized by a maximal value [16]. Therefore, this bay is defined by considerable diversity and high levels of productivity of bacteria and phytoplankton, which are valuable food objects for herring and provides high levels of storage lipids.

Specific hydrological conditions connected with the income of large amounts of fresh water from rivers, the presence of strong reversing tidal currents, and significant fluctuations of temperature, salinity, transparency and water color—especially in a warm-up period, develop in different parts of the relatively shallow-waters of Onega Bay [17,18,19]. As a result, different conditions can form for all organisms in the areas that are close to the bay.

The content of TAG and PL was lower among the fish of Onega Bay than among the herring from Kandalaksha Bay. The TAG/PL ratio was the same. According to Reference [20], absolute and relative contents are less important than the ratio of individual lipid classes. It may be considered that, in case of different levels of structural and storage lipids in fish in Kandalaksha and Onega bays, comparatively favorable conditions form for groups of herring during the feeding period.

## 3.2. Membrane Lipids

Among the herring from all of the studied water areas of the White Sea, the dominance of PC was determined. Among the lipids, this PL was the most significant in terms of quantity among different species of fish [21,22]. The greatest variations of concentration are found in relation to PC and minor PL (LysoPC, SFM), depending on herring habitat. These membrane PLs perform specific regulatory functions that depend on their concentration, and have a modulating effect on the activity of membrane-bound enzymes, in accordance with the living conditions of organisms [23,24]. Changing of the level of minor LysoPC and SFM cell membranes takes place under the influence of external signals (for example, fluctuation of temperature, salinity, and other stressful effects of the environment) [23,25].

According to the data of the discriminant analysis, the main factor load relates to CHOL. Using biochemical and discriminant methods, it was determined that CHOL levels in the herring from Kandalaksha Bay and the water area of the Tersksy Shore were maximal, while among the herring from the Dvina and Onega bays, it was minor. The revealed quantitative differences of CHOL and CHOL/PL ratio in the groups of herring from different habitats of the White Sea can be associated with different functional activities of the membrane lipids of fish, which are subject to a number of processes: Their intake of food; enhanced synthesis in the liver; modification of biomembranes of cells; and increasing or reducing of physiological functions (motion activity at different rates of flow, turbulence degree, pressure, environmental temperature, etc.). Variations in relation to CHOL and CHOL/PL ratios among the herring groups are one of the ways of regulating the state (fluidity) of cellular membranes in the process of the adaptation of organisms to changing environmental conditions. It is known that enhancement of cholesterol biosynthesis in the liver, as well as in the case of intake in food, is accompanied by acceleration of its esterification [26], which was shown in our research of the herring from Kandalaksha Bay and the water area of the Tersksy Shore, and may reflect an adaptive response to excessive intake of CHOL.

## 3.3. Fatty Acids

It is known that FA components of lipids are the most labile and are involved in the process of the development of adaptive responses [27,28,29]. One of the characteristics of FA composition of aquatic organisms of the northern latitudes is a high level of MUFA. In all of the groups of herring from the studied habitats of the White Sea, the prevalence of MUFA due to 18:1(n-9) and 16:1(n-7) was revealed. It is known that MUFA are food borne and they are biomarkers for specific groups of aquatic organisms: 18:1(n-9) FA—the main biomarker of dinophytes (Dinophyta division, Dinophyceae class) and bacterial plankton; and 16:1(n-7) FA—diatom (Ochrophyta division, Bacillariophyceae class) microalgae [30,31,32]. Dinophytes and diatom microalgae are major contributors to the White Sea phytoplankton biomass [16]. The levels of production of phytoplankton vary considerably in different habitats of the White Sea; they also change depending on seasons and years. According to the literature, in the Kandalaksha Bay, the biomass produced by planktonic bacteria is maximal in July–September. By the beginning of autumn (mid-August) in the White Sea, the complex of planktonic algae develops—where, at first, large dinophytes and then diatoms dominate. At the end of October (sampling time), the period of phytoplankton growth finishes and, in water, there are only large species of dinophytes and diatoms that completely disappear [33]. Apparently, this explains the low level of

16:1(n-7) FA among the herring in Onega Bay in comparison to the fish from other catch areas.

In the groups of herring from all the habitats of the sea, minor MUFA, individually, did not exceed 4.1% of total FA: 18:1(n-7) FA is synthesized (up to 53.6%) by green (Chlorophyta division, Chlorophyceae class) algae: 20:1(n-9) and 22:1(n-11) FA are synthesized de novo by only phytovorous species of *Calanus* ssp. [32,34]. In the formation of the zooplankton community, the *Copepoda* order plays a main role. Almost everywhere, they form the basis of biomass in the White Sea [35,36].

Relatively high content of SFA and 16:0 and 14:0 FA prevailing in fish is connected not only with synthesis de novo, but also with their accumulation due to consumption of phytoplankton and zooplankton, the presence and availability of which, in the food chains of the herring from various areas of the White Sea, differed. Low levels of 14:0 and 16:1(n-7) FA in the herring of Onega Bay, in comparison with those from other habitats, is possibly associated with the decrease of the productivity of diatoms in this period of biomarkers, of which they are. An index of 16:0/18:1(n-9) FA was higher among the herring of Onega Bay, which indicates a higher degree of intensity of lipid metabolism [37].

Low content of PUFA, especially 22:6(n-3) FA among White Sea herring, in comparison with other Arctic and sub-Arctic species of fish, indicates that these FA were not the factor in the functioning of biomembranes under these conditions. The ratio (n-3)/(n-6) PUFA plays a particular role due to competitive relations in the process of their metabolism [38]. In relation to the herring from Onega Bay, the ratio of (n-3)/(n-6) PUFA was higher than in the fish groups of other habitats. Changing of the degree of unsaturation of lipids is a common mechanism of regulation of activity of membrane enzymes under the conditions of fluctuation of environmental factors [29,39]. As a result of the application of discriminant analysis, it was shown that one of the directions of White Sea herring variability, and its division into groups, was formed by 18:2(n-6) and 18:3(n-3) FA (the second discriminant function).

In herring in various tropho-ecological conditions, high levels of particular FA varied within certain limits and depended mainly on the temperature and food regimes in this period of life. Determined fatty acid spectra of White Sea herring from different habitats of the White Sea indicate that the main source of food in the feeding period (October) was representative of diatoms and dinophytes, and the proportion of copepods was insignificant.

# 4. MATERIALS AND METHODS

The White Sea herring adults (females) were collected in autumn in Dvina (64°57′, 38°23′), Kandalaksha (67°02′, 32°23′), and Onega bays (64°59′, 36°37′) and in waters of the Tersky Shore of the White Sea, at depths of 50, 25 and 38 meters, and at temperatures of 6.5, 2.9 and 6.7 °C, respectively, according to the method described in Reference [40]. In total, 66 individual specimens (*n*) were analyzed. Characteristics, specifics, and geographical coordinates of sampling areas are presented in Table 4.

**Table 4.** Characteristics of sampling areas in the White Sea.

| Characteristic/Sampling | Dvina Bay | Kandalaksha Bay | Tarsky Shore | Onega Bay |
|---|---|---|---|---|
| Coordinates | 64°57′<br>38°23′ | 67°02′<br>32°23′ | 66°10′<br>37°03′ | 64°59′<br>36°37′ |
| Depth, m | 50.0 | 25.0 | 38.0 | 38.0 |
| Water temperature, °C | 6.5 | 2.9 | 5.9 | 6.7 |
| Salinity, ‰ | 26.5 | 27.1 | 27.0 | 26.4 |

The flesh of the fish was homogenized and fixed in a solvent system of chloroform:methanol (2:1, *v*/*v*). Samples were homogenated individually in glass vials and kept in a cold room at 4 °C. Total lipids (TL) were extracted by the method described in Reference [41]. The dry weights, recovered after the lipid extraction of the tissues, were dried over phosphoric anhydride until the samples reached a constant weight.

The lipid classes, such as PLs, TAGs, cholesterols (CHOL), wax esters and cholesterol esters (WE + CE), as one fraction, were analyzed using thin-layer chromatography. PLs, TAGs and WE + CE were quantified using spectophotometery [42], following the method of Sidorov [43], and CHOL by the method used in Reference [44]. In this research, we use a SF-2000 spectrophotometer (OKB Spectr, St. Petersburg, Russia) at a wave length of 540 and 50 nm, respectively. To identify the lipid classes on the plates, standards (Sigma Aldrich, St. Louis, MO, USA) for thin-layer chromatography were used.

High-performance liquid chromatography (HPLC, Stayer, AKVILON, Moscow, Russia) was used to identify six phospholipids: phosphatidylserine, phosphatidylethanolamine, phosphatidylinositol, phosphatidylcholine, lysophosphatidylcholine, and sphingomyelin following the method used in Reference [45]. Nucleosil 100–7 column with an acetonitrile:hexane:methanol:phosphorus acid (918:30:30:17.5 by volume) mobile phase was applied, and the detection was at a wave length of 206 nm (UV light). Phospholipid standards (Supelco-Analytical, Bellefonte, PA, USA) were used for the identification and quantification of the phospholipid compounds in the sample.

To analyze the fatty acid composition of the total lipid extracts, gas–liquid chromatography was used. Fatty acids were converted to fatty acid methyl esters (FAME) and were identified using a "Chromatec-Crystal-5000.1" (Yoshkar-Ola, Russia) gas chromatograph with a flame-ionization detector. A Zebron capillary gas chromatographic column (Phenomenex, Torrance, CA, USA) was installed into a chromatograph, and an isothermal column configuration was used (205 °C); the temperature of the detector and evaporator were 250 and 240 °C, respectively. The internal standard was 22:0 FA. Chromatec-Analytik-5000.1 software, version 2.6 (Yoshkar-Ola, Russia), was used for recording and interpreting the data. Standard mixtures of Supelco 37 FAME mix (Supelco-Analytical). We compared the equivalent the length of carbon chain and table constants, according to Jamieson [46].

The research was carried out using the equipment of the environmental biochemistry laboratory and the facilities of the Equipment Sharing Centre of the Institute of Biology, KarRC of RAS—"Multidisciplinary Basic and Applied Research on the Functioning of Living Systems in the North".

The data were analyzed to determine whether they exhibited a normal distribution. The significance of data differences was assessed using one-way analysis of variance (ANOVA). Multifactor analysis of variance (MANOVA) was used to study the influence of two characteristics: "fish length" (AB, cm) and "habitat" (complex of ecological factors) on total lipids; and phospholipids, triacylglycerol, cholesterol, and cholesterol esters (% of dry weight). Differences were considered significant at $p \leq 0.05$. In the study, StatGrafics 2.5 software, Excel, and Stadia statistical packages were used.

Differences in lipid and fatty acid content between the groups of White Sea herring, inhabiting different parts of the White Sea, were determined using the discriminant analysis, where four lipid parameters: Phospholipid, triacylglycerol, cholesterol ester and cholesterol (% of dry weight), and 11 indexes of fatty acids—14:0, 16:0, 18:0, 16:1(n-7), 18:1(n-9), 18:1(n-7), 20:1(n-9), 18:2(n-6), 20:4(n-6), 18:3(n-3), and 22:6(n-3) (% of total FA) were chosen as variables. As a method of multivariate statistics, the discriminant analysis allows to determine differences between groups of objects that differ in terms of a number of characteristics. It also makes it possible to define whether an object belongs to a particular group [47]. The analysis shows which variables make the most significant contribution to the separation of objects into groups, and how accurate the selected characteristics are for further analysis of such samples.

# 5. CONCLUSIONS

The evaluation of lipid and fatty acid status of White Sea herring, *Clupea pallasi maris albi* Berg (Clupeiformes, Clupeidae), from different bays of the White Sea (Dvina, Kandalaksha, and Onega), as well as the water area of the Terssky Shore of the sea basin (the mouth of the Varzuga River) were carried out in autumn. The group of herring inhabiting Kandalaksha Bay, which had a high lipid status (due to content of storage TAG), stands out. In the herring from Onega Bay, a low level of total lipids, including PL, was defined. Discriminant analysis identified two subgroups of herring, in the Dvina and Onega bays, which differed in terms of level of total lipids and PLs from Kandalaksha Bay, in terms of total lipids, TAG, and CHOL. Previous studies have shown that polar lipids have strong anti-inflammatory properties [48]. Inflammation is a trigger phenomenon for a vast array of diseases, including Cardiovascular Diseases (CVDs), cancer, and stroke. Hence, food with high levels of polar lipids are regarded as beneficial against the onset of these diseases [49].

Shown differences in lipid and fatty acid spectrums in the groups of herring from compared habitats of the White Sea can be explained by different trophic, mainly, ecological conditions in the studied cites. At the same time, the influence of genetic factors is not excluded. Defined lipid status is the most indicative criteria, reflecting the physiological state of individual specimens and groups as a whole. Variations of lipid composition in herring groups can be a reflection of qualitative and quantitative strategies of biochemical adaptations, which aim to compensate lipid metabolism of fish under the different tropho-ecological conditions of habitation in the White Sea.

# ACKNOWLEDGMENTS

The research was made in the frame of the budgetary theme No. 0221-2014-0003 and was supported by RFBR 14-04-00473 (biochemical analysis, fatty acids), RFBR 16-45-100162 (habitats analysis).

# REFERENCES


1. Kuznetsov, V.V. *The White Sea and Its Biological Features of Flora and Fauna*; Academy of USSR: Moscow, Russia, 1960; pp. 57–298.
2. Lapin, Y.E. *White Sea herring as a Biological Entity. Laws of Fish Population Dynamics of the White Sea and Its Basin*; Nauka: Moscow, Russia, 1966; pp. 5–28.
3. Lapin, Y.E.; Pohilyuk, V.V. Spatial dynamics generations of the White Sea herring *Clupea pallasi marisalbi* Berg. *J. Ichthyol.* **1993**, *33*, 367–371.
4. Galkina, L.A. *The Origin and the Locality of the White Sea Herring. VIII Session of the Scientific Council on the Issue of "Biological Resources of the White Sea and the Inland Waters of the European North" (Abstracts)*; KarRC RAS: Petrozavodsk, Russia, 1969; pp. 207–208.
5. Dushkina, L.A. The pigmentation of the larvae of sea herring *Clupea* in the early stages of development in relation to population dynamics. In *Biology of the White Sea Herring*; Elsevier Applied Science: London, UK, 1975; pp. 227–254.
6. Muhomedyarov, F.B. Migration ways of the small White Sea herring based on their labeling. In *Biology White Sea Herring*; Elsevier Applied Science: London, UK, 1975; pp. 109–125.
7. Layus, D.L. *Population Structure of the White Sea Herring: Data of Karyological Analysis. The Results of Study of the White Sea Herring*; ZIN RAS: St. Petersburg, Russia, 1990; pp. 113–125.
8. Vekhov, V.N. Current status of eelgrass in the White Sea. In *The White Sea. Biological Resources and Problems of Their Rational Use. Part 1*; ZIN RAS: St. Petersburg, Russia, 1995; pp. 176–187
9. Alimov, A.F. Production Hydrobiology and functioning of ecosystems. In *New Ideas in Oceanology*; Nauka: Moscow, Russia, 2004; Volume 1, pp. 264–279.
10. Shatunovskiy, M.I. *Ecological Features of Metabolism of Marine Fish*; Nauka: Moscow, Russia, 1980; p. 288.
11. Gershanowitsch, A.D.; Lapin, V.I.; Shatunovskij, M.I. Specific features of lipid metabolism in fish. *Usp. Sovrem. Biol.***1991**, *3*, 207–219.
12. Pavlov, D.S.; Nefedova, Z.A.; Veselov, A.E.; Nemova, N.N.; Ruokolainen, T.R.; Vasileva, O.B.; Ripatti, P.O. Age dynamics of lipid status of juveniles of Atlantic salmon (*Salmo salar* L.) from the Varzuga river. *J. Ichthyol.* **2009**, *49*, 1073–1080.
13. Sidorov, V.S. *Ecological Biochemistry of Fishes*; Nauka: Leningrad, Russia, 1983; p. 240.
14. Hochachka, P.W.; Somero, G.N. *Biochemical Adaptation: Mechanism and Process in Physiological Evolution*; Oxford University Press: New York, NY, USA, 2002; p. 466.
15. Iverson, S.J. Tracing aquatic food webs using fatty acids: From qualitative indicators to quantitative determination. In *Lipids in Aquatic Ecosystems*; Arts, M.T., Brett, M.T., Kainz, M.J., Eds.; Springer: New York, NY, USA, 2009; pp. 281–309.
16. Iljash, L.B.; Zitina, L.S.; Fedorov, V.D. *Phytoplankton of the White Sea*; Janus-K: Moscow, Russia, 2003; pp. 48–154.
17. Primakov, I.M. Morphometry and oceanographic characteristics. In *Biological Resources of the White Sea: Study and Use. The Fauna of the Seas*; ZIN RAS: St. Petersburg, Russia, 2012; Volume 69, pp. 7–13.
18. Sapozhnikov, V.V.; Arzhanova, N.V.; Mordasova, N.V. Hydrochemical bases productivity. In *Biological Resources of the White Sea: Study and Use. The Fauna of the Seas*; ZIN RAS: St. Petersburg, Russia, 2012; Volume 69, pp. 13–34.
19. Troshkov, V.A. Zooplankton. In *Biological Resources of the White Sea: Study and Use. The Fauna of the Seas*; ZIN RAS: St. Petersburg, Russia, 2012; Volume 69, pp. 67–87.

20. Karagezyan, K. *The Phospholipids and Their Role in the Life of the Organism*; Yerevan State University: Yerevan, Armenia, 1977; pp. 25–29.
21. Nefedova, Z.A.; Ruokolainen, T.R.; Vasilyev, O.B.; Nemova, N.N.; Kashulin, N.A. Some lipid metabolism of different age females of *Coregonus albula* from the lake. Syamozero (Karelia) and Lake. Kovdor (Murmansk region) during feeding. *J. Ichthyol.* **2005**, *45*, 689–694.
22. Murzina, S.A. The Role of Lipids and Their Fatty Acid Components in the Biochemical Adaptations of the Daubed Shanny *Leptoclinus maculatus* F. from the North-Western Coastal Waters of Spitsbergen. Ph.D. Thesis, Karelian State Pedagogical Academy, Petrozavodsk, Russia, 2010; p. 184.
23. Dyatlovitskaya, E.V.; Bezuglov, V.V. Lipids like bioeffectors. Introduction. *Biochemistry* **1998**, *63*, 67–74.
24. Kolomiytseva, I.K.; Perepelkina, N.I.; Patrushev, I.V.; Popov, V.I. The role of lipids in the endoplasmic reticulum and assembly dictyosomes of the neuronal cells of the cells of the cerebral cortex Yakut gopher *Citellus undulates* during hibernation. *Biochemistry* **2003**, *68*, 954–967.
25. Christie, W.W. Available online: http://lipidlibrary.aocs.org/ Primer/content.cfm?ItemNumber=39292&navItemNumber=19204 (accessed on 15 May 2016).
26. Lopuhin, Y.M.; Archakov, A.I.; Vladimirov, Y.A.; Kogan, E.M. *Cholesterinoz*; Medicine: Moscow, Russia, 1985; p. 350.
27. Nemova, N.N.; Nefedova, Z.A.; Murzina, S.A.; Veselov, A.E.; Ripatti, P.O.; Pavlov, D.S. The effect of environmental conditions on the dynamics of fatty acids in juveniles of the Atlantic salmon (*Salmo salar* L.). *Russ. J. Ecol.* **2015**, *46*, 267–271.
28. Yanes-Roca, C.; Rhody, N.; Nystrom, M.; Main, K.L. Effects of fatty acid composition and spawning season patterns on egg quality and larval survival in common snook (*Centropomus undecimalis*). *Aquaculture* **2009**, *287*, 335–340.
29. Kreps, E.M. *Lipids of Cell Membrane. Evolution of Brain Lipids*; Nauka: Leningrad, Russia, 1981; p. 339.
30. Viso, A.C.; Marty, J.C. Fatty acids from 28 marine microalgae. *Phytochemistry* **1993**, *34*, 1521–1533.
31. Viron, C.; Saunois, A.; Andre, P.; Perly, B.; Laffose, M. Isolation and identification of unsaturated fatty acid methyl esters from marinew micro-algae. *Anal. Chim. Acta* **2000**, *409*, 257–266.
32. Graeve, M.; Lundberg, M.; Boer, M.; Kattner, G.; Hop, H.; Falk-Petersen, S. The fate of dietary lipids in the Arctic ctenophore *Mertensia ovum* (Fabricius 1780). *Mar. Biol.* **2008**, *153*, 643–651.
33. Berger, V.J. *Production Potential of the White Sea*; ZIN RAS: St. Petersburg, Russia, 2007; pp. 66–131.
34. Falk-Petersen, S.; Hopkins, C.C.E.; Sargent, J.R. Trophic relationships in the pelagic, arctic food web. In Proceedings of the 24th European Marine Biology Symposium, Trondheim, Norway, 12–16 May 1990; pp. 315–333.
35. Primakov, I.M.; Berger, V.Y. Production of planktonic crustaceans in the White Sea. *Mar. Biol.* **2007**, *33*, 356–360.
36. Primakov, I.M.; Kuznetsov, E.A. The study, sustainable use and protection of natural resources of the White Sea. Articles XI All-Russian conference with international participation. In *Spatial Distribution of Summer Zooplankton of the White Sea and South-Eastern Part of the Barents Sea*; ZIN RAS: St. Petersburg, Russia, 2010; pp. 142–143.
37. Kalchenko, E.I.; Videnskaya, T.L.; Yurjeva, M.I. Feeding and fatty acid content of wild and captured juveniles of dog-salmon during keeping in river fishponds. In Proceedings of the International Conference Modern Problems of Physiology and Biochemistry of Aquatic Organisms, Petrozavodsk, Russia, 6–9 September 2004; KarRC RAS: Petrozavodsk, Russia, 2004; p. 58.
38. Sergeeva, M.G.; Varfolomeeva, A.T. *The Cascade of Arachidonic Acid*; Narodnoe Obrazovanie: Miscow, Russia, 2006; pp. 21–38.

39. Boldyrev, A.A.; Kaivarainen, E.I.; Ilykha, V.A. *Biomembranology*; KarRC RAS: Petrozavodsk, Russia, 2006; pp. 24–102.
40. Semushin, A.V.; Sherstkov, V.S.; Rukhlova, V.A. Diversity of fish species in the trawl catches in the southeastern Barents Sea in 1980–2008. *J. Ichthyol.* **2011**, *51*, 717–737.
41. Folch, J.; Lees, M.; Sloane Stanley, G.H. A simple method for the isolation and purification of total lipids from animal tissue. *J. Biol. Chem.* **1957**, *226*, 497–509.
42. Walsh, D.E.; Banasik, O.J.; Gilles, K.A. Thin-layer chromatographic separation and colorimetric analysis of barley or malt lipid classes and their fatty acids. *J. Chromatogr.* **1965**, *17*, 278–287.
43. Sidorov, V.S.; Lisenko, E.I.; Bolgova, O.M.; Nefedova, Z.A. Lipids of Fish. I. Methods of Analyses. The Tissues Specificity of Vebdace Lipids (*Coregonus albula* L.). In *Salmonidae of Karelia. Issue 1 Ecology Parasites Biochemistry*; Potapova, O.I., Smirnov, Y.A., Eds.; Karelian Branch of the USSR Academy of Sciences, Institute of Biology: Petrozavodsk, Russia, 1972; pp. 150–161.
44. Engelbrecht, F.M.; Mari, F.; Anderson, J.T. Cholesterol determination in serum. A rapid direction method. *S. Afr. Med. J.* **1974**, *48*, 250–356.
45. Arduini, A. High performance liquid chromatography of long-chain acylcarnitine and phospholipids in fatty acid turnover studies. *J. Lipid Res.* **1996**, *37*, 684–689.
46. Jamieson, G.R. GLC-identification techniques for longchain unsaturated fatty acids. *J. Chromatogr. Sci.* **1975**, *13*, 491–497.
47. Korosov, A.V.; Gorbach, V.V. *Computer Processing of Biological Data*; KarRC RAS: Petrozavodsk, Russia, 2007; p. 76.
48. Zabetakis, I. *Marine Oils and Diseases in Marine Oils (from Sea to Pharmaceuticals)*; NOVA Science Publishers: New, York, NY, USA, 2015; pp. 83–90.
49. Morphis, G.; Kyriazopoulou, A.; Nasopoulou, C.; Sioriki, E.; Demopoulos, C.A.; Zabetakis, I. Assessment of the in Vitro antithrombotic properties of sardine (*Sardina pilchardus*) fillet lipids and cod liver oil. *Fishes* **2016**, *1*, 1–15.

# CHAPTER 14

# Distribution, Spread, and Habitat Predictability of a Small, Invasive, Piscivorous Fish in an Important Estuarine Fish Nursery

**Marin F. D. Greenwood**

*ICF, 630 K Street, Sacramento, CA 95814, USA*

## ABSTRACT

Invasive species often cause negative ecological and economic effects. Florida has >20 established invasive fish species but only seven exist in saltwater. The present study examined *Belonesox belizanus* (Pike Killifish), a Central American euryhaline fish introduced to Tampa Bay (west-central Florida) in the early 1990s, which has quantifiably reduced populations of small-bodied native fishes and may compete with prized sportfish juveniles in estuarine nursery habitat. Long-term monitoring revealed that *B. belizanus* occurs in estuarine waterbodies along a 31-km stretch of the bay's eastern fringe, with a second, smaller population in two western tributaries. Spread rate was estimated to be 5.5–13 km $year^{-1}$, intermediate among invasive poeciliids. A novel implementation of boosted regression tree modeling to assess *B. belizanus* habitat predictability found greater probability of presence with decreasing water depth and pH, whereas presence tended to be greatest at polyhaline salinity. It is hypothesized that *B. belizanus* distribution in Tampa Bay is constrained by deep, seawall habitats acting as ecological barriers. Further *B. belizanus* spread therefore may be most likely to occur by human release (from aquaria or bait buckets) or bird carry-off. Newly restored tidal habitat within the current range probably will be invaded quite quickly by *B. belizanus*.

**Keywords:** *Belonesox belizanus*; nursery; Tampa Bay; invasive fish; ecological barrier; boosted regression tree; predation; estuary; habitat; prediction

# 1. INTRODUCTION

Invasive species, habitat degradation or loss, and pollution are the most important threats to native fishes in the USA [1]. Invasive species are defined by the United States government as "with regard to a particular ecosystem, a non-native organism whose introduction causes or is likely to cause economic or environmental harm, or harm to human, animal, or plant health" [2]. The number of established nonnative fishes in the USA more than doubled between 1970 and 2013 [3]. Negative ecological effects from invasive fishes can include predation, competition, and interference [4]. Economic losses from invasive fishes have been estimated at several billion US dollars annually [5]. Successful establishment of invasive fishes in new habitats, whether through intentional or unintentional releases, may depend on a variety of factors including broad physiological tolerance, rapid dispersal, and similarity of habitat to the native range [6].

Prediction of invasive fish presence and habitat use is one of the most important applications of species distribution modeling [7]. Establishing species-habitat relationships that are general and broadly applicable rather than idiosyncratic to a narrow set of conditions is desirable for allowing successful prediction of presence in unsampled areas [8]. Prediction serves as an important means of validating the generality of observed species-habitat relationships and was recently argued by Houlahan et al. [9] to be essential for demonstrating ecological understanding.

Peninsular Florida is likely to contain more invasive freshwater fish than any other similarly sized area in the world [10], probably because the warm climate facilitates survival of tropical fish that are released from freshwater aquaria or that escape from fish farms [11]. Although freshwater invasive fishes are prevalent, there are relatively few established invasive saltwater or estuarine fishes in Florida: Schofield and Loftus [12] described 22 permanently established freshwater fishes, whereas Schofield et al. [13] noted only two permanently established marine fishes (the lionfishes *Pterois volitans* and *Pterois miles*). Six of the permanently

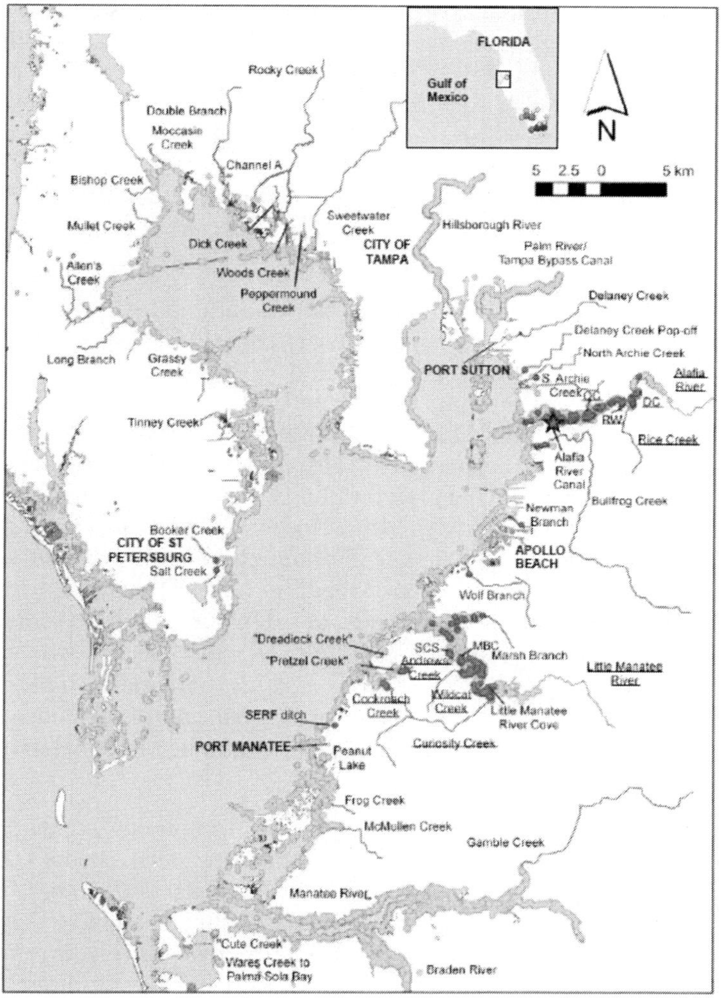


**Figure 1.** Presence of *Belonesox belizanus* in estuarine waterbodies of Tampa Bay, 1989–2007/2008, amended with permission from [24], with inset showing collections from elsewhere in Florida [15]. For main map, red dots represent presence (with a red star indicating the first occurrence, 1994), green dots represent absence. QC = Question Mark Creek; RW = Riverview West Creek; DC = Dogleg Creek; MBC = Mill Bayou Canal; SCS = Sun City Slough. Total *n* = 16,113 shoreline seines (see Supplementary Material). Underlined waterbodies were included in the boosted regression tree modeling of habitat predictability. Collections from elsewhere in Florida represent clustered records ranging from one specimen (white circle) to 20 or more specimens (dark blue circle).

established freshwater fishes are euryhaline and occur in saltwater habitat [13]. The Pike Killifish *Belonesox belizanus* Kner (Pisces: Poeciliidae) is among these species and was introduced to Miami-Dade County, south Florida, in 1957 when ~50 excess medical research specimens were released into an agricultural canal [14]. Although the main invasive distribution of *B. belizanus* is in south Florida (Figure 1), a secondary introduction occurred in the Tampa Bay watershed (west central Florida), possibly as a result of escape from a tropical aquaculture facility [15], and is the subject of the present study.

*B. belizanus* is native to Central American Atlantic drainages south of the Punta del Morro (Mexico) to northeastern Costa Rica [16,17] and although the largest species within the Poeciliidae, it is still relatively small (maximum size of 200-mm total length; [17]). All life stages of *B. belizanus* are carnivorous [17], with prominent teeth and a novel gape adaptation that allows the species to consume prey that are relatively large for its size [18]. The primary prey for all life stages (including neonates) is fish [14,19,20,21]. The ability to switch to invertebrate prey such as small shrimps is considered to allow *B. belizanus* to be a successful invader despite its specialized, invariant feeding mechanism [22]. In south Florida, *B. belizanus* was observed to deplete small resident fish populations after introduction [14,19] and the species was recently classified at the upper end of medium risk for potential negative effects [23]. Recognition of the ecological risk from *B. belizanus* is evidenced by the species being prohibited for possession in a number of American states, although not Florida.

There is strong quantitative evidence that the Tampa Bay population of *B. belizanus* has significantly reduced abundance of small-bodied resident fishes, in particular Sailfin Molly *Poecilia latipinna*, Eastern Mosquitofish *Gambusia holbrooki*, and Sheepshead Minnow *Cyprinodon variegatus* [24]. Although there is little evidence for direct effects on abundance or size distribution of estuary-dependent migratory species, which generally grow to larger sizes and recruit from larger source populations than resident fishes, the depletion of resident prey fishes by *B. belizanus* is of concern for its potential indirect effects on estuary-dependent species through competition [15]. In particular, Tampa Bay provides important nursery habitat for Common Snook *Centropomus undecimalis*, a prized sportfish supporting a substantial recreational fishery (>277,000 directed trips year$^{-1}$ worth $15.8 M year$^{-1}$; [25]), which experienced a considerable cold-kill-related decline in recent years [26] and occurs in estuarine areas in which *B. belizanus* is also found [27,28,29].

Knowledge of *B. belizanus* spread, distribution, and habitat associations in Tampa Bay is therefore of considerable importance both from the perspective of biodiversity (structuring of resident native fish assemblages) and potential effects on managed fisheries. Qualitative accounts of *B. belizanus* habitat in the native range describe occupation of a variety of shallow-water, low-flow habitats ranging from freshwater to hypersaline, including mangrove swamps, sinkholes, ponds, creeks, and river backwaters [17,30,31,32]. Quantitative evaluations of *B. belizanus* habitat have included multispecies analyses in the native range [33,34,35,36] and a detailed single-species evaluation of associations with physicochemical parameters from several datasets collected in south Florida [37]. The latter study found *B. belizanus* density was correlated with salinity and temperature at the southernmost, estuarine sites, whereas pH was most associated with changes in density at the northern, inland sites.

The present study used long-term monitoring and shorter duration sampling efforts [24] to ask three main questions. First, what is the distribution of *B. belizanus* in estuarine portions of the Tampa Bay watershed? Examination of the species' distribution allowed hypotheses to be developed for the potential factors constraining *B. belizanus* within Tampa Bay and, by extension, south Florida. Second, what is the spread rate of *B. belizanus*? Given the observed negative effects of the species, knowledge of spread rate is vital to understanding the expansion of the potential risk posed by *B. belizanus*, particularly because global climate change may provide more favorable temperatures over a greater spatial area of Florida and other states in the future [38]. Third, what are the habitat associations of *B. belizanus*? Following the philosophy of Houlahan et al. [9] that prediction should be used to aid ecological understanding, the present study employed a novel method to assess predictability of habitat use in test waterbodies based on boosted regression tree models developed from independent, training waterbodies. This allowed examination of the extent to which observed *B. belizanus*–habitat relationships were general and broadly applicable rather than idiosyncratic and limited to specific waterbodies. The present study is the first quantitative, single-species examination of *B. belizanus*–habitat relationships based solely on data collected in situ at the time of sampling, and demonstrates a method that can also be used for other species and locations. The main conclusions of the present study are that the principal Tampa Bay population of *B. belizanus* is distributed in eastern estuarine waterbodies, with a smaller, spatially restricted western population that is likely to have been a separate introduction. Although the rate of *B. belizanus* spread is comparable to that of some other invasive fishes, the distribution may be constrained by soft ecological barriers, which emphasizes the need to manage human-mediated spread. Success of *B. belizanus* in Tampa Bay appears facilitated by its spread rate, broad environmental tolerance, and relatively distinct ecological niche.

# 2. MATERIALS AND METHODS

All data used in the present study are provided as Supplementary Material.

## 2.1. Study Area and Sampling Techniques

Tampa Bay is Florida's largest open-water estuary (1031 km$^2$; [39]), a subtropical system that, although greatly altered by urbanization, still has relatively large areas fringed by mangroves, salt marshes, and other natural vegetation [29,40,41]. Four main tributaries (the Hillsborough, Alafia, Little Manatee, and Manatee rivers) provide freshwater inflow, with approximately 100 smaller tributaries also draining to the bay (Figure 1; [42]). The smaller tributaries and creeks range from relatively natural mangrove- or saltmarsh-lined waterbodies to urbanized systems with considerable amounts of seawall and riprap. Physicochemical conditions vary widely and salinity ranges from nearly fresh to polyhaline, depending on tides, precipitation, and wind. Tampa Bay is described in detail by Lewis and Estevez [39].

Analyses presented herein used data collected in 1989–2008 during diurnal shoreline seine sampling within estuarine waterbodies of the Tampa Bay watershed by the Florida Fish and Wildlife Conservation Commission, Fish and Wildlife Research Institute's Fisheries-Independent Monitoring program [24]. Data from trawling in deeper areas and seines deployed away from the shore were not considered because *B. belizanus* was never collected in these habitats. In all sampling, *B. belizanus* specimens were identified, counted, and measured (standard length, mm). Descriptive habitat data and physicochemical parameters were also recorded, as described further below. Secchi depth as a measure of water clarity was also measured, but is not included in analyses because sampling sites were sufficiently shallow that the Secchi disk was almost always visible on the substrate. Further details of the sampling procedures can be found elsewhere [24,43].

## 2.2. Data Analysis

### 2.2.1. Distribution and Spread Rate

Coordinates of sites at which *B. belizanus* was present or absent during seine sampling were overlaid on maps of the Tampa Bay watershed to assess the species' distribution. Long-term sampling in the lower Little Manatee River watershed was used to examine the rate of spread following first occurrence. This watershed was selected to examine rate of spread because it offered the most comprehensive, consistent longitudinal sampling and monthly sampling had already commenced before first occurrence (in contrast to the Alafia River, for example, wherein sampling was only seasonal at the time of first occurrence). Monthly 21.3-m shoreline seine sampling in the Little Manatee River watershed occurs in the lowermost 13 km of the river main stem and accessible adjacent habitats (Figure 2). Sampling is stratified into upstream and downstream zones 7 km upstream of the river mouth (Figure 2), with 4 samples per month in each zone. The temporal sequence of observations suggested that *B. belizanus* invaded the Little Manatee River from downstream to upstream. Following Ben Rais Lasram et al. [44], the rate of spread (S) was calculated:

$$S_t = \frac{(d_t - d_{t-1})}{(D_t - D_{t-1})} \quad (1)$$

where $d$ is distance upstream of the mouth of the Little Manatee River (km), $D$ is number of days from the first observation in the river, $t$ is a given observation of *B. belizanus*, and $t - 1$ is the previous downstream observation of *B. belizanus*.

### 2.2.2. Habitat Predictability

Habitat predictability of *B. belizanus* was investigated using data collected monthly in 2006 or 2007 with 9.1-m seines in 11 estuarine waterbodies within the species' range in the Tampa Bay watershed, consisting of small creeks, canals, and adjacent non-creek/canal habitats (river main-stems or bayous) (Figure 1, Table 1 and Table 2; [24]). The original purpose of sampling in these waterbodies was to assess the value of tidal creeks/canals as nekton habitat, with the adjacent non-creek/canal habitats generally also sampled to provide a contrast. Each waterbody was divided into longitudinal spatial strata (~200 m long) and daylight sampling was undertaken monthly at randomly chosen locations between each creek/canal's mouth and the upstream extent that could be reached within time or access constraints. Sites were chosen from a set of predefined geographic coordinates with a random number generator. The number of samples per waterbody was proportional to the extent of the waterbody that was sampled, with two samples per spatial stratum per month; thus, the number of seine samples ranged from four per month in smaller waterbodies (250–300 m long) to 10 per month in larger waterbodies (1000 m long). Adjacent waterbodies (rivers or bayou) also received two samples per month per stratum, with one stratum located upstream and downstream of the creek/canal mouths. A total of 935 shoreline seine samples was collected. Additional detailed description of the studied waterbodies is provided in [21,45].

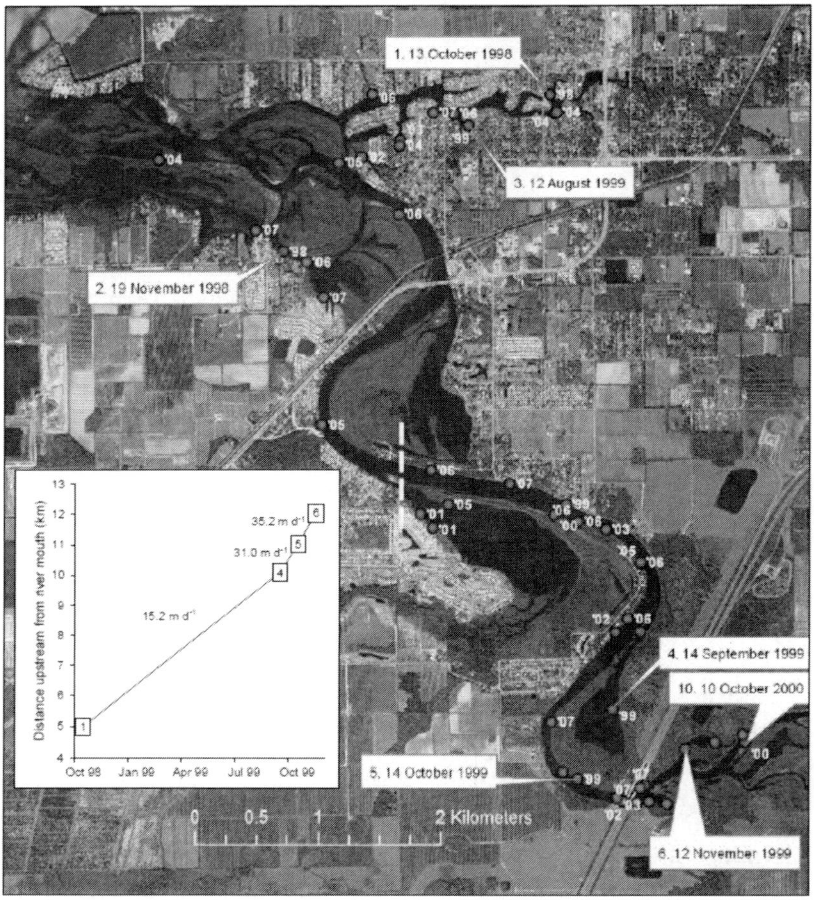


**Figure 2.** Presence (red dots with last two digits of year annotation) of *B. belizanus* in the lower Little Manatee River sampling area (km 0–13) from regular shoreline seine monitoring, 1997–2007. The first six presences are highlighted, along with the most upstream presence. The broken yellow line divides the upstream and downstream zones of monthly sampling. The inset graph shows rate of upstream spread from the first to sixth occurrence in 1997/1998.

**Table 1.** Tampa Bay estuarine waterbodies sampled to assess *Belonesox belizanus*habitat predictability, including mean and range of continuous habitat predictors.

| Waterbody Name | Waterbody Type | Year Sampled | Number of 9.1-m Seine Samples | Temperature (°C) | Salinity (Practical Salinity Scale) | Dissolved Oxygen (mg L⁻¹) | pH | Depth (m) | Gradient (%) |
|---|---|---|---|---|---|---|---|---|---|
| Alafia River | River | 2006 | 167 | 22.9 (13.9–30.3) | 3.5 (0.1–20.1) | 7.3 (3.4–16.5) | 7.7 (6.8–8.8) | 0.5 (0.1–1.0) | 20.6 (0.0–66.7) |
| Andrews Creek | Creek | 2007 | 72 | 25.1 (17.3–32.3) | 22.4 (6.3–32.8) | 5.5 (2.9–9.9) | 7.8 (7.5–8.1) | 0.5 (0.2–1.0) | 21.4 (0.0–66.7) |
| Cockroach Creek | Creek | 2007 | 120 | 23.8 (16.1–29.7) | 19.3 (0.4–33.0) | 3.8 (0.3–12.2) | 7.3 (6.8–8.1) | 0.5 (0.2–1.2) | 21.1 (0.0–60.0) |
| Curiosity Creek | Creek | 2006 | 120 | 22.5 (11.9–30.0) | 3.3 (0.2–17.2) | 6.5 (2.7–10.6) | 7.3 (6.4–7.9) | 0.7 (0.1–1.2) | 26.8 (0.0–66.7) |
| Dogleg Creek | Canal | 2006 | 48 | 22.6 (15.1–28.3) | 1.3 (0.2–4.0) | 6.9 (4.5–9.8) | 7.7 (7.3–8.3) | 0.7 (0.3–1.0) | 29.0 (6.7–53.3) |
| Hayes Bayou | Bayou | 2006 | 48 | 23.7 (12.8–30.5) | 11.6 (0.6–23.6) | 7.0 (3.6–10.3) | 7.6 (7.0–8.3) | 0.4 (0.1–1.0) | 14.3 (0.0–40.0) |
| Little Manatee River | River | 2006 | 96 | 23.8 (11.8–31.4) | 8.5 (0.1–27.0) | 6.7 (3.6–12.9) | 7.4 (6.3–7.9) | 0.5 (0.1–1.0) | 18.3 (0.0–53.3) |
| Question Mark Creek | Canal | 2006 | 48 | 23.4 (13.4–30.1) | 8.0 (1.0–16.4) | 6.5 (2.6–13.7) | 7.6 (6.9–8.4) | 0.5 (0.2–1.0) | 23.1 (0.0–46.7) |
| Rice Creek | Creek | 2006 | 48 | 21.7 (15.4–29.2) | 1.3 (0.1–8.1) | 8.1 (2.0–13.5) | 7.6 (7.1–8.8) | 0.5 (0.1–1.0) | 12.6 (0.0–66.7) |
| Riverview West Creek | Canal | 2006 | 48 | 23.3 (15.1–30.6) | 2.5 (0.2–8.9) | 7.8 (2.7–16.6) | 7.7 (6.9–8.7) | 0.5 (0.1–1.0) | 15.7 (0.0–40.0) |
| Wildcat Creek | Creek | 2006 | 120 | 22.7 (11.6–28.9) | 7.3 (0.2–23.4) | 6.5 (2.7–15.2) | 7.4 (7.0–8.1) | 0.5 (0.1–1.2) | 21.0 (0.0–66.7) |
| Overall | | | 935 | 23.2 (11.6–32.3) | 8.3 (0.1–33.0) | 6.4 (0.3–16.6) | 7.5 (6.3–8.8) | 0.5 (0.1–1.2) | 20.9 (0.0–66.7) |

**Table 2.** Shoreline and substrate type in Tampa Bay estuarine waterbodies sampled to assess *B. belizanus* habitat predictability.

| Waterbody Name | Shoreline Type | | | | | Substrate Type | | | |
|---|---|---|---|---|---|---|---|---|---|
| | Mangrove | Marsh | Shrubs/Trees | Small Terrestrial Veg. | Hardened | Mud | Mud-Sand | Sand | Oyster-Rock |
| Alafia River | 0.0% | 14.4% | 21.6% | 13.8% | 49.1% | 7.2% | 26.3% | 49.7% | 16.8% |
| Andrews Creek | 94.4% | 0.0% | 1.4% | 0.0% | 2.8% | 72.2% | 18.1% | 1.4% | 6.9% |
| Cockroach Creek | 99.2% | 0.0% | 0.0% | 0.0% | 0.0% | 69.2% | 20.8% | 1.7% | 8.3% |
| Curiosity Creek | 0.0% | 50.8% | 27.5% | 20.0% | 0.0% | 19.2% | 35.8% | 45.0% | 0.0% |
| Dogleg Creek | 0.0% | 0.0% | 39.6% | 29.2% | 4.2% | 33.3% | 41.7% | 20.8% | 4.2% |
| Hayes Bayou | 0.0% | 100.0% | 0.0% | 0.0% | 0.0% | 33.3% | 25.0% | 41.7% | 0.0% |
| Little Manatee River | 21.9% | 56.3% | 3.1% | 2.1% | 12.5% | 3.1% | 22.9% | 55.2% | 18.8% |
| Question Mark Creek | 2.1% | 10.4% | 66.7% | 4.2% | 14.6% | 22.9% | 41.7% | 33.3% | 2.1% |
| Rice Creek | 0.0% | 8.3% | 18.8% | 33.3% | 27.1% | 10.4% | 31.3% | 54.2% | 4.2% |
| Riverview West Creek | 0.0% | 2.1% | 0.0% | 25.0% | 72.9% | 68.8% | 18.8% | 6.3% | 6.3% |
| Wildcat Creek | 4.2% | 86.7% | 4.2% | 2.5% | 0.0% | 50.8% | 40.0% | 9.2% | 0.0% |

Relationships between *B. belizanus* probability of presence and a variety of habitat predictor variables were examined using Boosted Regression Trees (BRT; [46]). As described by Elith et al. [46], BRT is a machine learning method that combines regression trees (models that relate a response to their predictors by recursive binary splits) and boosting (an adaptive method for combining many simple models to give improved predictive performance), resulting in final models that are essentially additive regression models in which individual terms are simple trees, fitted in a forward, stagewise fashion. The advantages of BRT include no need for prior data transformation or elimination of outliers, ability to fit complex nonlinear relationships, and inclusion of interactions between predictors [46].

Habitat variables were assessed in the BRT modeling for their potential to influence the probability of presence of *B. belizanus* based on previously observed or hypothesized relationships. Julian day was included to represent seasonality [37]. Waterbody type was included to test for the potential importance of mesohabitat, given previous qualitative observations of habitat preference (Miller [30]: "swamps and the eddies of streams"; Turner and Snelson [20]: "...habitats with some degree of cover, such as mangrove swamps and narrow, shallow canals with dense aquatic vegetation...not found in large, open canals or borrow ponds"; Greenfield and Thomerson [31]: "...shallow water and areas of heavy vegetation", "temporary ponds that form during...flooding", "larger rivers and streams...usually found close to shore, in slower-moving backwater areas"; and Miller et al. [17]: "...ciénegas, lagoons, ponds, and slow-moving parts of creeks and rivers..."). Waterbodies were classified as creek (natural tributaries of larger rivers), canal (dredged inlets of larger rivers), bayou (large open embayment of a river), and river (mainstems of the two main rivers within the *B. belizanus* range, Alafia and Little Manatee). Dominant shoreline type at each sampling site was included because of qualitative observations of differences in *B. belizanus* presence between vegetation type [20], and was characterized as emergent marsh (primarily black rush, *Juncus roemarianus*), mangrove (red mangrove, *Rhizophora mangle*, or black mangrove, *Avicennia germinans*), small terrestrial vegetation (primarily leather fern, *Acrostichum danaefolium*), shrubs/trees, or artificially hardened (e.g., sea wall or rip rap). Associated with shoreline type were qualitative variables for whether or not the shoreline was inundated during sampling and whether or not the shoreline vegetation was overhanging the sampling site; these variables could influence the probability of presence along shoreline habitat, as shown for other fishes [47,48]. Substrate type was represented in the BRT models as mud, mud-sand, sand, or some oyster/rock. Water depth (at the seine's bag, i.e., 1.5 m from shore; with sampling limited to sites 1.2 m deep or less) and site gradient (calculated as the difference in water depth between the seine's bag and the seine's wing at the shore, divided by the distance between bag and shore, and expressed as a percentage) were included to address these variables' potential importance for predation risk [49]. Physicochemical variables previously shown to correlate with *B. belizanus* density in south Florida [37] were also included in the BRT models: salinity, temperature, dissolved oxygen, and pH. These physicochemical variables were measured at 0.2-m depth with a Hydrolab® multiprobe (OTT Hydromet, Kempten, Germany), reflecting *B. belizanus*'s tendency to inhabit surface waters [30].

In order to examine habitat predictability for *B. belizanus* presence, a novel approach based on creating two sets of BRT models was implemented. Hereinafter the models are referred to as the fixed subset and randomly selected subset models. For both types of models, training data are defined as the data used to formulate the model, whereas test data were not used for model formulation and instead were used to test the models' predictive performance, i.e., the test data were independent of the training data. For the fixed subset model, test data were represented by the data collected in three waterbodies (Andrews Creek, Curiosity Creek, and Question Mark Creek; $n$ = 240 samples, 25.7% of the 935 total samples), which were chosen on the basis of their habitat characteristics being within the range encompassed by the other eight waterbodies (Table 1 and Table 2). Training data thus consisted of the 695 samples (74.3%) from these eight other waterbodies. The fixed subset BRT model was implemented in the dismo package [50] of the R software environment [51] using a bag fraction (i.e., the proportion of data to be selected at each step of the model) of 0.5, a learning rate (i.e., the shrinkage parameter determining the contribution of each tree to the growing model) of 0.005, and tree complexity of 2 (i.e., allowing first-order interactions) in order to achieve Elith and coworkers' [46] recommendation of >1000 trees for reliable model predictions. Exploration of a higher order of tree complexity (5) did not yield substantial improvement in the model's predictive performance, so tree complexity of 2 was implemented for the final analysis. Given the stochasticity in BRT model formulation [46], 1000 iterations of the fixed subset model were undertaken in order to examine the variability in model predictive performance and the consistency in relationships between *B. belizanus* presence and habitat. Model fitting, tenfold crossvalidation, and prediction of the model to the test data were undertaken for each of the 1000 iterations of the fixed subset model.

The randomly selected subset BRT model consisted of randomly selecting (without replacement) 695 samples of the total 935 samples to use as training data, then repeating the steps described above for the fixed subset model: model fitting, crossvalidation, and prediction to the test data (i.e., the randomly remaining 240 samples). This whole procedure was repeated 1000 times, again to illustrate the variability in model predictive performance and consistency in *B. belizanus* presence–habitat relationships.

The predictive performance of the fixed and randomly selected subset BRT models in correctly modeling presence of *B. belizanus* was assessed using the area under the receiver operating characteristic curve (AUC). As noted by Huang and Frimpong [7], AUC values are commonly interpreted as follows: AUC > 0.9 is excellent; 0.8 < AUC < 0.9 is good; 0.7 < AUC < 0.8 is fair; 0.6 < AUC < 0.7 is poor; and 0.5 < AUC < 0.6 is failed (i.e., similar to random guessing). Huang and Frimpong suggested that this classification scheme is too demanding for prediction to test data, given that the scheme was developed for the purpose of assessing model fit rather than transferability (prediction); instead, they used a threshold of AUC < 0.6 to indicate limited transferability from training to test data for prediction [7]. This threshold was also considered appropriate for the present study. Similar to the approach used by Broennimann and Guisan [52], the variability in AUC over the 1000 iterations of the fixed subset and randomly selected BRT models allowed 95% intervals to be calculated, based on the 2.5th and 97.5th percentiles of AUC. The overlap of the 95% intervals was assessed to examine the evidence for differences in predictive performance between fixed subset and randomly selected BRT models for training, crossvalidation, and prediction to test data.

In addition to assessing predictive performance with AUC, the ability of the fixed subset BRT to predict presence in the three test waterbodies was assessed by summing the probability of presence over all samples for each waterbody to give an overall predicted % presence for each iteration of the model, with 95% intervals calculated based on the 2.5th and 97.5th percentiles of the 1000 iterations. The predicted estimates of % presence were compared to the actual % presence in each waterbody, for which 95% intervals were generated by bootstrapping—randomly sampling with replacement, for the same sample size as the original sample [53]—the presence/absence data 1000 times, then calculating the 2.5th and 97.5th percentiles of the % presence from each of the 1000 resamples.

The relative contribution [54] of the habitat predictor variables to *B. belizanus* presence was compared between the fixed and randomly selected subset BRT models using the 95% intervals from the 1000 model iterations. The consistency of *B. belizanus* presence–habitat relationships between the fixed and randomly selected subset models was assessed with partial dependence plots from R's gbm package [55]. The partial dependence plots illustrated the marginal effect of each habitat predictor on *B. belizanus* probability of presence after accounting for the average effects of all other predictors in the models, and were based on the 95% intervals of the marginal effects from the 1000 model iterations.

# 3. RESULTS

## 3.1. Distribution and Spread Rate

The first specimen of *B. belizanus* collected in the Tampa Bay watershed was made on 21 February 1994 from a canal off the Alafia River (at 3.2 km from the river mouth; Figure 1). During extensive sampling occurring in 2006/2007, *B. belizanus* was mostly collected in a number of waterbodies draining either directly to eastern Tampa Bay, to the Alafia River, or to the Little Manatee River (Figure 1). Data collected following this period (2008–2016), primarily as a continuation of long-term monitoring for status and trends in estuary-dependent sportfishes and nekton communities, found *B. belizanus* in similar areas within this range [56,57]. The straight-line distance from the northernmost to southernmost collection locations is approximately 31 km. *B. belizanus* was also collected in two small urban creeks in the city of St. Petersburg, on the western fringe of Tampa Bay (Figure 1).

The first collection of *B. belizanus* in the Little Manatee River watershed was made on 13 October 1998, approximately 5 km from the mouth of the river, with the farthest upstream collection within the sampling area at approximately river km 13 of the main stem in October 2000 (Figure 2). Four collections between October 1998 and November 1999 gave estimated spread rates of 15–35 m $day^{-1}$ (5.5–13 km $year^{-1}$; Figure 2).

## 3.2. Habitat Predictability

*B. belizanus* was found in 7% of shoreline seine samples from the 11 estuarine waterbodies studied during 2006/2007, with the greatest frequency of presence (12%–14%) in two polyhaline mangrove-dominated creeks (Andrews Creek and Cockroach Creek), a mesohaline marsh-dominated creek (Wildcat Creek), and a mesohaline canal mostly lined with shrubs/trees (Question Mark Creek) (Table 3). The species occurred along all shoreline and bottom types sampled with seines, across a broad range of physiochemical variables: salinity 0.2–29.3, temperature 11.8–30.1 °C, dissolved oxygen 0.5–14.7 mg $l^{-1}$, pH 6.9–8.4, depth 0.1–1.0 m, and gradient 0%–66.7%.

The predictive performance of the BRT models for *B. belizanus* presence was greatest for the training data, with slightly more variation in the area under the receiver operating characteristic curve (AUC) for the 1000 iterations of the randomly selected data subsets (95% interval: 0.92–0.98) than for the fixed subsets (0.92–0.96) (Figure 3). The 95% interval of AUC for the crossvalidation of the training data overlapped for the randomly selected and fixed subsets of data, again with more variation for the randomly selected subset. With respect to prediction of test data that were not included in development of the models, there was significantly lower AUC for the fixed subset model compared to the randomly selected model, with the latter having broad variation which overlapped the AUC of the crossvalidation models (Figure 3).

**Table 3.** Abundance and percentage presence of *B. belizanus* in Tampa Bay estuarine waterbodies sampled to assess habitat predictability.

| Waterbody Name | Number of 9.1-m Seine Samples | Number of Fish | % Presence |
|---|---|---|---|
| Alafia River | 167 | 2 | 1.2% |
| Andrews Creek | 72 | 14 | 12.5% |
| Cockroach Creek | 120 | 26 | 11.7% |
| Curiosity Creek | 120 | 10 | 5.0% |
| Dogleg Creek | 48 | 2 | 4.2% |
| Hayes Bayou | 48 | 3 | 6.3% |
| Little Manatee River | 96 | 6 | 5.2% |
| Question Mark Creek | 48 | 8 | 12.5% |
| Rice Creek | 48 | 0 | 0.0% |
| Riverview West Creek | 48 | 4 | 6.3% |
| Wildcat Creek | 120 | 22 | 14.2% |
| Overall | 935 | 97 | 7.2% |

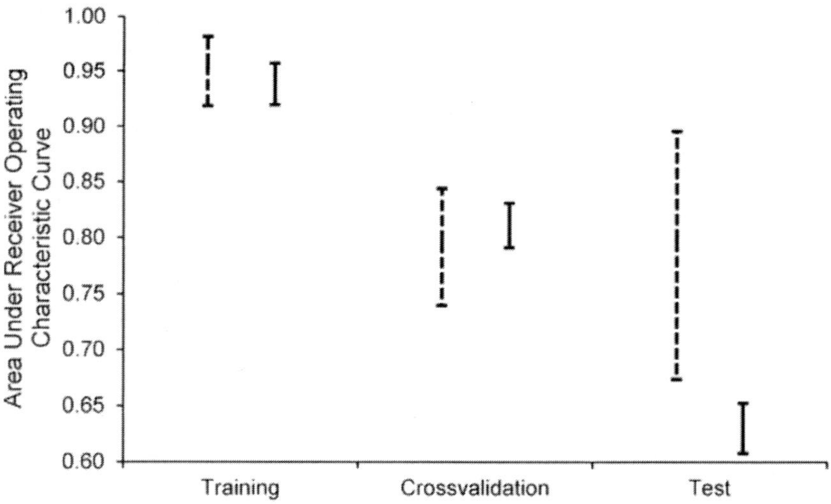


**Figure 3.** Predictive performance of boosted regression tree models for *B. belizanus* presence in Tampa Bay estuarine waterbodies for randomly selected (|- - -|) and fixed (|---|) subsets of 935 9.1-m shoreline seine samples, expressed as 95% intervals of the area under the receiver operating characteristic curve from 1000 model iterations. Randomly selected and fixed subset models included training data ($n$ = 695 samples) and prediction data ($n$ = 240 samples).

The BRT model for the fixed subset of the data provided a reasonable prediction of the presence of *B. belizanus* in one of the three test waterbodies: the 95% interval of the prediction for Curiosity Creek was

6.8%–7.4%, which was somewhat greater than the actual observed 5.0% presence, but well within the 95% interval for actual presence (Figure 4). In contrast, the BRT model underpredicted the presence in the two other waterbodies, and in the case of Andrews Creek, the 95% intervals for the predicted and actual presence did not overlap (Figure 4).

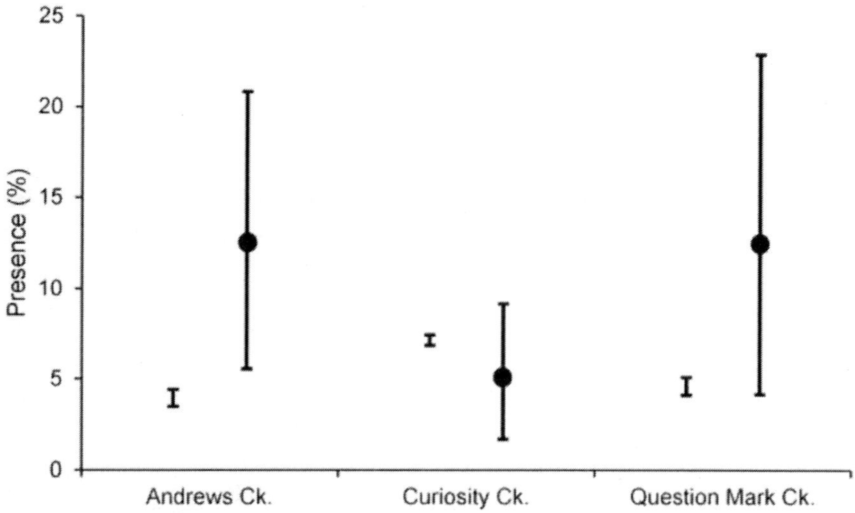


**Figure 4.** Predictions (|---|, 95% interval) of *B. belizanus* presence (% of samples) in three estuarine waterbodies of Tampa Bay, based on 1000 boosted regression tree model iterations for a fixed subset training dataset ($n$ = 695 samples), in comparison to actual occurrence (|—•—|, with 95% interval generated by 1000 bootstrapped resamples of the data).

In general, the BRT models based on the randomly selected and fixed subsets of the data suggested a similar relative ranking of habitat predictor variable importance in predicting *B. belizanus* presence (Figure 5). The range of variability in importance across the 1000 iterations of the two BRT models was considerably greater for the model based on randomly selected subsets of the data. Julian day was the most important predictor, followed by dissolved oxygen. Depth was suggested to be as important as pH, salinity, temperature and shoreline type in the model based on randomly selected subsets of the data, whereas shoreline type in particular had greater importance in the model based on the fixed subset of data. Substrate, gradient, and waterbody had relatively little importance (generally 5% or less), and shoreline inundation and vegetation overhang had minimal importance (Figure 5).

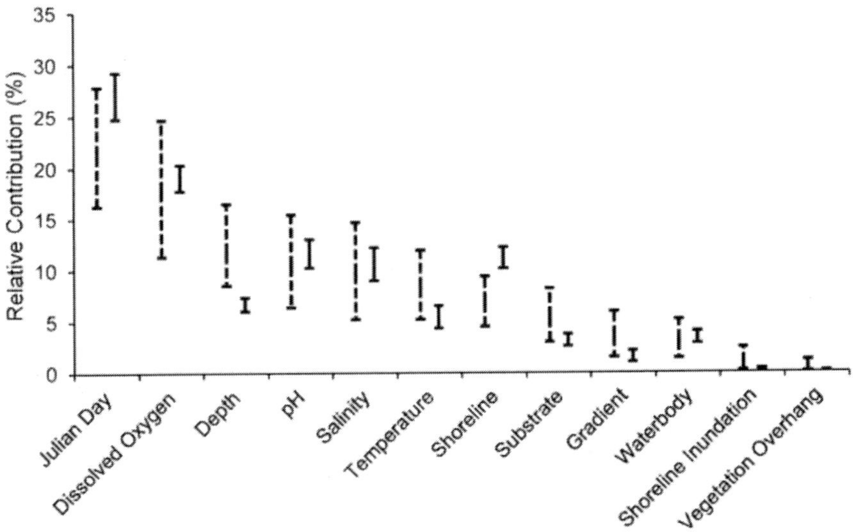


**Figure 5.** Relative contribution of habitat predictors from boosted regression tree models for *B. belizanus* presence in Tampa Bay estuarine waterbodies for randomly selected (|- - -|) and fixed (|---|) subsets of 935 9.1-m shoreline seine samples, expressed as 95% intervals from 1000 model iterations.

The relationships between *B. belizanus* probability of presence and habitat predictors showed some similarities and some differences between the BRT models based on the randomly selected and fixed subsets of the data. Both the randomly selected and fixed subset BRT models suggested that the probability of presence increased from around Julian day 240 (i.e., September) until the end of the year, although the increase was less pronounced for the randomly selected model (Figure 6). A major difference between the randomly selected and fixed subset models was the representation of the effect of dissolved oxygen in hypoxic conditions (<2 mg $L^{-1}$), for which the fixed subset model suggested a very high marginal effect on probability of presence (up to nearly 0.14) whereas the randomly selected subset model gave a marginal effect of around 0.04. This difference principally arose from high frequency of presence (0.6, i.e., 60% of samples) for *B. belizanus* during hypoxic conditions in Cockroach Creek in September 2007, with these observations being present in all iterations of the fixed subset model; the predicted effect of dissolved oxygen was more consistent between the randomly selected and fixed subset models over the rest of the variable's range.

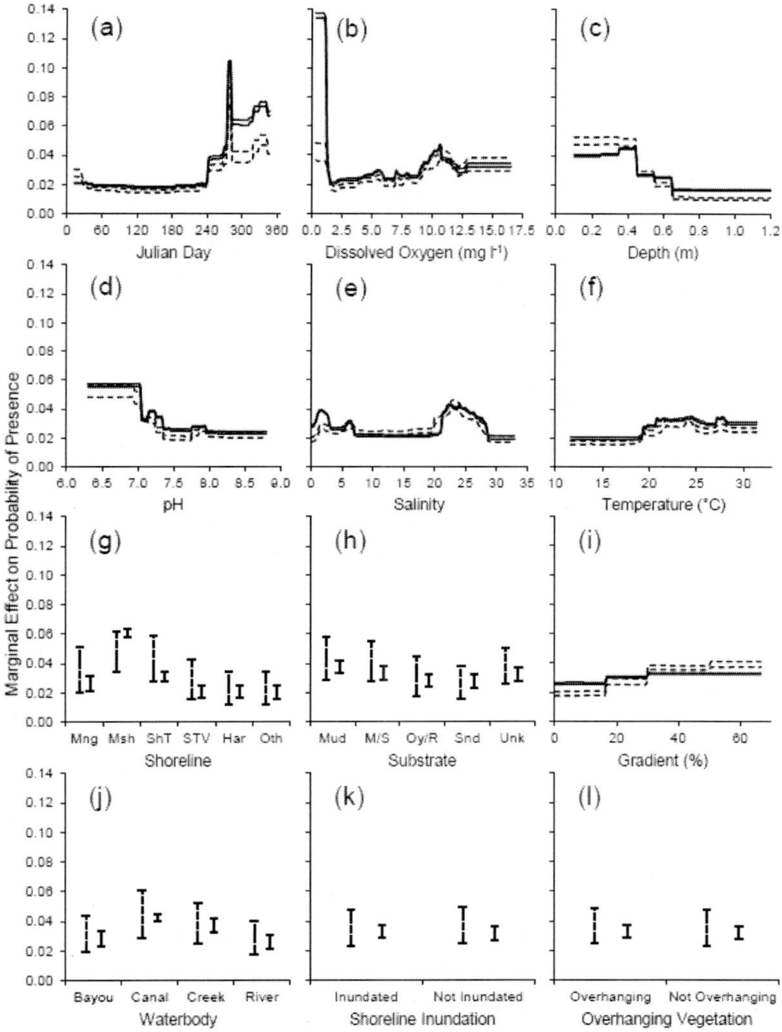


**Figure 6.** Partial dependence plots of habitat predictors from boosted regression tree models for *B. belizanus* presence in Tampa Bay estuarine waterbodies for randomly selected (|- - -|) and fixed (|–––|) subsets of 935 total 9.1-m shoreline seine samples, expressed as 95% intervals from 1000 model iterations. (a) Julian day; (b) dissolved oxygen; (c) water depth; (d) pH; (e) salinity; (f) temperature; (g) shoreline type: Mng (mangrove), Msh (marsh), ShT (shrubs/trees), STV (small terrestrial vegetation), Har (hardened), Oth (other); (h) substrate type: Mud, M/S (mud/sand), Oy/R (oyster/rock), Snd (sand), Unk (unknown); (i) gradient; (j) waterbody type; (k) shoreline inundation; (l) overhanging vegetation.

Other habitat predictors with relatively high contributions for which the predicted marginal effect on *B. belizanus* presence was reasonably consistent between randomly selected and fixed subset BRT models were depth and pH, for which the probability of presence decreased by two to three times over the range of each variable (Figure 6). The predicted marginal effect of salinity was not mononotic and predicted *B. belizanus* presence as a peak between salinity of about 20 and 30 for both randomly selected and fixed subset models, whereas somewhat higher probability of presence at low salinity (<7) was only suggested by the fixed subset model. The increased probability of presence with increasing temperature, particularly above ~19 °C, was consistent for fixed and randomly selected BRT models.

The fixed subset model suggested some significant differences between shoreline types (with marsh having greater probability of *B. belizanus* presence than other types), substrate (greater probability of presence over mud than oyster/rock or sand), and waterbody (greater probability of presence in canals and creeks than rivers, with bayou intermediate); however, the random subset model suggested a greater extent of overlap in the different levels of these predictors. A somewhat increasing probability of presence with increasing gradient was more pronounced in the random subset model than the fixed subset model (Figure 6).

# 4. DISCUSSION

## 4.1. Distribution and Spread Rate

The present study demonstrated that *B. belizanus* is well distributed in some estuarine portions of the Tampa Bay watershed, principally in the Alafia and Little Manatee River basins (Figure 1). From the first recorded presence in the Alafia River drainage in 1994, the species was subsequently detected in the Little Manatee River drainage in 1998. The mouths of these two rivers are approximately a straight-line distance of 17 km apart. Given the indentations along the Tampa Bay shoreline between the two rivers (Figure 1), as well as numerous convoluted potential inshore pathways (e.g., mosquito ditches and county road drainage ditches), the first presence of *B. belizanus* in the Little Manatee River nearly five years after the first recorded presence in the Alafia River is consistent with the estimated rate of spread in the Little Manatee River (5.5–13 km $year^{-1}$; Figure 2). Subsequent sampling in non-tidal freshwater ditches that are ultimately connected to estuarine tributaries has shown that *B. belizanus* has spread extensively inland in these drainages [56]. Collections of *B. belizanus* in south Florida following initial introduction indicates a spread rate of about 1.8 km in just over two months (from latitude 25.559516°, longitude −80.331945° in early December 1957 to approximately latitude 25.543571°, longitude −80.331424° on 18 February 1958; [14]). Extrapolation of this rate to a full year gives a value (9 km $year^{-1}$) within the range estimated in the present study. Additional study of *B. belizanus* spread rates in south Florida based on the extensive available collection database [15] was beyond the scope of the present study but would be valuable.

Invasion success is often linked with a rapid rate of spread [58], and spread rate has been shown to correlate with factors such as the availability of vacant ecological niches [59]. The shallow-water piscivore niche of *B. belizanus* has other members [28], although the fact that the species is immediately piscivorous after birth [20] and is resident year-round in such systems generally differentiates it from other piscivores. The niche (described as "piscivorous in open water habitat in the middle and high water column, principally using pectoral fins for motility") is recognized as being quite distinct from other fishes in the native range [60]. The spread rate of *B. belizanus* estimated in the present study is intermediate to those of other invasive poeciliids. Invasive mosquitofishes (*Gambusia affinis* and *G. holbrooki*) spread much more quickly than *B. belizanus*, with rates ranging from ~38 km $year^{-1}$ in the Colorado River watershed [59] to several hundred meters per day in small-scale studies [61,62]. These rapid rates have been attributed to relatively high boldness of invasive mosquitofishes that contributes to their classification as some of the world's worst invasive species [58]. In contrast, *B. belizanus* rate of spread is greater than the 1.6-km $year^{-1}$ spread of Trinidadian Guppy *Poecilia reticulata* [59], another invasive poeciliid known for negative ecological effects [63], and similar to the 7.9 km $year^{-1}$ spread of Sailfin Molly *P. latipinna* [59]. Among non-poeciliids, the *B. belizanus* spread rate is comparable to that of Round Goby *Neogobius melanostomus*, for which estimates range from 0.5 to 17 km $year^{-1}$ in the Great Lakes and Danube River [64,65,66]. Some larger invasive fishes of particular concern exhibit very rapid spread rates, e.g., Nile Perch *Lates niloticus* (150 km 6 $month^{-1}$ to 50 km $week^{-1}$; [67], Bighead Carp (1.7–2.4 km $day^{-1}$; [68]), and Northern Snakehead *Channa argus* (18 km $month^{-1}$; [69]). Mayan Cichlid *Cichlasoma urophthalmus* spread from south to east-central Florida was estimated to be 15 km $year^{-1}$ [70]; this species has been shown to negatively affect native fishes [71], with a range that includes Tampa Bay and is expanding [56,72]. Depletion of prey in a given area by *B. belizanus* may increase the likelihood of range expansion [19], as may agonistic intraspecific interactions such as territoriality [73] and avoidance of cannibalism [14].

The disjunct distribution of the western population of *B. belizanus* in Tampa Bay relative to the eastern population is not likely to have resulted from swimming because over 15 km of deeper, open water separates the two populations and the species has not been collected in offshore habitat or in the many km of shoreline around the bay between the two populations (Figure 1). Data are lacking to determine when the species first occurred in western Tampa Bay, but potential mechanisms facilitating occurrence include aquarium release [74], bait bucket release [65,75], or bird carry-off [11,76]. A single inseminated female *B. belizanus* released into a water body may be sufficient to establish a population because of large potential brood sizes (>300; [20]) and no evidence for deleterious phenotypic effects in poeciliid populations founded by a single individual [63].

Given the ability of *B. belizanus* to spread at up to 13 km year$^{-1}$, why then after over 20 years in Tampa Bay does the main, eastern population appear to be confined to a region that is a crow's-fly distance of ~30 km from north to south? Movement rates of fishes can be influenced by a number of factors, including the presence of suitable habitat [77]. The *B. belizanus* range on the eastern fringe of Tampa Bay is bounded by large port facilities to the north and south (Figure 1) which are characterized by dredged ship channels and several km of extensive seawalls creating deep-water habitat at the shoreline. As shown in the habitat predictability analysis, the probability of presence of *B. belizanus* decreased with increasing depth. On the basis of the present study's results, it is hypothesized that the deep port seawall habitats with little cover could function as "soft" ecological barriers (sensu Bond and Lake [78]) to *B. belizanus* because of avoidance of such areas as a result of habitat preference [77] or because predation is high in the deeper water and no individuals survive [79]. Between the two ports, the shoreline possesses a considerable amount of shallow-depth habitat including natural shorelines, mosquito ditches, and inland county drainage ditches, which probably facilitated spread of *B. belizanus*. The small western population may also be constrained by deep-water, seawalled habitats in the harbor adjacent to the creeks where the species was found in the present study. Spread of *B. belizanus* through the Everglades and south Florida was essentially unhindered by such potential soft barriers, at least to the south and west, and likely occurred through the extensive system of agricultural canals that include shallow nearshore areas. Thus, the species has reached the southwest coast to the first major urban area (near the city of Naples) but has not spread farther north (Figure 1) to other estuaries [80,81]. The BRT habitat predictability analysis indicated that *B. belizanus* does occur along hardened shorelines (such as seawalls) in shallower water, albeit at somewhat lower probability of occurrence than other shoreline types (Figure 6), emphasizing the hypothesis that the deep-water aspect of seawalls in ports and harbors may be more important than the shoreline type.

## 4.2. Habitat Predictability

The present study displayed a common feature of species distribution modeling: the performance of the BRT models fitted to training data was higher than the cross-validation BRT models, which in turn was higher than the test data prediction model performance [7,46]. Such patterns arise because prediction to test data is more difficult than simple interpolations to the sampled (training) regions, as a result of spatial habitat heterogeneity and varying ecological relationships [7]. The BRT model of *B. belizanus* probability of presence based on the fixed subset of data had some value in its transferability, per the >0.6 AUC criterion of Huang and Frimpong [7], but the comparison to the BRT model based on the randomly selected subset of training data demonstrated that some of the apparent habitat relationships were idiosyncratic as opposed to general [8]. The present study's emphasis on using prediction as a means of demonstrating ecological understanding [9] drove the need to consider model transferability not just in terms of cross-validation and prediction to test data, but also the consistency of observed species–habitat relationships through the comparison of the fixed and randomly selected subsets of data. Conducting 1000 iterations of the fixed and randomly selected BRT models allowed generation of 95% intervals, a novel approach that facilitated examination of the consistency of the observed relationships in order to evaluate which relationships appeared ecologically reasonable and therefore the predictability of habitat use. This approach is readily applicable to studies of other species where it is desirable to provide additional validation of observed species–habitat relationships.

Challenges in predictability of *B. belizanus* presence (Figure 4) may have reflected interannual differences in recruitment [37]—most waterbodies were sampled in 2006, but two were sampled in 2007—as well as differences between waterbodies in the equilibrium status of *B. belizanus* spread [82]. In addition, *B. belizanus* is comparatively rare, even in the waterbodies with the highest percentage presence (Table 3). Rarity can hinder prediction of species–habitat relationships as a result of overfitting few presences with many covariates, suggesting that exploration of techniques such as ensemble-based modeling would be worthwhile in the future [83]. Occupancy modeling based on repeated sampling at each site to account for imperfect detection may also provide additional insight [84]. Initial colonization of new waterbodies can result in high abundance of *B. belizanus*, with numbers declining thereafter [85]. Relatively low abundance generally appears to be typical of the species, even in the native range, and may reflect a conserved life history trait [37] which could be a function of some of the same factors influencing spread, e.g., territoriality [73], cannibalism [14], and resource depletion [19].

Dissolved oxygen was the second most important (after Julian day) of the habitat variables predicting *B. belizanus* presence in the studied estuarine waterbodies of Tampa Bay, although there was a considerable difference in the relationship between the BRT models based on the fixed and random subsets of data (Figure 6). The fixed data model showed occurrence was high at hypoxic conditions, reflecting a short-term event in Cockroach Creek during September 2007. Similar to other cyprinodontoids, *B. belizanus* may employ aquatic surface respiration in hypoxic conditions [86,87], possibly allowing the species to occupy such areas in relatively high numbers and prey successfully on species less adapted to low dissolved oxygen [88]. Alternatively, aquatic surface respiration may have allowed *B. belizanus* to remain within low dissolved oxygen habitat, albeit with greater susceptibility to capture during sampling, akin to greater susceptibility to predation observed in other fishes exhibiting aquatic surface respiration [88]. Although *B. belizanus* appears sufficiently physiologically flexible to cope with low dissolved oxygen, as observed in other invasive fishes in Florida [89], the relationships suggested in the present study were somewhat idiosyncratic rather than general and useful for prediction.

The greater probability of presence of *B. belizanus* with declining pH was consistent between fixed and random subset BRT models (Figure 6), and there are both similarities and differences with the patterns observed in previous studies. In freshwater habitat of south Florida's Everglades National Park, Kerfoot et al. [37] found a positive relationship between pH (range: ~7.4–8.0) and *B. belizanus* density at one site, a weak positive relationship at another site (pH range: ~7.2–7.4), and negative relationships at two other sites (pH ranges: ~7.3–7.6 and ~7.6–8.1). Multispecies ordination analyses within the native range do not demonstrate strong or consistent relationships between pH and *B. belizanus* populations [33,34,35,36]. As noted by Kerfoot et al. [37], changes in pH are known to directly affect ion uptake across gills and internal acid–base balance in fishes, ultimately affecting survival rates. For example, low concentration of calcium ions was found to negatively affect the predation functional response of invasive *N. melanogobius* [90]. Given potential future changes in pH as a result of climate change [91], future laboratory studies on *B. belizanus* could assess whether physiology and behavior (e.g., predation) are significantly affected by pH in a manner consistent with the lower probability of presence at higher pH observed in the present study.

The Environmental Matching Hypothesis predicts that the impacts of an invader decline as habitat conditions move further from its physiological optimum [90]. The present study found that *B. belizanus* is widely distributed across the salinity gradient, with evidence for greatest presence in polyhaline conditions (Figure 6). The observed effects of *B. belizanus* on small prey fish abundance in Tampa Bay estuarine waterbodies were consistent over a broad salinity gradient [24]. The present study contrasts with that of Kerfoot et al. [37], who found decreasing *B. belizanus* density with increasing salinity in south Florida red mangrove habitats; a negative relationship with salinity has also been found in the native range [33]. Water temperature in the coldest month is 2–3 °C cooler in the Tampa Bay watershed than south Florida [92], so that occupation of higher salinity by *B. belizanus* could ameliorate low-temperature effects. Such an effect was found for *Tilapia mossambica* (now *Oreochromis mossambicus*, a euryhaline freshwater species), which is restricted to estuaries near the southern limit of its range in Africa, possibly in order to maintain near-normal sodium and chloride ion concentrations during low winter temperatures [93]. A similar mechanism was not found for *C. urophthalmus*, however [94]. Nekton abundance often increases with increasing salinity because of the increasing contribution of marine-derived juveniles [43,95] and this may provide a greater potential pool of prey resources for *B. belizanus* at higher salinity. The estuarine focus of the sampling in the present study precludes any firm conclusions about the distribution or abundance of *B. belizanus* in the permanent, non-tidal freshwater reaches upstream of the study area; expansion of sampling in this habitat using similar seining methods would clarify the significance of the present study's findings with respect to salinity.

Lower lethal temperature (~9.5 °C) for *B. belizanus* is surprisingly low and rarely experienced in Tampa Bay [96]; feeding is, however, reduced (or ceases) at ~11–16 °C [92,97], which are more commonly observed winter temperatures in the area. During an 80-year cold front in south Florida, water temperature in some areas dropped below the laboratory-determined *B. belizanus* mean lethal temperature, yet the species persisted [98]. It would be informative to repeat Shafland and Pestrak's [92] and Kerfoot's [97] experiments, which were undertaken in freshwater aquaria, at a variety of salinity levels to assess potential ameliorating effects on low temperature effects. The present study found a consistent drop-off in *B. belizanus* probability of presence below about 19.5 °C (Figure 6), whereas Kerfoot et al. [37] found that density of *B. belizanus* in south Florida red mangroves increased as temperature decreased, although the lowest temperature in their dataset was 23 °C. Long-term temperature increases may favor the species' increased expansion [38], as noted for *C. uropthalmus* [71], producing effects on native species that may not yet have been manifested [99].

## 4.3. Management Implications

The range of shoreline habitats and physicochemical conditions at which *B. belizanus* was present in this study suggests that within certain parameters (i.e., nearshore, shallow-water habitat), the species has generally broad environmental tolerance, a key trait of successful invasive fishes [6]. The consistent importance of Julian day in predicting probability of presence suggests that future studies aiming to assess the occurrence of *B. belizanus* in Tampa Bay estuarine waterbodies should be focused on the period from September to December/January in order to maximize the likelihood of capture (Figure 6). The present study suggests that *B. belizanus* in estuarine areas of Tampa Bay spread at a modest rate compared to the most notable invasive fishes and that its distribution may be constrained by soft ecological barriers, i.e., deep seawall habitat. The recent classification as moderately invasive [23] appears appropriate. Nevertheless, strong evidence for negative effects on small-bodied fishes [24] and potential competition with *C. undecimalis* [15] indicates that consideration of possible management actions is warranted. Under current climate conditions, *B. belizanus* would be predicted to be able to occur 150–200 km farther north than Tampa Bay based on lower lethal temperature [92,97]. However, the present study's hypothesis that spread of *B. belizanus* is restricted by deep, anthropogenically altered habitats leads to the conclusion that movement north may be very much dependent on human-facilitated pathways, recently referred to as "jump dispersal" [100], for which the main vectors are releases from aquaria and bait buckets [75]. Internet-based sales of ornamental fish allow species to be spread easily within the U.S., suggesting that education of hobbyists and retailers is paramount [74] and particularly important in the case of *B. belizanus*, given that aggression by piscivores is one reason for release of fish from aquaria [101].

Regarding bait bucket releases, *B. belizanus* is not raised commercially as bait, but releases could occur as surplus fish from anglers cast-netting [102] and catching the species along with other small-bodied fishes. This is particularly true within the eastern portion of the *B. belizanus* Tampa Bay range, which is notably good habitat for sub-adults and adults of the primary inshore sportfish *C. undecimalis* [103]. One possible means of limiting spread by this vector would be to produce simple educational materials for anglers (e.g., posters at boat ramps) illustrating the current range of *B. belizanus*, emphasizing the need to not release it beyond this range, and giving warnings about its potential effects through competition with prized sportfishes that anglers may be targeting. Such materials could be combined with similar warnings for other spreading invasive fishes, such as *C. uropthalmus* [56]. Emphasis on not releasing *B. belizanus* from aquaria or bait buckets is also relevant for freshwater habitat because of the similarity in prey capture and potential competition with juveniles of another prized sportfish, Largemouth Bass *Micropterus salmoides* [104].

Restoration is being undertaken in the Tampa Bay watershed with the goal of approximately restoring the balance of estuarine habitat observed ca. 1950 [41]. Projections of habitat change as a result of sea level rise suggest that mangrove habitat will increase its percentage of coastal habitat coverage from ~74% to 85–89% by 2100, with a concomitant decrease in salt marsh from 25% to 10–14% [41]. The habitat predictability analysis demonstrated that *B. belizanus* is sufficiently flexible in terms of shoreline habitat occupation (Figure 6) that such changes would not be expected to result in considerable effects to the species. With a spread rate of several km $year^{-1}$, it would be expected that *B. belizanus* would colonize restored areas relatively quickly, particularly because some of the major restored sites are directly upstream of two of the waterbodies in the present study, Andrews Creek and Cockroach Creek [105]. This could disrupt typical successional fish assemblage dynamics that are observed in restored areas [106,107,108], as evidenced by effects on small-bodied resident fishes elsewhere in Tampa Bay [24] and potential competition for prey with transients using restored areas as a nursery, in particular *C. undecimalis* [15].

## ACKNOWLEDGMENTS

Thanks to personnel from the FWC-FWRI Fisheries-Independent Monitoring Program for collection and processing of fish samples. J. Buhrman and P. Shafland (FWC-FWRI), J. Krebs (University of South Florida), and C. Barker (University of California at Davis) procured several references. J. Sullivan and R. Dickerson (Hillsborough County Parks, Recreation and Conservation Department), and W. Anastasiou (Tampa Electric Company) facilitated access to several sampling sites. Data used in this study were derived from projects supported in part by funding from State of Florida Saltwater Fishing License sales; by the Department of the Interior, US Fish and Wildlife Service, Federal Aid for Sport Fish Restoration Grant Number F-43; by the Pinellas County Environmental Fund, Tidal Tributary Habitat Initiative Grant 2005-0003-003; by State Wildlife Grant SWG05-015; and by grants CM-254 and CM-280 from the Florida Office of Coastal Management, Department of Environmental Regulation, with funds provided by the US Office of Ocean and Coastal Resource Management, National Oceanic and Atmospheric Administration, under the Coastal Zone Management Act of 1972, as amended.

# REFERENCES


1. Wilcove, D.S.; Rothstein, D.; Dubow, J.; Phillips, A.; Losos, E. Quantifying Threats to Imperiled Species in the United States. *Bioscience* 1998, *48*, 607–615.
2. Obama, B. *Executive Order—Safeguarding the Nation from the Impacts of Invasive Species*; The White House, Office of the Press Secretary: Washington, DC, USA, 2016. Available online: https://www.whitehouse.gov/the-press-office/2016/12/05/executive -order-safeguarding-nation-impacts-invasive-species (accessed on 11 January 2017).
3. Matlock, G.C. Temporal Trends in Non-Native Fishes Established in the Continental United States. *Manag. Biol. Invasion* 2014, *5*, 349–355.
4. Mills, M.D.; Rader, R.B.; Belk, M.C. Complex Interactions between Native and Invasive Fish: The Simultaneous Effects of Multiple Negative Interactions. *Oecologia* 2004, *141*, 713–721.
5. Pimentel, D.; Zuniga, R.; Morrison, D. Update on the Environmental and Economic Costs Associated with Alien-Invasive Species in the United States. *Ecol. Econ.* 2005, *52*, 273–288.
6. Moyle, P.B.; Marchetti, M.P. Predicting Invasion Success: Freshwater Fishes in California as a Model. *Bioscience* 2006, *56*, 515–524.
7. Huang, J.; Frimpong, E.A. Limited Transferability of Stream-Fish Distribution Models among River Catchments: Reasons and Implications. *Freshw. Biol.* 2016, *61*, 729–744.
8. Wenger, S.J.; Olden, J.D. Assessing Transferability of Ecological Models: An Underappreciated Aspect of Statistical Validation. *Methods Ecol. Evol.* 2012, *3*, 260–267.
9. Houlahan, J.E.; McKinney, S.T.; Anderson, T.M.; McGill, B.J. The Priority of Prediction in Ecological Understanding. *Oikos* 2016, *126*, 1–7.
10. Shafland, P.L.; Gestring, K.B.; Stanford, M.S. Florida's Exotic Freshwater Fishes—2007. *Fla. Sci.* 2008, *71*, 220–245.
11. Tuckett, Q.M.; Ritch, J.L.; Lawson, K.M.; Hill, J.E. Landscape-Scale Survey of Non-Native Fishes near Ornamental Aquaculture Facilities in Florida, USA. *Biol. Invasions* 2017, *19*, 223–237.
12. Schofield, P.J.; Loftus, W.F. Non-Native Fishes in Florida Freshwaters: A Literature Review and Synthesis. *Rev. Fish Biol. Fish* 2015, *25*, 117–145.
13. Schofield, P.J.; Morris, J.A.; Akins, L. *Field Guide to Nonindigenous Marine Fishes of Florida*; NOAA Technical Memorandum NOS NCCOS 92; National Oceanic and Atmospheric Administration: Silver Spring, MD, USA, 2009. Available online: https://repository.library.noaa.gov/view/noaa/12675 (accessed on 13 February 2017).
14. Belshe, J.F. 1961. Observations of an Introduced Tropical Fish (*Belonesox belizanus*) in Southern Florida. Master's Thesis, University of Miami, Coral Gables, FL, USA, 1961.
15. Schofield, P.J.; Nico, L.; Neilson, M. *Belonesox belizanus*; USGS Nonindigenous Aquatic Species Database: Gainesville, FL, USA, 2013. Available online: https:// nas.er. usgs.gov/queries/factsheet.aspx?SpeciesID=843 (accessed on 19 January 2017).
16. Obregon-Barboza, H.; Contreras-Balderas, S.; Lozano-Vilano, M.L. The Fishes of Northern and Central Veracruz, Mexico. *Hydrobiologia* 1994, *286*, 79–95.
17. Miller, R.R.; Minckley, W.L.; Norris, S.M. *Freshwater Fishes of México*; University of Chicago Press: Chicago, IL, USA, 2007.
18. Ferry, L.A.; Paig-Tran, E.M.; Gibb, A.C. Suction, Ram, and Biting: Deviations and Limitations to the Capture of Aquatic Prey. *Integr. Comp. Biol.* 2015, *55*, 97–109.
19. Miley, W.W. 1978. Ecological Impact of the Pike Killifish, *Belonesox belizanus* Kner, (Poeciliidae) in Southern Florida. Master's Thesis, Florida Atlantic University, Boca Raton, FL, USA, 1978.

20. Turner, J.S.; Snelson, F.F.J. Population Structure, Reproduction and Laboratory Behavior of the Introduced *Belonesox belizanus* (Poeciliidae) in Florida. *Environ. Biol. Fishes* 1984, *10*, 89–100.
21. Greenwood, M.F.D.; Malkin, E.; Peebles, E.B.; Stahl, S.D.; Courtney, F.X. *Assessment of the Value of Small Tidal Streams, Creeks, and Backwaters as Critical Habitat for Nekton in the Tampa Bay Watershed. Report to the Florida State Wildlife Grants Program, Project SWG05-15*; Florida Fish and Wildlife Conservation Commission Fish and Wildlife Research Institute and University of South Florida College of Marine Science: St Petersburg, FL, USA, 2008; Available online: https://www.researchgate.net/publication/313398996_Assessment_of_the_value_of_small_tidal_streams_creeks_and_backwaters_as_critical_habitat_for_nekton_in_the_Tampa_Bay_watershed_Report_to_the_Florida_State_Wildlife_Grants_Program_Project_SWG 05–15 (accessed on 13 February 2017).
22. Harms, C.A.; Turingan, R.G. Dietary Flexibility Despite Behavioral Stereotypy Contributes to Successful Invasion of the Pike Killifish, *Belonesox belizanus* in Florida, USA. *Aquat. Invasions* 2012, *7*, 547–553.
23. Lawson, L.L., Jr.; Hill, J.E.; Hardin, S.; Vilizzi, L.; Copp, G.H. Evaluation of the Fish Invasiveness Screening Kit (Fisk V2) for Peninsular Florida. *Manag. Biol. Invasion* 2015, *6*, 413–422.
24. Greenwood, M.F.D. Assessing Effects of the Nonindigenous Pike Killifish on Indigenous Fishes in Tampa Bay, Florida, Using a Weighted-Evidence Approach. *Trans. Am. Fish Soc.* 2012, *141*, 84–99.
25. Fulford, R.; Yoskowitz, D.; Russell, M.; Dantin, D.; Rogers, J. Habitat and Recreational Fishing Opportunity in Tampa Bay: Linking Ecological and Ecosystem Services to Human Beneficiaries. *Ecosyst. Serv.* 2016, *17*, 64–74.
26. Stevens, P.W.; Blewett, D.A.; Boucek, R.E.; Rehage, J.S.; Winner, B.L.; Young, J.M.; Whittington, J.A.; Paperno, R. Resilience of a Tropical Sport Fish Population to a Severe Cold Event Varies across Five Estuaries in Southern Florida. *Ecosphere* 2016, *7*, e01400.
27. MacDonald, T.C.; Matheson, R.E.; Peebles, E.B.; McMichael, R.H.; Greenwood, M.F.D. Tidal Creeks as Nekton Habitat in the Tampa Bay Estuary. In Proceedings of the Tampa Bay Area Scientific Information Symposium (BASIS 5), St Petersburg, FL, USA, 20–23 October 2009; Cooper, S.T., Ed.; Tampa Bay Estuary Program and Tampa Bay Regional Planning Council: Tampa Bay, FL, USA, 2010.; pp. 109–122.
28. Krebs, J.M.; Bell, S.S. Risk of Predation Reflects Variation in the Reproductive Strategy of a Dominant Forage Fish in Mangrove Tidal Tributaries. *Oecologia* 2012, *168*, 737–748.
29. Krebs, J.M.; McIvor, C.C.; Bell, S.S. Nekton Community Structure Varies in Response to Coastal Urbanization near Mangrove Tidal Tributaries. *Estuar. Coasts* 2014, *37*, 815–831.
30. Miller, N. The Fishes of the Motagua River, Guatemala. *Bull. Am. Mus. Nat. Hist.* 1905, *23*, 95–123.
31. Greenfield, D.W.; Thomerson, J.E. *Fishes of the Continental Waters of Belize*; University Press of Florida: Gainesville, FL, USA, 1997.
32. Taylor, D.S.; Reyier, E.A.; McIvor, C.C.; Davis, W.P. An Assessment of Ichthyofaunal Assemblages within the Mangal of the Belize Offshore Cays. *Bull. Mar. Sci.* 2007, *80*, 721–737.
33. Lopez-Lopez, E.; Sedeno-Diaz, J.E.; Romero, F.L.; Trujillo-Jimenez, P. Spatial and Seasonal Distribution Patterns of Fish Assemblages in the Rio Champoton, Southeastern Mexico. *Rev. Fish Biol. Fish* 2008, *19*, 127–142.
34. Esselman, P.; Allan, J. Relative Influences of Catchment-and Reach-Scale Abiotic Factors on Freshwater Fish Communities in Rivers of Northeastern Mesoamerica. *Ecol. Freshw. Fish* 2010, *19*, 439–454.
35. Vega-Cendejas, M.E.; Santillana, M.H.d.; Norris, S. Habitat Characteristics and Environmental Parameters Influencing Fish Assemblages of Karstic Pools in Southern Mexico. *Neotrop. Ichthyol.* 2013, *11*, 859–870.

36. Jourdan, J.; Krause, S.T.; Lazar, V.M.; Zimmer, C.; Sommer-Trembo, C.; Arias-Rodriguez, L.; Klaus, S.; Riesch, R.; Plath, M. Shared and Unique Patterns of Phenotypic Diversification Along a Stream Gradient in Two Congeneric Species. *Sci. Rep.* 2016, *6*, 38971.
37. Kerfoot, J.R., Jr.; Lorenz, J.J.; Turingan, R.G. Environmental Correlates of the Abundance and Distribution of *Belonesox belizanus* in a Novel Environment. *Environ. Biol. Fishes* 2011, *92*, 125–139.
38. Rahel, F.J.; Olden, J.D. Assessing the Effects of Climate Change on Aquatic Invasive Species. *Conserv. Biol.* 2008, *22*, 521–533.
39. Lewis, R.R.; Estevez, E.D. *The Ecology of Tampa Bay, Florida: An Estuarine Profile*; U.S. Department of the Interior, Fish and Wildlife Service: Washington, DC, USA, 1988.
40. Raabe, E.A.; Roy, L.C.; McIvor, C.C. Tampa Bay Coastal Wetlands: Nineteenth to Twentieth Century Tidal Marsh-to-Mangrove Conversion. *Estuar. Coasts* 2012, *35*, 1145–1162.
41. Sherwood, E.T.; Greening, H.S. Potential Impacts and Management Implications of Climate Change on Tampa Bay Estuary Critical Coastal Habitats. *Environ. Manag.* 2014, *53*, 401–415.
42. Lewis, R.R., III; Robison, D. *Setting Priorities for Tampa Bay Habitat Protection and Restoration: Restoring the Balance*; Tampa Bay National Estuary Program: St. Petersburg, FL, USA, 1996
43. Greenwood, M.F.D.; Matheson, R.E., Jr.; McMichael, R.H., Jr.; MacDonald, T.C. Community Structure of Shoreline Nekton in the Estuarine Portion of the Alafia River, Florida: Differences Along a Salinity Gradient and Inflow-Related Changes. *Estuar. Coast. Shelf Sci.* 2007, *74*, 223–238.
44. Ben Rais Lasram, F.; Tomasini, J.A.; Guilhaumon, F.; Romdhane, M.S.; Do Chi, T.; Mouillot, D. Ecological Correlates of Dispersal Success of Lessepsian Fishes. *Mar. Ecol. Prog. Ser.* 2008, *363*, 273–286.
45. Tampa Bay Tidal Tributary Research Team. *Tampa Bay Tidal Tributary Habitat Initiative Project: Final Report and Management Recommendations. Tampa Bay Estuary Program Technical Report #02-08*; Report to the Pinellas County Environmental Fund; Sherwood, E.T., Ed.; Tampa Bay Estuary Program: St. Petersburg, FL USA, 2008; Available online: https://www.tbeptech.org/TBEP_TECH_PUBS/2008/TBEP_02_08_Tidal_Tributary_Habitat_Initiative_Integrated_Summary_Report.pdf (accessed on 13 February 2017).
46. Elith, J.; Leathwick, J.R.; Hastie, T. A Working Guide to Boosted Regression Trees. *J. Anim. Ecol.* 2008, *77*, 802–813.
47. Rozas, L.P.; Minello, T.J. Estimating Densities of Small Fishes and Decapod Crustaceans in Shallow Estuarine Habitats: A Review of Sampling Design with Focus on Gear Selection. *Estuaries* 1997, *20*, 199–213.
48. Cocheret de la Moriniere, E.; Nagelkerken, I.; van der Melj, H.; van derVelde, G. What Attracts Juvenile Coral Reef Fish to Mangroves: Habitat Complexity or Shade? *Mar. Biol.* 2004, *144*, 139–145.
49. Ellis, W.L.; Bell, S.S. Conditional Use of Mangrove Habitats by Fishes: Depth as a Cue to Avoid Predators. *Estuaries*2004, *27*, 966–976.
50. Hijmans, R.J.; Phillips, S.; Leathwick, J.; Elith, J. Dismo: Species Distribution Modeling. R Package Version 1.1-1. 2016. Available online: https://CRAN.R-project.org/package=dismo (accessed on 22 October 2016).
51. R Core Team. *R: A Language and Environment for Statistical Computing*; R Foundation for Statistical Computing: Vienna, Austria, 2016; Available online: https://www.R-project.org/ (accessed on 22 October 2016).
52. Broennimann, O.; Guisan, A. Predicting Current and Future Biological Invasions: Both Native and Invaded Ranges Matter. *Biol. Lett.* 2008, *4*, 585–589.
53. Brown, M.L.; Allen, M.S.; Beard, T.D. Data Management and Statistical Techniques. In *Fisheries Techniques*, 3rd ed.; Zale, A.V., Parrish, D.L., Sutton, T.M., Eds.; American Fisheries Society: Bethesda, MD, USA, 2012; pp. 15–77.

54. Friedman, J.H.; Meulman, J.J. Multiple Additive Regression Trees with Application in Epidemiology. *Stat. Med.* 2003, *22*, 1365–1381.
55. Ridgeway, G. with Contributions from others. gbm: Generalized Boosted Regression Models. R Package Version 2.1.1. 2015. Available online: https://CRAN.R-project.org/package=gbm (accessed on 12 October 2016).
56. Lawson, K.M.; Tuckett, Q.M.; Ritch, J.L.; Nico, L.G.; Fuller, P.L.; Matheson, R.E.; Gestring, K.; Hill, J.E. Distribution and Status of Five Non-native Fish Species in the Tampa Bay Drainage (USA), a Hot Spot for Fish Introductions. *Bioinvasions Rec.* (under review).
57. MacDonald, T. Personal Communication, Florida Fish and Wildlife Conservation Commission, Fish and Wildlife Research Institute: St. Petersburg, FL, USA, 2016.
58. Rehage, J.S.; Sih, A. Dispersal Behavior, Boldness, and the Link to Invasiveness: A Comparison of Four *Gambusia* Species. *Biol. Invasions* 2004, *6*, 379–391.
59. Olden, J.D.; Poff, N.L.; Bestgen, K.R. Life-History Strategies Predict Fish Invasions and Extirpations in the Colorado River Basin. *Ecol. Monogr.* 2006, *76*, 25–40.
60. Córdova-Tapia, F.; Zambrano, L. Fish Functional Groups in a Tropical Wetland of the Yucatan Peninsula, Mexico. *Neotrop. Ichthyol.* 2016, *14*.
61. Armstrong, R.C. Dispersal Rates of Mosquitofish through Cattails in Drainage Ditches. *Proc. Calif. Mos. Control Assoc.*1977, *45*, 103–104.
62. Alemadi, S.D.; Jenkins, D.G. Behavioral Constraints for the Spread of the Eastern Mosquitofish, *Gambusia Holbrooki* (Poeciliidae). *Biol. Invasions* 2008, *10*, 59–66.
63. Deacon, A.E.; Ramnarine, I.W.; Magurran, A.E. How Reproductive Ecology Contributes to the Spread of a Globally Invasive Fish. *PLoS ONE* 2011, *6*, e24416.
64. Bronnenhuber, J.E.; Dufour, B.A.; Higgs, D.M.; Heath, D.D. Dispersal Strategies, Secondary Range Expansion and Invasion Genetics of the Nonindigenous Round Goby, *Neogobius melanostomus*, in Great Lakes Tributaries. *Mol. Ecol.*2011, *20*, 1845–1859.
65. Kornis, M.; Mercado-Silva, N.; Vander Zanden, M. Twenty Years of Invasion: A Review of Round Goby *Neogobius melanostomus* Biology, Spread and Ecological Implications. *J. Fish Biol.* 2012, *80*, 235–285.
66. Brandner, J.; Cerwenka, A.F.; Schliewen, U.K.; Geist, J. Bigger Is Better: Characteristics of Round Gobies Forming an Invasion Front in the Danube River. *PLoS ONE* 2013, *8*, e73036.
67. Ligtvoet, W.; Mkumbo, O.C. Synopsis of ecological and fishery research on Nile perch (*Lates niloticus*) in Lake Victoria, conducted by HEST/TAFIRI. In Proceedings of the CIFA. Report of the 5th session of the Sub-Committee for the Development and Management of the Fisheries in Lake Victoria, Mwanza, Tanzania, 12–24 September 1989; FAO Fisheries Report 430. Food and Agriculture Organization of the United Nations: Rome, Italy, 1990.; pp. 35–74.
68. Peters, L.M.; Pegg, M.A.; Reinhardt, U.G. Movements of Adult Radio-Tagged Bighead Carp in the Illinois River. *Trans. Am. Fish Soc.* 2006, *135*, 1205–1212.
69. Lapointe, N.W.R.; Odenkirk, J.S.; Angermeier, P.L. Seasonal Movement, Dispersal, and Home Range of Northern Snakehead *Channa argus* (Actinopterygii, Perciformes) in the Potomac River Catchment. *Hydrobiologia* 2013, *709*, 73–87.
70. Nico, L.G.; Beamish, W.H.; Musikasinthorn, P. Discovery of the Invasive Mayan Cichlid Fish "*Cichlasoma*" *urophthalmus* (Günther 1862) in Thailand, with Comments on Other Introductions and Potential Impacts. *Aquat. Invasions* 2007, *2*, 197–214.
71. Harrison, E.; Lorenz, J.J.; Trexler, J.C. Per Capita Effects of Non-Native Mayan Cichlids (*Cichlasoma urophthalmus*; Gunther) on Native Fish in the Estuarine Southern Everglades. *Copeia* 2013, *2013*, 80–96.
72. Paperno, R.; Ruiz-Carus, R.; Krebs, J.M.; McIvor, C.C. Expansion of the Mayan Cichlid, *Cichlasoma urophthalmus* (Pisces, Cichlidae), above the 28° N Latitude in Florida. *Fla. Sci.* 2008, *71*, 293–304.
73. Bisazza, A.; Novarini, N.; Pilastro, A. Male Body Size and Male-Male Competition: Interspecific Variation in Poeciliid Fishes. *Ital. J. Zool.* 1996, *63*, 365–369.

74. Padilla, D.K.; Williams, S.L. Beyond Ballast Water: Aquarium and Ornamental Trades as Sources of Invasive Species in Aquatic Ecosystems. *Front. Ecol. Environ.* 2004, *2*, 131–138.
75. Rahel, F.J. Biogeographic Barriers, Connectivity and Homogenization of Freshwater Faunas: It's a Small World after All. *Freshw. Biol.* 2007, *52*, 696–710.
76. Frederick, P.C.; Spalding, M.G.; Sepulveda, M.S.; Williams, G.E.; Nico, L.; Robins, R. Exposure of Great Egret (*Ardea albus*) Nestlings to Mercury through Diet in the Everglades Ecosystem. *Environ. Toxicol. Chem.* 1999, *18*, 1940–1947.
77. Gilliam, J.F.; Fraser, D.F. Movement in Corridors: Enhancement by Predation Threat, Disturbance, and Habitat Structure. *Ecology* 2001, *82*, 258–273.
78. Bond, N.R.; Lake, P.S. Local Habitat Restoration in Streams: Constraints on the Effectiveness of Restoration for Stream Biota. *Ecol. Manag. Restor.* 2003, *4*, 193–198.
79. Mclvor, C.C.; Odum, W.E. Food, Predation Risk, and Microhabitat Selection in a Marsh Fish Assemblage. *Ecology*1988, *69*, 1341–1351.
80. Idelberger, C.F.; Stafford, C.J.; Erickson, S.E. Distribution and Abundance of Introduced Fishes in Florida's Charlotte Harbor Estuary. *Gulf. Caribb. Res.* 2011, *23*, 13–22.
81. Idelberger, C.F.; Stevens, P.W.; Weather, E. Fishes of the Lemon Bay Estuary and a Comparison of Fish Community Structure to Nearby Estuaries Along Florida's Gulf Coast. *Fla. Sci.* 2016, *79*, 159–177.
82. Elith, J.; Leathwick, J.R. Species Distribution Models: Ecological Explanation and Prediction across Space and Time. *Annu. Rev. Ecol. Evol. Syst.* 2009, *40*, 677–697.
83. Breiner, F.T.; Guisan, A.; Bergamini, A.; Nobis, M.P. Overcoming Limitations of Modelling Rare Species by Using Ensembles of Small Models. *Methods Ecol. Evol.* 2015, *6*, 1210–1218.
84. Hegel, T.M.; Cushman, S.A.; Evans, J.; Huettmann, F. Current State of the Art for Statistical Modelling of Species Distributions. In *Spatial Complexity, Informatics, and Wildlife Conservation*; Cushman, S.A., Huettmann, F., Eds.; Springer: New York, NY, USA, 2010; pp. 273–311.
85. Trexler, J.C.; Loftus, W.F.; Jordan, F.; Lorenz, J.J.; Chick, J.H.; Kobza, R.M. Empirical Assessment of Fish Introductions in a Subtropical Wetland: An Evaluation of Contrasting Views. *Biol. Invasions* 2000, *2*, 265–277.
86. Nordlie, F.G. Physicochemical Environments and Tolerances of Cyprinodontoid Fishes Found in Estuaries and Salt Marshes of Eastern North America. *Rev. Fish Biol. Fish* 2006, *16*, 51–106.
87. Stevens, P.W.; Blewett, D.A.; Casey, J.P. Short-Term Effects of a Low Dissolved Oxygen Event on Estuarine Fish Assemblages Following the Passage of Hurricane Charley. *Estuar. Coasts* 2006, *29*, 997–1003.
88. Domenici, P.; Lefrancois, C.; Shingles, A. Hypoxia and the Antipredator Behaviours of Fishes. *Philos. Trans. R. Soc. Lond. B Biol. Sci.* 2007, *362*, 2105–2121.
89. Schofield, P.J.; Loftus, W.F.; Brown, M.E. Hypoxia Tolerance of Two Centrarchid Sunfishes and an Introduced Cichlid from Karstic Everglades Wetlands of Southern Florida, U.S.A. *J. Fish Biol.* 2007, *71*, 87–99.
90. Iacarella, J.C.; Ricciardi, A. Dissolved Ions Mediate Body Mass Gain and Predatory Response of an Invasive Fish. *Biol. Invasions* 2015, *17*, 3237–3246.
91. Duarte, C.M.; Hendriks, I.E.; Moore, T.S.; Olsen, Y.S.; Steckbauer, A.; Ramajo, L.; Carstensen, J.; Trotter, J.A.; McCulloch, M. Is Ocean Acidification an Open-Ocean Syndrome? Understanding Anthropogenic Impacts on Seawater Ph. *Estuar. Coasts* 2013, *36*, 221–236.
92. Shafland, P.L.; Pestrak, J.M. Lower Lethal Temperatures for Fourteen Non-Native Fishes in Florida. *Environ. Biol. Fishes* 1982, *7*, 149–156.
93. Allanson, B.R.; Bok, A.; van Wyk, N.I. The Influence of Exposure to Low Temperature on *Tilapia Mossambica* Peters (Cichlidae). II. Changes in Serum Osmolarity, Sodium and Chlorida Ion Concentrations. *J. Fish Biol.* 1971, *3*, 181–185.
94. Stauffer, J.R.J.; Boltz, S.E. Effect of Salinity on the Temperature Preference and Tolerance of Age-0 Mayan Cichlids. *Trans. Am. Fish Soc.* 1993, *123*, 101–107.

95. Peterson, M.S.; Ross, S.T. Dynamics of Littoral Fishes and Decapods Along a Coastal River-Estuarine Gradient. *Estuar. Coast. Shelf Sci.* 1991, *33*, 467–483.
96. Gilmore, R.G., Jr.; Bullock, L.H.; Berry, F.H. Hypothermal Mortality in Marine Fishes of South-Central Florida, January, 1977. *Northeast Gulf. Sci.* 1978, *2*, 77–97.
97. Kerfoot, J. Thermal Tolerance of the Invasive *Belonesox Belizanus*, Pike Killifish, Throughout Ontogeny. *J. Exp. Zool. A Ecol. Genet. Physiol.* 2012, *317*, 266–274.
98. Boucek, R.E.; Rehage, J.S. Examining the Effectiveness of Consumer Diet Sampling as a Nonnative Detection Tool in a Subtropical Estuary. *Trans. Am. Fish Soc.* 2014, *143*, 489–494.
99. Strayer, D.L.; Eviner, V.T.; Jeschke, J.M.; Pace, M.L. Understanding the Long-Term Effects of Species Invasions. *Trends Ecol. Evol.* 2006, *21*, 645–651.
100. Everman, E.; Klawinski, P. Human-Facilitated Jump Dispersal of a Non-Native Frog Species on Hawai'i Island. *J. Biogeogr.* 2013, *40*, 1961–1970.
101. Gertzen, E.; Familiar, O.; Leung, B. Quantifying Invasion Pathways: Fish Introductions from the Aquarium Trade. *Can. J. Fish Aquat. Sci.* 2008, *65*, 1265–1273.
102. Stevens, P.W. Sampling Fish Communities in Saltmarsh Impoundments in the Northern Indian River Lagoon, Florida: Cast Net and Culvert Trap Gear Testing. *Fla. Sci.* 2006, *69*, 135–147.
103. Winner, B.L.; Blewett, D.A.; McMichael, R.H.; Guenther, C.B. Relative Abundance and Distribution of Common Snook Along Shoreline Habitats of Florida Estuaries. *Trans. Am. Fish Soc.* 2010, *139*, 62–79.
104. Kerfoot, J.R., Jr.; Turingan, R.G. Similarity and Disparity in Prey-Capture Kinematics between the Invasive Pike Killifish (*Belonesox Belizanus*) and the Native Florida Largemouth Bass (*Micropterus Floridanus*). *Fla. Sci.* 2011, *74*, 137.
105. Henningsen, B.F.; Norton, N.T.; Young, B.E. The Rock Ponds Ecosystem Restoration Project—A True Mosaic of Coastal Habitats for Tampa Bay. Proceedingso of the Tampa Bay Area Scientific Information Symposium (BASIS 6), St. Petersburg, FL, USA, 28–30 September 2015; Burke, M., Ed.; Tampa Bay Estuary Program and Tampa Bay Regional Planning Council: St. Petersburg, FL, USA, 2015.; p. 60. Available online: https://www.tbeptech.org/BASIS/BASIS6/BASIS6_Proceedings_FINAL.pdf#page=67 (accessed on 12 February 2017).
106. Motta, P.J. *Fisheries Evaluation of Restoration of Peanut Lake, Tampa Bay, Florida*; Department of Biology, University of South Florida: Tampa, FL, USA, 1995.
107. Kurz, R.C.; Fenwick, R.W.; Davis, K.A. *A Comparison of Fish Communities in Restored and Natural Salt Marshes in Tampa Bay, Florida*; Southwest Florida Water Management District: Tampa, FL, USA, 1998. Available online: http://images.library.wisc.edu/EcoNatRes/EFacs/Wetlands/Wetlands24/reference/econatres.wetlands24.rkurz.pdf (accessed on 13 February 2017).
108. Bahk, B.M. *Succession of Ichthyofauna in a Constructed Tidal Creek Salt Marsh in Cockroach Bay, Florida*; Resource Management Department, Southwest Florida Water Management District: Brooksville, FL, USA, 1999.

# Index